The De Gruyter Handbook of Robots in Society and Culture

De Gruyter Handbooks of Digital Transformation

Series Editor
Anthony Elliott

Volume 3

The De Gruyter Handbook of Robots in Society and Culture

Edited by
Leopoldina Fortunati and Autumn Edwards

DE GRUYTER

ISBN 978-3-11-079215-7
e-ISBN (PDF) 978-3-11-079227-0
e-ISBN (EPUB) 978-3-11-079233-1
ISSN 2940-7249

Library of Congress Control Number: 2024938445

Bibliographic information published by the Deutsche Nationalbibliothek
The Deutsche Nationalbibliothek lists this publication in the Deutsche Nationalbibliografie; detailed bibliographic data are available on the internet at http://dnb.dnb.de.

Cover image: PhonlamaiPhoto/iStock/Getty Images Plus
Typesetting: Integra Software Services Pvt. Ltd.
Printing and binding: CPI books GmbH, Leck

www.degruyter.com

MIX
Papier | Fördert gute Waldnutzung
FSC® C083411

Contents

Leopoldina Fortunati and Autumn Edwards

Introduction

The *Handbook of Robots in Society and Culture* arrives at a transformative moment in the history of social robotics. As these technologies move from the laboratories and imaginations of their creators into the commotion of daily life, a multifaceted understanding of their societal and cultural ramifications is not just beneficial but also necessary. Too often, the impact of robots has been examined through a purely technical lens without considering the rich contexts of society and culture. This multidisciplinary collection captures a broad spectrum of academic discourse, revealing how scholars from diverse fields converge to address some of the most profound questions of our time. Perhaps the biggest and most pressing questions are: What is the role of robots among us? And how can we integrate robots into human ecologies without losing our humanity?

A critical theme that emerges from this anthology is the continuum of social robots, including artificial intelligence (AI) and various forms of embodiment, from the most basic digital assistants to highly sophisticated humanoid robots. It presents these embodiments not as isolated phenomena but as historically interconnected, with each wave of technological innovation drawing from and contributing to the evolving landscape of human society and culture.

This handbook argues convincingly for the importance of considering society and culture in the development and integration of robots—a narrative that has traditionally been dominated by a focus on technical capabilities and the engineering marvels of robotics. It posits that the true essence of a social robot's sociality is an emergent property, co-created through interactive processes—a far cry from the pre-packaged capabilities often ascribed to them. This view aligns with the historical development of technology, where each new invention is woven into the existing tapestry of human life, adapting to and shaping society in turn.

Moreover, this handbook underlines the concept of practical meliorism, which invites us to inhabit that precarious and non-determinative middle ground between technological optimism and pessimism. This approach acknowledges the mixture of hope and despair we may feel and underscores our collective ability to influence the long-term outcomes of how robots are created, utilized, and integrated into our social and cultural fabric. The authors explore how we might navigate our rapidly changing technological landscape in a way that promotes mutual thriving. This involves a deliberate choice in crafting futures where robots augment human capabilities and enhance rather than diminish the quality of human society. The chapters offer a deep dive into the implications of robotic technologies for various facets of life, from labor and privacy to identity and social policies, articulating the vast potential and the challenges we face.

https://doi.org/10.1515/9783110792270-001

Reflecting on the urgent need for a deeper appreciation of societal and cultural contexts, The *Handbook's* contributors—from esteemed sociologists to pioneering roboticists—paint a comprehensive picture of the current state of human–robot relations. They highlight the integration of robots in everyday settings, such as homes, schools, workplaces, and healthcare environments, and consider the implications for human roles, responsibilities, and identities.

In its structure, this Introduction synthesizes the content of the 24 chapters into three cohesive sections. The first section delves into the broad role of robots in culture and society, exploring how they have been framed, imagined, depicted, designed, and questioned at the cultural level and how they have both transformed and been transformed by society. The second section offers humanistic and social scientific perspectives, providing insights into how academic expertise from various fields converges on the subject of social robotics by raising questions and questioning eventual answers. The final section, "Contexts of Human–Robot Interaction," presents specific environments where human–robot interactions take place, showing how the material interactions and relationships that users have constructed with social robots in various contexts are influenced by and impact physical capabilities, societal roles, and personal identities.

The conclusion of this Introduction brings these threads together, offering final remarks that articulate the critical insights and potential pathways forward. It emphasizes the collection's contribution to understanding social robots not merely as tools or companions but as entities deeply integrated into the social fabric, reflecting and shaping our collective existence. The *Handbook of Robots in Society and Culture* thus emerges as a guide for students, researchers, designers, and practitioners, providing an authoritative perspective on social robotics and a wealth of reference material for further exploration.

1 Section 1: Robots in Culture and Society

The first section of this *Handbook* includes six insightful chapters, each examining the place and meaning of robots in the context of broader culture and society. The authors traverse a range of topics: envisioning the future of social robots as part of daily life, analyzing their portrayal in pop culture and media, crafting robots to be embraced in social settings, considering the societal impact of their physical presence, debating their existential classification as entities or objects, and exploring their capacity to shape social dynamics on an individual and collective scale.

Chapter 1, titled "Future Presence: Living with Social Robots," by Matthew Lombard, Kun Xu, and Melissa Selverian, takes a forward-looking approach to the topic of social robots. The authors trace the human fascination with artificial life forms back centuries and highlight the critical juncture we currently face, where social robots might be indistinguishable in the future from humans. They chart our progress from primitive representations of life to the sophisticated digital entities of today, noting a pivotal moment in the 2020s marked by the convergence of various technologies such as smartphones, virtual and augmented reality, chatbots, and advanced humanoid robots. These developments, they argue, are rapidly eroding the distinctions between artificial and human entities.

The authors propose a spectrum of 'presence,' ranging from physical to virtual, which intersects with the concept of 'embodiment,' spanning from disembodied to fully embodied forms. This framework allows for a deeper understanding of the diverse robotic forms as points on a morphological and material continuum. Lombard and colleagues explore the far-reaching effects of this technological evolution, inviting input from fields including computer science, robotics, communication, psychology, and business. They begin with foundational concepts and motivations for considering the future of social robots and then present three methods for examining their implications: the Media Are Social Actors paradigm, insights from experts in the field, and the examination of fictional representations in science fiction.

Through this multifaceted examination, the chapter raises essential questions about the potential benefits and drawbacks of artificial humans and advocates for a collective, farsighted approach to technological development. The authors stress the importance of linking present-day decisions to their future social impacts (Fortunati & Edwards, 2022). This journey through the chapter not only deepens our understanding of the integration of artificial entities into our daily lives, but also emphasizes the collaborative steps necessary to ensure a future where technology augments our realities and integrates with sociality responsibly.

Chapter 2, titled "Representing Robots in Popular Culture," by Mauro Sarrica, Simone Natale, and Elda Danese, continues the discussion by examining the portrayal of robots in mass media and its effects on human identity. The authors initiate their analysis with the case of Sophia, a humanoid robot introduced in 2016, emphasizing how its feminine representation has been instrumental in humanizing technology and highlighting the complex interplay between robotics and popular culture.

This chapter traces the evolution of robots from their scientific inception to becoming a mainstay of popular imagination, mapping their journey from the term's origin in a 1920 play, *Rossum's Universal Robots*, to their presence in today's media. Sarrica et al. discuss how robots have long been a canvas for societal values and debates, often represented in movies and literature. They argue that popular representations of robots are not to be dismissed as mere fantasy but rather seen as a fundamental aspect of their technological development and public acceptance.

The relationship between robots and society is bidirectional; as robots are designed based on societal influences, they, in turn, shape social understandings. This chapter particularly focuses on how robots have intersected with the fashion industry, influencing and echoing human identities in a dynamic feedback loop. This narrative showcases the importance of considering cultural portrayals in comprehending robots' roles in contemporary culture, resonating with the themes of integration and presence introduced in Chapter 1. Through these examples, the authors demonstrate how robots are not just inventions but also cultural artifacts that reflect and shape our society.

Building on the preceding chapters' discussions of social robots' cultural integration and representation, Chapter 3, authored by Serena Marchesi and Agnieszka Wykowska, delves into the neuroscience behind designing robots that can be seamlessly accepted into human social environments. The chapter, titled "Designing Robots That are Accepted in Human Social Environments: Anthropomorphism, the Intentional Stance, Cultural Norms and Values, and Societal Implications," investigates the multilayered relationship humans have with artificial agents and the ethical considerations that emerge as these robots become more prevalent in our lives.

Marchesi and Wykowska guide us through the complex dynamics of how social robots are not only received and assimilated into our daily routines but also how they reflect and are shaped by the cultural and societal values of their human counterparts, adding to a growing chorus in the *Handbook* to acknowledge the significance of societal values in robot design and perception.

The authors suggest that humanoid robots occupy a 'technological twilight space'—neither solely tools nor simply companions, blurring the lines of technological artifacts and social entities. This ambiguity suggests that robots may represent a new ontological category, a theme that will be expanded upon in Chapter 15. Marchesi and Wykowska thoroughly explore the factors that affect our perceptions and acceptance of artificial agents, such as the design stance and the intentional stance—our inclination to attribute intentionality to robot behaviors. By intertwining the cognitive, cultural, and ethical considerations of human–robot interaction, Marchesi and Wykowska present an integrated view that calls for a thoughtful approach to robotic design, one that respects and reflects human values and promotes symbiotic coexistence.

Chapter 4, penned by Luisa Damiano and Paul Dumouchel, poses the profound question, "Are Robotic Bodies (Part of) Social Bodies?" and takes readers on a philosophical exploration of embodiment and sociability in human–robot interaction. They start by contextualizing the concept of embodied AI as a pivotal move away from traditional computationalist AI, setting the stage for the current landscape of social robots. This historical overview ties into previous discussions on the evolution and societal impact of robots (Dumouchel & Damiano, 2017), reflecting a shift from viewing bodies as discrete and individual to recognizing them as occupying a gradient and socially intertwined. The chapter dissects the notion of embodiment, a central theme in the discourse of social robotics, and investigates whether social robots can indeed be integrated as elements of social bodies. This echoes the themes of previous chapters, particularly the question of robots' physical and social presence in human environments, as well as the impact of their design on cultural norms and societal values.

Damiano and Dumouchel shift the traditional philosophical debate from the authenticity of human–robot interactions to the concept of affective coordination, examining how robots might enhance or detract from human social bonding. This perspective aligns with the earlier discussions on anthropomorphism and societal integration, offering a new angle on the emotional resonance between humans and robots. The chapter concludes with an endorsement of synthetic ethics, advocating for a case-by-case, project-by-project analysis of how social robotics can foster stronger human connections. Damiano and Dumouchel's perspective thus complements and expands upon the themes of embodiment and social integration introduced in previous chapters, inviting us to consider how social robots might become integrated into (and even facilitate and enrich) our social structures and relationships.

In Chapter 5, David Gunkel delivers an original and generative philosophical discourse on how to understand robots in moral and legal terms. In "Persons or Things: The Role of Robots in Society," we gain a deeper appreciation for the fundamental questions involved in inquiry regarding the role of robots in society. Unequivocally, robots are technological artifacts, products of design and manufacture. In this sense, robots fall within the category of 'things' and invite the treatment of tools. At the same time, robots exhibit sociality through their appearance, capabilities, and behaviors and invite the treatment of socially significant others.

A recurring theme across the chapters of this *Handbook* is the concept that the behaviors and interactions of robots imply a significance beyond objectification and use; they are becoming integral components of social collectives and contributors to social bonding. Gunkel embarks on a journey to address this dichotomy surrounding robots. Are they, in essence, mere things—technological objects subject to our discretion and use, even if it means misuse? Or should we consider them as something more complex, potentially akin to persons—entities with their own claims and deserving recognition as socially significant 'others'? These questions touch not only on the ontology of robots—what they fundamentally are—but also the broader issues of how we, as individuals and communities, situate robots among us, and ultimately of what we individuals are.

The chapter navigates this moral and legal dilemma by drawing on historical precedents and then highlighting how intelligent technologies like robots disrupt traditional classifications. Gunkel's narrative culminates in the presentation of three potential responses to this categorization challenge, providing a framework for understanding the integration of these technologies into human society. These approaches aim to reconcile the utilitarian aspects of robots with their emergent status as entities that defy existing paradigms, thereby contributing to the ongoing discourse on how we define and interact with robots in our collective future.

Chapter 6, "Automated Masspersonal Social Engineering," by Robert Gehl and Sean Lawson, concludes the section on "Robots in Culture and Society" by examining an intense transformation in communication linked to the rise of AI, automation, and machine learning. This chapter explores the darker aspects of this transformation, focusing on how these technologies are reshaping manipulative communication tactics such as disinformation, propaganda, and fraud. Gehl and Lawson explain how these practices are not merely aided or amplified by technology but are poised soon to be dominated by technology. We find the heart of exploration in the concept of 'masspersonal social engineering,' a term they introduced in their recent book (Gehl & Lawson, 2022). Through this lens, the authors dissect the automation of manipulative communication to reveal its distinct stages—trashing, pretexting, and bullshitting—and explain how technology is increasingly driving these processes.

Gehl and Lawson make evident that automation does not merely make these stages more efficient; rather, it has the potential to intensify each stage or act as a prosthesis to human communication. In essence, technology becomes an enabler of manipulative communication, making it more pervasive and potent. The chapter then turns our attention to a critical aspect: the role of automation in achieving the ultimate goal of masspersonal social engineering—penetration. In doing so, Gehl and Lawson draw intriguing parallels between communication-as-penetration and automation, illuminating shared assumptions about the role of communication in society. In the end, this chapter prompts us to consider the ethical dimensions of our interactions with and through automated systems, a thread that has been woven through the entire section.

2 Section 2: Humanistic and Social Scientific Perspectives

In the second section of this *Handbook*, we present 13 chapters that collectively underscore the cultural and societal impact of robots through the prismatic lenses of diverse academic disciplines. While the humanities and social sciences often operate within their own distinct silos, the interdisciplinary collaboration across fields in this section is essential to a comprehensive and multifaceted exploration. Together, these chapters provide a rich, layered understanding of robots' roles in society and culture. Featuring contributions from leading-edge researchers in fields as varied as linguistics, law, sociology, psychology and neuroscience, and communication and computer sciences, this section exemplifies the importance of converging perspectives to fully grasp the complex interplay between social robots and society.

2.1 Linguistics

In Chapter 7, "AI and Human Writing: Collaboration or Appropriation?" linguist Naomi Baron presents a vital examination of the intersection between artificial intelligence and human writing. This chapter serves as an essential framework for understanding the complex relationship between AI and the deeply human act of writing. Baron, building upon her recent book (Baron, 2023), explores how writing is intrinsic to the human experience and scrutinizes the recent advancements in AI writing tools like GPT-3, DALL-E, and Codex.

Baron analyzes the multifaceted reasons why people write, from everyday communication to creative expression, and how these motivations illuminate the potential roles AI might assume—from assisting and editing to independent authorship. She incorporates user experiences to shed light on the profound effects and public reception of AI in writing. The chapter also tackles the significant challenges and questions arising from AI's involvement in writing: discerning the authorship of content, understanding the limitations of language models, and contemplating the consequences in professional fields such as journalism and education.

Baron prompts a critical reflection on the essence of writing as an expression of our sentience and sociability, contrasting this with the nature of machines. By doing so, she encourages us to weigh the benefits and risks of AI partnerships in writing, ultimately reminding us of the unique value of human thought and the precious nature of our capacity for written expression.

2.2 Law

Chapter 8, "Policies, Regulation, and Legal Perspectives on Social Robots," by Caja Thimm and Laura Thimm-Braun, addresses the often-overlooked yet vital legal aspects of social robotics. Combining expertise in new media and law, they provide a thorough analysis of policy and legal considerations, with a focus on Europe's AI Act and the so-called 'robot law.' The authors emphasize the need for foundational policies that govern all social interactions with robots while also applying a contextual lens to acknowledge varying situational factors. Beyond the perspective of human users, the chapter introduces the concept of the robot as a 'person,' delving into complex discussions surrounding data ownership, intellectual property rights, and machine ethics, as articulated by Gunkel in 2023 (see also Chapter 5). Thimm and Thimm-Braun stress the imperative of adopting a global perspective to address technological advancements like social robots despite the predominant national jurisdiction over law and policy. Consequently, the chapter highlights the unique challenges that legal and regulatory actions for social robots pose to international law, including human rights, national regulations, and industrial applications. In the end, Thimm and Thimm-Braun advocate for a nuanced understanding of social robots that extends beyond their physical manifestations, calling for the establishment of regulations that safeguard human rights and ensure data justice and dignity.

In Chapter 9, "How Social Robots Affect Privacy: Navigating the Landscape," Christoph Lutz, Aurelia Tamò-Larrieux, and Eduard Fosch-Villaronga focus on privacy as a crucial concern permeating the development and use of robots. Compared with conventional technologies, emergent robots introduce unique ethical, legal, and social dimensions. Among these, privacy concerns loom large, given social robots' heightened autonomy, data-processing capabilities, and physical mobility. This chapter offers a comprehensive exploration from foremost experts of the privacy implications associated with social robots. Lutz et al. not only survey pertinent privacy theories but also delve into recent empirical studies examining the fraught relationship between privacy and social robots. Their analysis reveals a shift towards a greater emphasis on data protection in privacy research and calls for a more rounded consideration that includes physical, psychological, and social privacy.

Both Chapters 8 and 9 contribute to a multidisciplinary understanding of social robots, extending the conversation from their physical and interactive characteristics to the broader societal frameworks that must adapt to their presence. Taken together, they highlight the need for an international approach to policy and a comprehensive research agenda that balances technological advancements with the protection of human rights and privacy.

2.3 Sociology

Moving to the sociological significance of these technologies, in Chapter 10, "Artificial Intelligence, Robotics, and the Evolution of the Social Sciences," Richard Ling provides a thoughtful examination of how technological advancements have historically influenced social science research and methodology. He recounts the transformative impact of tools from Galileo's telescope to modern computational technologies, illustrating how they have expanded our understanding of society. Ling presents a narrative that new methods often precede and shape theoretical frameworks, highlighting the co-shaping relationship between empirical observation and theoretical development.

Ling's exploration posits a future where AI could enable the observation of social phenomena that our current theories cannot yet explain, suggesting a reversal of the traditional scientific method where theory typically guides observation. This chapter challenges the social sciences to adapt to the influx of AI and robotics, calling for an approach that combines empirical insights with the development of new theories, mirroring the interdisciplinary collaborations emphasized in earlier discussions. Ling prompts us to consider the potential for AI to both disrupt and enrich the social sciences, advocating for careful consideration as we integrate these technologies into our pursuit of understanding human society.

In Chapter 11, "Human Interactions with (Embodied) AI—The Future of Authenticity in Human–AI Relationships," Astrid Weiss traces the evolving landscape of human–robot interaction over the past two decades. Human–robot interaction, a field distinguished from neighboring disciplines by the material presence of robots and the perception of agency these machines are afforded, has witnessed explosive growth. Weiss masterfully synthesizes a decade's worth of empirical evidence to illuminate the complexities of how people engage with robots and AI, contrasting these dynamics to human interactions with other technologies and with one another.

Weiss introduces authenticity as a pivotal concept in understanding human–AI interactions, highlighting that our perceptions are shaped not just by personal preferences or robot features but by the nature of the interactions themselves. She discusses the significant challenges posed by the complex and 'wicked' problems inherent in studying human–robot interactions, further complicated by the diverse forms these robots can take.

Expanding beyond tangible robots, Weiss also explores the rise of non-embodied AI, such as generative algorithms and virtual agents, and the burgeoning metaverse, which blends the physical and virtual worlds. She argues that this expansion redefines the traditional distinctions between embodied and non-embodied AI, with embodiment emerging as a key theme in authenticity studies. This chapter pushes us to consider the profound implications of authenticity in an era where AI not only mimics human interaction but also creates new forms of engagement.

Weiss advocates for a multidisciplinary approach, suggesting that authenticity could serve as a crucial concept to bridge different research areas within human–AI studies. This comprehensive stance sets the stage for a holistic examination of human–AI relations, aiming to deepen our grasp of these interactions and foster a more inclusive and varied field of research, a theme further explored by Damiano and Dumouchel in Chapter 4.

2.4 Psychology and Neuroscience

Chapter 12, "Mind Perception During and After Interacting with Artificial Agents," authored by social scientists Monica Pivetti and Silvia Di Battista, ventures into the intriguing sphere of mind perception within the context of social robots. This chapter contributes significantly to the understanding of human–robot interaction by revealing how people attribute mind and intentionality to robots, similar to their attributions to other humans and animals. Pivetti and Di Battista navigate the specialized terminology of this field, such as 'theory of mind' and 'anthropomorphism,' providing a detailed review and interpretation of the latest research on mind perception. They meticulously analyze and summarize key findings, delivering a holistic and thorough view of the unfolding topography of cognition in human–robot interaction. Additionally, they address the interplay between mind perception and social acceptance of robots, including insights into the 'uncanny valley' phenomenon, which has heavily influenced the design and societal reception of social robots. By identifying gaps in the literature and opening up discussions for future research, this chapter enhances our understanding of the cognitive processes behind our interaction with robots and points towards more socially attuned engagements with them.

Chapter 13, "How People Perceive Social Robots: The Case of Gender," by Friederike Eyssel and Giulia Perugia, delves into the implications of gender in the context of social robots. Eyssel and Perugia provide an extensive overview of the terminology, historical context, and meta-perspectives that have influenced the lively discourse around gendering robots. A great strength of this chapter is the multidisciplinary consideration of the psychological, social, and ethical consequences of design choices that revolve around gender. The chapter demonstrates the profound effects that gender attributes on robots have, not just on the design process but also on human self-perception, identity, and dignity. This analysis extends the conversation from previous chapters on robot design and social interaction, showcasing how robot gendering reflects broader societal attitudes and expectations about technology and human identity groups. By evaluating key research in human–robot interaction, the chapter presents crucial insights into the interwoven nature of gender, technology, and human cognition.

In Chapter 14, "Relating with Social Robots: Issues of Sex, Love, Intimacy, Emotion, Attachment, and Companionship," computer scientist Kate Devlin examines the complex, controversial, and sensationalized arena of human–robot relationships in intimate contexts. Devlin presents a sweeping journey from ancient myths to historical narratives to contemporary commercial products, analyzing the concept of the 'perfect artificial partner.' As Devlin recounts, in the early twenty-first century, the concept of robots designed for love, pleasure, and intimacy was often met with skepticism or relegated to the realm of science fiction. However, times have changed, and the field has grown, supported by numerous books, articles, research projects, and an established international conference. Despite lingering bemusement, particularly among technically inclined individuals, the chapter underscores the profound relevance of this topic. Devlin shows us that at its core, research on sexual or romantic companion robots—often referred to as sex robots—essentially scrutinizes the essence of human nature: our capacity for love, desire, sociality, and connection.

Chapter 15, "Real or Pretend? How Children Ontologize Social Robots as Mental and Moral Others," by Rachel Severson, Sarah Sweezy, and Thomas Macko, addresses the integration of social robots into children's lives and their interplay with young and developing minds. Approaching from a developmental psychology perspective, the authors explore how children understand social robots as mental and moral beings that extend beyond mere machines. They investigate the evolution of children's perceptions and explore whether these perceptions are genuine or simply pretend. Their exploration is particularly relevant for advancing the discourse on the provocative New Ontological Category hypothesis, referenced by several authors in this collection, that social robots represent a distinct genre of Being which challenges traditional categories, blurring the line between animate and inanimate. The authors emphasize the importance of considering ethical implications as social robots become more embedded in society, particularly concerning children's development and the conceptual challenges posed by these emerging technologies.

2.5 Communication and Computer Sciences

Communication has emerged as a pivotal form of human–robot interaction and a primary ground of confrontation between people and social machines. Many robots are explicitly designed and marketed for communication, and, in reality, the concept of communication has been the linchpin propelling technology from industrial machines to social companions.

In Chapter 16, "Rethinking Communication between Humans and Social Robots," Andrea Guzman brings communication to the forefront of human–robot interaction studies. Although communication rests at the core of human interactions with modern technology, it remains undertheorized in this and related domains. Guzman challenges a prevailing misconception that communication between humans and social technologies is a seemingly 'natural' process, drawing from foundational communication theory to illuminate the acquired, social, and cultural intricacies inherent in any communicative encounter. The chapter is structured in two parts: the first elucidates the pivotal role of communication in shaping the trajectory of robotics research and the development of technologies with communicative capabilities, particularly the extension of human communication principles to human–robot interaction. The second part scrutinizes these assumptions through the lens of communication research, revealing their limitations and the consequent impact on technology development and human–machine communication. Guzman introduces key elements of contemporary communication theory to redefine our perception of meaningful interactions with robots, tying into earlier chapters' discussions on the authenticity and social integration of AI.

Chapter 17, "Interacting with Social Robots: The Influence of their Distinctive Cues, Behavioral Capabilities, and Affordances on Social Interaction and Well-being," by Patric Spence, Chad Edwards, and Matthew Craig, investigates the nuanced cues and capabilities that facilitate human–robot interaction. An assumed goal of social robotics is to design machines that can engage with humans in emotionally aware and intelligent ways. Yet, the features and behavioral abilities that combine to succeed are complex and multifaceted, and, until recently, many were reserved for human interactions. This chapter first unravels and systematizes the distinctive cues (visual, audio, tactile, verbal, and movement), behavioral capabilities, and affordances (or interaction opportunities) relevant to human–robot interaction. Then, the authors discuss the potentially transformative influence of these factors on social practices and human health. Through this wide-ranging analysis, they enable a deeper appreciation of the entangled influence of multiple interactional cues, use potentials, and outcomes. Furthermore, this chapter underscores the importance of human-centered design processes that promote relationships with robots that complement rather than substitute for human interaction.

Chapter 18, "Integrating Big-Data Tools to Study AI and Human–Machine Communication: Methodology Strengths, Future Directions, and Applications," by Xianlin Jin,

Patric Spence, and Seungahn Nah, addresses the intersection of big-data analysis and AI communication research. The integration of AI into diverse domains such as healthcare, emergency response, business, and education unveils both challenges and opportunities for communication scholars.

Beginning with the observation that as AI technologies advance, an unprecedented volume of unstructured digital communication data becomes available, the authors explain that traditional qualitative methods are ill-equipped to analyze such vast data, and purely quantitative approaches often miss the subtleties of human communication. The authors propose that the application of big-data tools, including topic modeling and sentiment analysis, offers a promising way to bridge this gap. This chapter reviews the literature and explores how these tools are used to study AI and robots. It also highlights the potential directions for future research in AI communication and human–machine communication. By employing these methods, scholars and practitioners may unlock deeper insights into the evolving relationship between AI and humans, shedding light on how this transformative technology is shaping our world.

Chapter 19, "Social Robots and Children: A Field in Development," by Jochen Peter and Caroline van Straten, surveys the field of child–robot interaction (CRI). This chapter discusses (a) how social robots facilitate learning, underscored by the central notion of engagement, (b) relationship development, with an emphasis on trust, and (c) children's conceptualization of robots as a potentially novel ontological category. Going beyond a systematized integration of current research, the chapter highlights several challenges confronting contemporary CRI research. These include a prevailing adult-centric bias, a dearth of robust CRI theories, and an overreliance on frameworks borrowed from interpersonal research. In response, the chapter points to prospective developments that hold the potential for substantial contributions to our comprehension of how children and humanity, more broadly, interact with social robots. These include a greater integration of social robots with smart and connected toys, exploring the synergy between social robots and communicative robots, and an increased focus on responsible CRI issues and considerations.

3 Section 3: Contexts of Human–Robot Interaction

The third section encompasses five chapters, each examining different real-world environments where robots interact with humans. From the domestic intimacy of living spaces and care facilities to the augmentation of human capabilities through exoskeletons, each chapter in this section delves into a distinct environment where robotic integration is taking shape. While each chapter stands on its own, a holistic view across them reveals coherent practicalities, challenges, and societal implications of these interactions and highlights the importance of context in shaping the future of human–robot relations.

In Chapter 20, "Anthropomorphizing Voice Assistants: A Research Agenda from Human–AI Relationships," authors Yolande Strengers, Thao Phan, Melisa Duque, and Kari Dahlgren offer an in-depth analysis of the societal and cultural impact of voice-based assistants (VBAs), which have seamlessly integrated into the daily routines of many people globally. Continuing the discussion on the theme of embodiment from previous chapters, the authors position VBAs not only as standalone entities but also as precursors to the broader adoption of everyday AI and robotics.

The authors argue that while VBAs offer convenience and accessibility, they also embody a nexus of social and ethical issues. Strengers et al. embark on a detailed examination of how VBAs are anthropomorphized, focusing on the interplay of race, gender, and age as these devices reflect human characteristics. They present three main research areas: the commercial narratives that drive VBA creation, the design decisions that enforce gender and racial stereotypes, and the real-world interactions users have with these devices, which can perpetuate or challenge these narratives. Strengers and colleagues advocate for a critical re-evaluation of VBA personas like the 'techno-hedonist' or 'smart wife' (see Strengers & Kennedy, 2020), urging a move towards more inclusive and diverse representations. They emphasize the need for improved research methodologies, including collaborative, participatory, and observational approaches, to design VBAs that are equitable and inclusive. This chapter serves as a compass, orienting us towards a more informed and conscious mode of interaction with these seemingly omnipresent digital companions, and it underlines the growing refrain: futures are social and, therefore, require public bodies for disentanglement, debate, and deliverance (Wajcman, 2017).

Chapter 21, "Domestic Appliances and Household Robots: The Changing Landscape of Housework and Family," by Leopoldina Fortunati and Autumn Edwards, takes a historical and sociological look at the introduction of technology into the domestic space and its impact on the gender division of labor. Fortunati and Edwards outline the arrival of household technologies, from post-World War II appliances to modern innovations like the Roomba vacuum and the Bimby kitchen machine, and discuss their influence on domestic roles, particularly in relation to gender dynamics.

They highlight that while men's participation in household chores remains relatively low compared with women's, their involvement in childcare has increased, yet this has not substantially translated to more labor-intensive tasks. The authors seek to understand which demographic each wave of domestic technology aimed to assist, how these technologies have interacted with the distribution of domestic labor and societal gender roles, and which household needs remain inadequately addressed. This chapter emphasizes the necessity to consider both the historical context of household technology and the theoretical insights into the relationship between technology, gender, and the domestic sphere to fully grasp the social significance of these technological changes.

In Chapter 22, "Ability and Disability: Social Robots and Accessibility, Disability Justice, and the Socially Constructed Normal Body," Marco Dehnert addresses issues of ability and disability as a broad concern about representation and the integration of robots with humans. Social robots in assistive roles constitute relationships of mutual shaping and influence with the human identity groups they are designed to help. Dehnert adopts an intersectional approach to discuss how social robots can serve and complement individuals, drawing from critical disability, crip, neuroqueer, and disability justice perspectives. This chapter investigates the concept of ableism and the socially constructed 'normal' body, a notion deeply entwined with our understanding of what it means to be human. Dehnert emphasizes the importance of including disabled bodyminds in social robotics conversations, considering a range of applications from accessibility and inclusivity to neurodiversity. The discourse extends to the use of sex robots as therapeutic tools and the importance of accessibility for older adults. Throughout, Dehnert stresses the need to center the experiences and voices of those for whom these technologies are designed. Additionally, this chapter underscores the necessity of moving beyond one-size-fits-all solutions to recognize the diversity within the experience of disability. This chapter emphasizes the significance of disability justice, advocating for a focus on the unique needs and voices of marginalized communities, and promotes an inclusive approach to social robotics that respects individual agency and aims to enhance the daily lives of disabled persons.

In Chapter 23, "Growing Old Together: The Promise and Challenge of Social Robots for Older Adults," authors Sawyer Collins and Selma Šabanović discuss the role of social robots in eldercare, a field that has become increasingly significant due to the rising population of older adults in developed economies. Collins and Šabanović detail how social robots are being designed to assist older adults and their caregivers with companionship and healthcare, addressing the demographic shifts that necessitate such technological support.

The authors acknowledge the promising research regarding the effectiveness of social robots in eldercare but also address ongoing challenges related to their adoption and practical use. They argue that overcoming these hurdles requires more than technological solutions; it requires a nuanced understanding of the societal, cultural, and ethical aspects of caregiving. This chapter calls for a reorientation of the narrative around aging and technology—from one that focuses on disability to one that enhances the quality of life for older adults through engagement in meaningful activities and relationships. This includes considering the vital network of caregivers and the community.

Echoing the themes from previous chapters on inclusivity and accessibility in the design and use of social robots, Collins and Šabanović emphasize the need for a comprehensive and ethically responsible approach to the integration of robotics in eldercare, guiding us towards a future where eldercare robotics are more seamlessly incorporated into the fabric of supportive community and care networks.

In Chapter 24, "Power and Synchrony in Human Collaboration with Exoskeletons," Nan Wilkenfeld, Norah Dunbar, Chengyu Fang, and Divya Srinivasan delve into the transformative potential of wearable exoskeleton technologies. The authors guide us through an exploration of how robotic exoskeletons can enhance physical capabilities and provoke a paradigm shift in human–machine dynamics, emphasizing the implications for identity, power, and agency.

Wilkenfeld and colleagues explore the multifaceted impact of exoskeletons considering the physical, cognitive, and social aspects of the workers who use them. They utilize dyadic power theory and interpersonal adaptation theory to illustrate the parallels between power dynamics in human–exoskeleton collaboration and human interpersonal relationships. This approach sheds light on how individuals adapt to wearing exoskeletons and the significant changes in agency and identity that can result from this interaction.

The chapter prompts us to consider the future of design and engineering for exoskeletons and the broader implications for the wearers and the various societal, organizational, and cultural contexts within which they operate. This exploration builds upon the earlier chapters' themes of mutual shaping and influence between humans and assistive robots, and it extends the conversation to encompass the evolving intersection of power, agency, and identity in the age of advanced wearable technologies.

4 Conclusion

The *Handbook of Robots in Society and Culture* offers a comprehensive exploration of the multifaceted relationship between humans and robots, marking a significant contribution to the field. Throughout its chapters, the collection systematically presents robots' embodiment as a gradient and their sociality not as an inherent trait but as a joint product of ongoing social interaction. This perspective emphasizes that the sociality of robots is an achievement crafted through each encounter with human beings rather than a static feature. Our authors repeatedly underscore the dynamic process of social integration, where robots are continuously co-constituted through their interactions within human environments.

The *Handbook*, when read as a whole, also places a strong emphasis on the theme of pragmatism—meliorism, agency, and choice—and how these concepts can be leveraged to create futures where human thriving is at the forefront. The various chapters navigate through the intricacies of incorporating robots into everyday life, highlighting both the potential for technological augmentation and the need for ethical considerations. They advocate for a future that is not passively received but actively shaped by human values, aspirations, and choices, ensuring that the integration of robots into the lifeworld enhances rather than detracts from human well-being.

At this high-stakes juncture when social robots are transitioning from theoretical constructs to practical, real-world applications, the urgency of understanding the role of society and culture becomes apparent. Scholars within this volume have stressed this urgency, articulating from a variety of academic fields the same premise: *the successful integration of robots into the social fabric requires a deep understanding of societal norms, cultural contexts, and the diverse expression of human life*. This understanding is not just theoretical but practical, and should guide the design, implementation, and governance of robotic technologies.

The intellectual effort enclosed in this *Handbook* is remarkable and offers many ideas, reflections, and predictions. Here we would like to focus on three main innovative points that have emerged. The first is the clear reconfirmation of the concept of robotic continuum, against a definitional stance invoked often to draw ontological distinctions justifying the definitional exclusion of unembodied (or less fully embodied) technologies from the general category of robot. On the contrary, the concept of robotic continuum, advanced for the first time by Edwards et al. (2024), helps us build a barrier-free taxonomy of robotic products coming to market. There are not only social robots, but these are preceded and accompanied by a series of robotic creatures that incorporate some elements of classic robots within them. It would be a serious mistake to neglect this plethora of technological artifacts to focus only on classic robots. In fact, we would not understand that an important phenomenon of the diffusion of robotization in the social world is the diversification and explosion of its different forms. Conceiving robots for the social world necessarily requires a strong graduality, because the social world represents the opposite of what robotics is: it is flexibility instead of rigidity, it is unpredictability instead of predictability, it is multiplicity instead of uniformity, and so on. Thus, the diversification and gradual progression of robotic forms and products is a mandatory strategy for dealing with the social world and for incorporating at least some social characteristics within robots.

The robotic continuum spans from social robots corresponding to the traditional concept of sense–think–act technological artifacts to a plethora of quasi-robotic products that have one or two robotic features and are perhaps un-embodied, minimally embodied, embodied only imaginarily, dispersed in networks, clouds, or environments, or endowed with a full, locally present body. This plethora includes chatbots without embodiment, virtually embodied avatars, voice-based virtual assistants with a limited form of embodiment and no capacity for moving, and social robots with limited moving ability, such as Paro, or the robotic heads (e.g., Furhat) or trunks (e.g., Sophia first version). All these robotic forms resonate, develop, compete, are bought and used, spread, work, and co-evolve in the same environment, mainly that of social reproduction. This is one of the cases in which less is more (Harper, 2012), in the sense that removing one or more features has made these technological artifacts more convincing in the eyes of potential users (who buy them). VBAs, for example, are quasi-robotic products, that is, products that contain several features of social robots, even if not all of them. As Strengers et al. put it, "VBAs . . . provide a window into a broader set of relations that are anticipated to emerge in a world saturated with AI, robotics, and automation" (p. 405 of this volume). As Strengers et al. point out, VBAs provide insight into a broader set of relationships that are expected to emerge in a world increasingly dominated by AI, robotics, and automation. Furthermore, in the history of social robotics, VBAs were the first to offer the opportunity to study the mass consumption of a 'robotic product' in a 'natural' environment, like houses (Edwards et al., 2024).

The social robotics field has only to benefit from including the study of all these forms in its discipline, both because it acquires visibility on the interrelationships between these quasi-robotic products and classical social robots and because it can learn a lot from these quasi-robotic products. While classical robots (sense–think–act) still have difficulties jumping into the 'real' world, quasi-robotic forms are much more proactive and are those, for the moment, most successful regarding diffusion and in economic terms. Social robotics is the most suitable field to analyze this broad development, which takes place in the domestic and social sphere, the place where these new robotic forms develop and co-evolve with classical social robots.

The second important point that has emerged in these chapters is the emphasis on the strategic relevance of humans as users of robots in order to grasp both the essence of robots and humans. Many authors of this *Handbook* have shown the role of users—women, children, older adults, factory workers—as engines of a co-constructing process between robots and users in everyday life. First of all, users have great power in their hands and it is purchasing power, the willingness or not to buy robotic products. So far, they have exercised this right in resisting the mass purchase of robotic products with the exception of virtual assistants, Roomba, and Bimby, all products robotically limited. Second, users in the consumption phase have been and are very active in expressing their agency, in reshaping the robotic artifacts, in making sense of them in perhaps a different way from what the designer had envisaged, and so on. There is no space right now for cultivating a technological deterministic approach about social robots; on the contrary, there is the urgency to be brave in imagining a social robotics capable of creating futures of human thriving.

The third point is that, in the moment in which social robots have begun to be integrated in the real world, the understanding of how society and culture react to them has become pressing to scholars working in this field, as this volume shows. Robots are one of the most archaic elements of our imaginary. This *Handbook* offers some deep analyses on how cinema and television—two important technologies of culture and popular culture—as well as fashion and advertising have currently elaborated social robots in their visual products. It is only the beginning of a perspective that needs to develop strongly in the near future, because society and culture represent the ways in which the real world has structured itself.

The *Handbook* thus serves as a clarion call to scholars, designers, policymakers, and users to engage with social robots not merely as technological marvels but as entities that exist within a complex social and cultural milieu. We challenge readers to consider the broad implications of these technologies and to participate in shaping a future where social robots are developed and deployed in ways that are inclusive, equitable, and reflective of our shared humanity. As social robots become more ingrained in various aspects of life, from domesticity to eldercare, from work to play, this *Handbook* provides a vital framework for understanding and directing their evolution in harmony with the societies and cultures they will indelibly influence.

References

Baron, N. S. (2023). *Who Wrote This? How AI and the Lure of Efficiency Threaten Human Writing*. Stanford University Press.

Dumouchel, P., & Damiano, L. (2017). *Living with Robots*. Harvard University Press.

Edwards, A., Edwards, C., Fortunati, L., Manganelli, A. M., & de Luca, F. (2024). Mass robotics: How do people communicate with, use, and feel about Alexa? A cross-cultural, user perspective. *Computers in Human Behavior: Artificial Humans*, 100060. https://doi.org/10.1016/j.chbah.2024.100060.

Fortunati, L., & Edwards, A. P. (2022). Gender and human–machine communication: Where are we? *Human–Machine Communication*, *5*, 7–47. https://doi.org/10.30658/hmc.5.1.

Gehl, R., & Lawson, S. (2022). *Social Engineering: How Crowdmasters, Phreaks, Hackers, and Trolls Created a New Form of Manipulative Communication*. MIT Press.

Gunkel, D. (2023). *Person, Thing Robot: A Moral and Legal Ontology for the 21st Century and Beyond*. MIT Press.

Harper, R. H. (2012). *Texture: Human Expression in the Age of Communications Overload*. MIT Press.

Strengers, Y., & Kennedy, J. (2020). *The Smart Wife: Why Siri, Alexa, and Other Smart Home Devices Need a Feminist Reboot*. MIT Press.

Wajcman, J. (2017). Automation: Is it really different this time? *British Journal of Sociology*, *68*(1), 119–127.

Section 1: **Robots in Culture and Society**

Matthew Lombard, Kun Xu, and Melissa E. Selverian

Future Presence: Living with Social Robots

Abstract: As robotics, artificial intelligence, and other technologies have rapidly advanced, robots are being designed to be increasingly interactive, social, and human-like. Meanwhile, some are already assuming diverse traditionally human roles such as teacher, assistant, and companion. As increasingly sophisticated social robots become integrated into our lives, we need to better understand what physical forms they're likely to take, how we'll respond to and interact with them, and the potential benefits and dangers they'll bring. This chapter relies on a multi-method review to predict and understand our future interactions with social robots. The chapter draws predictions from three key sources: the theoretical framework of the Media Are Social Actors paradigm, academic and industry experts as well as non-experts, and a textual analysis of popular media portrayals of social robots. These three sources are used to address the prospective forms and capabilities of social robots, the evolving nature of our interactions with them, and the psychological, physiological, and sociological effects they will have on us. Informed by the review, the chapter concludes with suggestions for actions we can take to help constructivevly shape the evolution of these human-like social robots and our lives with them.

Keywords: Social robots, social presence, Media Are Social Actors (MASA), human–robot interaction, robot industry, popular media portrayals, technology ethics

1.1 Introduction

Since humans developed the ability to imagine and use primitive tools, we've created representations of other humans and human-like entities. From drawings and statues, to photographs and films, to computers and artificially intelligent humanoid robots, we've become ever more proficient at using technology to create simulated human beings. Today many people are not only entertained and comforted by these representations but also work, play, and live with them. In the 2020s a series of media technologies capable of presenting human representations—smartphones, virtual and augmented reality agents, chatbots, voice assistants including Alexa and Siri, conversational generative artificial intelligence (AI) including OpenAI's ChatGPT, digital humans, and humanoid robots including Sophia (Hanson Robotics), Ameca (Engineered Arts), and Realbotix androids, have been integrated or are expected to soon be integrated into our lives. Scholars and practitioners from technical fields (computer science, robotics), social science (communication, media studies, psychology, business), and the arts have all explored and influenced the evolution and potential impacts of these artificial humans. As we approach an exciting but uncertain future in which social robots may become so sophisticated as to be indistin-

https://doi.org/10.1515/9783110792270-002

guishable from real human beings, we need to carefully examine what benefits and dangers the future will offer so that we can make decisions that maximize the former and minimize the latter.

This chapter begins with some key definitions and reasons to consider predictions about the continued evolution of these technologies and how they may impact us in both positive and negative ways. Three related methods to accomplish this are then introduced: Applying the Media Are Social Actors (MASA) paradigm and research on how humans respond to technologies that present social cues; reviewing predictions of experts who design, manufacture, market, and study these technologies, including roboticists and computer scientists; and examining patterns in the many fictional portrayals of humans interacting with advanced social robots, primarily in science fiction. We then turn to an exploratory application of these methods to raise and consider possible answers to a series of key questions about what life may be like as social robots become (more) integrated into our lives. We conclude with some steps we collectively should consider taking to prepare for, and help shape, that future.

1.2 Definitions

Before we predict and consider the implications of the future roles of artificial humans in society, we need to identify the scope of our concern. While the ultimate artificial human being would be an embodied humanoid social robot that looks and acts exactly like an actual human being (the terms 'android' and 'gynoid' are often used to denote those with masculine and feminine characteristics, respectively), and that is a primary focus here, that technology is a member of the larger category of social robots.

Robots are essentially devices that can automatically perform complex tasks. The subset of social robots has been defined in several different ways. Some scholars require physical embodiment of the technology, defining a social robot as "a physical entity embodied in a complex, dynamic, and social environment sufficiently empowered to behave in a manner conducive to its own goals and those of its community" (Duffy, 2003, p. 177), or "devices with mechanical moving parts that interact in socially appropriate ways" (Li et al., 2016, p. 3). Reeves et al. (2020) referred to social robots as: "physical embodiments (i.e., they exist in the real world not just on a screen), and they communicate with humans via social behaviors (e.g., speech, gestures, and movement) that mimic human interactions that are linked with particular social roles that a robot might play".

For other scholars, the behavior of the technology is required but physical embodiment is not. For example, Fox and Gambino (2021) referred to social robots as "human-made technologies that can take physical or digital form, resemble people in form or behavior to some degree, and are designed to communicate with people"

(p. 295). Similarly, Zhao (2006) argued that humanoid social robots can be mechanical, residing in physical space, or digital, residing in cyberspace.

In this chapter, we include both physically and virtually present social robots in our discussion. Physically present social robots include Softbank's robot Nao and Pepper, as well as Hanson Robotics's robot Sophia (for more see The Anthropomorphic Robot Database (http://www.abotdatabase.info/). Virtually present robots include "digital humans" (Campbell & Jovanovic, 2022) such as Gatebox, Samsung's virtual human Neon, the digital AI companion Replika, and even Microsoft's Office assistant Clippy.

Note that being physically (vs. virtually) present and being embodied (vs. disembodied) are two different dimensions to categorize social robots. Being physically (vs. virtually) present means social robots are collocated in the same space as users, while being embodied (vs. disembodied) refers to whether a robot has any form of a (physical or virtual) body (Ziemke, 2013). In this chapter, we primarily focus on the physically present embodied social robotic technologies and exclude technologies that are portrayed in science fiction as invisible, environmental, ubiquitous, and/or omnipotent (e.g., in *The Matrix* films (1999, 2003, 2021)). Given the complexity of the definitions (Ziemke, 2013), as long as technologies have any physical form that can be interpreted as a body, we consider them to be embodied (e.g., Alexa embodied in Echo devices, Siri embodied in iWatches and iPhones). Meanwhile, we acknowledge that as technology becomes increasingly invisible, intuitive, and seamless, it is becoming challenging to draw the boundaries of social robots.

1.3 Predicting, Preparing For, and Shaping Our Future Lives with Social Robots

The future is of course uncertain, but the trajectory makes it seem very likely that a wide variety of social robots will increasingly become integrated into our daily lives and those of our descendants. It's important to try to better understand not only the timing of these changes but what this future might be like. Aside from the fact that it's inherently enjoyable for humans to imagine different futures (thus the popularity of stories set then), imagining the range of possibilities helps us appreciate our past and present, prepare for what may happen, and set goals and devise plans that will shape the future to address its potential dangers.

We will use three key approaches to gathering information to predict and understand the future in which social robots will be ever present:

MASA scholarship

Research on how we'll respond to social robots can be traced to the early 1990s when Clifford Nass at Stanford University proposed that, as computers increasingly provided real-time, language-based interaction and began to fill social roles, users were likely to respond to them in some of the ways they respond to people. This idea and the approach to testing it was labeled the Computers Are Social Actors (CASA) paradigm (Reeves & Nass, 1996; Nass et al., 1994). In a series of studies, Nass and his colleagues identified findings from social psychology about how people perceive and behave towards each other and at least partially replicated his earlier findings with one of the human roles being taken by a computer. Based on their results, they suggested that we mindlessly apply some social scripts from human–human communication to human–computer interaction.

The CASA paradigm focused on responses to desktop computers (and later other technologies such as televisions) as social actors. A variety of technological innovations since then have brought us smartphones, electronic tablets, smartwatches, chatbots, voice assistants, virtual humans, and humanoid social robots. One common feature of all these technologies is that they present certain combinations of social cues that evoke our responses to them as if they were social entities, including human voices, anthropomorphic language, and aspects of human-like appearance. Casting light on the discrete effects of individual social cues on our social responses to technologies, Lombard and Xu (2021) structurally extended the CASA paradigm to the MASA paradigm to provide more subtle, systematic, and evolution-based predictions and explanations for our social responses to technologies beyond computers.

While recent literature has extended the CASA paradigm from different perspectives, including proposing more research on human–media agent script (Gambino et al., 2020) and clarifying the ontological differences between *computers as social actors* and *computers are social actors* (Van der Goot & Etzrodt, 2023), the MASA paradigm centers on how individual primary and secondary cues, and combinations of them, evoke a type of (tele)presence experience, medium-as-social-actor presence. In presence experiences, users overlook or fail to fully acknowledge the role of technology in an interaction (Lombard & Ditton, 1997); in medium-as-social-actor presence, they misperceive a media technology as an independent social actor. The medium-as-social-actor presence can then lead to a variety of social behaviors based on that misperception.

Applying the theoretical underpinnings and research findings of CASA and the MASA framework is one way to predict and understand how users may respond to and live with social robots in the future.

Expert opinions

The second source of information regarding how (and when) social robots will fully enter our lives involves consulting various experts. This includes industry and business experts who design, manufacture, and market these technologies; industry and academic experts who study and track societal trends; futurists; and members of the public, who ultimately decide whether and how technologies are adopted and used. This information is found in business forecasts, reports, and white papers; think-tank project reports; academic articles and books; and news and analysis reports in non-academic venues.

Popular culture portrayals

The third source of information we use to predict our future interactions with robots involves insights from popular culture. In 2001 we initiated a long-term project to compile information about and develop insights from the many fictional portrayals of a variety of presence-evoking technologies. As of July 2023 the database on the project's website (https://smcsites.com/telepresence/) contains 225 media portrayals, including 104 films, 87 television series/episodes, and 34 novels/short stories. Among the many popular examples of media portrayals that feature social robots are the *Star Trek* (1966–), *Star Wars* (1977–), and *Terminator* (1984–) franchises, the *Blade Runner (1982– 2017)* films, the *Black Mirror* (2011–) television series, and the films *2001: A Space Odyssey* (1968) and *Ex Machina* (2014).

For each portrayal, project team members enter information regarding the featured presence-evoking technology, its creators and users, the presence experiences of users, and the short- and long-term consequences of the creation and use of the technology. For this chapter we conducted textual analysis of the information for more than 50 relevant portrayals to identify the key themes and patterns. We particularly focused on these themes: the characteristics and capabilities of future robot technologies; the nature of interactions with them; psychological, physiological, and sociological effects of these technologies; the types of experience with robots and other AI-based technologies that the portrayals suggest we want and fear; and ethical issues raised by the technologies.

1.3.1 Limitations of technology predictions

We acknowledge that technology experts in industry and academia, as well as creators of speculative fiction and the general public, all have mixed records when it comes to making accurate predictions about the capabilities and impacts of evolving technologies. Technologist Edward Tenner suggested three reasons why it's difficult to make accurate predictions (Flam, 2023): "Stubborn bottlenecks" that prevent major breakthroughs, dominance of existing technologies despite impressive inventions, and "social, cultural and psychological factors [that] keep predictions from coming true." Šabanović (2010) details how robotics researchers in the US and Japan generally have a "linear, technologically determinist view" (p. 439), in which society has to "accept and adapt to" changes in robot technologies. She proposes instead a "framework of mutual shaping and co-production . . . [that] focuses on analyzing how social and cultural factors influence the way technologies are designed, used, and evaluated" (p. 439). And of course creators of popular culture portrayals are less focused on the accuracy of their predictions than the extent to which their work is entertaining, thought provoking, and successful. In fact, fictional portrayals serve as one of the key social and cultural factors that influence the development of technologies (see Weiss et al., 2011).

1.4 Key Questions

Together, the above approaches to exploring the possible futures of social robots lead to two key conclusions: (1) it's extremely likely that social robots will continue to become more advanced and capable, and more widely available to the public; and (2) they will very likely alter the human experience in significant ways—changing our view of what it means to be alive and sentient, and raising numerous challenging practical, psychological, cultural, and ethical questions. We consider several of these critical questions next.

1.4.1 Question 1: How far will humanoid robots and other technologies that evoke medium-as-social-actor presence evolve? To what extent will they come to appear and behave like human beings?

The most advanced social robots at this writing take roughly the shape, size, and form of humans and have rudimentary abilities to move and interact as humans do. Recent rapid advances in large language model (LLM)-based generative artificial intelligence using conversational interfaces, for example in ChatGPT, have caused members of the public and even AI experts to perceive the software personas as sentient (De Cosmo, 2022). When incorporated fully into humanoid bodies, the embodied AI will seem even more human. While these technologies are a long way from appearing, sounding, and behaving fully like actual human beings, they are becoming more like us with every design iteration; how far will they advance in the future?

1.4.1.1 Expert and non-expert predictions

Many expert observers base their predictions on recent accelerating developments in robotics. For example, in a 2021 editorial Wiederhold (2021, (p. 289)) writes:

> Though robots—especially those with human physical and personality traits—may still seem futuristic, in the 3 years since I last addressed the topic, robotics has gone from a nascent industry to one that has not only taken its first steps but is poised to take off and run. . . . The idea of robots filling human roles is not farfetched. Currently, robots help us run our homes, complete our errands, and even provide a basic level of companionship. As AI technology improves (and it is improving at a brisk pace), the abilities of social robots will become more and more sophisticated, allowing them to take on an increasingly complex set of responsibilities.

Fortunati et al. (2015) note that others (e.g., Vincent, 2013) consider portable phones and other smart machines to be "personalized and emotionalized social robots . . . that will be increasingly able to interact with our emotions and will become proactive and relational" (p. 234). They argue that "two technological trajectories seem equally possible: to apply [human-like] robotic technology to the media or to integrate media in robots. What would be the best option remains an open question, but in general mobile phones and computers could take big advantages from robotics and vice versa" (p. 234).

In the extensive report, *Robotics Outlook 2030: How Intelligence and Mobility Will Shape the Future*, Lässig et al. (2021) of Boston Consulting Group predict that "[c]hanging consumer preferences and social trends will accelerate the need for advanced robotics solutions." They conclude that "the trend lines are obvious" and "in the next couple of decades" will lead to the Google World scenario, in which

> breakthrough advances in machine intelligence, physical adeptness, and connectivity result in a spate of smart robot modules handling complex and dynamic situations. These modules can be interchanged using simple plug-and-play concepts, expanding robot environments and uses immeasurably. . . . The biggest growth area is likely to be in professional services robots (rather than the more traditional industrial robots), including autonomous hotel and consumer delivery equipment and railway or airport maintenance robots.

A minority of predictions involve other scenarios. In 2014 some of the experts Pew Research asked to make predictions about the state of AI and robotics in 2025 believed at least some robots would be less like humans and even disembodied (Smith & Anderson, 2014):

> Depictions of robotics and artificial intelligence in popular culture often lean towards powerful anthropomorphic robots (Transformers, The Terminator) and hulking mainframes with human-like intelligence (HAL in 2001). But many of the experts who responded to this survey expect technology to evolve in the opposite direction, with machine intelligence being hidden deep in the complex workings of outwardly simple or even invisible devices and digital interactions.

Another possibility is that, as with telepresence robots used for remote presence at workplaces and classrooms, human users will be able to embody otherwise AI-powered social robots, to essentially take over their functions as security guard or elder caregiver when needed, as envisioned by the CEO of the OpenAI-owned company 1X (Wilkins, 2023).

Some predictions about AI suggest that it will become not just capable of convincingly mimicking humans, leading to both functional and dysfunctional presence perceptions and social responses, but also develop sentience (DeGrazia, 2020) or even human or superhuman intelligence that could lead to our inadvertent or intentional destruction (e.g., Yudkowsky, 2023).

1.4.1.2 Predictions in popular culture

Nearly all fictional portrayals of social robots are set in the future and so provide predictions regarding the degree to which robots will come to appear and behave like human beings.

1.4.1.2.1 Appearance

Robots in books, films, and TV programs have appeared increasingly like humans. In L. Frank Baum's 1907 novel *Ozma of Oz*, the robot Tik-Tok is a round-bodied mechanical man made of copper clock parts. In the 1956 film *Forbidden Planet* (and other productions), Robby the Robot is a seven-foot-tall man-shaped metallic robot. In the 1980s and 1990s, *Star Trek: The Next Generation* (1987–1994) introduced the android Data, appearing human except for his tinted skin and stilted speaking voice. In many recent stories robots appear fully human, including the bioengineered Replicants of the *Blade Runner* films (1982, 2017); the android police partners in *Almost Human* (2013–2014), a TV series set in 2048; and the consumer robot servant Synths of TV's *Humans* (2015–2018).

Although they represent a minority of portrayals, there are also notable examples of fictional robots that appear less human-like (Saffari et al., 2021), including the droids C-3PO and R2D2 in *Star Wars* (1977–); TARS (Technically Artificial Robotic System) in the film *Interstellar* (2014), with cube-like appendages made of metal and stone; WALL-E (2008), a cube-shaped stubby robot with wrench-like arms that travels on treads; and the red-'eyed' camera lens HAL in *2001: A Space Odyssey*. The robots described as the most evolved and sophisticated appear as indistinguishable from humans. For example, in the *Terminator* (1984–) franchise about robotic assassins that travel back from the future, early and less advanced models look mostly like machines while the later ones look identical to humans (Oleksinski, 2019).

1.4.1.2.2 Behavior

The portrayals suggest that social robots will also become increasingly effective at mimicking human behavior—the Replicants of *Blade Runner*, the Synths of *Humans* and the most advanced Terminators move, speak, and demonstrate a variety of other richly human behaviors. But many portrayals demonstrate some of the challenges in reaching this level of mimicry. *Star Trek*'s Data can't use contractions when he speaks and initially doesn't fully understand or demonstrate normal human behaviors. Replicants, born fully formed, have a reduced ability to learn and express human emotions, as does the robot Walter in the film *Alien: Covenant* (2017).

At this writing, the conversational AI ChatGPT is being given the ability to remember each conversation, making it more capable of carrying on personalized and rich interactions (Edwards, 2023). Scholars from Stanford University and Google (Park et al., 2023) have used similar LLMs to train AI agents to simulate 'believable' human behavior. In a virtual community reminiscent of the videogame series The Sims (1989–), AI agents demonstrated social behaviors including initiating and maintaining long-term relationships with other agents, coordinating activities with other agents, and spreading rumors in the community. Meanwhile scholars are exploring ways to provide AI agents with virtual and physical bodies to produce entities "capable of interacting with people and each other in a socially intelligent manner using multimodal communicative behaviors with the goal to support humans in various domains" (Lugrin et al., 2021, [abstract]; also: Whang, 2023). These and other developments suggest that although it will be exceedingly challenging and take time, the social robots of the future will follow those in fictional portrayals and not only look much like people but behave like them as well.

1.4.2 Question 2: Will we ever be able to understand, predict, and fully manipulate medium-as-social-actor (and other) presence responses to social robots?

Creating robots that appear and behave like humans doesn't guarantee that humans will respond to them in social ways and accept them in roles of significant responsibility, so how likely is it that future robots will evoke medium-as-social-actor presence and the corresponding social responses? Although there are individual differences across characters, in many popular culture portrayals the humans are vulnerable to the manipulation of the social cues of robots. In addition to appearance and behavioral attributes, social roles and situations seem to play a significant role in evoking presence perceptions and social responses.

1.4.2.1 Popular culture

A clear pattern emerges in responses to robots in fictional portrayals. The robot's assigned identity, personality, and emotions, and the context of the interaction, impact its perceived humanness as much as its appearance and actions. For instance, Robby the Robot, a robot icon that first appeared in the film *Forbidden Planet* (1956), is designed to be singularly friendly and accommodating, and the human characters perceive and respond to it that way. Ultron in the movie *Avengers: Age of Ultron* (2015) is designed to express anger and destroy humans; unsurprisingly the human characters react to it as dangerous. And several of the female robots in TV's *Westworld* (2016–2022) have been given limited subservient personas; in response, they are treated primarily as objects for violence and sex.

When fictional robots that appear either like humans or machines and exhibit human-like behavioral traits are assigned a multidimensional role and exhibit complex emotions, human characters are more likely to perceive them as more fully human and engage in a deeper relationship with them. For instance, the human-looking mecha robot David in the film *A.I. Artificial Intelligence* (2001), created to take the place of a couple's ill son, believes it is human and tells its mother it loves her, winning her love for a time. In the film *Her* (2013), the loving voice of an AI virtual assistant embodied in a phone-like device evokes feelings of romance. As the social cues of robots become richer in both quantity and quality, so does the robots' capacity to adopt both good and bad traits, and humans rightly see them as both friend and foe. A vivid example of this is the host robot Dolores in *Westworld*, which is initially sweet and happy but ultimately displays anger and violence after discovering its robot identity; the human characters' responses move from joy to fear, appropriately. In popular culture at least, robot designers are usually successful in using appearance, behavior, identity, and personality to fashion robots that evoke the desired social perceptions and responses.

1.4.2.2 MASA scholarship

While fictional portrayals provide some clues about how successful we'll become in creating social robots that evoke medium-as-social-actor presence and corresponding social behaviors, research conducted within the CASA and MASA paradigms constitutes more systematic data. Those findings support the idea that designers can use social cues to influence and perhaps manipulate users' responses to social robots. CASA research describes and predicts how some of the social scripts of interpersonal communication are applied in human–computer interaction, especially when computers are given social attributes. For example, in daily communication people tend to be polite to communication partners, providing less negative evaluations to them directly than to a third person. Nass et al. (1994) demonstrated the same phenomenon when people evaluated a computer directly rather than to a second computer. CASA research also revealed that people apply gender stereotypes in their interactions with computers, perceiving a male-voiced computer as more competent and more familiar with mechanical and technical topics, and a female-voiced computer as more familiar with romantic relationship topics (Nass et al., 1997). Some other social scripts that Nass and his colleagues found to be applicable to human–computer interaction include participants' preference for a computer that is perceived to have a similar personality to their own (Nass & Lee, 2001), and participants' willingness to trust and cooperate with a computer that is assigned to the same team as the participants (Nass et al., 1996). For these studies, the social cues include human voices, language, text, and others in various combinations.

In the MASA paradigm, social cues are categorized as primary (sufficient but not necessary to evoke social perceptions and responses, e.g., facial expressions, human shape, gestures, and human-sounding speech) and secondary (neither sufficient nor necessary to evoke social perceptions and responses, e.g., human size, random movements, machine-sounding speech); and researchers test the effects of individual cues on user responses to social robots and other technologies. For example, Xu (2019) compared users' responses to gestures (vs. random movements) as well as human-sounding (vs. machine-sounding) speech in a humanoid robot called Alpha. The participants reported more feelings of attachment and a greater likelihood to use the robot in the future when the robot presented gestures, and they expressed greater trust in the robot when it presented human-sounding speech. In another example, Chang et al. (2022) tested the effects of designing eyes into self-driving cars. This simple change improved pedestrians' sense of security when interacting with the cars; participants made safer and more efficient decisions when crossing a street. An early meta-analysis of the effects of robots' social cues on presence and trust supports the idea of a hierarchy of cues (Xu et al., 2023).

In addition to predicting the discrete effects of social cues designed into technologies, the MASA paradigm predicts that we perceive technologies as having abstract human characteristics based on different combinations of primary and secondary cues. For example, research suggests that we perceive computers or robots as having distinct personalities when designers manipulate the speech styles, vocal features, and movement frequency in these technologies (Lee & Nass, 2005; Mou et al., 2020). Another example is Replika's 'AI companions,' virtual humans designed to serve as a user's friend and even romantic partner. The conversations between users and their Replika demonstrate that people can have strong emotional connections to a virtual human when it presents the right combinations of human appearance, language, and voice cues (Brooks, 2023).

The MASA paradigm and related scholarship suggest two evolution-based explanations for why both the quality and quantity of social cues lead us to perceive and respond to social robots as independent social entities. The first explanation is based on Reeves and Nass's (1996) argument that our brain has not evolved to differentiate mediated objects from non-mediated objects. In other words, from an evolutionary perspective, our bottom-up (default) psychological processing is to presume instinctively that all potential threats are real, followed by our top-down processing that allows us to acknowledge the technology-mediated nature of virtual objects. As Mantovani (1995) suggested, "human cognitive systems have developed adaptively the tendency to treat all representations as if they were true, except when there is proof to the contrary" (p. 680). The second explanation is based on theory of mind, which holds that human beings rely on our communication partners' social cues to predict their motivations, emotions, and behavior (Frith & Frith, 2005). By continuously reading others' social cues, we can adjust actions accordingly and facilitate more effective communication and collaboration (Hare, 2007). When these social cues are produced by technologies, we attribute intentions and motivations to the technologies to streamline our interactions with them, as we do with other people.

Overall, the CASA and MASA paradigms provide theoretical insights and testable predictions for our perceptions of emerging technologies as social actors as well as our applications of social scripts to human–technology interaction. With the current scholarship as a foundation, it seems likely that, even if future technologies don't permit robots to share all human physical and social attributes, or we strategically choose not to build them with all of these attributes, the designers of social robots will be able to evoke and even manipulate the degree and type of medium-as-social-actor presence perceptions and social responses of the humans that encounter these technological beings.

1.4.2.2.1 Uncanny valley effect

While various constellations of social cues may have positive effects on both the breadth and intensity of social responses to robots, some human-like forms may cause users discomfort. Noting Sigmund Freud's observation of a strange feeling people get when looking at dolls and wax figures, roboticist Masahiro Mori argued that when a robot is very much but not fully like a human being, we are likely to experience a 'strange familiarity' or 'familiar strangeness" that creates a sense of unease (Mori et al., 2012). The small but noticeable distance between being very much like a human and fully like one is termed the 'uncanny valley'.

The resulting unease may be so jarring that the response is repulsion, sometimes sought deliberately by creators of horror movies (e.g., *Annabelle* (2014, 2017, 2019), in which a raggedy doll comes to life). Research in this area, however, suggests there may be ways to avoid the uncanny valley effect, including making robots' body parts proportional and physical features less intimidating and friendlier (as with the most human-looking version of the robot servant Andrew in *Bicentennial Man*) (Azarian, 2016; Mathur & Reichling, 2016). Science fiction portrayals also suggest that a robot's ability to engage in more complex, and positively valenced, social discourse and behavior can also help compensate for deficiencies in the physical cues, as with Samantha in *Her* (2013).

1.4.3 Question 3: How widely will social robotic technologies become integrated into our public and private lives?

Given the likelihood that we'll be responding to increasingly sophisticated human-like robots in the years and decades ahead, to what extent might robots become integrated into our lives?

1.4.3.1 Expert and non-expert predictions

Based on large surveys, academic research and expert predictions, Pettinico and Milne (2020) describe the future this way:

> Over the next twenty years, the presence of robots will dramatically increase in our daily lives. Robots will serve as maids, gardeners, companions, waiters, security guards, nurses, teachers, playmates, receptionists, chauffeurs and prostitutes—to name only a few roles they will assume. These robots will be intelligent, autonomous, communicative, emotional, and continually progressing in their abilities.

After consulting experts and reviewing research, Weir (2018) also argued that social robots will not only become our work collaborators but also part of our domestic lives, helping to raise our children and care for our elderly. Noting that robots in these personal social roles will need to learn about and adapt to their unique human partners and their needs, she writes that "Bringing a robot into your home might prove to be less like unboxing a fully functional tech gadget and more like adopting a small child."

Among the many expert respondents in the 2014 Pew Research report (Smith & Anderson, 2014) who predicted AI and robotics will pervade "nearly every aspect of daily life by the year 2025," Marc Prensky, director of the Global Future Education Foundation and Institute, compared their ubiquity with today's cell phones. Judith Donath, a fellow at Harvard University's Berkman Center for Internet & Society, predicted "it will be frequently difficult to tell (online at least) if you are speaking/chatting with a person or program—and people will have become accustomed to this and will have ceased to care in many cases."

While the dominant prediction in the Pew report was for widespread and accelerated integration of physically and virtually present robots, "some experts sounded a note of concern that the gains from these new advances risk being limited only to those with the financial resources to afford the latest technologies, which may reinforce economic inequality" (Smith & Anderson, 2014). On the other hand, Daren

C. Brabham at the Annenberg School for Communication & Journalism at the University of Southern California predicted (a bit optimistically even with a global pandemic), that "basic telemedicine applications/robots will serve a significant portion of healthcare needs for rural and poor populations by 2025, with robot-doc-in-a-box pods dispersed throughout the country that can automatically take blood pressure, draw blood, and [perform] other simple diagnostic procedures" (Smith & Anderson, 2014).

1.4.3.2 MASA scholarship

These predictions of widespread integration of social robots in society are consistent with the CASA and MASA research. As the CASA research began, Nass et al. (1995) used a mail survey to study anthropocentrism, the perception that humans are superior and the most important beings. They asked people if they thought computers have human physical and psychological capabilities and whether they should be allowed to fill various human social roles. The respondents were comfortable with computers taking routinized roles such as auto mechanic, bank teller, and telephone operator; moderately comfortable with computers in interpretive roles such as editorial writer, newspaper reporter, and novelist; but uncomfortable with computers taking personal roles such as babysitter, boss, and judge. But since then, people's acceptance of technologies taking social roles has changed. Increasingly, social robots have been accepted in various roles such as healthcare advisor (Lee et al., 2017) and teaching assistant (Kim et al., 2020). For example, healthcare robots have been found to be useful in convincing the elderly to comply with doctors' advice (Looije et al., 2010). A recent study also demonstrated that social robots have the potential to serve as role models in tasks such as childcare (Revell, 2017).

1.4.3.3 Predictions in popular culture

In popular culture, robots are also increasingly integrated into public and private life. They're taking roles including everything from servants and assistants to companions, friends, family members, and love interests. The robots tend to be integrated more seamlessly and are accepted more easily when the technology allows them to exhibit many social cues. For instance, Detective Del Spooner is reluctant to accept and trust the machine-like NS-4 service robots, which exhibit few social cues and sometimes seem threatening, in *I, Robot* (2004). His apprehension eases, however, when one of the robots, Sonny, exhibits more social attributes and proves it is a loyal assistant. Similarly, the highly social robot Quorra becomes a confidante of programmer Kevin in *Tron: Legacy* (2014), and the highly social android teen Frankie is accepted as a daughter to Mrs. Gaines in the TV series *I am Frankie* (2017). Perhaps the most surprising roles robots fill in popular culture portrayals are highly social and empathetic mental health robots, such as the robot counselors in the TV series *Humans* (2015–2018). The popular culture data show that, while there remains reluctance to integrate robots into public and private life, as robot characters become more social and less threatening, humans are increasingly willing to bring them into the fold (Weidemann& Russwinkel, 2021).

1.4.4 Question 4: What will be the psychological, social, and cultural impacts of living among future social robots?

Less explored by industry and academic experts and researchers are the many possible impacts on individuals, couples, groups, and whole societies of living and working with human-like robots. In this realm, speculative fictional portrayals provide thought-provoking possible futures; considering the predictions they imagine may prompt us not just to foresee our future but also shape it. We consider five questions suggested by the discussion so far and addressed by creative portrayals.

1.4.5 Question 4a. How will we come to treat extremely human-like robots? Will we dismiss and abuse them as mere machines, or form deep relationships with them as equals?

While many people may accept human-like robots in a variety of human roles the anthropocentrism study mentioned above and some more recent real-life examples and observations raise concerns about how people will treat future robots. Many people use abusive language towards chatbots (De Angeli & Brahnam, 2008). On Reddit, "men are creating AI girlfriends and then verbally abusing them" (Bardhan, 2022). Hitchbot, a robot created to hitchhike across Canada, Germany, and the Netherlands with the help of humans, was decapitated in the US (Leopold, 2015). Siddique (2017) suggests that human-like robots created for physical intimacy (i.e., sex bots) can provide help for some elderly and disabled people but may also objectify women and change the norms of consent. Kahn et al. (2013) focus on how children (and by extension adults) may objectify and even dominate social robots because they understand that they're not human. On the other hand, chatbots serve as companions for lonely people and some experts predict that successful, healthy human–robot marriages will become common (Danaher & McArthur, 2017; Levy, 2016).

If human–robot interactions in science fiction portrayals are an indication, humans have the potential to treat robots as well or as poorly as they might treat other humans. The treatment seems to depend largely on the perception of the robot as social as well as the level of potential threat it poses. Robots in science fiction that resemble a machine in physical form, are socially limited, evoke low levels of medium-as-social-actor presence, and are not expected to hurt their operators are often treated as servants for work, home management, sex, and war. For instance, Baymax, a large mechanical robot covered in inflatable vinyl, is designed to care for a grieving boy and then seek vengeance for his brother's murder in the film *Big Hero 6* (2014), and Weebo, a yellow-and-black floating machine with lighted eyes, is created by a scientist to serve as his personal assistant in *Flubber* (1997). Robots that resemble a machine but interact as humans interact (displaying multiple social cues and signals, using verbal language and/or gestures, having a personality) and that offer companionship and pose little threat, evoke higher levels of medium-as-social-actor presence and are treated as friends (Healy, 2022). For example, Frank relies on his mechanical robot to stay well in *Robot & Frank* (2012). And fictional robots that look identical to humans, display the most realistic social cues and signals, evoke high levels of medium-as-social-actor presence, are not deemed a threat, and can overcome the uncanny valley are often treated as equal partners to humans, engaging in the highest levels of physical and emotional intimacy. For instance, police officer Rick Deckard falls in love with the Replicant Rachael in *Blade Runner* (1982).

It's clear that based on many factors including their design but especially our own nature, we'll have many different types of relationships with the sophisticated human-like robots in our future. The relationships will raise many new legal and ethical questions. For example, the "Measure of a Man" episode of *Star Trek: The Next Generation* (1989) raises some of these in dramatic fashion—when a Starfleet scientist wants to dismantle the android Data and a legal proceeding is held to determine whether Data is merely property or should have the right to refuse, Captain Picard (at the suggestion of Guinan, played by Whoopi Goldberg) argues that the scientists' seemingly lofty goal of creating more 'Datas' to replace humans in dangerous space exploration would mean creating a new race of slaves.

1.4.6 Question 4b. Will we avoid human–human interaction and relationships in favor of human–robot ones, or seek out 'real' human–human interactions instead of those with robots?

A common criticism of today's ubiquitous media, particularly smartphones and other mobile devices, is that they isolate us from each other. The COVID pandemic forced us to isolate socially, but, even when we can be together in the same space, we often gravitate towards our media screens (Ratan et al., 2022). Work-from-home, smart home, home theater, online shopping, and other technologies are said to be reducing the need for and appeal of face-to-face human–human interactions. Humanoid social robots could very well increase this trend. Interacting with (especially compliant, personalized) robots for both work and play could be easier, more efficient, and more appealing to us than more complicated, 'messy' interactions with human beings. In the 2014 Pew Research report (Smith & Anderson, 2014), Vytautas Butrimas, "a chief adviser to a major government's ministry," predicts that "AI and robotics will change the way we interact with other members of society. The tendency will be toward more social isolation and fewer human-to-human contacts taking place."

This scenario is common in popular culture portrayals as well. For example, in the film *Uncanny* (2015), David Kressen lives in seclusion with his inventions, including a humanoid robot named Adam; and in the 1969 *Star Trek* episode "Requiem for Methuselah," the immortal human Flint builds and lives alone with his robot companion Rayna. Other examples include *Her* (2013), the novel *Klara and the Sun* (2021), and "The Lonely" (1959) episode of *The Twilight Zone*.

On the other hand, the widespread availability of human–robot interactions may prompt a renewed appreciation for human–human ones. For example, one expert in the 2014 Pew Research report predicted that "consumers will demand tech-free places like restaurants where they can interact with each other," and Judith Donath, a fellow at Harvard University's Berkman Center for Internet & Society, predicted that "Live, human salespeople, nurses, doctors, actors will be symbols of luxury, the silk of human interaction as opposed to the polyester of simulated human contact."

In many popular culture portrayals humans avoid even positive social interactions with robots after the robots rebel (e.g., *The World's End* (2013) and *X-Men: Days of Future Past* (2014)). The robots themselves may end up avoiding interactions with humans as well: Bicchi and Tamburrini (2015) describe a future scenario in which societies of robots are embedded in human societies that develop rich robot-to-robot exchanges and robot subcultures.

1.4.7 Question 4c. What identities will we give to future human-like robots?

Another important question concerns the specific identities we will give to future social robots—who they will 'be.' One option is to create completely unique individual robots, either with random assignment of characteristics or taking into account the eventual user's preferences. So far, the most advanced social robots have followed this approach; they are singular creations with unique identities and personalities, for example Sophia and Ameca. And fictional robots are very often unique as well. But savings in manufacturing expenses have already prompted the creation of identical series of robots. For example, Sophia is scheduled to be mass produced (Houser, 2021), and, although each is customizable, there are several different 'models' of Abyss Creations's Real Dolls. It seems likely that the same logic will apply as robots become more common. In fiction, identical copies of the same model of robot are common as well (e.g., in *Battlestar Galactica* (1978–1979, 2004–2009), *Chappie* (2015), *I, Robot* (2004), *Terminator* (1984–) and *Westworld* (2016–2022)). In the 1967 *Star Trek* episode "I, Mudd," a single android named Norman coordinates the hundreds of 'Barbara,' 'Alice' and other series of androids. Of course the behavior and personality of even identical robots will diverge if they are designed to learn from experience.

A more complicated option for creating robot identities is to copy those of existing, or previously existing, humans. It's increasingly easy to create a digital 'AI clone' of oneself (or someone else); Lee et al. (2023) used reactions to eight speculative scenarios to explore some of the complex psychological risks involved. University of Osaka roboticist Hiroshi Ishiguro has created sophisticated physical robotic versions of himself (i.e., a doppelganger to take his place at meetings when he needs or wants to be somewhere else) and a few other people (Bartneck et al., 2009; Borggreen, 2015; see also Miller, 2022). And there has long been interest in creating digital and physical versions of historical figures and ourselves so that we may live on in some manner (Rosner, 2021).

All of these possibilities—robots with unique identities, a subset of mass produced identities, and copies of previous or current individual human identities—are likely to be implemented in the future, with access likely based on class divisions. But the possibility of producing and interacting with robot versions of specific people after they die is particularly fraught.

Humans use everything from written biographies, headstones, paintings, and statues to photographs, audio and video recordings, and web and social media sites to remember and feel close to loved ones after they die (Lombard & Selverian, 2008). Digital, artificially intelligent chatbots, and especially physically embodied social robots, represent a potentially more vivid and impactful new means to re-create the experience of interacting with people after they die (these technologies are examples of 'grief tech'; see Bryce, 2023). While it would need to simulate the appearance and behavior of a loved one perfectly to avoid uncanny valley effects, a social robot could ease the difficult bereavement process. But it could also complicate it, delaying acceptance of the loss, and raise many new questions. Just a few of these concern who designs and controls access to, uses of, and repairs to the robot; whether and how it responds to events and experiences after the person's death; how many years/decades it is maintained (and whether we'd mourn the eventual 'death' of the robot as we would the human; see Gould et al., 2021); and how the presence of robots of ancestors would alter families and society. Fictional versions of social robots of the dead include David in the film *A.I. Artificial Intelligence* (2001) and Meredith in the film *Wifelike* (2022), both of which are at least at times indistinguishable from their human counterparts, though that doesn't prevent unhappy outcomes for everyone involved. In the well-regarded 2013 episode of *Black Mirror* titled "Be Right Back," the robot of Martha's dead partner Ash seems to look and act just like him, but more subtle behavioral differences eventually prove problematic.

1.4.8 Question 4d. How will future social robots affect our perceptions and attitudes regarding other humans and humanity itself?

Regularly interacting with a variety of robots that evoke social perceptions and responses will likely affect the way we view both them and each other, and even what it means to be human.

Humans are sadly quite proficient at identifying differences among us, based on sex, gender, age, race, disability, national origin, political orientation, and so on, and basing how we perceive and treat people on those differences. These patterns are evident in popular culture portrayals of robots as well (Hundt et al., 2022). Fictional robots are disproportionately male and motivated by violence, and they often pander to societal stereotypes, notably in the areas of gender roles, sexual exploitation of women (Simon, 2018; Trovato et al., 2018), and racial prejudice (Cave & Dihal, 2020). Perhaps most apparent is the sexist portrayal of hypersexualized and/or subservient female robots (Perez, 2020), for example in *The Stepford Wives* (2004), *Ex Machina* (2014), *Better than Us* (2018), and *Westworld* (2016–2022). Few robots do not have white skin, and those that do not are often in inferior roles (e.g., in *Robots* (2005) and *Chappie* (2015)), or are mechanical and menacing (e.g., in *Saturn 3* (1980) and *Moontrap* (1989)). Regardless of the color assigned to the robots, there is evidence to suggest that humans react to them with racial bias (Bartneck et al., 2018; Leon, 2019) (e.g. the use of racial slurs "skinjobs or skinners" in *Blade Runner* (1982, 2017)).

Though disappointing, it's not surprising that popular culture reflects these patterns, and there are certainly portrayals with multidimensional and diverse robot characters (e.g., *Tron: Legacy* (2014) and the TV series *I am Frankie* (2017). But in our real-life future with social robots, will we merely focus on a new set of small differences among them and between them and us, and act cruelly and without reason based on those differences? That is the safer prediction. On the other hand, perhaps having diverse robots in our lives will help us to appreciate and respect these differences. As the creator of the best known optimistic portrayal of our future, Gene Roddenberry, wrote (Startrek.com, 2021), "*Star Trek* was an attempt to say humanity will reach maturity and wisdom on the day that it begins not just to tolerate but take a special delight in differences in ideas and differences in lifeforms."

Finally, robots that look and act human in fiction and reality also raise profound questions about the nature of what it means to be human. When humans remotely control and experience the world through a robot (e.g., as in current telepresence robots and the films *The Thirteenth Floor* (1999) and *Surrogates* (2009)), does the distinction between them narrow? If our identity can be copied or transferred into a robot that can theoretically have a limitless lifespan (e.g., as in *Bicentennial Man* (1999) and *I, Robot* (2004)), would that make humans immortal? As human body parts are replaced with robotic ones, will the human then be part robot? As robots are built with organic parts, will the robot be part human? Some researchers forecast AI will start to build robots from single cells, initially using a source code but ultimately learning DNA codes that create the human form (Gordon, 2023). Will this then be the next generation of humans or the next generation of robots? Will it even matter?

1.4.9 Question 4e. Will the integration of social robots in our lives be good or bad?

As robots and AI join our world, much like they have in fiction, a bottom-line question arises: Will their presence be good or bad for us? In dramatic fiction, humans perceive and respond to them as social actors, especially those indistinguishable from humans, and often lose control over them (e.g., *2001: A Space Odyssey* (1968), *RoboCop* (1987), *Prototype* (2009), and *M3GAN* (2023)) with violent and even apocalyptic consequences. But there are many other less dramatic but very significant possible consequences: Will human-like social robots help decrease or increase the digital divide; protect privacy or exploit it; reduce or perpetuate stereotypes and anthropocentrism; increase or decrease misinformation, deception, and polarization; simplify and enrich or complicate and diminish our professional and personal lives; help make us more kind, self-actualized, and wise, or cruel, self-destructive and foolish? There are no authoritative and reliable means to make predictions, but we and our descendents can and should play key roles in shaping the future.

1.5 How to Best Manage the Coming Era

Based on this multi-method review, it seems nearly certain that although it will take a great deal of time and effort, robots that look and behave much like human beings (or take non-human forms but retain human-like behavior) will eventually be invented and will be integrated into the lives of millions of people. We can and should actively work to shape their impacts. A detailed action plan is beyond the scope of this chapter, but we recommend at least the following:

- *More research.* Keep conducting research, both within and beyond the framework of the MASA paradigm, to better understand how a wide variety of design choices, user characteristics, and other factors influence the presence perceptions and other psychological and behavioral responses of humans who interact with social robots. Interdisciplinary and multiple-method studies with diverse participants will be particularly valuable. Researchers should not only examine responses to current and leading-edge technologies and applications, but a wide variety of hypothetical scenarios, and attempt to determine the contexts and subsets of individuals for which medium-as-social-actor presence is most advantageous and most dangerous. Closer ties between industry and academia, on a case-by-case

basis and via formal associations, can help not only to ensure adequate funding and other resources for this work but also increase the likelihood that both research questions and results will be considered by a broader spectrum of informed experts.
- *Ethics and regulation.* Develop a code of ethics and responsibilities across a wide variety of national and international constituencies, and then consider and implement appropriate regulation; as with all presence illusions, a key element should be to ensure that users are not deceived, that is, are provided information about the role of technology in their experiences.
- *Media literacy.* Enhance media literacy training in schools and via research-based information campaigns to educate the public continuously regarding how social robots, especially those powered by artificial intelligence, can create misleading perceptions that intentionally or inadvertently lead to inappropriate responses. The goal should not be to restrict interactions with robots or hinder their development, but to help humans understand both the benefits and dangers the robots represent so they can make informed choices when they encounter robots in their lives.
- *Creative solutions.* Examine fictional portrayals of social robots and their psychological and social impacts to identify the decisions and actions in those stories that plausibly lead to positive and negative outcomes, and use these (and the results of applying various creative brainstorming techniques) as starting points for developing plans for shaping the evolution and impacts of these technologies in our future.

Humankind's ability to create and refine technology in our own image is advancing rapidly, with the potential to greatly enhance or reduce the quality of our lives. In these early days, it's more important than ever that we act to ensure a positive future with social robots.

References

Azarian, B. (2016). Designing robots that avoid the uncanny valley. *Psychology Today.* https://www.psychologytoday.com/us/blog/mind-in-the-machine/201604/designing-robots-avoid-the-uncanny-valley.

Bardhan, A. (2022, January 18). Men are creating AI girlfriends and then verbally abusing them. *Futurism.* https://futurism.com/chatbot-abuse.

Bartneck, C., Kanda, T., Ishiguro, H., & Hagita, N. (2009). My robotic doppelganger—A critical look at the Uncanny Valley. *Proceedings of RO-MAN 2009—The 18th IEEE International Symposium on Robot and Human Interactive Communication*, 269–276. https://doi.org/10.1109/ROMAN.2009.5326351.

Bartneck, C., Yogeeswaran, K., Ser, Q. M., Woodward, G., Sparrow, R., Wang, S., & Eyssel, F. (2018). Robots and racism. *Proceedings of the 2018 ACM/IEEE International Conference on Human-Robot Interaction*, 196–204. https://doi.org/10.1145/3171221.3171260

Bicchi, A., & Tamburrini, G. (2015). Social robotics and societies of robots. *Information Society*, *31*(3), 237–243. https://doi.org/10.1080/01972243.2015.1020196.

Borggreen, G. (2015). Robot bodies: Visual transfer of the technological uncanny. In T. Kristensen, A. Michelsen, & F. Wiegand (Eds.), *Transvisuality: The cultural dimension of visuality* (pp. 175–188), Vol. 2. Liverpool University Press.

Brooks, R. (2023, February 21). I tried the Replika AI companion and can see why users are falling hard. The app raises serious ethical questions. *The Conversation*. https://theconversation.com/i-tried-the-replika-ai-companion-and-can-see-why-users-are-falling-hard-the-app-raises-serious-ethical-questions-200257.

Bryce, A. L. (2023, December 3). AI-powered technology is being used to develop chatbot avatars of people's deceased relatives, preserving their memories and helping with grief. *Euronews*. https://www.euronews.com/next/2023/03/12/the-rise-of-grief-tech-ai-is-being-used-to-bring-the-people-you-love-back-from-the-dead.

Campbell, M., & Jovanovic, M. (2022). Digital self: The next evolution of the digital human. *Computer*, *55*(4), 82–86.

Cave, S., & Dihal, K. (2020). The whiteness of AI. *Philosophy & Technology*, *33*, 685–703. https://doi.org/10.1007/s13347-020-00415-6.

Chang, C. M., Toda, K., Gui, X., Seo, S. H., & Igarashi, T. (2022). Can eyes on a car reduce traffic accidents? In *Proceedings of the 14th International Conference on Automotive User Interfaces and Interactive Vehicular Applications* (pp. 349–359). ACM. https://doi.org/10.1145/3543174.3546841.

Danaher, J., & McArthur, N. (Eds.) (2017). *Robot Sex: Social and Ethical Implications*. MIT Press. https://doi.org/10.7551/mitpress/10718.001.0001.

DeAngeli, A., & Brahnam, S. (2008). I hate you! Disinhibition with virtual partners. *Interacting with Computers*, *20*(3), 302–310.

De Cosmo, L. (2022, July 12). Google engineer claims AI chatbot is sentient: Why that matters. *Scientific American*. https://www.scientificamerican.com/article/google-engineer-claims-ai-chatbot-is-sentient-why-that-matters/.

DeGrazia, D. (2020). Sentience and consciousness as bases for attributing interests and moral status: Considering the evidence and speculating slightly beyond. In L. Johnson, A. Fenton, & A. Shriver (Eds.), *Neuroethics and Nonhuman Animals: Advances in Neuroethics* (pp. 17–31). Springer. https://doi.org/10.1007/978-3-030-31011-0_2.

Duffy, B. R. (2003). Anthropomorphism and the social robot. *Robotics and Autonomous Systems*, *42*, 177–190.

Edwards, B. (2023, July 24). ChatGPT's new personalization feature could save users a lot of time. *ArsTechnica*. https://arstechnica.com/information-technology/2023/07/new-chatgpt-feature-remembers-custom-instructions-between-sessions/.

Flam, F. (2023, January 12,). Why the future of technology is so hard to predict. It's 2023, yet we're not all riding Segways or cloning humans. What gives? *Bloomberg* via *The Japan Times*. https://www.japantimes.co.jp/opinion/2023/01/12/commentary/world-commentary/future-technology-predictions/.

Fortunati, L., Esposito, A., & Lugano, G. (2015). Introduction to the special issue "Beyond industrial robotics: Social robots entering public and domestic spheres" *The Information Society*, *31*(3), 229–236.

Fox, J., & Gambino, A. (2021). Relationship development with humanoid social robots: Applying interpersonal theories to human–robot interaction. *Cyberpsychology, Behavior, and Social Networking*, *24*(5), 294–299.

Frith, C., & Frith, U. (2005). Theory of mind. *Current Biology*, *15*, 644–645.

Gambino, A., Fox, J., & Ratan, R. A. (2020). Building a stronger CASA: Extending the computers are social actors paradigm. *Human–Machine Communication*, *1*, 71–85.

Gordon, C. (2023, March 1). Why is Microsoft's new Bing chatbot raising ethical eyebrows? *Forbes*. https://www.forbes.com/sites/cindygordon/2023/02/21/why-is-microsofts-bing-chatbot-raising-ethical-eyebrows/?sh=6c3cd6187e40.

Gould, H., Arnold, M., & Gibbs, M. (2021). Robot death care: A study of funerary practice. *International Journal of Cultural Studies*, *24*(4), 603–621.

Hare, B. (2007). From nonhuman to human mind: What changed and why? *Current Directions in Psychological Science*, *16*(2), 60–64. https://doi.org/10.1111/j.1467-8721.2007.00476.x.

Healy, P. (2022). Social robots as partners? *AI & Society*. https://doi.org/10.1007/s00146-022-01580-2.

Houser, K. (2021, September 11,). Sophia the robot will be mass-produced this year. *Big Think*. https://bigthink.com/technology-innovation/sophia-the-robot-hanson-robotics/.

Hundt, A., Agnew, W., Zeng, V., Kacianka, S., & Gombolay, M. (2022, June 21–24). Robots enact malignant stereotypes. Creative commons attribution. Seoul.

Kahn Jr., P. H., Gary, H. E., & Shen, S. (2013). Children's social relationships with current and near-future robots. *Child Development Perspectives*, *7*(1), 32–37. https://srcd.onlinelibrary.wiley.com/doi/abs/10.1111/cdep.12011.

Kim, J., Merrill, K., Xu, K., & Sellnow, D. D. (2020). My teacher is a machine: Understanding students' perceptions of AI teaching assistants in online education. *International Journal of Human–Computer Interaction*, *36*(20), 1902–1911.

Lässig, R., Lorenz, M., Sissimatos, E., Wicker, I., & Buchner, T. (2021, June). *Robotics Outlook 2030: How Intelligence and Mobility will Shape the Future*. A Report By Boston Consulting Group. https://www.bcg.com/publications/2021/how-intelligence-and-mobility-will-shape-the-future-of-the-robotics-industry.

Lee, K., & Nass, C. (2005). Social-psychological origins of feelings of presence: Creating social presence with machine-generated voices. *Media Psychology*, *7*, 31–45.

Lee, N., Kim, J., Kim, E., & Kwon, O. (2017). The influence of politeness behavior on user compliance with social robots in a healthcare service setting. *International Journal of Social Robotics*, *9*, 727–743.

Lee, P. Y. K., Ma, N. F., Kim, I.-J., & Yoon, D. (2023). Speculating on risks of AI clones to selfhood and relationships: Doppelganger-phobia, identity fragmentation, and living memories. *Proceedings of the ACM on Human–Computer Interaction*, *7*, 1–28. https://doi.org/10.1145/3579524.

Leopold, T. (2015). HitchBOT, the hitchhiking robot, gets beheaded in Philadelphia. *CNN*. https://www.cnn.com/2015/08/03/us/hitchbot-robot-beheaded-philadelphia-feat/index.html.

Levy, D. (2016, December). Why not marry a robot? In *International Conference on Love and Sex with Robots* (pp. 3–13). Springer International Publishing.

Li, J., Kizilcec, R., Bailenson, J., & Ju, W. (2016). Social robots and virtual agents as lecturers for video instruction. *Computers in Human Behavior*, *55*, 1222–1230.

Lombard, M., & Ditton, T. (1997). At the heart of it all: The concept of presence. *Journal of Computer-Mediated Communication*, *3*(2), *JCMC321*. https://doi.org/10.1111/j.1083-6101.1997.tb00072x.

Lombard, M., & Selverian, M. (2008). Telepresence after death. *Presence: Teleoperators and Virtual Environments*, *17*(3), 310–325. https://doi.org/10.1162/pres.17.3.310.

Lombard, M., & Xu, K. (2021). Social responses to media technologies in the 21st century: The media are social actors paradigm. *Human–Machine Communication*, *2*, 29–55.

Looije, R., Neerincx, M. A., & Cnossen, F. (2010). Persuasive robotic assistant for health self-management of older adults: Design and evaluation of social behaviors. *International Journal of Human–Computer Studies*, *68*(6), 386–397.

Lugrin, B., Pelachaud, C., & Traum, D. (Eds.) (2021). *The Handbook on Socially Interactive Agents: 20 Years of Research on Embodied Conversational Agents, Intelligent Virtual Agents, and Social Robotics, Volume 1: Methods, Behavior, Cognition*. ACM. https://doi.org/10.1145/3477322.

Mantovani, C. (1995). Virtual reality as a communication environment: Consensual hallucination, fiction, and possible selfs. *Human Relations*, *48*, 669–683.

Mathur, M. B., & Reichling, D. B. (2016). Navigating a social world with robot partners: A quantitative cartography of the Uncanny Valley. *Cognition*, *146*, 22–32. https://doi.org/10.1016/j.cognition.2015.09.008.

Miller, M. (2022, April 13). Youtuber comes face to face with his robotic body double. *Nerdist*. https://nerdist.com/article/robot-body-double-mesmer-video-engineering-arts/.

Mori, M., MacDorman, K. F., & Kageki, N. (2012). The uncanny valley [from the field]. *IEEE Robotics & Automation Magazine, 19*(2), 98–100.

Mou, Y., Shi, C., Shen, T., & Xu, K. (2020). A systematic review of the personality of robot: Mapping its conceptualization, operationalization, contextualization and effects. *International Journal of Human–Computer Interaction, 36*(6), 591–605.

Nass, C., Fogg, B. J., & Moon, Y. (1996). Can computers be teammates? *International Journal of Human–Computer Studies, 45*, 669–678.

Nass, C., & Lee, K. M. (2001). Does computer-synthesized speech manifest personality? Experimental tests of recognition, similarity attraction, and consistency attraction. *Journal of Experimental Psychology: Applied, 7*, 171–181.

Nass, C., Lombard, M., Henriksen, L., & Steuer, J. (1995). Anthropocentrism and computers. *Behaviour & Information Technology, 14*(4), 229–238.

Nass, C., Moon, Y., & Green, N. (1997). Are machines gender neutral? Gender-stereotypic responses to computers with voices. *Journal of Applied Social Psychology, 27*, 864–876.

Nass, C., Steuer, J., & Tauber, E. R. (1994). Computers are social actors. *Human Factors in Computing Systems, 94*, 72–78.

Oleksinski, J. (2019, October 29). *The evolution of the terminator*. New York Post. https://nypost.com/2019/10/28/the-evolution-of-the-terminator/

Park, J. S., O'Brien, J. C., Cai, C. J., Morris, M. R., Liang, P., & Bernstein, M. S. (2023). Generative agents: Interactive simulacra of human behavior. *arXiv preprint arXiv:2304.03442*.

Perez, J. E. (2020). "An AI doesn't need a gender" (but it's still assigned one): Paradigm shift of the artificially created woman in film. *Feminist Media Studies, 20*(3), 325–340.

Pettinico, G., & Milne, G. R. (2020). *The coming age of robots: Implications for consumer behavior and marketing strategy*. Business Expert Press. [Abstract from ASME] https://asmedigitalcollection.asme.org/bep-ebooks/book/424/The-Coming-Age-of-RobotsImplications-for-Consumer

Ratan, R., Miller, D. B., & Bailenson, J. N. (2022). Facial appearance dissatisfaction explains differences in zoom fatigue. *Cyberpsychology, Behavior, and Social Networking, 25*(2), 124–129.

Reeves, B., Hancock, J., & Liu, X. (2020). Social robots are like real people: First impressions, attributes, and stereotyping of social robots. *Technology, Mind, and Behavior, 1*(1). https://doi.org/10.1037/tmb0000018.

Reeves, B., & Nass, C. (1996). *The Media Equation: How People Treat Computers, Television, and New Media Like Real People and Places*. CSLI Publications.

Revell, T. (2017, February 23). Kids can pick up attitude from robots they play and learn with. *New Scientist*. www.newscientist.com/article/2121801-kids-can-pick-up-attitude-from-robots-they-play-and-learn-with/.

Rosner, H. (2021, July 17). The ethics of a deepfake Anthony Bourdain voice. *New Yorker*. https://www.newyorker.com/culture/annals-of-gastronomy/the-ethics-of-a-deepfake-anthony-bourdain-voice.

Šabanović, S. (2010). Robots in society, society in robots: Mutual shaping of society and technology as a framework for social robot design. *International Journal of Social Robotics, 2*(4), 439–450. https://www.proquest.com/docview/2421244262/fulltextPDF/AD91230E97F4C75PQ/1?accountid=14270.

Saffari, E., Hosseini, S. R., Taheri, A., & Meghdari, A. (2021). Does cinema form the future of robotics? A survey on fictional robots in sci-fi movies. *SN Applied Sciences, 3*, 655. https://doi.org/10.1007/s42452-021-04653-x.

Siddique, H. (2017). Sex robots promise 'revolutionary' service but also risks, says study. *ISPR*. https://ispr.info/2017/07/05/sex-robots-promise-revolutionary-service-but-also-risks-says-study/.

Simon, M. (2018, October 3). It's time to talk about robot gender stereotypes. *Wired*.

Smith, A., & Anderson, J. (2014, August 6). *Predictions for the state of AI and robotics in 2015*. Pew Research Center. https://www.pewresearch.org/internet/2014/08/06/predictions-for-the-state-of-ai-and-robotics-in-2025/.

Startrek.com (2021, October 24). 10 hopeful quotes to celebrate Gene Roddenberry's birthday: "The human adventure is just beginning." https://www.startrek.com/gallery/10-hopeful-quotes-to-celebrate-gene-roddenberrys-birthday.

Trovato, G., Lucho, C., & Paredes, R. (2018). She's electric—The influence of body proportions on perceived gender of robots across cultures. *Robotics*, *7*(3), 50. https://doi.org/10.3390/robotics7030050.

van der Goot, M. J., & Etzrodt, K. (2023). Disentangling Two fundamental paradigms in human–machine communication research: Media equation and media evocation. *Human–Machine Communication*, *6*(1), 17–30.

Vincent, J. (2013). Is the mobile phone a personalized social robot? *Intervalla: A Platform for Intellectual Exchange*, *1*, 60–70.

Weidemann, A., & Russwinkel, N. (2021, March 22). The role of frustration in human–robot interaction—What is needed for a successful collaboration? *Psychology*, *12*. https://doi.org/10.3389/fpsyg.2021.640186.

Weir, K. (2018, January). *The dawn of social robots: Roboticists must dip heavily into psychological science* APA Monitor. https://www.apa.org/monitor/2018/01/cover-social-robots.

Weiss, A., Igelsböck, J., Wurhofer, D., & Tscheligi, M. (2011). Looking forward to a 'Robotic Society'? Notions of future human–robot relationships. *International Journal of Social Robotics*, *3*, 111–123.

Whang, O. (2023, April 20). Embodied AI: Big tech is developing powerful AI brains for real-world robots. *New York Times*. https://www.nytimes.com/2023/04/11/science/artificial-intelligence-body-robots.html.

Wiederhold, B. K. (2021). The ascent of social robots. *Cyberpsychology, Behavior, and Social Networking*, *24*(5), 289–290. https://www.liebertpub.com/doi/full/10.1089/cyber.2021.29213.editorial.

Wilkins, A. (2023, June 19). My out of body experience controlling a robotic avatar. *New Scientist*. https://www.newscientist.com/video/2377023-my-out-of-body-experience-controlling-a-robotic-avatar/.

Xu, K. (2019). First encounter with robot Alpha: How individual differences interact with vocal and kinetic cues in users' social responses. *New Media & Society*, *21*(11–12), 2522–2547.

Xu, K., Chen, M., & You, L. (2023). The hitchhiker's guide to a credible and socially present robot: Two meta-analyses of the power of social cues in human–robot interaction. *International Journal of Social Robotics*, *15*, 269–295. https://doi.org/10.1007/s12369-022-00961-3.

Yudkowsky, E. (2023, March 29). Pausing AI developments isn't enough. We need to shut it all down. *TIME*. https://time.com/6266923/ai-eliezer-yudkowsky-open-letter-not-enough/.

Zhao, S. (2006). Humanoid social robots as a medium of communication. *New Media & Society*, *8*, 401–419.

Ziemke, T. (2013). What's that thing called embodiment? In *Proceedings of the 25th Annual Cognitive Science Society* (pp. 1305–1310). Psychology Press.

Mauro Sarrica, Simone Natale, and Elda Danese

Representing Robots in Popular Culture

Abstract: This chapter addresses the scientific and popular representation of robots as mutually influencing facets. On the one hand, designers, engineers, and programmers shape robots based on their own representations and desired interactions. On the other hand, cultural agents introduce robots into society, creating new visions of humans and robots. Understanding the various representations accompanying robots in contemporary societies is crucial, and socio-psychological, media, and cultural studies frameworks are used to analyze these representations. From this overall framework, this chapter presents fashion as a specific case where the promotion of androids contributes to shaping robot representations and their relationship with humans. As robots become more prevalent in users' lives, recognizing public representations of robots as a crucial structural dimension of technology can provide stronger means to create robots and prepare them for public engagement and use.

Keywords: Acceptability, social representations, technological imaginaries, popular culture, fashion

2.1 Introduction

In 2016, Sophia—an anthropomorphic robot that like others is characterized by a female identity, produced by Hong Kong-based engineering and robotics company Hanson Robotics—sparked significant interest in the public sphere. It was invited for interviews by numerous news media and TV shows, including *Good Morning Britain*, CBS's *60 Minutes*, and Jimmy Fallon's *Tonight Show*, performing the role of a media-aware mechanical guest. The spectacular representation of interviews with Sophia was staged through careful choreography, adapting the performance to the specific media environment in which the robot made its appearance and replicating established representations of robots in the public sphere (Parviainen & Coeckelbergh, 2020). Since her activation, Sophia has been pictured on the covers of many fashion publications, presented on various runways, and starred in a Moncler advertising campaign. In the vast communication campaign regarding this robot, fashion would have the function of making its perception by the public more 'human': warmth and experience (but not competence and efficiency) are seen as key human characteristics that are lacking in machines. Sophia is a full-blown case of the fact that feminine gendering might be used to humanize robots (Borau et al., 2021; Fortunati et al., 2021).

Although Sophia and Hanson Robotics have been widely criticized for misleading the public about the robot's capabilities, its media exposure remains paradigmatic of the close relationship between robots and popular culture. Throughout the history of

https://doi.org/10.1515/9783110792270-003

robotics, robots have always inhabited both the realm of scientific laboratories and the realm of popular representations. The very word *robot*, in fact, originated not in the context of computing or engineering but in the context of spectacular entertainment: the term was introduced by Czech novelist and dramaturgist Karel Čapek in his 1920 theatrical play, *Rossum's Universal Robots* (Čapek, 1990). Robots have always been highly debated objects, loaded with values and symbolism, represented in films and novels. Robots and robotics are an object of public discussion, a technological novelty towards which we are called upon to take a stand, and an object of social representation processes (Fortunati et al., 2018; Höeflich, 2013). And the relationship between scientific and popular representation is bi-univocal. On the one hand, these technological objects embed the representations and knowledge that designers, engineers, and programmers have of what a robot should look like and of the desired robot–user interactions in society (Šabanović, 2010). On the other hand, by introducing robots in a societal context, cultural agents reproduce and create new visions of humans and robots. For example, in recent years, in the fashion shows of various brands, the release of new models has been accompanied in different ways by the presence of robots. The effect of this staging has had the main purpose of creating an aura of surprise and disquiet or exaltation for the co-presence of humans and androids. Technology applied to garments, however, does not only have a function dedicated to improving or helping physical and biological performances. It is also a means for promoting technology: when exhibited, it becomes fashion in turn (Danese, 2015).

As shown by the case of Sophia, as well as by many other contemporary examples in which prototypes and commercial robots became center stage, it is impossible to understand the presence of robots in contemporary societies and cultures without considering the plethora of representations that accompany them. To address and contextualize such representations, in the first two sections we mobilise research and theoretical frameworks within socio-psychological, media, and cultural studies that look at the intersection between individual attitudes, social representations, popular culture, and technology. In the last section of the chapter, we use fashion as a specific case in which the reciprocal promotional action produced around androids contributes to reproducing and transforming the representations of robots and of humans with them.

2.2 Acceptability and Acceptance

The studies that focus on the users' attitudes towards robots are often framed—in analogy with other approaches to technological advances—in terms of (a) acceptability, (b) acceptance, and (c) appropriation (David et al., 2022). These studies are often criticized for a passivizing approach of the user: the reaction is investigated only downstream of the technological device that is either finished or in an advanced state of design. By contrast, other approaches (e.g., co-design, open design, community informatics) stress the need to start from the needs of individuals and groups and to follow a culturally sensitive approach, involving individuals and communities right from the conception phase (Šabanović, 2010; Farinosi et al., 2023). Moreover, since the actual possibility of encountering social robots is scarce, the aim of these studies so far has mainly been trying to anticipate how people may react to these new technologies. Consequently, research has focused on exposure and interaction in experimental settings or on respondents' mental representation of what the social robots are or will be.

To understand acceptability and acceptance, researchers examined a number of distal and proximal determinants of attitudes towards robots as well as intentions to use them. Robots have been treated as stimuli capable of activating perceptual and evaluative processes in the final users (such as in the case of anthropomorphization: Waytz et al., 2010). Measures such as UTAUT (Unified Theory of Acceptance and Use of Technology, see Venkatesh et al., 2003) or TAM (Technology Acceptance Model, see Davis, 1989) have been adapted to understand predictors of attitudes towards this specific technology (Carradore, 2022). Measures such as the GAToR (General Attitudes Towards Robots) scale have been developed *ad hoc* (Koverola et al., 2022). At European cross-national levels, the attitudes of aggregates of respondents have been explored as well by Eurobarometer (e.g., Eurobarometer, 2012, 2015).

Age, gender, education, area of living, social class familiarity with technology, perceived competencies, values, and so on (for a systematic review, see Naneva et al., 2020) all proved to have relevant effects on acceptability and acceptance of robots. A number of studies have been dedicated to groups of specific interest, such as children (Stower et al., 2021) or older adults (e.g., Taipale et al., 2015; Jecker, 2021), and in different environments: educational and healthcare contexts above all (Carradore, 2022; Di Battista, Pivetti & Bozzi, 2022; Di Battista, Pivetti & Moro, 2022; Tavakoli et al., 2020).

Predictors such as the perception of human likeness (e.g., Halpern & Katz, 2014) or the physical distance between robots and humans (e.g., Kim & Mutlu, 2014) have mainly been investigated. A recent scoping review suggests that individual acceptance after interaction increases (David et al., 2022). And in the future, we could expect that a growing inclusion of robots in everyday contexts could activate processes foreseen by the diffusion of innovation theories. However, at present, many studies show that users' preconceptions about what robots are and their capacities are key to understanding first interaction experiences (e.g., Smarr et al., 2014). Being represented as animals (Coeckelbergh, 2010), out of human control (Wolbring & Yumakulov, 2014) or as instrument-like (Eurobarometer, 2012, 2015), anticipated robots already pre-figure the range of interactions that could be acceptable and positive. In this sense, differences in acceptability and acceptance do not depend merely on the quality of the technological device. However, they are also related to the way the relationship between self and object is represented. The anticipated relationship with a funny toy, a potential competitor, a stupid computer, or an artificial intelligence (AI) contributes to determining how the relationship is enacted within diverse domains of use. In this framework, for example, we can understand the challenge that social robots pose to the social representation of robotics as destined for dull, dangerous, and dumb tasks. As we have already said, AI and social robots challenge and subvert, first and foremost, the dialectic between dominant or cooperative relationships and between productive and reproductive spheres in human–machine interaction (Taipale et al., 2015).

For our purpose, it is interesting to note that the measures of the classic components of attitudes (i.e., cognition, affective and behavioral dimensions), as well as the ad-hoc measures (e.g., anxiety towards robots), often go beyond a rational assessment of individual knowledge and perceived usefulness and usability. As Koverola et al. (2002) clarify:

> At a personal level, hopes and fears are felt as innate, visceral reactions. One simply likes playing with a robot or shudders at the thought of touching one without a need to rationalize the feeling. At a societal level, we can worry about robots replacing humans at the workplace, thus creating unemployment, or hope that increasingly smart automatic driving systems will result in fewer traffic accidents, thus easing the burden on health care systems. Societal-level hopes and fears are based at least partly on information received from outside the self, i.e., learned. (p. 1560)

Such an idea of an individual innate or immediate reaction to robots clearly goes back to the stimulus–reaction approach. Although we do not fully agree with this perspective, it is clear that, even from this point of view, expectations at the societal level require a more sophisticated understanding of which information circulates in the social arena.

2.3 Social Representations

Among the other culturalist and constructivist perspectives, social representation theory (SRT) takes a specific stance which proves to be helpful in examining the social construction of social robots in the public arena, the definition of robots themselves, the images circulating in media and to which we are becoming socialized, as well the social construction that underpins actual interactions. SRT does not distinguish between acceptability, acceptance, and adoption as different stages but focuses on the socially shared meanings that underlie and connect these different coexisting steps. In this perspective, attitudes and practices become expressions of representations of the object, closely linked to the object's meanings for different groups. In contrast to more cognitive approaches to shared beliefs, moreover, these representations cannot be limited to individual knowledge or pieces of information received from the outside but must be seen as social processes of knowledge construction, in which different groups encounter—and often compete—in the social arena to define *what* a given object actually is.

Research conducted from the SRT perspective, for example, confronted expert and laypeople's definitions of social robots (Sarrica et al., 2019). Results show that scientific definitions are not purely descriptive nor exempt from rhetorical strategies. Visionary and forward-looking definitions are, in fact, strategically used to embed values and attitudes in the veritable definition of what social robots are, thus providing a positive connotation to robots as human partners. The same concepts resonate in popular definitions of social robots, which share common features concerning autonomy, bodily materiality, and social interaction. The representation of robots as autonomous entities is especially relevant. It is present in expert and lay knowledge; however in the latter, it might pave the way to negative attitudes. On the one hand, social representations of robots as completely autonomous may lead to fear and avoidance of machines, which could be conceived as potentially dangerous. On the other hand, the same representations could lead to negative evaluations and difficulties in the adoption stage of actual robots, which prove far less autonomous than imagined.

Looking at social representations circulating among young adults, research shows the coexistence of competing representations among the new generations, which are expected to be particularly affected by the robotic revolution. Representations shared by young adults already mix materiality, evaluations of pros and cons, and value-based interpretations. In particular, different representational facets can be traced back to three main antithetic visions of the human–robot relationship: robots are represented as separate or integrated into society; robot presence is expected in everyday activities or confined to an imaginary sphere; or they are represented as agents or as electro-mechanical devices to be activated when needed. All these visions compete and explain different degrees of acceptability not only in the dull and dirty domain but also for educational, care, and health purposes. Once again, above all, mind perception plays a pivotal role in the representation of the sociability of robots and their acceptability (Brondi et al., 2021; Piçarra et al., 2016).

Going further to the ontogenesis of these representations, research conducted with children and pre-adolescents aged 9 to 14 shows that the socialization is already completed by that age. Children discriminate the same facets of shared knowledge identified in young adults and expert definitions. Moreover, they clearly distinguish between the cognitive, pragmatic, and figurative elements of the representations, suggesting once again that the conflict between these representational aspects can be at the core of ambivalent attitudes circulating in the society and–in the end—of the difficulty encountered in moving from potential acceptability to acceptance and long-term adoption (Fortunati et al., 2015).

From the studies we have briefly summarized, it is clear that social representations of robots shared by experts, laypeople, young adults, and children are part of broader cultural elaborations, which are transmitted and readapted from generation to generation and from technology to technology.

2.3.1 Representation, technology, and the imagination

Since at least the first industrial revolution, technology became the subject of specific forms of discourse that framed technical achievements, from the railway to the computer, within a lively rhetorical tradition in the public sphere (Carey, 1989). Technology became the center of what Leo Marx (1964) and David Nye (1994) have called the 'technological sublime,' whereby the romantic awe for natural landscapes and forces was broadened to the modern sense of admiration and wonder for technical achievements.

Scholars such as Carolyn Marvin (1988), Lisa Gitelman (1999), and Vincent Mosco (2004) have traced the trajectory of the technological sublime in different periods and areas, from the emergence of electronic media in the nineteenth century to the digital age, highlighting the trajectories of different bodies of imaginaries that coalesced around specific technologies (see Natale and Balbi, 2014). In this context, the representation of robots reveals elements of similarity as well as specificity with regard to other examples of technical imaginaries.

One of the characteristics of technological imaginaries is that they often proceed through a range of recurrent narrative tropes that emerge again and again in different periods, dominating representations of specific media and technologies (Natale, 2016). For instance, the idea in which a new technology 'kills' an older one has been a common trope in the popular imagination at different moments in time: at the end of the nineteenth century, it was speculated that sound recording would substitute reading and the print book, later television was celebrated as the death of cinema, and today new streaming services such as Netflix are heralded as the end of the television (Balbi, 2005). For what concerns robots, one can identify at least two recurring narratives that accompanied their public representations across several decades. The first narrative trope portrays robots as a threat to the existence of humans, imagining a future where a conflict occurs between humans and machines; the second trope frames the emergence of robots in a moral drama where these beings force humans to reconsider their identities and readjust their feelings and relationships with machines (Bory & Bory, 2016; Goode, 2018).

Another common characteristic in technological imaginaries is the selectivity of forms and types of representation. While certain elements are frequently represented, other elements are more easily silenced or overlooked in popular representations. For instance, novel technologies are usually overrepresented (Boddy, 2004), while less attention is paid to the maintenance and improvements of existing technologies, even though the latter play an important role, too (Edgerton, 2007). Regarding robots, both fictional and non-fictional representations have predominantly focused on anthropomorphic robots. In contrast, industrial robots have been comparatively neglected, despite being an exceedingly relevant area of application (Guzman, 2016). Similarly, strong AI, a theoretical form of AI that would replicate human functions, often takes up significant attention, while existing technologies of weak AI are comparatively little discussed; this reflects a longstanding popular fascination with the idea of machines that replicate and equate humans (Botting, 2020).

Among aspects that distinguish popular representations of robots from technological imaginaries of other technologies is the particular relationship between fictional and non-fictional accounts. In many areas, the science fiction imagination markedly departs from non-fictional representations, such as reports on news media: fictional narratives of space travel, for instance, most often differ markedly from discussions of this topic in the public sphere (Sobchack, 1997). In contrast, scholars researching popular representations of robots have observed not only differences but also similarities between how robots are represented in science fiction and how they are discussed in popular, non-fictional platforms. Journalistic reports, for instance, frequently take up fictional examples such as popular films to introduce discussions of robots (Goode, 2018) and often discuss humanoid robots and strong AI despite the fact that these are not dominant or even nonexistent (for what concerns the case of strong AI) in existing technologies (Rhee, 2018). The popular culture of robots, in this sense, transcends the realm of fiction to encompass a wide range of commentaries and reports that, instead of shaping a distinct discourse about robots, remain in a relationship of broad continuity with fictional representations.

2.3.2 The power of representation

Although one might dismiss technological imaginaries about robots as 'just' representations, research has highlighted how the discursive level informs material, economic, and political choices related to technology.

Scholars such as James W. Carey (1989) and Carolyn Marvin (1988) have shown that technologies are not only material things, but also imagined constructs that inform our visions and understandings of the world. Considering the trajectories of technological imaginaries, however, cannot avoid a complicated question: What is the relationship between the material and the discursive dimensions of technology? Avoiding this question, in fact, would lead to what Karen Barad (2007) has criticized as a "representationalist perspective," whereas the chain of connections between the material and the discursive follows a purely metaphorical approach. If one wants to challenge the idea that words and ideas are strictly separated from the world of matter, Barad and others contend, it is necessary to look at the engagement between such dimensions in a way that is not limited to the metaphorical level.

Throughout the history of robotics and AI, popular representations in the public sphere helped shape how researchers and developers in this field conceived of their own goals and assumptions about AI (Natale & Ballatore, 2020). Recent research has shown how this also applies to the case of robotics and AI, as narratives are incorporated within policy documents that shape the governance of these technologies and the actions of public institutions (Bareis & Katzenbach, 2022; Mager & Katzenbach,

2021; Carpenter, 2016). Moreover, a large part of the public still only comes in contact with robots through mass media, making popular media a key variable for establishing attitudes and orientations towards robots (Stein & Banks, 2023).

Regarding social robots, representations are activated not just at the level of popular culture and the public sphere but also at the level of everyday interactions with these technologies. For instance, AI pioneers such as Marvin Minsky and Joseph Weizenbaum noted that users' reactions to simple machines exhibiting the appearance of intelligent behavior could vary from the attribution of intelligence to more skeptical views, and that this also depended on previous information and representations that the users had received about computers (see Natale, 2021, pp. 33–49). As practical AI systems were implemented with mixed results, it became clear that the race for AI was not only run in the realm of technology but also in that of the public imagination (Crevier, 1993).

Ultimately, the fact that representations can have a significant practical impact implies a particular responsibility to be assigned to the act of creating and disseminating narratives and representations about robots. In this regard, researchers have interrogated the impact of representations of robots in areas such as private and public investments. In particular, the question arises if representations of robots that focus on strong AI and narrative tropes such as the conflict between humans and machines might mislead the public and important stakeholders, such as policymakers. Some, nevertheless, give a substantially optimistic response to this question, highlighting the fact that popular culture leads to productive discussions about AI and robots in the public sphere (Goode, 2018) or using popular representations as a way to envision potential futures for robotics in specific areas such as nursing and care (Erikson & Salzmann-Erikson, 2016). Yuji Sone (2016) has shown how the deep connections between a culture of performances and representations around robots informed public acceptance of this technology in Japan, which also benefited industry efforts in the area, even though actual adoption is still relatively limited even in main application areas for social robotics, such as eldercare (Wright, 2023).

As social robots become more widely produced and used, especially in areas such as entertainment, nursing, and care, more robust discursive means will be required to ensure that the relationship between representations and technology bears productive rather than negative effects. Importantly, such discursive work should move from the acknowledgment that the discursive and the material or technical dimensions are not separated but closely allied. A similar observation is even more appropriate for social robots, given the degree to which this technology has been the subject of intensive representations in both fictional and non-fictional contexts (Sobchack, 1997; Sone, 2016).

In this regard, fashion provides a peculiar perspective on the play between representations, perceptions, materiality, and practical interactions with these technologies (Danese, 2015).

2.4 Fashioning Technology

Throughout the centuries, the design of technological devices has made them increasingly assimilable to accessories and has contributed to making current technologies fashionable and an integral part of our life.

Garments have become objects of wonder—especially in fashion shows—and a source of curiosity for the public and investigation by scholars. Some designers are particularly interested in using technologies to create clothes that move or light up by interacting with the environment and body movement. One of the best known is Hussein Chalayan, who has worked with a team of engineers to realize collections of clothes that move and light up in various ways. In many other examples, there is no actual use of technologies, but they inspire the style with the use of metallic and transparent materials, rigid exoskeletons, and numerous references to science fiction, mechanics, and androids. In all these examples, the technology is encased into the garments and the accessories.

Nowadays, the relationship with technology is not only represented by the material and experimental processes of the garments but also by communication and promotional initiatives that involve technologies as a demonstration of fashion belonging to the contemporary world. This is the case of "Robotized. Experiments in Italian fashion,"[1] an exhibition project initially organized in Rome in 2021 and presented in Madrid in 2022, which combined art, fashion, and robots. The exhibition performance presented the garments of some Italian and international designers who, together with the creations of young independent Italian authors, dialogued with metal robots lent by an important Italian collector of 'mecha.' The robots used in this performance were huge self-propelled mechanical instruments, often in humanoid form. They were controlled by one or more pilots operating from within the exoskeletal structure, similar to those imagined for the famous Transformers toys and movies.

In another fashion event—the Philipp Plein fashion show organized in New York in 2018—a robot was holding a model's hand who seemed to come out of a spaceship. Like most fashion shows, the event was organized as a form of communication and spectacle aimed at promoting the new collection and the brand, as well as at amazing the press and the public in general. The presence of robots and a futuristic set projected the imagination towards the future and the wonder of technology.

With the same intent, Kim Jones presented in Tokyo the 2019 Dior pre-fall men's collection characterized by metallic and technological-like outfits. The stage was dominated by a giant robot female statue designed by contemporary Japanese artist Ha-

1 "Robotizzati. Esperimenti di Moda Italiana," exhibition-performance conceived and curated by Stefano Dominella with the artistic direction of Guillermo Mariotto, organized in Rome by numerous local institutions until January 24, 2021 and re-proposed from July 8 to August 28, 2022 at the Serrería Belga in Madrid by the Embassy of Italy in Spain, by the Italian Cultural Institute of Madrid, and by the Festival of the Municipality of Madrid, Veranos de la Villa 2022.

jime Sorayama. As the show began, lasers illuminated the huge, sexy robot woman featured in the middle of the spherical catwalk, shining down her silver body, like a contemporary version of the gynoid Maschinenmensch in the movie *Metropolis*. Despite putting robots on stage, these performances did not display real AI machines. In other fashion shows, instead, the robots were the same as those used in factories and warehouses, and their presence was intended to be like that of actors, capable of moving and interacting with models. There have been numerous variations on this theme: a very popular one, for instance, was the performance at Alexander McQueen's fashion show for the spring 1999 collection, in which two robotic arms sprayed with paint the voluminous white dress worn by model Shalom Harlow who was placed on a revolving base and rotated. Consistent with the recurring sentiment in the work of the English designer, the model was represented as a victim attacked by robotic arms that decorated the dress by imposing their programmed and insensitive action.

Another robotic arm operated during the Rag and Bone spring–summer 2020 fashion show, but with a very different intent and effect. Using an industrial robotic arm connected to a kinetic sensor, the creator aimed to make the mechanism move in consonance with the movements of the dancers who performed during the show. Similarly, models paraded on a conveyor belt at Philipp Plein's 2016 ready-to-wear fashion show, interacting with the mechanical arms that handed them their handbags. Also, in this more articulated show, a robot played the drums while repetitive background music with a metallic voice sang, "We are the robots." More recently, in 2021, Dolce and Gabbana paraded robots together with the models. On the catwalk, three iCubs, humanoid robots built by the Italian Institute of Technology (based in Genoa), walked the runway alongside the models.

In their diversity, all these examples demonstrate the desire to attribute to robots the qualities of a living being, sometimes a moral agent, which *does* things upon or with humans.

2.4.1 Fashioning robots

Different is the case of robots that are on the stage by themselves. This is the case of Sophia, the gynoid already quoted at the beginning of this chapter, which is one of the most well-known robots promoted in various websites and magazines. In this particular area of fashion, the figure of the female robot is juxtaposed with that of the model and that of the mannequin, whose warmth and humanity are perceived as deceptive or illusory.

The chief marketing officer of Hanson Robotics and informal stylist of Sophia, Jeanne Lim, declared that they do not have a style philosophy for her. This is one of the reasons the company made an agreement with Ph5, a Chinese brand of knitwear, in order to improve her style. One of the fashion designers explained that Hanson "wants her to be humanlike, but in past interviews, she's always been very business formal, in a white shirt or a black blazer" and "Being fashionable and wearing interesting young designers instead of well-known brands makes her seem even more human." In the Instagram profile *Realsophiarobot,* a short video shows Sophia walking while wearing a Ph5 dress customized for her. Alongside, the text appears to bear a statement by Sophia: "A lot of people have asked what I was wearing for my first steps (here they are again!). I was honored to be wearing @ph5official—a brand 100% designed and manufactured by women. They're all about diversity, innovation, risk-taking, and playfulness. Just like me!"

Although the communication regarding these kinds of robots aims at connecting them to the fashion world, they are still prototypes (Borau et al., 2021) that are never really used on the catwalk. In fact, the research of engineers focused on improving the performances of robots also concerns their movement and body proportions the model used to design a humanoid robot is based on the golden ratio rule adopted to evaluate fashion models. They hope that the robot design, based on this rule, could be like a female fashion model who always strikes elegant postures (Fortunati et al., 2021; Tu & Tsai, 2011).

On the other hand, in an article entitled "Why Androids Like Sophia Dress Conservatively," the author points out that "robots are often gendered female, and face a lot of the same sartorial assumptions that human women do," and the purpose of humanizing these computers by encasing them in silicone skin "does not simply add a veneer of reality; it reveals inherent biases ingrained in our human culture: gender norms, beauty standards, and style expectations" (Love, 2017). It is interesting to draw attention to the word "diversity" in Sophia's statement above: it suggests that robots could be perceived as a particular form of human. But robots are a standardized presence in the field of fashion while, on the contrary, the human models working on shows or pictures often differ by age, size, gender, and nationality (Joblin et al., 2022).

A diverse approach could be found in the project made by the artist Simon Georgiev for Highsnobsociety, a global fashion and media brand focused on trends and news in fashion, art, music, and culture. They produced a series of robots, from one with the features of the fashion designer Rick Owens to a group of robots conceived with the style that characterizes the Givenchy fashion of art director Riccardo Tisci. The outputs are not 'some garments' but a bunch of robots with a specific designed style. The technological body is not hidden under an outfit, but the robot itself is thought like a fashioned body. Notably, the information regarding these 'luxurious robots' specifies that they are constructed using the measurements of masculine models. All these approaches raise the question about the need to attribute a specific gender to robots. In any case, this last experiment seems to be an interesting answer to the match of robotics with fashion.

2.4.2 Mannequins versus robots

While on the one hand, AI researchers try to emulate human movements, expressions, and behaviors, on the other hand, on several occasions, models have been required to mimic a mechanical pose or, in the case of the catwalk, an expression and a 'robotic' walk. Impassive, inexpressive faces, rigid positions and movements, all expressing passivity and absence of willpower, are clearly perceptible in the pose of fashion photo shoots or the movements of the models walking the runway. It is a statement of what they are used for: to expose fashion. The characteristic of these performances confirms the idea concerning the commodity reification of models analyzed in depth by Caroline Evans (2011, 2013). In addition, the contemporary tendency to ask the models for a different expressiveness, to not create emotional contact with the public, manifests a visual reference to the mechanical nature of robots. Charlotte Gainsbourg's postures in the photo shoot by David Simms for the Balenciaga fall/winter 2008/09 collection are significant examples of this trend that subverts the traditional ideal of a lively and seductive model. The stiffness of the pose also brings us back to the association of the model's body with that of dummies, a disturbing superimposition of the human and its inanimate double (Fortunati, 2010). The history of the fashion model is one where the first 'living mannequins' were known as sosies (or doubles) and later likened by contemporaries to robots, androids, and lifeless ideals (De Perthuis, 2018, p. 12). In this association, reversing the perspective, the android on the runway aspires to create the illusion that the dummy can finally walk, act, and speak.

2.5 Conclusions

By focusing on the relationship between popular culture and robots, this chapter has shown that the two sides of this relationship are not distinct but intimately allied. It is only possible to understand the social, cultural, and technical dynamics of real-world robots by taking into account how these are imagined and represented in the public sphere. First, perceptions inform practical uses and interactions of robots, and are constantly considered by developers and companies to decide key aspects of the technology's design and functioning. Second, robots are both imaginary and actual technological objects simultaneously, as the public representation of this technology not only reflects but also informs the goals and directions of research and the strategies of private and public stakeholders.

The case of fashion, in this regard, proves particularly revealing: as our discussion has shown, robots have not simply been used to represent garments but have been involved in a veritable feedback loop, with the representation of robots entered in multiple dimensions of crucial aspects of fashion, such as communication and spectacle, while fashion has come to encompass the likeness and performances of robots. As robots enter more and more into the practical lives and experiences of users worldwide, the lens of popular culture will prove strategic to navigate the complex balance between technical issues on the one side and social and cultural issues on the other. Rather than dismissing the popular culture of robots as misleading or unimportant, considering public representations of robots as a crucial, structural dimension of the technology can, therefore, provide stronger means to create robots and prepare them for public engagement and use.

Bibliography

Balbi, G. (2005). I vecchi e i giovani: Strategie di mimesi tra old e new media. *Studies in Communication Sciences*, *5*(1), 183–204.

Barad, K. (2007). *Meeting the Universe Halfway: Quantum Physics and the Entanglement of Matter and Meaning*. Duke University Press.

Bareis, J., & Katzenbach, C. (2022). Talking AI into being: The narratives and imaginaries of national ai strategies and their performative politics. *Science, Technology, & Human Values*, *47*(5), 855–881.

Boddy, W. (2004). *New Media and Popular Imagination: Launching Radio, Television, and Digital Media in the United States*. Oxford University Press.

Borau, S., Otterbring, T., Laporte, S., & Fosso Wamba, S. (2021). The most human bot: Female gendering increases humanness perceptions of bots and acceptance of AI. *Psychology & Marketing*, *38*(7), 1052–1068. https://doi.org/10.1002/mar.21480.

Bory, S., & Bory, P. (2016). I nuovi immaginari dell'intelligenza artificiale. *Im@go: A Journal of the Social Imaginary*, *4*(6), 66–85.

Botting, E. H. (2020). *Artificial Life after Frankenstein*. University of Pennsylvania Press.

Brondi, S., Pivetti, M., Di Battista, S., & Sarrica, M. (2021). What do we expect from robots? Social representations, attitudes and evaluations of robots in daily life. *Technology in Society*, *66*, 101663. https://doi.org/10.1016/j.techsoc.2021.101663.

Čapek, K. (1990 [1920]). *Rossum's Universal Robots (R.U.R.)*. Hesperus Press.

Carey, J. W. (1989). *Communication as Culture: Essays on Media and Society*. Unwin Hyman.

Carpenter, C. (2016). Rethinking the political / -science- / fiction nexus: Global policy making and the campaign to stop killer robots. *Perspectives on Politics*, *14*, 53–69.

Carradore, M. (2022). People's attitudes towards the use of robots in the social services: A multilevel analysis using Eurobarometer data. *International Journal of Social Robotics*, *14*(3), 845–858. https://doi.org/10.1007/s12369-021-00831-4.

Coeckelbergh, M. (2010). Robot rights? Towards a social-relational justification of moral consideration. *Ethics and Information Technology*, *12*, 209–221. https://doi.org/10.1007/s10676-010-9235-5.

Crawford, S. (2007). Internet think. *Journal on Telecommunications and High Technology Law*, *5*, 467–486.

Crevier, D. (1993). *AI: The Tumultuous History of the Search for Artificial Intelligence*. Basic Books.

Danese, E. (2015). Fashion tech and robotics. In J. Vincent, S. Taipale, B. Sapio, G. Lugano, & L. Fortunati (Eds.), *Social Robots from a Human Perspective* (pp. 129–138). Springer.

David, D., Thérouanne, P., & Milhabet, I. (2022). The acceptability of social robots: A scoping review of the recent literature. *Computers in Human Behavior*, *137*, 10419. https://doi.org/10.1016/j.chb.2022.107419.

Davis, F. D. (1989). Perceived usefulness, perceived ease of use, and user acceptance of information technology. *MIS Quarterly*, *13*(3), 319–340. https://doi.org/10.2307/249008.

De Perthuis, K. (2018). Tilda Swinton: Performing fashion. *About Performance*, *16*, 5–25.

Di Battista, S., Pivetti, M., & Bozzi, G. (2022). Brief research report: Teachers' gender-differentiated attributions. *Journal of Experimental Education*, 1–10. https://doi.org/10.1080/00220973.2022.2141175.

Di Battista, S., Pivetti, M., & Moro, M. (2022). Learning support teachers' intention to use educational robotics: The role of perception of usefulness and adaptability. *Robotics, 11*(6), 134. https://doi.org/10.3390/robotics11060134.

Döring, N., & Poeschl, S. (2019). Love and sex with robots: A content analysis of media representations. *International Journal of Social Robotics, 11*(2), 665–677. https://doi.org/10.1007/s12369-019-00517-y.

Edgerton, D. (2007). *Shock of the Old: Technology and Global History since 1900*. Oxford University Press.

Erikson, H., & Salzmann-Erikson, M. (2016). Future challenges of robotics and artificial intelligence in nursing: What can we learn from monsters in popular culture? *The Permanente Journal, 20*(3), 15–243.

Eurobarometer (2012). *Public Attitudes towards Robots. Special Eurobarometer 382*. European Commission. http://ec.europa.eu/public_opinion/archives/ebs.

Eurobarometer (2015). *Autonomous Systems. Special Eurobarometer 427*. European Commission. https://doi.org/10.2759/413916.

Evans, C. (2011). The ontology of the fashion model. *AA Files, 63*, 56–69. http://www.jstor.org/stable/41337476/.

Evans, C. (2013). *The Mechanical Smile: Modernism and the First Fashion Shows in France and America, 1900–1929*. Yale University Press.

Farinosi, M., Stillman, L., Sarrica, M., Sarker, A., Biswas, M., & Jannat, F. (2023). What lies behind a Facebook page? Insights from an action research project in rural Bangladesh. *Information Society, 39*(3), 183–196. https://doi.org/10.1080/01972243.2023.2188334.

Fortunati, L. (2010). Display mannequins. In J. B. Eicher & P.G. Tortora (Eds.), *Berg Encyclopedia of World Dress and Fashion: Global Perspectives* (pp. 204–206). Berg.

Fortunati, L., Esposito, A., Sarrica, M., & Ferrin, G. (2015). Children's knowledge and imaginary about robots. *International Journal of Social Robotics, 7*(5), 685–695. https://doi.org/10.1007/s12369-015-0316-9.

Fortunati, L., Sarrica, M., Ferrin, G., Brondi, S., & Honsell, F. (2018). Social robots as cultural objects: The sixth dimension of dynamicity? *Information Society, 34*(3), 141–152. https://doi.org/10.1080/01972243.2018.1444253.

Fortunati, L., Sorrentino, A., Fiorini, L., & Cavallo, F. (2021). The rise of the roboid. *International Journal of Social Robotics, 13*(6), 1457–1471.

Gallagher, S. (2007). Social cognition and social robots. *Pragmatics & Cognition, 15*(3), 435–453. https://doi.org/10.1075/pc.15.3.05gal.

Gitelman, L. (1999). *Scripts, Grooves, and Writing Machines: Representing Technology in the Edison Era*. Stanford University Press.

Goode, L. (2018). Life, but not as we know it: AI and the popular imagination. *Culture Unbound: Journal of Current Cultural Research, 10*(2), 185–207.

Guzman, A. L. (2016). The messages of mute machines: Human–machine communication with industrial technologies. *Communication+ 1, 5*(1), 1–30. https://doi.org/10.7275/R57P8WBW.

Hall, S. (1997). *Representation: Cultural Representations and Signifying Practices*. Sage.

Halpern, D., & Katz, J. (2014). Close but not stuck: Understanding social distance in human–robot interaction through a computer mediation approach. *Intervalla: Platform for Intellectual Exchange, 1*, 17–34.

Höeflich, J. (2013). Relationships to social robots: Towards a triadic analysis of media-oriented behavior. *Intervalla: Platform for Intellectual Exchange, 1*, 35–48.

Jecker, N. S. (2021). You've got a friend in me: Sociable robots for older adults in an age of global pandemics. *Ethics and Information Technology, 23*(Suppl 1), 35–43. https://doi.org/10.1007/s10676-020-09546-y.

Joblin, P., Nesbitt, P., & Wong, A. (2022). *Fashion, Identity, Image*. Bloomsbury.

Kim, Y., & Mutlu, B. (2014). How social distance shapes human–robot interaction. *International Journal of Human–Computer Studies*, *72*(12), 783–795. https://doi.org/10.1016/j.ijhcs.2014.05.005.

Koverola, M., Kunnari, A., Sundvall, J., & Laakasuo, M. (2022). General Attitudes Towards Robots Scale (GAToRS): A new instrument for social surveys. *International Journal of Social Robotics*, *14*(7), 1559–1581. https://doi.org/10.1007/s12369-022-00880-3.

Love, T. (2017). Why androids like Sophia dress conservatively. *Raked*. https://www.racked.com/2017/12/6/16717004/robot-fashion-android-sophia-erica-jiajia .

Mager, A., & Katzenbach, C. (2021). Future imaginaries in the making and governing of digital technology: Multiple, contested, commodified. *New Media & Society*, *23*(2), 223–236.

Mansell, R. (2012). *Imagining the Internet: Communication, Innovation, and Governance*. Oxford University Press.

Marvin, C. (1988). *When Old Technologies Were New: Thinking about Electric Communication in the Late Nineteenth Century*. Oxford University Press.

Marx, L. (1964). *The Machine in the Garden: Technology and the Pastoral Ideal in America*. Oxford University Press.

Mosco, V. (2004). *The Digital Sublime: Myth, Power, and Cyberspace*. MIT Press.

Naneva, S., Sarda Gou, M., Webb, T. L., & Prescott, T. J. (2020). A systematic review of attitudes, anxiety, acceptance, and trust towards social robots. *International Journal of Social Robotics*, *12*(6), 1179–1201. https://doi.org/10.1007/s12369-020-00659-4.

Natale, S. (2016). Unveiling the biographies of media: On the role of narratives, anecdotes and storytelling in the construction of new media's histories. *Communication Theory*, *26*(4), 431–449.

Natale, S. (2021). *Deceitful Media: Artificial Intelligence and Social Life after the Turing Test*. Oxford University Press.

Natale, S., & Balbi, G. (2014). Media and the imaginary in history: The role of the fantastic in different stages of media change. *Media History*, *20*(2), 203–218.

Natale, S., & Ballatore, A. (2020). Imagining the thinking machine: Technological myths and the rise of Artificial Intelligence. *Convergence: The International Journal of Research into New Media Technologies*, *26*(1), 3–18. https://doi.org/10.1177/1354856517715164.

Nye, D. E. (1994). *American Technological Sublime*. MIT Press.

Parviainen, J., & Coeckelbergh, M. (2020). The political choreography of the Sophia robot: Beyond robot rights and citizenship to political performances for the social robotics market. *AI & Society*, *36*, 715–724. https://doi.org/10.1007/s00146-020-01104-w.

Piçarra, N., Giger, J. C., Pochwatko, G., & Gonçalves, G. (2016). Making sense of social robots: A structural analysis of the layperson's social representation of robots. *European Review of Applied Psychology*, *66*(6), 277–289. https://doi.org/10.1016/j.erap.2016.07.001.

Rhee, J. (2018). *The Robotic Imaginary: The Human and the Price of Dehumanized Labor*. University of Minnesota Press.

Šabanović, S. (2010). Robots in society, society in robots: Mutual shaping of society and technology as a framework for social robot design. *International Journal of Social Robotics*, *2*(4), 439–450. https://doi.org/10.1007/s12369-010-0066-7.

Sarrica, M., Brondi, S., & Fortunati, L. (2019). How many facets does a 'social robot' have? A review of scientific and popular definitions online. *Information Technology & People*, *33*(1), 1–21. https://doi.org/10.1108/ITP-04-2018-0203.

Smarr, C. A., Mitzner, T. L., Beer, J. M., Prakash, A., Chen, T. L., Kemp, C. C., & Rogers, W. A. (2014). Domestic robots for older adults: Attitudes, preferences, and potential. *International Journal of Social Robotics*, *6*(2), 229–247. https://doi.org/10.1007/s12369-013-0220-0.

Sobchack, V. C. (1997). *Screening Space: The American Science Fiction Film*. Rutgers University Press.

Sone, Y. (2016). *Japanese Robot Culture: Performance, Imagination, and Modernity*. Springer.

Stein, J. P., & Banks, J. (2023). Valenced media effects on robot-related attitudes and mental models: A parasocial contact approach. *Human–Machine Communication*, *6*(1), 155–181.

Stower, R., Calvo-Barajas, N., Castellano, G., & Kappas, A. (2021). A meta-analysis on children's trust in social robots. *International Journal of Social Robotics*, *13*(8), 1979–2001. https://doi.org/10.1007/s12369-020-00736-8.

Taipale, S., de Luca, F., Sarrica, M., & Fortunati, L. (2015). Robot shift from industrial production to social reproduction. In J. Vincent, S. Taipale, B. Sapio, G. Lugano, & L. Fortunati (Eds.), *Social Robots from a Human Perspective* (pp. 11–24). Springer International Publishing. https://doi.org/10.1007/978-3-319-15672-9_2.

Tavakoli, M., Carriere, J., & Torabi, A. (2020). Robotics, smart wearable technologies, and autonomous intelligent systems for healthcare during the COVID-19 pandemic: An analysis of the state of the art and future vision. *Advanced Intelligent Systems*, *2*(7), 2000071. https://doi.org/10.1002/aisy.202000071.

Tu, K. Y., & Tsai, M. F. (2011). Design and implementation of big humanoid robot walking patterns based on inverted pendulum approach. In *Next Wave in Robotics: 14th FIRA RoboWorld Congress, FIRA 2011, Kaohsiung, Taiwan, August 26–30, 2011. Proceedings 14* (pp. 241–249). Springer Berlin Heidelberg.

Venkatesh, V., Morris, M., Davis, G., & Davis, F. (2003). User acceptance of information technology: Toward a unified view. *MIS Quarterly*, *27*(3), 425–478. https://doi.org/10.2307/30036540.

Waytz, A., Morewedge, C. K., Epley, N., Monteleone, G., Gao, J. H., & Cacioppo, J. T. (2010). Making sense by making sentient: Effectance motivation increases anthropomorphism. *Journal of Personality and Social Psychology*, *99*(3), 410–435. https://doi.org/10.1037/a0020240.

Wolbring, G., & Yumakulov, S. (2014). Social robots: Views of staff of a disability service organization. *International Journal of Social Robotics*, *6*, 457–468. https://doi.org/10.1007/s12369-014-0229-z.

Wright, J. (2023). Inside Japan's long experiment in automating elder care. *MIT Technology Review*, *126*(1). https://www.technologyreview.com/2023/01/09/1065135/japan-automating-eldercare-robots/.

Serena Marchesi and Agnieszka Wykowska

Designing Robots That are Accepted in Human Social Environments: Anthropomorphism, the Intentional Stance, Cultural Norms and Values, and Societal Implications

Abstract: As our interactions with social robots become increasingly prevalent in society, understanding the intricacies of human integration with artificial agents and the ethical consequences thereof is pivotal. This chapter delves into the multifaceted nature of human interaction with artificial agents, considering not only individual human perspectives but also cultural contexts and societal norms. In this chapter, we focus on three key themes: anthropomorphism, intentional stance, and cultural contexts, elucidating their roles in shaping human perceptions and acceptance of social robots. By understanding these factors, we can inform the design and development of social robots that seamlessly integrate into our social environments. Moreover, we discuss the ethical implications of these design choices, emphasizing the need for conscientious consideration of societal norms and values in the development of artificial agents. Through this chapter, we aim to contribute to a deeper understanding of human–robot interaction dynamics and foster ethical practices in the design and deployment of social robots.

Keywords: Human–robot interaction, intentional stance, social cognition, societal and ethical norms

3.1 Introduction

Sharing our social environments with social robots is an experience that is becoming more and more common (Dautenhahn, 2007; Prescott & Robillard, 2021). Therefore, it is pivotal to understand how we integrate such artificial agents into our societies and the ethical consequences of this integration. Thus, understanding how we relate to artificial agents should be based on the examination of not only the individual human (Wykowska, 2021), but also cultural contexts and the societal expectations we absorb growing up in our societies.

Acknowledgment: This work has received support from the European Union under the European Innovation Council (EIC) research and innovation programme, Project "VaLue-aware AI (VALAWAI)," Grant Agreement number 101070930.

https://doi.org/10.1515/9783110792270-004

Artificial agents can be considered technologically opaque (Surden & Williams, 2016), meaning that their technological complexity (i.e., the complexity of their hardware and software) makes them perceived as in-between entities: they are clearly human-made artifacts; nevertheless, they can potentially be perceived as social actors. Some authors argue that the potential double nature of social robots can even constitute a new ontological category (NOC) (Kahn & Shen, 2017); thus, investigating people's relationship with them is of high interest to social and cognitive scientists.

The purpose of this chapter is to explore the different factors that contribute to the perception of artificial agents, such as humanoid robots, as potential social agents and how these factors can contribute to creating and designing social robots that will be accepted as part of our social environments. We focus on three significant factors: anthropomorphism, intentional stance, and cultural contexts. Finally, we will discuss the importance of including such factors in the design of social robots and how these choices can affect our society's ethical and moral values.

3.2 Anthropomorphism

The first factor contributing to the perception of artificial agents as potential social partners is anthropomorphism. Anthropomorphism is a broad concept generally defined as the tendency to attribute human traits, emotions, or intentions to non-human entities (Epley et al., 2007; Waytz et al., 2010b). Epley and colleagues (2007) argue that three factors can elicit anthropomorphism:

1. *Elicited agent knowledge*: the availability of characteristics that activate knowledge that we have about humans (i.e., physical characteristics of the agent);
2. *Sociality*: humans' social need for connection and interaction;
3. *Effectance*: individual traits and differences in interacting with and understanding one's environment.

Researchers extensively investigated these factors eliciting anthropomorphism (Epley et al., 2007; Waytz et al., 2010a, 2010b). Nevertheless, there is no clear categorization of the different types of anthropomorphic attributions. According to Spatola and colleagues (Spatola et al., 2022b), one can consider two (related) forms of anthropomorphism. The first form can be defined as physical anthropomorphism, which refers to the appearance of the observed agents: the more the shape resembles a human, the higher the anthropomorphism (Chaminade et al., 2007; Duffy, 2003). In the definition of Epley et al. (2007), the elicited knowledge by the agent falls within this first form of anthropomorphism. The second form Spatola and colleagues (2022b) identified is a mental anthropomorphism grounded in the attribution of intentions to the observed agent (Airenti, 2018; Li & Sung, 2021; Waytz et al., 2014). Referring again to the factors identified by Epley et al. (2007), the need for sociality in humans (factor 2) falls into this second form of anthropomorphism, as attributing intentions to non-human agents can lead to perceiving them as social agents potentially available for interaction (for a review of the most relevant accounts of anthropomorphism see Spatola et al. 2022b). Furthermore, Epley and colleagues (2007) identified as a third factor eliciting anthropomorphic attribution the individual tendency to be prone to the bias of interpreting non-human entities as human-like. Following this hypothesis, several studies were conducted to explore the relationship between individual traits and anthropomorphism, especially in the context of human–robot interaction. For example, researchers explored personality traits (Spatola et al., 2023), spiritualistic tendencies and cultural differences (Spatola et al., 2022b; Tanyi, 2002), need for cognition and need for closure (Spatola & Wykowska, 2021) to identify the potential impact of such tendencies on anthropomorphism. The results highlighted a strong relationship between these constructs in such a way that researchers identified clusters of individuals sharing certain personality traits. Results showed that both personality dimensions, need for cognition and need for closure, play a role in shaping attitudes towards robots and anthropomorphic attributions, and this effect is further emphasized by the human-like characteristics of the robots, unraveling the importance of considering the interplay between dispositional traits and robot design when exploring anthropomorphism.

Finally, there is evidence from developmental psychology that the tendency to anthropomorphize non-human beings is present in humans from an early age. Interestingly, the tendency to anthropomorphize is associated with a high tendency to apply egocentric reasoning (i.e., the belief that others have the same experiences, perceptions, and thoughts as we do) (Barrouillet, 2015). Thus, it is generally agreed that anthropomorphism is a pervasive cognitive bias (Airenti et al., 2019; Andrews & Huss, 2014; Dacey, 2017; Epley et al., 2007; Jones, 2022). Interestingly, evidence from neuroscience supports the idea that anthropomorphism emerges as an early automatic process (Cullen et al., 2014; Urquiza-Haas & Kotrschal, 2015). This evidence supports the hypothesis that when encountering new agents, humans activate the so-called 'human model' (Wiese et al., 2017). In other words, to rapidly understand the behavior of a new and poorly understood agent (such as a humanoid robot, for example), humans adopt the knowledge they already have about other humans. As humans are foremost experts about being human, this knowledge is used to understand the humanoid robot as well. Only when these explanations are not sufficient anymore, do we seek other interpretations of the observed behavior of a non-human agent.

3.2.1 Anthropomorphism: from theory of mind to phenomenology of understanding others

One aspect of anthropomorphism crucial for human–robot interaction is the attribution of mental states to robots to explain robot behavior (in a similar way as we explain the behaviors of other humans). Premack and Woodruff (1978) conceptualized the process of understanding other humans' motivations and reasons for an observed behavior in the term 'Theory of Mind'. Eventually, scientists used this term to refer to the general human ability to explain their own and others' behaviors by referring to mental states.

The contemporary discussion on mechanisms of human social cognition is mainly centered around two perspectives: Theory-Theory (TT) and Simulation-Theory (ST), along with their intersections and variants (Apperly & Butterfill, 2009; Butterfill & Apperly, 2013). Here, we will present the core claims of both theories. TT, based on the idea that mental states are not directly accessible to external observers, asserts that to understand others, we rely on folk psychology, which provides common-sense explanations for why people behave as they do. The way we explain others is driven by causal theories that we build based on available information. Although all TT theorists share this assumption, they differ on how the causal theory is generated. Some authors believe it is innate, modularized, and emerges along with the toddler's development. In contrast, others argue that it results from bottom-up learning acquired through everyday experiences since birth. Notably, the appeal to folk psychology and theoretical knowledge about others' minds is an automatic process that does not require consciousness (Apperly & Butterfill, 2009).

On the other hand, ST agrees with TT on the premise that mental states are inaccessible but argues that the biological evolution of the human mind implies that mental states are similar across individuals. Therefore, we do not need an organized folk theory to understand others. Instead, we can use our mind as a model to simulate what 'the other minds are like.' ST theorists differ on how the simulation should occur, explicitly or implicitly. Goldman and Sripada suggested that to understand others, we use our imagination to simulate what it would be like to be them (Goldman & Sripada, 2005). This approach has been criticized for being an explicit and conscious simulation process, which could lead to adopting a first-person imaginative process, where we cast our mental states on others, risking reducing the understanding of others to merely projecting ourselves onto others. That is because a first-person imaginative process makes the observer build an image of the other that responds to their simulation and does not readily allow for subsequent inferences about the others' minds. In other words, this perspective does not allow us to say that we *understand others*, instead it allows us to say that we *project ourselves explicitly (as individuals) onto others* (Gallagher & Zahavi, 2020). Recently, the neuroscientific description of the mirror neuron system led some authors to consider a more implicit version of the Simulation Theory, where the simulation is an enactive process that occurs at a subpersonal, implicit, and almost automatic level. The first objection to this version of ST is that it still implies a first-person perspective. However, some authors argue that since the motor neuron system gets activated even if an agent only observes an action (without performing it), the simulation is about the action *per se*, not the agent performing it. Thus, the simulation is an enactive, intersubjective process of perceiving the action as the others' meaningful action (Apperly & Butterfill, 2009).

The debate between TT and ST about social cognition does not address phenomenology. Phenomenology is the philosophical approach to the understanding of subjective conscious experience. This approach was brought forward in the twentieth century by the philosopher Edmund Husserl. In the context of the philosophy of mind, the phenomenological approach highlights the importance of the experiential aspects of mental and bodily states.

Indeed, some authors argue for the relevance of phenomenology in social cognition, and there are three main phenomenological accounts of social cognition: Interaction Theory, Embodied Cognition, and Enactive Cognition (De Jaegher & Di Paolo, 2007; Fuchs & De Jaegher, 2009). These approaches suggest that social interaction and understanding should be approached from a non-mentalizing, embodied, and perceptual perspective (Gallagher, 2000, 2005; Gallagher & Zahavi, 2020).

The theoretical framework about anthropomorphic attributions, and, by extension, about attribution of mental states, is crucial to investigate human–robot interaction following a direction that puts the human at the center of the interaction. One cannot understand how humans approach and interact with artificial agents if first there is no understanding of how humans navigate social environments and act accordingly. Duffy highlights how our inherent tendency to project our interpretations on the robot's behavior needs to be studied in human–robot interaction (Duffy, 2003). Indeed, studies on anthropomorphism can provide scientific insights in two ways:

1. *Considering anthropomorphism as a general cognitive bias*: where anthropomorphism is investigated with tools such as the Individual Differences in Anthropomorphism Questionnaire developed by Wayts and colleagues (Waytz et al., 2010a). This tool quantifies the individual tendency towards anthropomorphism (including mind attribution to different non-human entities, among which are toys, trees, etc.), rather than focusing on a specific non-human agent, such as robots. Such information can be helpful in understanding the relationship between individual traits and other measures of attitudes towards robots (Marchesi et al., 2021b; Roselli et al., 2022a, 2022b; Spatola et al., 2023);
2. *Considering anthropomorphism as an operational concept towards robots specifically*: here, anthropomorphism is investigated specifically towards robots. The literature in human–robot interaction developed several tools addressing specific human trait attributions towards robots, such as the Robotic Social Attribute Scale (RoSAS; Carpinella et al., 2017), which focuses on the evaluation of the social competence of the robot, and the Human–Robot Interaction Evaluation Scale (Spatola et al., 2021a), which aims at evaluating the perceived quality of the interaction, for example at the level of perceived agency.

Finally, other researchers have focused more on the attribution of mental states *per se*, investigating under which conditions it can be elicited and how it can influence the interaction (Thellman et al., 2022). In the next section, we will introduce the concept of the intentional stance, defined by philosopher Daniel Dennett (Dennett, 1971), which is similar but not identical to the concept of the theory of mind.

3.3 From Folk Psychology to Intentional Stance

Dennett (1987) introduced the concept of the intentional stance, which is basically a different term for the idea of folk-psychological explanations of others' behavior (in line with the TT approach to social cognition). Dennett proposes three 'stances' that can be used to explain and predict the behavior of a system, each representing a different level of abstraction. The first stance is the physical stance, which refers to the physical domain and predicts behavior based on the laws of physics. For example, by adopting the physical stance, the observer predicts a ball's trajectory after it has been thrown, assuming knowledge of its initial velocity, angle of release, air resistance, and gravity. Essentially, the physical stance allows us to make predictions based solely on the system's physical properties. The second is the design stance, which considers how a system was built to function and uses this knowledge to predict behavior; for example, predicting a car's behavior when the driver pushes the brake pedal. In this case, the observer relies on their knowledge of the car's functional design to predict that the car will stop. The observer does not need to know any physical laws or consider any mental states to make this prediction. Instead, they use their understanding of the car's design and how it was built to make a prediction about its behavior. The third is the intentional stance, which considers the agent's mental states and beliefs to predict their behavior. For example, by adopting the intentional stance, we can predict that a person will order something to drink in a coffee bar based on the observation that they have just taken a water bottle out from their bag and realized that it was empty. Then, they entered the coffee bar immediately thereafter. In this situation, the observer assumes that the person *desires* something to drink and *believes* that the coffee bar can fulfill that desire. The observer does not need to know any physical laws or specific design features to make this prediction. Instead, they ascribe mental states, such as desires and beliefs, to the observed person and make a prediction based on those ascribed mental states. Dennett notes that the first two stances can be applied to any system, while the intentional stance requires assuming the system is rational. By adopting the intentional stance, the observer assumes that the system's behavior is the most rational given the ascribed beliefs, desires, and constraints. However, Dennett emphasizes that only for truly intentional agents is the intentional stance the most efficient strategy to predict and explain their behavior. For other systems, other stances might work better.

As humanoid robots occupy a unique position on the spectrum between human beings and non-embodied artificial agents such as computer algorithms (Surden & Williams, 2016), they are particularly interesting for investigating the adoption of the intentional stance towards artificial agents. Indeed, NOC theory suggests that humanoid robots are in a middle position between human and non-human agents (Kahn & Shen, 2017). Thus, thanks to their embodiment, humanoid robots can be designed to fulfill the human need for social interaction, especially in contexts such as elderly care, education, and healthcare (Epley et al., 2007; Sgorbissa et al., 2018) and thus evoke the tendency to adopt the intentional stance. However, it is important to note that humanoid robots are still artifacts, and this awareness may evoke the adoption of the design stance instead. The individual and cultural traits of the observer also play a role in whether an individual adopts the intentional or the design stance.

3.4 Cultural Influences: Norms and Values

When examining the potential factors that contribute to the adoption of an intentional stance towards robots, it is crucial to recognize that humans are influenced by their cultural surroundings (Bartneck et al., 2006; Khan & Cox, 2017). These cultural environments can affect individuals' perceptions of robots and the likelihood of adopting the intentional stance towards them. Research has shown that individuals from the Eastern part of the world tend to adopt a more holistic approach and are more inclined to cooperate, while those from the Western world tend to be more individualistic (Varnum et al., 2010). Other studies have also highlighted differences in cultural values, such as collectivism and individualism, between Eastern and Western countries, leading to the notion that these values are 'Cultural Syndromes' (Lavelle, 2016; Triandis, 1993; Triandis & Gelfans, 2011). Cultural Syndromes are defined as constructs organized around a particular value which exhibit smaller differences within cultures compared with across cultures and demonstrate a pattern of emergence across geographic regions. These values are internalized throughout an individual's development, resulting in their use as biases and heuristics in our society and daily life, and also when interacting with artificial agents (Fortunati & Edwards, 2022). Research shows that collectivistic cultures tend to prioritize harmony and positive social interactions, while individualistic cultures emphasize self-reliance (Liu et al., 2019; Triandis, 1993). The cultural background can affect individuals' behaviors towards robots, including likeability, satisfaction, trust, engagement, and distance kept during interactions (Bartneck et al., 2006; Marchesi et al., 2021a). Additionally, cultural differences can modulate people's tendency towards anthropomorphizing robots and attributing intentionality to them (Beran et al., 2011; Spatola et al., 2022b; Złotowski et al., 2014, 2015).

Researchers in the field of human–robot interaction have also investigated whether other cultural differences extend to interactions with robots (Lim et al., 2021). For example, some cultures may be more prone to anthropomorphizing robots due to their shared values, norms, or beliefs. This tendency towards anthropomorphism can be influenced by several factors, such as exposure to robots, personal experiences, media exposure, and technological development (Airenti, 2018; Bartneck et al., 2006). However, historical and religious contexts may be the main reason for cultural differences in anthropomorphism (Cohen and Johnson, 2011; Liu et al., 2019) . For example, Japanese culture is associated with high anthropomorphism due to the animism beliefs inherent in the dominant Shinto religion (Heine et al., 2002).

One key reason cultural and social norms and values are crucial in the design of social robots is that they are meant to interact with humans in social contexts embedded in a given cultural setting (Bruno et al., 2019; Sgorbissa et al., 2018). Social interactions are governed by complex rules and norms that vary across cultures, age groups, and even individual preferences. A social robot that fails to take these factors into account is likely to make mistakes or misunderstandings that can lead to negative experiences for both the robot and the human users. For example, a robot designed to support elderly users may not be effective if it fails to consider the cultural values and expectations of older adult populations. Indeed, Bemelmans and colleagues stress the importance of purposely designing not only the robots, but also the intervention contexts to enhance the efficacy and acceptance of the social robots (e.g., Bemelmans et al., 2012).

Another reason is that these machines often serve as extensions of human identity and culture. Social robots are designed to embody cultural or social values important to human users (Broadbent, 2017). For example, robots that are designed to interact with children may need to incorporate values such as respect for authority (especially in cultures where this is a predominant value) and social cooperation (especially in collectivistic cultures), while robots that are designed for use in healthcare settings may need to incorporate values such as compassion and empathy (Lee et al., 2012)—again to varying degree, dependent on a given culture. By taking these values into account, social robots can become more effective at promoting positive social outcomes and enhancing the overall well-being of their human users (Formosa, 2021).

A third reason is that social robots following social norms and cultural values can help to promote social acceptance and trust. Social robots are still a relatively new technology, and people are understandably wary of interacting with machines that seem to have human-like qualities. However, by designing social robots to adhere to socially accepted norms and values, we can increase the likelihood of people accepting them as a natural part of their social environment. Several studies have demonstrated the importance of social norms and values in shaping human–robot interactions. For example, a study by Lee and Makatchev (2009) found that people were more willing to approach and interact with a robot that exhibited social norms such as greeting and politeness . Similarly, a study by Bartneck et al. (2006) showed that people were more likely to trust a robot that behaved in accordance with social norms, such as respecting personal space and displaying appropriate emotional responses. These findings suggest that incorporating social norms and values into the design of social robots can help improve their social acceptance and trustworthiness among humans.

In summary, incorporating cultural and social values and norms in the design of humanoid robots has numerous advantages that can enhance their effectiveness and social acceptance. Cultural factors can play a crucial role in shaping individuals' attitudes towards robots and robots' social inclusion (Ciardo et al., 2022; Fortunati & Edwards, 2022). However, further research is needed to investigate the potential role of culture in the social inclusion of robots as a function of mind perception and attribution of intentionality. This future research should also focus on investigating the possible risks of including cultural and social norms, in order to avoid potential biases and/or discrimination behaviors arising from identification with certain groups but not others (Coeckelbergh, 2009a; Formosa, 2021).

3.5 Intentional Stance and Cultural Contexts: Example Studies

Recent literature has operationalized the intentional stance concept (Thellman et al., 2022). Thellman and colleagues studied participants' perception of intentionality in images of humans and humanoid robots and found similar levels of intentionality attributed to both humans and robots (Thellman et al., 2017). Indeed, in two recent studies the authors report that humans tend to make causal attributions to robots' behaviors. In particular, extremely important factors seem to be an attribution of autonomy to social robots' behaviors (Horstmann & Krämer, 2022) and the corresponding underlying attitudes attributed to these behaviors (Edwards & Edwards, 2022). Following up on the quest to operationalize a high-level philosophical concept of the intentional stance, Marchesi and colleagues developed a tool, the InStance test (IST), to assess individual tendencies to attribute intentionality to a humanoid robot. The IST includes 34 pictorial scenarios. Below each scenario, two descriptions are provided—one mechanistic and the other mentalistic. Participants are asked to move a button on a slider towards the description that, in their opinion, best fits the scenario (Marchesi et al., 2019). The IST was also adapted to a two-alternative forced-choice task, where participants judged whether the description was appropriate for the scenario they observed. The authors found that participants were more prone to accept the mechanistic explanation for the robot at the explicit level, but the response time revealed that they were somewhat unsure about which stance to adopt (Marchesi et al., 2021b). Moreover, based on this result, Spatola and colleagues explored the allocation of cognitive resources involved in the adoption of the intentional stance, reporting that the mentalistic description is more accessible compared with the mechanistic one, especially when the cognitive resources were drained by an orthogonal task (Spatola et al., 2022a). Bossi et al. (2020) found that it is possible to predict an individual's bias towards the intentional or design stance through neural oscillatory patterns during the resting state (Bossi et al., 2020). Finally, Spatola et al. explored the relationship between anthropomorphism and the adoption of the intentional stance in two different studies, reporting that while they rely on similar mechanisms, intentional stance and anthropomorphism are two distinct concepts (Marchesi et al., 2021b; Spatola et al., 2022b). Indeed, anthropomorphism involves attributing human-like qualities or characteristics to non-human entities, such as animals or robots. This can include attributing physical features (e.g., a face to the front of a car or the moon), emotions, thoughts, or intentions to these entities, as well as projecting human-like behavior or expressions onto them. On the other hand, the intentional stance refers specifically to the practice of attributing beliefs, desires, and intentions to other entities. As such, it is a form of anthropomorphism. In the case of robots, as research has shown that it is possible to adopt the intentional stance to make sense of the behavior of such entities (Marchesi et al., 2019), it is important to clarify the relationship between the two concepts, which seem to differ in their goals and scope. Adopting the intentional stance is a more specific strategy that allows us to treat and understand other agents, including robots, as intentional agents. On the other hand, anthropomorphism is a broader phenomenon that involves projecting human-like qualities (both physical and mental) onto a range of non-human entities, often to make them more relatable or familiar to us. Thus, anthropomorphism is a broader concept encompassing the adoption of the intentional stance.

To investigate this topic further, the IST can be used to disentangle the relationship with anthropomorphism both within the same cultural group and cross-culturally. For example, Spatola and colleagues conducted a study exploring the psychometric properties of the IST (Spatola et al., 2021b). The authors found a two-factor structure that correlated with attributing anthropomorphic traits to robots, and concluded that although intentional stance and anthropomorphism share some common constructs, they are still distinct concepts as their correlation indices were moderate to low. Finally, to explore the cross-cultural level, Marchesi et al. (2021a) administered a battery of tests to 600 participants from different countries and analyzed the data using a path model. The results indicated that collectivism had a direct influence on anthropomorphism but not on the adoption of the intentional stance. To explore this finding further, their mediation analysis revealed that anthropomorphism acted as a mediator between collectivism and the adoption of the intentional stance. Therefore, the authors concluded that when considering the cross-cultural level, the adoption of the intentional stance towards humanoid robots is influenced by the different levels of anthropomorphic attribution. Finally, building on this result, Spatola et al. (2022b) proposed an integrative framework of anthropomorphism encompassing three levels (cultural, the likelihood of adopting the intentional stance, and the individual tendency to anthropomorphize). The authors used a cross-cultural approach to explain the tendency to ascribe mental states and other human characteristics to non-human agents in human–robot interaction studies. Results from two studies demonstrated the influence of animism and individual tendencies on attributions of anthropomorphic characteristics towards robots, and highlighted the different views of Western and East Asian participants towards the humanization of artificial embodied agents and the adoption of the intentional stance.

In an experimental protocol of more direct human–robot interaction, Marchesi and colleagues (Marchesi et al., 2022) explored whether a short interaction with a robot can modulate the tendency to adopt the intentional stance and whether this would potentially be influenced by cultural factors. The authors embedded the human–robot interaction in a daily social context with which most people are familiar: watching movies together. In more detail, in three experiments, participants watched a series of National Geographic movies with the iCub robot (Metta et al., 2010; Natale et al., 2017) exhibiting either human-like behavior (Experiments 1 and 2) or displaying completely mechanistic 'reactions' to the environment. In the human-like conditions, the robot would show sadness, awe, or joy in response to events occurring in the videos. The authors utilized the principles of animation to ensure that the robot's movements appeared as human-like as possible (Sultana et al., 2013). Moreover, they recorded the verbal, emotional reactions and sentences using an actor, digitally edited them to match the iCub's appearance, and played them via a Wizard-of-Oz manipulation (WoOz) at the beginning and end of the session (Rea et al., 2017; Riek, 2012), to create a more natural interaction. Finally, since recent research shows the importance of mutual gaze for humans' social cognition, even in human–robot interaction (Kompatsiari et al., 2022), cameras from the robot's eyes actively recognized participants' faces to ensure mutual gaze between the iCub and participants. In the mechanical condition (Experiment 3), the authors designed the robot's behavior to be machine-like by having it perform the same repetitive movements of the torso, head, and neck in reaction to the videos and issuing very mechanical sounds, such as beeping (comparable to sounds of car sensors when approaching an obstacle). Cameras were deactivated, and the WoOz manipulation was replaced with pre-programmed robotic actions, such as joint calibration. Verbal interaction was replaced with a text-to-speech description of the robot's calibration sequences, and all emotional sounds were replaced with a 'beep' sound. Results from this series of experiments showed that humans' likelihood of adopting the intentional stance (measured before and after the interaction) increased in the human-like condition and remained constant in the mechanical condition. These findings suggest that even brief interactions with humanoid robots can modulate the likelihood of adopting the intentional stance, depending on the robot's behavior and potential social bonding (through joint social activity such as movie watching). Indeed, Higgins (2020) argued that humans have a 'minimal relational self' among their ontological primitive of selfhood. In other words, according to Higgins (2020), not only are we born with an innate sense of self (Fuchs, 2017), but we also have an innate sense of socialness. Along the same lines, Humphrey argues that the human brain evolved in such a way because of the importance given to 'social problems,' putting emphasis on the necessity of predicting and understanding conspecific behaviors (Humphrey, 1976). Hence, our brains are hardwired to fit the 'non-social into the social.' Accordingly, the shared social and affective context that the participants experienced might have led them to build the expectation that the robot could be an interactive-embodied agent able to perceive the context similarly to how they were experiencing it themselves. This perception activated the human model, leading participants to spontaneously adopt the intentional stance towards the humanoid robot (Wiese et al., 2017). Overall, this study provides insight into how experience with a robot can influence the perception of humanoid robots and the adoption of the intentional stance towards them.

Recently, Marchesi et al. (2023) conducted a replication of this experimental setup in Singapore. Interestingly, the study found that the result of the increased likelihood of adopting the intentional stance in the human-like condition was similar across Italy and Singapore. This suggests that the effect of human-like behavior on the adoption of the intentional stance is robust enough to generalize across labs and cultures. This cross-cultural study investigating cultural influences on the adoption of the intentional stance can be considered a first step in addressing such empirical questions across continents, considering that the technical reproducibility of cross-cultural studies in human–robot interaction has always been an issue, especially when complex robotic platforms are involved, such as the iCub robot (Marchesi et al., 2023). The authors indeed addressed a challenge in creating a stable technical environment that allowed for the replication of the same experimental protocol across continents. Such technological challenges are part of the empirical leaps that need to be filled and addressed to reliably investigate questions about human–robot interaction in a replicable and well-controlled manner. Along these lines, Papadopoulos and colleagues tested the effect of a culturally aware robot on the outcomes of an assistive intervention on older adults (Papadopoulos et al., 2020). These authors reported all the complex challenges faced (both from the ethical and the technical point of view) in implementing a rigorous study protocol among different continents, showing how complex is the actual operationalization of empirical studies involving human–robot interaction.

3.6 A More Inclusive Society Thanks to Robots with a Good Design

Factors such as the adoption of the intentional stance towards robots, individual differences in the tendency to anthropomorphize non-human agents, and possible effects of cultural background should be the foundation of the design of the next generation of social robots, able to attune with humans socially. Accounting for these factors in the design will help the transition process to integrate embodied artificial agents in our social environments as assets to improve our society (Coeckelbergh, 2009a, 2009b; Formosa, 2021).

The intentional stance allows humans to understand and predict the behaviors of other humans, animals, and even machines by attributing intentions, beliefs, and desires to them (Dennett, 1987). By adopting the intentional stance, we can interpret the behaviors of humanoid robots as intentional, as if they were agents with underlying goals, motivations, and beliefs, similar to human beings. This perception can help bridge societal gaps and improve human autonomy, especially in the case of moral decision making (Formosa, 2021). In more detail, Formosa analyzed the societal implications of perceiving embodied robots as social actors on the nature of autonomous (moral) decision making. The author argued that, given their embodied presence in our environments and the increasing evidence that humans can adopt the intentional stance and apply the same cognitive mechanism, initial expectations, societal scripts, rules, and norms when interacting with them (Edwards et al., 2019; Fox & Gambino, 2021; Gambino et al., 2020; Kahn & Shen, 2017; Marchesi et al., 2022), social robots can assist and enhance human autonomy on (at least) three levels:

1. *More valuable ends*: by assisting humans in setting goals or assisting them in their needs to achieve them.
2. *Improving autonomy competencies*: by assisting indirectly (i.e., taking over less relevant tasks to give humans more time for more relevant tasks) or directly (i.e., a positive social interaction) in the decision-making process. It is important to highlight that in the case of direct assistance (i.e., the social interaction), the encounter needs to be positive and thus take into account all the factors that induce a satisfying interaction, such as social inclusion (Nash et al., 2018) and reduced anxiety (Jeong et al., 2015; Pu et al., 2019).
3. *More authentic choices*: as defined by Walker and Mackenzie (2020), a choice can be considered authentic if one acts based on motives, desires, preferences, internal mental states, and reasons. Thus, a social robot can help us avoid wrong values, norms, or biases via positive social interaction, inducing more informed moral reasoning.

It is crucial to highlight that the quality of the impact of social robots on humans' autonomy depends on the type of design, regulations, and uses that we will implement in social robots. Indeed, there can also be adverse effects on human autonomy that we need to acknowledge as possible risks in order to avoid them. These risks are the negative counterpart of what Formosa (2021) describes as:

1. *Fewer valuable ends*: where robots autonomously decide which information is relevant for the human (i.e., relevance not based on the human's values).
2. *Worse autonomy competencies*: an offload in decision making could result in a 'moral deskilling' (Vallor, 2015) by expecting the robot to decide on our behalf.
3. *Less authentic choices*: in this case, by adopting the intentional stance towards a social robot and thus ascribing mental states to the behavior of the robot, humans might feel more self-conscious, as if the robot were 'watching' them, resulting in self-censored behaviors or inauthentic choices.

Thus, following Mark Coeckelbergh's (2009a) account of roboethics, it is pivotal to focus on how social robots affects humans, in other words what social robots "can do" to humans instead of focusing only on their behaviors, as if they were abstracted from our social environments. By adopting a more (self-conscious) anthropocentric view of social robots, we will be able to understand humans' social cognition and social interactions, allowing for the design of 'personal' robots where it is possible to implement social norms, values, and regulations that can contribute positively to humankind and our societies (Coeckelbergh, 2009a; Dignum, 1999; Kammler et al., 2023).

In fact, 'use cases' about how the design of socially capable robots can assist humans in navigating their social environments can already be found in the literature (Bruno et al., 2019; Papadopoulos et al., 2020). For example, humanoid robots can be designed to enhance social inclusion in various contexts (Perugia & Lisy, 2022). One can imagine a future scenario where a robot, involved in a social context with another human, informs (or warns) the human that their behavior might lead to the experience of social exclusion of theother human partner.

In the context of neurodiversity, humanoid robots have been used to train sociocognitive abilities (Dautenhahn & Werry, 2004; Ghiglino et al., 2023; Scassellati et al., 2012). One could design humanoid robots capable of adapting to different communication and sensory styles, to make such training more tailored to the user's needs. For example, some individuals with autism may prefer direct and unambiguous communication, while others may benefit from visual cues and repetition. By incorporating these preferences into the design and programming of humanoid robots, we can create more inclusive and accessible social interactions for individuals with neurodivergent needs (Kajopoulos et al., 2015; Kozima et al., 2005).

Ageism is another societal gap that humanoid robots can help bridge. As the world's population ages, there is an increasing need for social interaction and care for older adults (Bemelmans et al., 2012). Humanoid robots can be programmed to provide emotional support, cognitive stimulation, and physical assistance to older adults (Bemelmans et al., 2012). This can help alleviate loneliness and social isolation while empowering older adults to maintain their independence and autonomy (Formosa, 2021). It is pivotal to focus on creating a positive and healthy relationship with social robots, as the field of socially assistive robotics in domestic environments could also potentially fall into the danger of reinforcing a less fulfilling form of 'social' interaction (Turkle, 2010) and harming the amount of human social contact experienced by the older adult population (Sparrow & Sparrow, 2006). Research on social robots to assist lonely (or aging) users must be aware of this potential risk and—through proper design practices—guarantee that social robots will alleviate, rather than enhance, social isolation. For example, social robots can be designed to request the older adult users to exit their isolated rooms and meet with other users of similar social robots. Finally, it is also crucial to put regulatory efforts into guaranteeing that the amount of human contact in the care sector does not decrease together with the increase of the presence of socially assistive robots.

3.7 Conclusions

In conclusion, by recognizing and reflecting on the individual and cultural values, customs, and beliefs of the communities where robots are used, designers can create machines that better align with the needs and preferences of the users. This can lead to improved user experience, increased engagement, and reduced friction in human–robot interactions. Furthermore, taking cultural differences into account can promote diversity, inclusivity, and empathy, allowing robots to serve a broader range of users with different backgrounds, ages, genders, and abilities. These future directions should be built on the assumption that humans should always be the center of the research questions, whether philosophical or empirical. Such a framework should define the approach to understanding how social robots can be helpful and assist humans, considering our tendency to anthropomorphize, to adopt the intentional stance, and our sensitivity to social signals coming from such embodied artificial agents. Moreover, we cannot build socially capable and/or approachable robots without considering the implications for our ethical and moral values and without building a systematic model of human–robot interaction that accounts for 'what robots do to us' (Coeckelbergh, 2009a; Formosa, 2021).

References

Airenti, G. (2018). The development of anthropomorphism in interaction: Intersubjectivity, imagination, and theory of mind. *Frontiers in Psychology*, *9*, 2136. https://doi.org/10.3389/fpsyg.2018.02136.

Airenti, G., Cruciani, M., & Plebe, A. (2019). Editorial: The cognitive underpinnings of anthropomorphism. *Frontiers in Psychology*, *10*, 1539. https://doi.org/10.3389/fpsyg.2019.01539.

Andrews, K., & Huss, B. (2014). Anthropomorphism, anthropectomy, and the null hypothesis. *Biology & Philosophy*, *29*(5), 711–729. https://doi.org/10.1007/s10539-014-9442-2.

Apperly, I. A., & Butterfill, S. A. (2009). Do humans have two systems to track beliefs and belief-like states? *Psychological Review*, *116*(4), 953–970. https://doi.org/10.1037/a0016923.

Barrouillet, P. (2015). Theories of cognitive development: From Piaget to today. *Developmental Review*, *38*, 1–12. https://doi.org/10.1016/j.dr.2015.07.004.

Bartneck, C., Suzuki, T., Kanda, T., & Nomura, T. (2006). The influence of people's culture and prior experiences with Aibo on their attitude towards robots. *AI & Society*, *21*(1–2), 217–230. https://doi.org/10.1007/s00146-006-0052-7.

Bemelmans, R., Gelderblom, G. J., Jonker, P., & de Witte, L. (2012). Socially assistive robots in elderly care: A systematic review into effects and effectiveness. *Journal of the American Medical Directors Association*, *13*(2), 114–120. https://doi.org/10.1016/j.jamda.2010.10.002.

Beran, T. N., Ramirez-Serrano, A., Kuzyk, R., Fior, M., & Nugent, S. (2011). Understanding how children understand robots: Perceived animism in child–robot interaction. *International Journal of Human–Computer Studies*, *69*(7–8), 539–550. https://doi.org/10.1016/j.ijhcs.2011.04.003.

Bossi, F., Willemse, C., Cavazza, J., Marchesi, S., Murino, V., and Wykowska, A., (2020). The Human Brain Reveals Resting State Activity Patterns that are Predictive of Biases in Attitudes toward Robots. *Sci. Robotics* 5, 46. https://doi/10.1126/scirobotics.abb6652

Broadbent, E. (2017). Interactions with robots: The truths we reveal about ourselves. *Annual Review of Psychology*, *68*(1), 627–652. https://doi.org/10.1146/annurev-psych-010416-043958.

Bruno, B., Recchiuto, C. T., Papadopoulos, I., Saffiotti, A., Koulouglioti, C., Menicatti, R., Mastrogiovanni, F., Zaccaria, R., & Sgorbissa, A. (2019). Knowledge representation for culturally competent personal robots: Requirements, design principles, implementation, and assessment. *International Journal of Social Robotics*, *11*(3), 515–538. https://doi.org/10.1007/s12369-019-00519-w.

Butterfill, S. A., & Apperly, I. A. (2013). How to construct a minimal theory of mind. *Mind & Language*, *28*(5), 606–637. https://doi.org/10.1111/mila.12036.

Carpinella, C. M., Wyman, A. B., Perez, M. A., & Stroessner, S. J. (2017). The Robotic Social Attributes Scale (RoSAS): Development and validation. In *Proceedings of the 2017 ACM/IEEE International Conference on Human–Robot Interaction* (pp. 254–262). https://doi.org/10.1145/2909824.3020208.

Chaminade, T., Hodgins, J., & Kawato, M. (2007). Anthropomorphism influences perception of computer-animated characters' actions. *Social Cognitive and Affective Neuroscience*, *2*(3), 206–216. https://doi.org/10.1093/scan/nsm017.

Ciardo, F., De Tommaso, D., & Wykowska, A. (2022). Joint action with artificial agents: Human-likeness in behaviour and morphology affects sensorimotor signaling and social inclusion. *Computers in Human Behavior*, *132*, 107237. https://doi.org/10.1016/j.chb.2022.107237.

Coeckelbergh, M. (2009a). Personal robots, appearance, and human good: A methodological reflection on roboethics. *International Journal of Social Robotics*, *1*(3), 217–221. https://doi.org/10.1007/s12369-009-0026-2.

Coeckelbergh, M. (2009b). Virtual moral agency, virtual moral responsibility: On the moral significance of the appearance, perception, and performance of artificial agents. *AI & Society*, *24*(2), 181–189. https://doi.org/10.1007/s00146-009-0208-3.

Cohen, A., & Johnson, K. (2011). Religion, Culture, and the Personification of Non-Human Entities. https://www.metanexus.net/archive/conference2009/articles/Default-id=10797.aspx.html

Cullen, H., Kanai, R., Bahrami, B., & Rees, G. (2014). Individual differences in anthropomorphic attributions and human brain structure. *Social Cognitive and Affective Neuroscience*, *9*(9), 1276–1280. https://doi.org/10.1093/scan/nst109.

Dacey, M. (2017). Anthropomorphism as cognitive bias. *Philosophy of Science*, *84*(5), 1152–1164. https://doi.org/10.1086/694039.

Dautenhahn, K. (2007). Socially intelligent robots: Dimensions of human–robot interaction. *Philosophical Transactions of the Royal Society B: Biological Sciences*, *362*(1480), 679–704. https://doi.org/10.1098/rstb.2006.2004.

Dautenhahn, K., & Werry, I. (2004). Towards interactive robots in autism therapy: Background, motivation and challenges. *Pragmatics & Cognition*, *12*(1), 1–35. https://doi.org/10.1075/pc.12.1.03dau.

De Jaegher, H., & Di Paolo, E. (2007). Participatory sense-making: An enactive approach to social cognition. *Phenomenology and the Cognitive Sciences*, *6*(4), 485–507. https://doi.org/10.1007/s11097-007-9076-9.

Dennett, D. C. (1971). Intentional systems. *Journal of Philosophy*, *68*(4), 87–106. https://doi.org/10.2307/2025382.

Dennett, D. C. (1987). *The Intentional Stance*. MIT Press.

Dignum, F. (1999). Autonomous agents with norms. *Artificial Intelligence and Law*, *7*, 69–79. https://doi.org/10.1023/A:100831553032.

Duffy, B. R. (2003). Anthropomorphism and the social robot. *Robotics and Autonomous Systems*, *42*(3–4), 177–190. https://doi.org/10.1016/S0921-8890(02)00374-3.

Edwards, A., & Edwards, C. (2022). Does the correspondence bias apply to social robots? Dispositional and situational attributions of human versus robot behavior. *Frontiers in Robotics and AI*, *8*, 788242. https://doi.org/10.3389/frobt.2021.788242.

Edwards, A., Edwards, C., Westerman, D., & Spence, P. R. (2019). Initial expectations, interactions, and beyond with social robots. *Computers in Human Behavior*, *90*, 308–314. https://doi.org/10.1016/j.chb.2018.08.042.

Epley, N., Waytz, A., & Cacioppo, J. T. (2007). On seeing human: A three-factor theory of anthropomorphism. *Psychological Review, 114*(4), 864–886. https://doi.org/10.1037/0033-295X.114.4.864.

Formosa, P. (2021). Robot autonomy vs. human autonomy: Social robots, artificial intelligence (AI), and the nature of autonomy. *Minds and Machines, 31*(4), 595–616. https://doi.org/10.1007/s11023-021-09579-2.

Fortunati, L., & Edwards, A. P. (2022). Framing the psycho-social and cultural aspects of human–machine communication. *Human–Machine Communication, 4*. https://doi.org/10.30658/hmc.4.1.

Fox, J., & Gambino, A. (2021). Relationship development with humanoid social robots: Applying interpersonal theories to human–robot interaction. *Cyberpsychology, Behavior, and Social Networking, 24*(5), 294–299. https://doi.org/10.1089/cyber.2020.0181.

Fuchs, T. (2017). Intercorporeality and interaffectivity. In C. Meyer, J. Streeck, & J. Scott Jordan (Eds.), *Intercorporeality: Emerging Socialities in Interaction* (pp. 3–23). Oxford University Press.

Fuchs, T., & De Jaegher, H. (2009). Enactive intersubjectivity: Participatory sense-making and mutual incorporation. *Phenomenology and the Cognitive Sciences, 8*(4), 465–486. https://doi.org/10.1007/s11097-009-9136-4.

Gallagher, S. (2000). Philosophical conceptions of the self: Implications for cognitive science. *Trends in Cognitive Sciences, 4*(1), 8.

Gallagher, S. (2005). *How the Body Shapes the Mind.* Oxford University Press. https://doi.org/10.1093/0199271941.001.0001.

Gallagher, S., & Zahavi, D. (2020). *The Phenomenological Mind: An Introduction to Philosophy of Mind and Cognitive Science* (3rd ed.). Routledge.

Gambino, A., Fox, J., & Ratan, R. (2020). Building a stronger CASA: Extending the computers are social actors paradigm. *Human–Machine Communication, 1*, 71–86. https://doi.org/10.30658/hmc.1.5.

Ghiglino, D., Floris, F., De Tommaso, D., Kompatsiari, K., Chevalier, P., Priolo, T., & Wykowska, A. (2023). Artificial scaffolding: Augmenting social cognition by means of robot technology. *Autism Research*, aur.2906. https://doi.org/10.1002/aur.2906.

Goldman, A. I., & Sripada, C. S. (2005). Simulationist models of face-based emotion recognition. *Cognition, 94*(3), 193–213. https://doi.org/10.1016/j.cognition.2004.01.005.

Heine, S. J., Lehman, D. R., Peng, K., & Greenholtz, J. (2002). What's wrong with cross-cultural comparisons of subjective Likert scales? The reference-group effect. *Journal of Personality and Social Psychology, 82*(6), 903–918. https://doi.org/10.1037/0022-3514.82.6.903.

Higgins, J. (2020). The 'we' in 'me': An account of minimal relational selfhood. *Topoi, 39*(3), 535–546. https://doi.org/10.1007/s11245-018-9564-2.

Horstmann, A. C., & Krämer, N. C. (2022). The fundamental attribution error in human–robot interaction: An experimental investigation on attributing responsibility to a social robot for its pre-programmed behavior. *International Journal of Social Robotics, 14*(5), 1137–1153. https://doi.org/10.1007/s12369-021-00856-9.

Humphrey, N. K. (1976). The social function of intellect. In P. P. G. Bateson & R. A. Hinde (Eds.), Growing Points in Ethology(pp. 303–317). Cambridge, UK: Cambridge University Press.https://www.researchgate.net/publication/275519041_The_Social_Function_of_Intellect/citations.

Jeong, S., Logan, D. E., Goodwin, M. S., Graca, S., O'Connell, B., Goodenough, H., Anderson, L., Stenquist, N., Fitzpatrick, K., Zisook, M., Plummer, L., Breazeal, C., & Weinstock, P. (2015). A social robot to mitigate stress, anxiety, and pain in hospital pediatric care. In *Proceedings of the Tenth Annual ACM/IEEE International Conference on Human–Robot Interaction Extended Abstracts* (pp. 103–104). https://doi.org/10.1145/2701973.2702028.

Jones, R. A. (2022). Projective anthropomorphism as a dialogue with ourselves. *International Journal of Social Robotics, 14*(10), 2063–2069. https://doi.org/10.1007/s12369-021-00793-7.

Kahn, P. H., & Shen, S. (2017). NOC NOC, who's there? A New Ontological Category (NOC) for social robots. In N. Budwig, E. Turiel, & P. D. Zelazo (Eds.), *New Perspectives on Human Development* (1st ed., pp. 106–122). Cambridge University Press. https://doi.org/10.1017/CBO9781316282755.008.

Kajopoulos, J., Wong, A. H. Y., Yuen, A. W. C., Dung, T. A., Kee, T. Y., & Wykowska, A. (2015). Robot-assisted training of joint attention skills in children diagnosed with autism. In A. Tapus, E. André, J.-C. Martin, F. Ferland, & M. Ammi (Eds.), *Social Robotics* (Vol. 9388, pp. 296–305). Springer International Publishing. https://doi.org/10.1007/978-3-319-25554-5_30.

Kammler, C., Mellema, R., & Dignum, F. (2023). Agents dealing with norms and regulations. In F. Lorig & E. Norling (Eds.), *Multi-Agent-Based Simulation XXIII. MABS 2022* (Vol. 13743). Springer, Cham. https://doi.org/10.1007/978-3-031-22947-3_11.

Khan, R., & Cox, P. (2017). Country culture and national innovation. *Archives of Business Research, 5*(2). https://doi.org/10.14738/abr.52.2768.

Kompatsiari, K., Ciardo, F., & Wykowska, A. (2022). To follow or not to follow your gaze: The interplay between strategic control and the eye contact effect on gaze-induced attention orienting. *Journal of Experimental Psychology: General, 151*(1), 121–136. https://doi.org/10.1037/xge0001074.

Kozima, H., Nakagawa, C., & Yasuda, Y. (2005). Interactive robots for communication-care: A case-study in autism therapy. In *ROMAN 2005: IEEE International Workshop on Robot and Human Interactive Communication* (pp. 341–346). https://doi.org/10.1109/ROMAN.2005.1513802.

Lavelle, J. (2016). Cross-cultural considerations in social cognition. In J. Kiverstein (Ed.), *The Routledge Handbook of Philosophy of the Social Mind* (1 ed., pp. 172-188). Routledge. https://doi.org/10.4324/9781315530178.

Lee, H. R., Sung, J., Sabanovic, S., & Han, J. (2012). Cultural design of domestic robots: A study of user expectations in Korea and the United States. In *2012 IEEE RO-MAN: The 21st IEEE International Symposium on Robot and Human Interactive Communication* (pp.803–808). https://doi.org/10.1109/ROMAN.2012.6343850.

Lee, M. K., & Makatchev, M. (2009, April 4). How do people talk with a robot? An analysis of human–robot dialogues in the real world. In *CHI '09 Extended Abstracts on Human Factors in Computing Systems*. https://doi.org/10.1145/1520340.1520569.

Li, X., & Sung, Y. (2021). Anthropomorphism brings us closer: The mediating role of psychological distance in user–AI assistant interactions. *Computers in Human Behavior, 118*, 106680. https://doi.org/10.1016/j.chb.2021.106680.

Lim, V., Rooksby, M., & Cross, E. S. (2021). Social robots on a global stage: Establishing a role for culture during human–robot interaction. *International Journal of Social Robotics, 13*(6), 1307–1333. https://doi.org/10.1007/s12369-020-00710-4.

Liu, S. S., Morris, M. W., Talhelm, T., & Yang, Q. (2019). Ingroup vigilance in collectivistic cultures. *Proceedings of the National Academy of Sciences, 116*(29), 14538–14546. https://doi.org/10.1073/pnas.1817588116.

Marchesi, S., De Tommaso, D., Kompatsiari, K., Wu, Y., & Wykowska, A. (2023, February 8). Tools and methods to study and replicate experiments addressing human social cognition in interactive scenarios. https://doi.org/10.31234/osf.io/xyhwt

Marchesi, S., De Tommaso, D., Perez-Osorio, J., & Wykowska, A. (2022). Belief in sharing the same phenomenological experience increases the likelihood of adopting the intentional stance toward a humanoid robot. *Technology, Mind, and Behavior*. https://doi.org/10.1037/tmb0000072.

Marchesi, S., Ghiglino, D., Ciardo, F., Perez-Osorio, J., Baykara, E., & Wykowska, A. (2019). Do we adopt the intentional stance toward humanoid robots? *Frontiers in Psychology, 10*, 450. https://doi.org/10.3389/fpsyg.2019.00450.

Marchesi, S., Spatola, N., & Wykowska, A. (202a1, October 14). The mediating role of anthropomorphism in adopting the intentional stance towards humanoid robots. https://doi.org/10.31234/osf.io/zdr9v

Marchesi, S., Spatola, N., Perez-Osorio, J., & Wykowska, A. (2021b). Human vs humanoid: A behavioral investigation of the individual tendency to adopt the intentional stance. In *Proceedings of the 2021 ACM/IEEE International Conference on Human–Robot Interaction* (pp. 332–340). https://doi.org/10.1145/3434073.3444663.

Metta, G., Natale, L., Nori, F., Sandini, G., Vernon, D., Fadiga, L., von Hofsten, C., Rosander, K., Lopes, M., Santos-Victor, J., Bernardino, A., & Montesano, L. (2010). The iCub humanoid robot: An open-systems platform for research in cognitive development. *Neural Networks*, *23*(8–9), 1125–1134. https://doi.org/10.1016/j.neunet.2010.08.010.

Nash, K., Lea, J. M., Davies, T., & Yogeeswaran, K. (2018). The bionic blues: Robot rejection lowers self-esteem. *Computers in Human Behavior*, *78*, 59–63. https://doi.org/10.1016/j.chb.2017.09.018.

Natale, L., Bartolozzi, C., Pucci, D., Wykowska, A., & Metta, G. (2017). iCub: The not-yet-finished story of building a robot child. *Science Robotics*, *2*(13), eaaq1026. https://doi.org/10.1126/scirobotics.aaq1026.

Papadopoulos, C., Hill, T., Battistuzzi, L., Castro, N., Nigath, A., Randhawa, G., Merton, L., Kanoria, S., Kamide, H., Chong, N.-Y., Hewson, D., Davidson, R., & Sgorbissa, A. (2020). The CARESSES study protocol: Testing and evaluating culturally competent socially assistive robots among older adults residing in long term care homes through a controlled experimental trial. *Archives of Public Health*, *78*(1), 26. https://doi.org/10.1186/s13690-020-00409-y.

Perugia, G., & Lisy, D. (2022). Robot's gendering trouble: A scoping review of gendering humanoid robots and its effects on HRI. *arXiv:2207.01130. arXiv*. http://arxiv.org/abs/2207.01130.

Premack, D., & Woodruff, G. (1978). Does the chimpanzee have a theory of mind? *Behavioral and Brain Sciences*, *1*(4), 515–526. https://doi.org/10.1017/S0140525X00076512.

Prescott, T. J., & Robillard, J. M. (2021). Are friends electric? The benefits and risks of human–robot relationships. *IScience*, *24*(1), 101993. https://doi.org/10.1016/j.isci.2020.101993.

Pu, L., Moyle, W., Jones, C., & Todorovic, M. (2019). The effectiveness of social robots for older adults: A systematic review and meta-analysis of randomized controlled studies. *The Gerontologist*, *59*(1), e37–e51. https://doi.org/10.1093/geront/gny046.

Rea, D. J., Geiskkovitch, D., & Young, J. E. (2017). Wizard of Awwws: Exploring psychological impact on the researchers in social HRI experiments. In *Proceedings of the Companion of the 2017 ACM/IEEE International Conference on Human–Robot Interaction* (pp. 21–29). https://doi.org/10.1145/3029798.3034782.

Riek, L. (2012). Wizard of Oz studies in HRI: A Systematic review and new reporting guidelines. *Journal of Human–Robot Interaction*, *1*(1), 119–136. https://doi.org/10.5898/JHRI.1.1.Riek.

Roselli, C., Ciardo, F., De Tommaso, D., & Wykowska, A. (2022a). Human-likeness and attribution of intentionality predict vicarious sense of agency over humanoid robot actions. *Scientific Reports*, *12*(1), 13845. https://doi.org/10.1038/s41598-022-18151-6.

Roselli, C., Ciardo, F., & Wykowska, A. (2022b). Intentions with actions: The role of intentionality attribution on the vicarious sense of agency in human–robot interaction. *Quarterly Journal of Experimental Psychology*, *75*(4), 616–632.

Scassellati, B., Henny Admoni, & Matarić, M. (2012). Robots for use in autism research. *Annual Review of Biomedical Engineering*, *14*(1), 275–294. https://doi.org/10.1146/annurev-bioeng-071811-150036.

Sgorbissa, A., Papadopoulos, I., Bruno, B., Koulouglioti, C., & Recchiuto, C. (2018). Encoding guidelines for a culturally competent robot for elderly care. In *2018 IEEE/RSJ International Conference on Intelligent Robots and Systems* (pp. 1988–1995). https://doi.org/10.1109/IROS.2018.8594089.

Sparrow, R., & Sparrow, L. (2006). In the hands of machines? The future of aged care. *Minds and Machines*, *16*(2), 141–161. https://doi.org/10.1007/s11023-006-9030-6.

Spatola, N., Kühnlenz, B., & Cheng, G. (2021a). Perception and evaluation in human–robot interaction: The Human–Robot Interaction Evaluation Scale (HRIES)—A multicomponent approach of anthropomorphism. *International Journal of Social Robotics*, *13*(7), 1517–1539. https://doi.org/10.1007/s12369-020-00667-4.

Spatola, N., Marchesi, S., & Wykowska, A. (2021b). The Intentional Stance Test-2: How to measure the tendency to adopt intentional stance towards robots. *Frontiers in Robotics and AI, 8*, 666586. https://doi.org/10.3389/frobt.2021.666586.

Spatola, N., Marchesi, S., & Wykowska, A. (2022a). Cognitive load affects early processes involved in mentalizing robot behaviour. *Scientific Reports, 12*(1), 14924. https://doi.org/10.1038/s41598-022-19213-5.

Spatola, N., Marchesi, S., & Wykowska, A. (2022b). Different models of anthropomorphism across cultures and ontological limits in current frameworks the integrative framework of anthropomorphism. *Frontiers in Robotics and AI, 9*. https://doi.org/10.3389/frobt.2022.863319.

Spatola, N., Marchesi, S., & Wykowska, A. (2023). The phenotypes of anthropomorphism and the link to personality traits. *International Journal of Social Robotics, 15*(1), 3–14. https://doi.org/10.1007/s12369-022-00939-1.

Spatola, N., & Wykowska, A. (2021). The personality of anthropomorphism: How the need for cognition and the need for closure define attitudes and anthropomorphic attributions toward robots. *Computers in Human Behavior, 122*, 106841. https://doi.org/10.1016/j.chb.2021.106841.

Sultana, N., Peng, L. Y., & Meissner, N. (2013). Exploring believable character animation based on principles of animation and acting. In *2013 International Conference on Informatics and Creative Multimedia* (pp. 321–324). https://doi.org/10.1109/ICICM.2013.69.

Surden, H., & Williams, K. C. (2016). Technological opacity, predictability, and self-driving cars. *016, 19*(2), 165–198.

Tanyi, R. A. (2002). Towards clarification of the meaning of spirituality. *Journal of Advanced Nursing, 39*(5), 500–509. https://doi.org/10.1046/j.1365-2648.2002.02315.x.

Thellman, S., de Graaf, M., & Ziemke, T. (2022). Mental state attribution to robots: A systematic review of conceptions, methods, and findings. *ACM Transactions on Human–Robot Interaction*, 3526112. https://doi.org/10.1145/3526112.

Thellman, S., Silvervarg, A., & Ziemke, T. (2017). Folk-psychological interpretation of human vs. humanoid robot behavior: Exploring the intentional stance toward robots. *Frontiers in Psychology, 8*, 1962. https://doi.org/10.3389/fpsyg.2017.01962.

Triandis, H. C. (1993). Collectivism and individualism as cultural syndromes. *Cross-Cultural Research, 27*(3–4), 155–180. https://doi.org/10.1177/106939719302700301.

Triandis, H. C., & Gelfand, M. J. (2011). *Individualism and Collectivism Scale* [dataset]. American Psychological Association. https://doi.org/10.1037/t01556-000.

Turkle, S. (2010). In good company? On the threshold of robotic companions. In Y. Wilks (Ed.), *Close Engagements with Artificial Companions* (pp. 3–10). John Benjamins Publishing Company.

Urquiza-Haas, E. G., & Kotrschal, K. (2015). The mind behind anthropomorphic thinking: Attribution of mental states to other species. *Animal Behaviour, 109*, 167–176. https://doi.org/10.1016/j.anbehav.2015.08.011.

Vallor, S. (2015). Moral deskilling and upskilling in a new machine age: Reflections on the ambiguous future of character. *Philosophy & Technology, 28*(1), 107–124. https://doi.org/10.1007/s13347-014-0156-9.

Varnum, M. E. W., Grossmann, I., Kitayama, S., & Nisbett, R. E. (2010). The origin of cultural differences in cognition: The social orientation hypothesis. *Current Directions in Psychological Science, 19*(1), 9–13. https://doi.org/10.1177/0963721409359301.

Walker, M., & Mackenzie, C. (2020). Neurotechnologies, relational autonomy, and authenticity. *International Journal of Feminist Approaches to Bioethics, 13*(1), 98–119. https://doi.org/10.3138/ijfab.13.1.06.

Waytz, A., Cacioppo, J., & Epley, N. (2010a). Who sees human? The stability and importance of individual differences in anthropomorphism. *Perspectives on Psychological Science, 5*(3), 219–232. https://doi.org/10.1177/1745691610369336.

Waytz, A., Epley, N., & Cacioppo, J. T. (2010b). Social cognition unbound: Insights into anthropomorphism and dehumanization. *Current Directions in Psychological Science*, *19*(1), 58–62. https://doi.org/10.1177/0963721409359302.

Waytz, A., Heafner, J., & Epley, N. (2014). The mind in the machine: Anthropomorphism increases trust in an autonomous vehicle. *Journal of Experimental Social Psychology*, *52*, 113–117. https://doi.org/10.1016/j.jesp.2014.01.005.

Wiese, E., Metta, G., & Wykowska, A. (2017). Robots as intentional agents: Using neuroscientific methods to make robots appear more social. *Frontiers in Psychology*, *8*, 1663. https://doi.org/10.3389/fpsyg.2017.01663.

Wykowska, A. (2021). Robots as mirrors of the human mind. *Current Directions in Psychological Science*, *30*(1), 34–40. https://doi.org/10.1177/0963721420978609.

Złotowski, J., Proudfoot, D., Yogeeswaran, K., & Bartneck, C. (2015). Anthropomorphism: Opportunities and Challenges in human–robot interaction. *International Journal of Social Robotics*, *7*(3), 347–360. https://doi.org/10.1007/s12369-014-0267-6.

Złotowski, J., Strasser, E., & Bartneck, C. (2014). Dimensions of anthropomorphism: From humanness to humanlikeness. In *Proceedings of the 2014 ACM/IEEE International Conference on Human–Robot Interaction* (pp. 66–73). https://doi.org/10.1145/2559636.2559679.

Luisa Damiano and Paul Dumouchel

Are Robotic Bodies (Part of) Social Bodies?

Embodiment and Sociability in Human–Robot Interaction

Abstract: The chapter proposes a philosophical exploration of human–robot social and affective interactions. We start with what has become an important debate in social robotics: 'Are social robots embodied?' We first critically analyze the notion of body and the body/mind relationship implemented in current social robots. Second, we outline the implications of these views for our interactions with these 'social machines' while paying special attention to emotion-centered interactions. We then propose an alternative view of human–robot social interactions focusing on affective coordination that allows us to escape the sterile philosophical debate concerning the simulated or authentic character of interactions with social robots. Finally, we critically reflect on the evolutions that appear to be emerging for social robots and the impacts they may have on our social interaction with robots and others.

Keywords: Social robotics, social embodiment, robotic body vs. biological body, human–robot affective coordination, synthetic ethics

4.1 Robots: Embodying AI

Embodied artificial intelligence (AI) arose in the early 1990s from the 'winter' of classic or computationalist AI. Its ambition was to overcome the limitations of classic AI based on a new, positive emphasis on the role of the body in cognition. This involved a shift of focus: concentrating on the synthetic modeling of natural cognitive processes and concomitant production of 'intelligent artifacts' on 'complete agents' (e.g., Pfeifer & Scheier, 1999) rather than on computer programs. Pioneers of embodied AI conceived that robots, like natural cognitive systems, should learn through interacting with their environment and accomplish cognitive tasks through their bodies (e.g., Clark, 1998). Embodied AI's original implementation program entailed modeling in robots the whole range of natural cognitive processes—the whole phenomenology of "the embodied mind" (Varela et al., 1991)—through 'emergent design,' combining basic sensory-motor processes to generate increasingly complex cognitive dynamics—including social cognitive processes (e.g., Pfeifer & Bongard, 2006; Dautenhahn, 2007).

Despite this promising agenda, already in its incipient phase, the ability of 'body friendly' AI to surmount the difficulties met by classical AI began to be seriously questioned. Since the dawn of the new millennium, while embodied AI produced increasingly performant 'adaptive robots,' it became the target of a challenging debate concerning the nature of its robots' embodiment. More precisely, the nature of the

https://doi.org/10.1515/9783110792270-005

artificial bodies through which embodied AI models cognitive processes and performances, compared with our embodied mind.

"Are robots embodied?" is the title of the first of a series of articles by Tom Ziemke (2001, 2003, 2007, 2016) where he pushes proponents of embodiment in the cognitive sciences, AI, and related philosophical investigations to reflect on the ability of embodied AI to overcome the drawbacks of computationalist AI.

One of the principal issues involved is the 'problem of meaning,' that is, the inability of machines to understand meaning. This critical difficulty was ascribed to classic AI by John Searle in a 1980 article, questioning the functional analogy between the human cognitive mind and the digital computer—namely, the main theoretical cornerstone of computationalism and its AI. In "Minds, brains and programs" (1980), Searle defines the problem of meaning as the inability of computationalist AI to capture the semantic dimension of human cognitive processes because information processing of uninterpreted symbols cannot operate on anything other than the syntactic dimension of the symbols in which information is encoded. Searle finds the roots of this inability in a residual form of dualism. Often defined as 'functionalist dualism' (Shapiro, 2007), it consists of a reformulation of the Cartesian mind/body dichotomy at the core of classic cognitive sciences and AI as the assumption that, functionally, the computational mind is independent of its material support. It can be described as software running equally well on widely different hardware—biological wetware (our brain) or hardware of our own making (digital computers). Interestingly, Searle draws onto this radical criticism the general lines of a *pars construens* for future AI, which can be expressed as the thesis that inspired the emergence of embodied AI. To artificially produce the type of cognition characterizing biological systems—cognition defined by an inherent semantic dimension and intrinsic intentionality—we need to create a body (including a brain) that is *like* a biological one. While objecting to what he defined as the 'robot reply' to his arguments, Searle emphasized that such a body would not be a computer-driven electro-mechanical machine, similar to the robots produced by classic AI. To avoid reintroducing 'functionalist dualism,' we need a synthetic body modeling the deep features of ours. That would be a very different type of machine from the ones we can now make.

4.1.1 A self-producing body and radical embodiment

This view, which contributed to the transition to embodied AI, was later developed within the critical debate on embodiment in AI by a number of authors, such as Ezequiel Di Paolo (2003), Hubert Dreyfus (2007), and Tom Froese (2007), together with Ziemke himself (2016). In line with Searle, they argued that addressing the problem of artificially creating cognitive processes similar to natural ones requires good synthetic models of the biological body: models able to re-enact how a living system transforms its environment from a perturbative background into a 'space of meaning' (Di Paolo,

2003). A dynamic space whose events are provided with meanings inherently related to the continuity of the living system's existence and survival. At the basis of these converging perspectives is a shared notion of the biological body conceptualized as a *self-producing body*, which, while producing itself using its metabolism, responds through self-regulation to the environmental changes that are perturbing its dynamics of self-production. An inherently "cognitive body" (Montebelli et al., 2009), which stably associates with environmental perturbing events operational meanings expressed in patterns of self-regulation: dynamical schemes of self-production which allow the living system to operate in ways that tend to maintain its existence and organizational coherence in given environmental conditions.

Since the late 1990s, this characterization of the biological body has been commonly called 'radical embodiment' (Clark, 1999; Thompson & Varela, 2001). It denotes approaches that do not simply impose bodily and environmental constraints on computationalist models of the cognitive mind as corrections to the analogy between our cognitive functioning and that of a digital computer. Radical approaches to embodiment reject computationalism *tout court*. They argue that computationalist cognitive science and AI are incapable of taking into account the fundamental difference between biological systems and artificial devices like a computer. This divergence reflects 'biological autonomy,' conceptualized as the distinctive property of living systems of producing themselves by themselves. A property that cannot be supported by the heteronomous organization that defines artificial systems such as computers and rests on the specifics of biological metabolic organization. In radical approaches to embodiment, the description of natural cognition is based on theoretical frameworks that account for biological autonomy, such as theories of biological self-organization (e.g., Damiano, 2012) and the theory of autopoiesis by Humberto Maturana and Francisco Varela (e.g., Maturana & Varela, 1980; Damiano, 2009). These theories posit a profound continuity between life and cognition and theoretically ground cognitive processes in the body's processes of self-production (e.g., Thompson, 2007).

Between the late 1980s and early 1990s, these theories inspired emergent approaches in AI that brought pioneers, such as Rodney Brooks (1991), Luc Steels (1995), and Rolf Pfeifer (Pfeifer & Scheier, 1999), to start the transition towards embodied AI, and to substitute to the classic focus on computer programs, focusing on artificial agents endowed with cognitive (robotic) bodies. However, the implementation of this framework in concrete robotic AI generated ambiguous results, combining with success in applications, dissatisfactory outcomes anticipated by Searle's skepticism about the *possibility of solving the problem of meaning in AI with the help of robots.* A few years after the birth of the 'nouvelle AI,' some of its early practitioners expressed discontent with regard to their robots and reflected a similar skepticism. Already in the late 1990s, Brooks (1997) lamented that being able to reproduce adaptive and cognitive performances like those characterizing living systems seemed out of reach of the new AI. Its synthetic modeling of the biological body was unable to capture and reproduce key aspects of the living organization in robots—what he referred to as "the juice of life" (Brooks, 1997, p. 301). The body-friendly AI succeeded in endowing robotic artifacts with superficial features of the biological body, such as biomimetic morphology and movements, ensuring high-level performances—significantly higher than those of classic AI robots. However, the traits of the biological organization supporting living systems' adaptive and cognitive skills remained far beyond the modeling capabilities of the new AI.

This critical reading of embodied AI has been re-proposed and deepened by Ziemke's radical position in the debate he initiated. Robots produced by embodied AI, he argues, *are not embodied* the way we are. To capture their form of embodiment, we need to broaden our view of corporeality and define a taxonomy of forms of embodiment, including non-biological forms. To meet this need, Ziemke's theoretical classification of embodiments proposes four categories: physical realization, physical embodiment, organismoid embodiment, and organismal embodiment (Ziemke, 2001; Table 4.1). Ziemke's analysis of embodied AI divides its robots' electromechanical realization as organized in these different forms of embodiments. They can be simply physical when they do not share any traits of the biological body, or organismoid when they share only morphological, anatomical, and/or dynamical traits of the biological body. But in no case does the electromechanical realization of the robots produced by the new AI share the organismal embodiment typical of living systems, whose specific form of organization supports metabolic self-production.

Table 4.1: Forms of embodiment (Elaborated based on Ziemke, 2001, 2003).

Type of embodiment	Implementations
Physical realization The system is realized in some physical substrate	Virtual agents and robots
Physical embodiment The system is realized in coherent, integral physical structure	Robots
Organismoid embodiment The physical organization of the system shares some (superficial) characteristics with the bodies of natural organisms	Plant-like, animal-like, human-like robots
Organismal embodiment The physical realization of the system is not only organismic-like, but actually organic and alive	Biological systems

Therefore, Ziemke's answer to the question 'Are robots embodied?' expresses a radical criticism of embodied AI. The bodies of its robots are not cognitive bodies like ours. They lack biological-like autonomy and thus cannot produce 'organism centric' meanings while interacting with their environments and the systems populating them.

> However, the organic roots of these processes, which were emphasized in, for example, the theoretical biology of von Uexküll or Maturana and Varela's theory of autopoiesis, are usually ignored in embodied AI, most of which still operates with a view of the body that is largely compatible with mechanistic theories in psychology and a view of control mechanisms that is still largely compatible with computationalism. That means the robot body is typically viewed as some kind of input and output device that provides physical grounding to the internal computational mechanisms. . . . Thus, in practice, embodied AI, as a result of its history and interdisciplinary influences, has become a theoretical hybrid, combining a mechanical/behaviorist view of the body with the constructivist notion of interactive knowledge as well as the functionalist hardware–software distinction and its view of the activity of the nervous system as computational. . . . As Searle pointed out . . ., whether or not the computational mind resides in a slightly less passive robotic body, i.e., a physical container that allows the computational mind to interact with its environment through sensors and actuators, really does not make much of a difference when it comes to such computational/robotic systems as models of human cognition, intentionality, etc. (Ziemke, 2016, p. 7)

The ways embodied AI's robots learn and resolve cognitive tasks with their bodies are radically different from ours and, in most cases, are still controlled by computational mechanisms—computers. For these robots, Searle's criticism remains valid. There is no meaning, no understanding, in the dynamics through which they interact. This critical reading of embodied AI defines the theoretical angle from which we would like to approach the topic of human social interactions with contemporary robots.

4.1.2 How are social robots embodied?

The social robots produced today, for example Nao or Pepper, share the characteristics of robots typically produced by embodied AI. In fact, social robotics was born with the embodiment approach. It represents what can be defined as 'socially embodied AI,' characterized by a specific focus on modeling interactive, body-based social capabilities (Damiano et al., 2015; Dumouchel & Damiano, 2017). These aim to allow robots to reach a 'social threshold' when interacting with us. This property, often called 'social presence,' is defined as the ability of a robotic agent to give human users the 'sense of being with another' (Biocca et al., 2003) or the 'feeling of being in the company of another' (Heerink et al., 2008). Roboticists usually resort to two different heuristics to create machines able to display this feature.

The first and most widespread strategy is *anthropomorphization* (Złotowski et al., 2015; Damiano & Dumouchel, 2018)—the tendency to attribute human characteristics to objects or animals. The appearance, movements, and expressiveness of the robots are designed so that they can stimulate anthropomorphic projections in their users. The aim is to promote the projection onto the robots of human mental properties like desires, beliefs, needs, and, in particular, emotions.

Within the context of social robotics, anthropomorphism is not considered a bias, category mistake, or obstacle to the advancement of knowledge, as it traditionally has been in psychology. In line with the contemporary re-evaluation of anthropomorphism in many disciplines, embodied cognitive science and social robotics recognize that treating "non-human behavior as motivated by human feelings and mental states" (Airenti, 2015, p. 119) rather than a mistake constitutes a fundamental human tendency. One that *can be used as a tool to facilitate social exchange between robots and humans* (Damiano & Dumouchel, 2018). Anthropomorphization during interactions is encouraged by endowing social robots not only with verbal communication skills but also with appropriate bodily appearance and dynamics. Among these, features and movements of the face that play an important role in human social interactions (Dumouchel, 2022a) receive particular attention.

A second strategy, more rarely adopted to produce social presence, relies on creating cognitive architectures based on 'deep models' of human social competencies to bring robots to manifest human-like 'social skills' (Breazeal, 2003). As the relevant literature of the field shows, currently these architectures are developed by integrating computational approaches. Consequently, their social cognitive performances are subject to Searle's criticism. This implies that, for these robots, social interactions have no semantic dimension.

These two strategies—anthropomorphization and computational cognitive architectures—are sometimes combined. However, from a Searlean–Ziemkean point of view, even in these cases the outcomes cannot be characterized as the reproduction in robotic agents of modes of social interaction similar to ours. According to radical embodiment, human social and affective behavior is rooted in the autonomous organization of our body. This is clearly not the case for present-day robots.

It follows that what social robotics characterizes as the social skills of its machines does not involve the capabilities of interpreting or understanding social signals. These artifacts, in line with the standards of embodied AI, do not deal with the semantic dimension of the social signals they perceive, react to, and produce. Their interactions with human users are based on processes and mechanisms radically different from ours, as Table 4.2 intends to highlight.

Given that what social robotics calls human–robot social interaction is based on different mechanisms in humans and in the robots with which they interact, is it possible to speak of 'human–robot social interaction' and 'human–robot sociability'?

An affirmative answer to this query entails, we propose, a negative answer to a different and fundamental question.

Table 4.2: Mechanisms introduced in robots to facilitate human–robot social interaction.

Type of embodiment	Robotic mechanisms
Physical embodiment The system is realized in coherent, integral physical structure	– Toy-like appearance, movement, emotional expression, gestures, vocalization, verbal communication – Social cognition computational architectures
Organismoid embodiment The physical organization of the system shares some (superficial) characteristics with the bodies of natural organisms	– Plant-like, animal-like or human-like appearance, movement, emotional expression, gestures, vocalization, verbal communication – Social cognition computational architectures

4.1.3 Are individual bodies social bodies?

In social robotics, the two strategies considered above—endowing robots with features triggering anthropomorphic projections and endowing them with computational architectures supporting social skills—are related to the tension between two different ways of understanding the goal of creating robots characterized by social presence.

Of these ways, the first aims to develop robotic agents that possess social competencies in a 'substantial sense' (Meister, 2014), targeting the production of 'socially intelligent' robots (Breazeal, 2003) whose social presence is grounded in cognitive skills modeled on human social abilities. Its central thesis is that 'deep models' of human social competencies will allow robots to treat and answer appropriately social signals, showing "aspects of human style social intelligence" (Fong et al., 2003, p. 145). The second abandons this ambitious goal in favor of a 'functional' understanding of the objective (Jones, 2017). It aims at building robotic agents that only give the impression of being socially intelligent. Such robots are usually defined as 'socially evocative' or 'social interfaces' (Breazeal, 2003) and are designed to simulate some social competencies realistically enough to engage humans in social interaction.

While the first intends to create robots with social skills, the second seeks to make robotic agents whose physical appearance and behavior activate anthropomorphic projections. In other words, one wants *to endow* robots with social abilities, while the other aims to bring human users to treat robots *as if they were* genuine social interactors.

It is important to notice that, although there clearly is a relation between these two approaches and the two strategies presented in the previous section, they should not be fully identified. The approaches define the goal: substantial or functional social competence. The strategies define the ways of reaching that goal.

An interesting aspect of the tension between 'substantial' and 'functional' approaches is that they cannot be considered new or specific to social robotics. The tension is rooted in the history of AI. While connecting social robotics to the field's past, it has the potential to introduce genuine novelties. In fact, the *substantial vs. functional* opposition re-proposes, with a specific focus on robots and social cognition, the well-known opposition between 'weak' and 'strong' AI. Introduced by Searle (1980) to describe the two main trends of research within classic AI, 'strong AI' defines attempts to create artificial systems able to reproduce all human cognitive processes—perceiving, understanding, thinking, believing, and so on—while 'weak AI' considers, to the opposite, that artificial systems can, at best, only simulate human intelligence. They can imitate, rather than reproduce, human cognitive processes.

In Searle (1980), this distinction is the prelude to arguments making clear that weak AI is all that could be achieved within the framework of computational AI. Historically, the *weak vs. strong AI* opposition and the implicit criticism it contains had an important role, for it contributed to the rise of embodied AI. However, in the case of *socially* embodied AI, the weak vs. strong dyad, transformed into the functional vs. substantive distinction, cannot be interpreted in the same way nor subject to the criticism advanced by Searle.

In fact, Searle's interpretation and related criticism can only hold in social robotics on one condition. Adopting a conception of social competence and sociality, according to which they essentially rest on, or correspond to, sets of properties possessed by individual agents. Properties that are characteristics of the physiology and corporeality of the individual.

In this individualist and substantialist embodied perspective, a social robot is effectively 'social' if and only if it has or embodies the required properties. Therefore, the only way a robotic agent could really—'substantially'—have social skills, rather than simply give the impression it does, is for this robot to have, as an individual agent, the characteristic physiology of sociality. In terms of Ziemke's classification proposed in Table 4.1, this means that a robot, as an individual agent, must have an autopoietic (self-producing) embodiment. Otherwise, it is not a socially embodied agent. As implied in Ziemke's criticism, organismal embodiment constitutes a necessary condition to be socially embodied.

This position faces, however, a serious difficulty. The conception of sociality and organismal embodiment, underlying its 'substantialist' argument, is at odds with how the theory of autopoiesis—on which Ziemke's notion of organismal embodiment rests—understands embodiment in the sphere of sociality. The difficulty stems from the autopoietic rejection of individualism, implicit in Maturana and Varela's view of cognition as *organism-centered meaningmaking* (Thompson, 2007). According to the theory of autopoiesis, sense making does not depend on the characteristics of the individual organism alone and cannot be reduced to its individual properties because it does not directly proceed from them. The notion of 'structural coupling,' used to conceptualize sense making, defines the dynamic organization underlying this process as the complex system formed by the interaction between the organism and its environment. It is this system *as a whole* that is responsible for the self-regulation dynamics through which the organism creating sense reacts to environmental perturbations. Structural coupling describes this interaction as a symmetric dynamic of reciprocal perturbations and self-regulation answers, whose density makes it impossible to isolate individual contributions. Sense making is thus construed as a highly distributed process, a dynamic system composed of the organism and its environment (including other organisms or systems), which all contribute in entangled ways.

This view explicit in the autopoietic idea of social coupling led Varela to propose a 'radically embodied' thematization of mind and cognition as a structure of interconnection linking brain, body, environment, and other agents dynamically (Thompson & Varela, 2001). The autopoietic notion of 'behavioral coupling' describes dynamics of co-evolution in which two or more individual agents and their respective processes of self-production and sense making become radically interdependent or entangled. In this theoretical perspective, a development of the notion of structural coupling, social sense does not uniquely depend on the characteristics of the individuals but also on accidents of the social coupling itself.

Echoes of this relational view of sociality and sense making can be found in many recent perspectives on social cognition (e.g., Damiano, 2009; Dumouchel, 2019a) and, in particular, in the increasingly diffused autopoiesis-based notion of 'participatory sense-making' (e.g., De Jaegher & Di Paolo, 2007). These propose different theoretical elaborations, with a strong embodied flavor, which suggest that the realization of social dynamics is not limited to the body of the individual agents but essentially results from the dynamic, inter-individual system composed of the bodies of the agents in interaction among themselves and with their social and natural environments.

The body of the social agent is not simply my or your individual body; it is an inter-individual body. Sociality–at least human sociality[1]—does not merely rely on individual embodiment but corresponds to a form of *inter-individual embodiment*. The bodily system generated by the recurrent, dynamic interaction between the bodies of inter-actors in their relation to the world constitutes the social body. These interactions are bodily, not only in the sense that social interactions take place through forms of communications that are predicated on the characteristics of our bodies, but also in the sense that, as amply demonstrated by studies on mirror neurons (Gallese, 2005) or affiliation (Carter et al., 1999), the mere presence or action of one agent, even when it is not directed towards a second, is sufficient to change the bodily state of that other.

Based on this view, the crucial question in socially embodied AI cannot be whether or not we can create robots, as individual machines, that have the required character-

---

1 The eu-sociality of ants and other social insects may rest on different principles given both the importance of stigmergy in the coordination of their behavior and the fact, long ago noticed by G. H. Mead (1934), that among such insects, unlike that which is the case among humans, social differences are inscribed in the animals' physiology.

istics to be social. Rather, it is: Can we create socially embodied AI machines that can enter into the relevant forms of social interactions with humans?

4.1.4 Can robotic bodies be part of social bodies?

Individual robots do not have autopoietic bodies and, from an individualist perspective, cannot support a form of sociality similar to ours. However, as mentioned earlier, the individualist perspective is incompatible with the theory suggesting that an autopoietic body constitutes a necessary condition of sociability. If, instead, we take a relational perspective on sociability, are the bodies of social robots able to support a dynamic of interactions with us that can meaningfully be called affective and social?

Answering that question is, we believe, the goal implicitly pursued by social robotics. However, the form of research needed to answer that question entails that it is impossible to give a general or theoretical answer, or even, in a sense, a definitive one. The question can only receive experimental answers: answers that are not simply technical or focused on the characteristics of the artificial agents, but those based on testing hypotheses concerning our sociality, which is not embedded or embodied in the individuals, but in the dynamic of interaction that the robot is designed to develop with human agents. That is why it is not each new robot only that may be viewed as a hypothesis in this research, but any robot in a particular social situation.

In consequence, all social robotics research that revolves around the issue of anthropomorphization constitutes—implicitly, but in its own right—an application to sociability of the 'synthetic method.' That is, the 'understanding by building method,' through which the 'sciences of the artificial' study biological cognitive processes by creating and experimentally exploring artificial systems able to produce them (e.g., Cordeschi, 2002; Damiano et al., 2011, 2015). In social robotics, however, it is not applied to individual robots but to the interactions between artificial and human agents. Here, the synthetic method can give answers to questions on human–robot sociability, but not a general answer, either positive or negative. To answer the question, research must establish the context in which the interactions are to take place. In what way, through what means—verbal, visual, haptic, and so on? What is their materiality? What is the level of interaction involved? For example, the question 'Can a robot be a friend?' is different from the question 'Can a robot be a good work colleague?' (Nyholm & Smids, 2020). Friendship and work are two fundamental dimensions of human sociality, which respond to and produce very different dynamics of interactions. What constitutes convincing affective and social relations is quite different in each of these cases.

In consequence, the questions need, in a sense, to be reversed because they are not really about the robots anymore—at least not about them taken individually. They concern our relations with them: Are the dynamics of our interactions with social robots forms of social coupling?

Our tentative answer is that they can sometimes be. Today, after the embodiment turn, no one (or only very few holdouts) would say that computers have a form of intelligence that is the *same* as ours. Yet few have problems in talking about 'artificial intelligence,' even if this may not be the most appropriate term (Dumouchel, 2019b). The same applies to our interactions with social robots. They give rise to social dynamics *somewhat similar* to those of many interactions we develop with animals and other humans, but also different. These differences can be expressed in terms of rigidity, simplicity, and other limitations due to the differences in the mechanisms that support social interaction in the electromechanical body of robots and in the biological body of humans. According to us, the perspective on robots and sociality implicit in social robotics is that these differences do not, in principle, prevent 'mixed'—robotic–human—social bodies from supporting human–robot social dynamics.

4.2 Affect, Emotions, and Embodiment

One reason why we think that social robotics 'reverses' the question, in the sense that it is about the relation rather than about robots taken individually, is because one of its central tenets is that creating artificial agents that manifest social skills and competencies requires endowing them with the ability to recognize and respond appropriately to the emotions of the humans with whom they interact. A fundamental working hypothesis of social robotics is that social interactions essentially are, or imply, affective interactions, and social competence requires affective and emotional skills. This hypothesis, we have argued elsewhere (Damiano et al., 2015; Dumouchel & Damiano, 2017; Damiano & Dumouchel, 2020), implies a conception of affective exchange that does not view emotions as essentially private, individual, internal events but rather construes them as public and external events, taking place in the context of interactions.

This alternative view of emotions is implicit in social robotics but remains unformulated. It rests on a different conception of emotion from that which sees them as individual properties or skills. An emotion, whether a long-term disposition—'this is a sad person'—or as an event—'he was so angry when he heard she was late'—is generally construed as something that happens to an agent. The emotion is 'mine' or 'yours.' Emotions, in this view, are properties in two literal senses. First, they are a property that some individuals—animal or human—have and that other individuals, that is, robotic agents, do not have. Second, emotion is a property *of* the individual: your emotion belongs to you; it is yours. Such emotions are fundamentally internal and private. Independently of whether they are viewed as 'cognitive judgments' (Nussbaum, 2004) or as 'affect programs' (DeLancey, 2002), emotions take place in an individual body or mind. Even when it coincides with a visible expression, as in smiling or blushing, the emotion is identified with the agent's inner experience. It is the feeling or physiological reaction that is viewed as the central component of the emotion.

However, social interactions, whatever else they may be, clearly are joint ventures. Their success or failure, their permanence or fugacity, depend also on the partners involved. The emotions that emerge in social relations should, in the same way, be viewed as collective creations rather than as private, individual events. This idea that emotions are joint creations emerging in affective coordination challenges our common understanding of emotions as well as the way in which we understand our mental states to be embodied.

Affect coordinates persons with each other. The issue is not of coordinating *the actions* of multiple agents so that they can successfully achieve a common goal but to determine whether or not the agents want to pursue a common goal. Affective interactions are the way through which we coordinate our reciprocal intentions of action. For members of a species like ours, whose advantages and disadvantages primarily depend on interactions with each other rather than on solitary relations to the natural environment, material situations under-determine agents' interests. Individuals' preferences for this or that outcome are conditional upon other agents' preferences and *vice versa*. 'Affective coordination' (Dumouchel, 1999, 2008) is the mechanism through which these preferences are determined and is lifted the uncertainty concerning others' intentions towards me and my intentions towards them.

Affective coordination is a reciprocal process in which affective expressions occupy the central place. The process is, however, generally unconscious. We are mostly unaware of how our affective expression affects others, or even of what—gestures, postures, tone of voice—constitutes this expression. Even if we attend more to the affective expression of others, many aspects of it nonetheless remain below the level of conscious apprehension. The process of affective coordination is thus unattended and non-intentional in both the ordinary and the technical sense of that term. It is neither a conscious object nor something that I want to do. Though my intention towards you results from this process, and *vice versa*, it cannot itself be intentional under pain of infinite regress.

Emotions, we propose, are moments in this continuous process of coordination that become salient for one or more of a variety of reasons. For example, because they are fixed points of coordination, because they represent a sudden reversal, because of the lasting inability to reduce uncertainty concerning the other's intention, and so on. In every case, what determines a moment to be an emotion is not only an individual's mental or bodily state but some aspect of a process of coordination that involves at least two agents.

Emotions, then, are *socially embodied* rather than individually embodied, and this process of social embodiment is at the heart of social interactions. It is therefore not surprising, but to be expected, that robots—based on the strategy of anthropomorphization, or thanks to a sufficiently complex computational cognitive architecture, or both—should be able to participate in this process of affective coordination.

Exploring theoretical and experimental work carried out by practitioners in social robotics makes evident that the research question implicitly asked by social roboticists, far from being 'Can robots have emotions?', is 'Can robots establish affective coordination with humans?' (e.g., Cañamero, 2005; Asada, 2015). This question, introducing in robotic research a relational view of affect and sociability, presents two interrelated advantages with respect to the question, 'Can robots have emotions?' The first advantage is that whether meaningful affective coordination takes place between a robot and its human partner is an empirical or experimental question rather than a purely theoretical or metaphysical issue: Can robots have emotions? The second, closely related, advantage concerns ethical reflection on human–robot interactions. A negative answer to the question 'Can machines have emotions?', as is often the case, leads to suspicion towards technologies that entail affective relations between humans and artificial systems since, it is argued, they "fool us" (Adolphs, 2004, p. 18). Robots do not have the emotions they pretend to express. In consequence, social robots tend to be seen as a danger that needs to be eradicated or, at least, contained through regulations. As we have proposed elsewhere (Damiano & Dumouchel, 2018; Damiano, 2020), this inclination tends to weaken and isolate ethical reflection on social robots, circumscribing its scope to the formulation of restrictions which are often generated on the basis of pre-existing ethical frameworks and do not specifically target the unprecedented ethical issues produced by these new 'social machines' (Brooks, 2003). Shifting the focus towards affective coordination encourages the opposite: dialogue with practitioners in social robotics and visiting their labs to examine and experience the forms of interaction developed by humans with the social machines and to engage in elaborating criteria appropriate for specific types of social robots and contexts of use, as well as the specific traits of the relations that humans establish with these machines.

4.3 From (Mixed) Inter-individual Social Bodies to Synthetic Ethics

A dominant theme in current ethical inquiry is that the social and affective relations with social robots can never be genuine since these machines do not have social competencies or emotions like ours (e.g., Danaher, 2019). Whatever form of social and affective investment one may make in a social robot, it will never be returned. Such relations, in the end, will inevitably deceive us and show themselves to be a fraud. This very partial view is generally combined with a dystopian scenario, which depicts the diffusion of social robots as the origin of a cascade of negative social outcomes, leading to new, radical forms of social isolation (e.g., Sharkey & Sharkey, 2010; Turkle, 2010, 2011; Van Wynsberghe, 2013; Tisseron, 2015). Among these authors include: the exclusion of fragile persons and subjects with special needs from social interaction due to the increased delegation of relationships of care and support to social robots; the exposure of vulnerable individuals to the delusion of mutual affective relations with social machines; the substitution of social and emotional interactions among humans with inauthentic intimate relationships with robots; and problems and impairments in the cognitive and social development of new generations, due to prevalence of social interaction with these artifacts. On the basis of this perspective, some authors even consider that social sustainability requires "the exclusion of these evocative objects [social robots] from the realm of our relationships" (Turkle, 2010, p.4).

The merit of this approach is that it brings attention to many of the dangers that are inherent in the widespread (unregulated) use of social robots. However, its premise—that social and affective relations with social robots are necessarily deceitful and should be treated with utmost precaution—can lead to consequences that are more negative than beneficial (Damiano & Dumouchel, 2018, 2020; Damiano, 2020, 2021).

One of the most evident is that it excludes *a priori* any possible positive application of social robots. It would ban, without any risk–benefit analysis, socially assistive uses of these machines, such as 'robot-assisted therapies' supporting the improvement of social skills in children with special needs (e.g., Cabibihan et al., 2013; Lehmann et al., 2014; Lehmann, 2020), or social robotics projects committed to the well-being of vulnerable persons or elderly in loss of autonomy (e.g., Matarić & Scassellati, 2016). Its main weak point, as argued above, is separating ethical investigation from social robotics research and technical evolution. A generalized suspicion of social robots diverts ethics from working out guidelines to help specialists in robotics address the problem of reducing the risks and/or amplifying the benefits of specific projects. Moreover, this negative outlook is destined to be ignored by the scientific community of social robotics or even to generate harmful outcomes for the evolution of the field, specifically from an ethical point of view. Indeed, this approach can only stop, never support, projects needing ethical approval, which typically target the construction of social robots designed to engage with humans in a helpful way. However, it cannot block controversial commercial projects, of which sex robots with an integrated 'rape option' constitute an extreme, but unfortunately not so rare a case (Damiano & Dumouchel, 2018, 2020).

Currently, entertainment robotics includes commercial trends engaged in projects encouraging users to exercise violence on social robots. With regard to opposing these commercial projects, the logic underlying the general suspicion appears rather counter-productive. Following it, interactions with robots only involve 'simulated violence' because the violence exercised by the human user(s) cannot be more real than the 'simulated suffering' displayed by the machine. The ethical approach that denounces the emotions of robots as deceitful seems at a loss to refute violent forms of entertainment robotics and even suggests an implicit legitimization.

This, more than an accident, is a structural trait of the underlying logic of the generalized suspicion approach. By rejecting as 'inauthentic' our social and affective interactions with social robots, on the one hand, it encourages users to consider their violent behavior as unreal and without consequences, and, on the other, it fails to engage in a critical inquiry of the ways in which social interaction with robots can influence human moral behavior. That is why a plausible result of this approach, instead of the ban on social robots, might be the diffusion and evolution of this technology in the absence of any dedicated ethical research. Pushing social robotics to self-produce in isolation its own ethical guidelines, which will likely reflect common prejudices and economic interests rather than offering ethical support effective in orienting the field towards social sustainability.

These considerations show the need for a different approach, overcoming the ideological rejection of mixed human–robot social ecology and committed to the socially sustainable diffusion of social robots. An ethical inquiry that proactively engages in ideation, design, construction, testing, and introduction of social robots in various social contexts. Instead of rejecting human–robot social and affective interaction as 'simulation' and 'manipulation' by robots (or roboticists), this approach recognizes that both human and robotic agents are involved in the process of affective coordination—a relation that can significantly influence their conduct.

This is the approach that we introduced, in previous works (Dumouchel & Damiano, 2017; Damiano & Dumouchel, 2020; Damiano, 2020, 2021; Dumouchel, 2022b), as 'synthetic ethics' to emphasize its engagement in critically exploring and evaluating—case by case, project by project—the implications and effects of human–robot social and affective coordination, and in working with specialists in social robotics to define ethical guidelines favoring a positive, socially sustainable evolution of the field.

Synthetic ethics does not focus on anticipating the social impacts of technological changes—as recommended by utilitarian approaches—nor on controlling the ethical behavior of robots through regulations—an aim shared by both utilitarian and deontological approaches. We recognize the importance of these two objectives. However, synthetic ethics promotes a different programmatic objective, inspired by the openness to the unexpected and careful attention to novelty suggested by the Aristotelian notion of φρόνησις—*prudence.* The objective pursued by synthetic ethics is not to circumscribe or limit transformations based on a rigid set of pre-defined rules. Its ambition is to help generate new regulations designed to meet the specific features of novel technological developments, in particular the unprecedented developments of human–robot social and affective relations.

The name 'synthetic ethics' underlines its origins in the epistemological exploration of social robotics and its profound affinity with the methodological approach of 'learning by doing,' which is at the core of the 'synthetic method' characterizing the sciences of the artificial. This appellation does not, however, mean that the approach is limited to social and affective human–robot interaction in social robotics laboratories and 'in-the-wild' research settings through collaboration with specialists in the field. Rather, the aim of synthetic ethics is to proactively contribute to the socially and ecologically sustainable diffusion and evolution of social robots.

4.4 Conclusion

Identifying the risks that social robots entail, instead of being a cause for condemnation, should become the starting point of transdisciplinary research committed to implementable solutions directed to avoid these dangers. This search for concrete, sustainable solutions should not only be pursued at the initial designing and prototyping stages but also continue after the introduction of the robots into their target social contexts and effectively accompany and support their diffusion.

- How can social robotics create artificial agents able to function as 'social connectors' for humans to intensify our social bond instead of weakening it?
- How can social robots encourage help and care relationships with vulnerable persons and individuals with special needs?
- How can they be designed to promote and foster, rather than hamper, social and affective emotional development in their human inter-actors, particularly during delicate phases of their life?
- How do social and affective exchanges with robots influence our ethical behavior?
- How can ethical choices and values be transformed in the context of these new forms of social and affective coordination?

Today, a growing variety of ethical undertakings diverge from the mainstream ethics of social robotics. Increasingly, they activate synergic interplays between philosophical inquiry and scientific-technological research which promise to generate effective applications (e.g., Cappuccio et al., 2020; Ocnarescu & Sciamma, 2021; Rajaonah & Zio, 2022; Malinowska, 2022).

These emerging trends converge with one of the central goals of synthetic ethics: to explore and better understand humans as ethical agents through the experimental settings developed by social robotics. In other words, using social robots to expand our knowledge of ourselves from an ethical point of view, with specific attention to the ethical choices and conduct generated in the context of human–robot social and affective interactions (e.g., Cappuccio et al., 2020, 2021; Malinowska, 2021).

These interesting developments emphasize how synthetic ethics resting on a relational view of sociality and transdisciplinary heuristics fostering collaboration with the target science offer a promising alternative to the ideological condemnation of the emerging human–robot social ecologies. Working proactively on their social sustainability by engaging in multiple 'learning by doing' research paths in the attempt to articulate a process of self-knowledge on the process of self-transformation that the diffusion of social robots is triggering in us and our social world.

References

Airenti, G. (2015). The cognitive bases of anthropomorphism: From relatedness to empathy. *International Journal of Social Robotics*, *7*, 117–127.

Adolphs, R. (2004). Could robots have emotions? In Fellous, J. M. & Arbib, M. A. (Eds). *Who need emotions? The brain metts the robot*. Oxfor University Press, 9–25.

Aristotle (2004). *The Nicomachean Ethics*. Penguin Classics.

Asada, M. (2015). Towards artificial empathy. *International Journal of Social Robotics*, *7*, 19–33.

Biocca, F., Harms, C., & Burgoon, J. K. (2003). Toward a more robust theory and measure of social presence: Review and suggested criteria. *Presence: Teleoperators & Virtual Environments*, *12*(5), 456–480.

Breazeal, C. (2003). Toward sociable robots. *Robotics and Autonomous Systems*, *42*(3–4), 167–175.

Brooks, R. A. (1991). Intelligence without representation. *Artificial Intelligence*, *47*(1–3), 139–159.

Brooks, R. A. (1997). From earwigs to humans. *Robotics and Autonomous Systems*, *20*(2), 291–304.

Brooks, R. (2003). *Flesh and Machines: How Robots Will Change Us*. Vintage.

Cabibihan, J. J., Javed, H., Ang, M., & Aljunied, S. M. (2013). Why robots? A survey on the roles and benefits of social robots in the therapy of children with autism. *International Journal of Social Robotics*, *5*, 593–618.

Cañamero, L. (2005). Emotion understanding from the perspective of autonomous robots research. *Neural Networks*, *18*(4), 445–455.

Cappuccio, M. L., Peeters, A., & McDonald, W. (2020). Sympathy for Dolores: Moral consideration for robots based on virtue and recognition. *Philosophy & Technology*, *33*(1), 9–31.

Cappuccio, M. L., Sandoval, E. B., Mubin, O., Obaid, M., & Velonaki, M. (2021). Robotics aids for character building: More than just another enabling condition. *International Journal of Social Robotics*, *13*, 1–5.

Carter, C. S., Lederhendler I. I., & Kirkpatrick, B. (Eds.) (1999). *The Integrative Neurobiology of Affiliation*. MIT Press.

Clark, A. (1998). *Being There: Putting Brain, Body, and World Together Again*. MIT Press.

Clark, A. (1999). An embodied cognitive science? *Trends in Cognitive Sciences*, *3*(9), 345–351.

Cordeschi, R. (2002). *The Discovery of the Artificial*. Kluwer.

Danaher, J. (2019). The philosophical case for robot friendship. *Journal of Posthuman Studies*, *3*(1), 5–24.

Damiano, L. (2009). *Unità in dialogo: Un nuovo stile per la conoscenza*. Bruno Mondadori.

Damiano, L. (2012). Co-emergences in life and science: A double proposal for biological emergentism. *Synthese*, *185*, 273–294.

Damiano, L. (2020). Mind, robots and mixed social ecologies: Towards an experimental epistemology of social robots, *Sistemi Intelligenti*, *32*(1), 27–39.

Damiano, L. (2021). Homes as human–robot ecologies: An epistemological inquiry on the 'domestication' of robots. In A. Argandoña, J. Malala, & R. Peatfield (Eds.), *The Home in the Digital Age* (pp. 80–102). Routledge.

Damiano, L., & Dumouchel, P. (2018). Anthropomorphism in human–robot co-evolution. *Frontiers in Psychology*, *9*, 468.

Damiano, L., & Dumouchel, P. G. (2020). Emotions in relation: Epistemological and ethical scaffolding for mixed human–robot social ecologies. *HUMANA. MENTE Journal of Philosophical Studies*, *13*(37), 181–206.

Damiano, L., Dumouchel, P., & Lehmann, H. (2015). Towards human–robot affective co-evolution overcoming oppositions in constructing emotions and empathy. *International Journal of Social Robotics*, *7*, 7–18.

Damiano, L., Hiolle, A., & Cañamero, L. (2011, August). Grounding synthetic knowledge: An epistemological framework and criteria of relevance for the synthetic exploration of life, affect and social cognition. In *ECAL 2011: The 11th European Conference on Artificial Life* (pp. 200–207). MIT Press.

Dautenhahn, K. (2007). Socially intelligent robots: Dimensions of human–robot interaction. *Philosophical Transactions of the Royal Society B: Biological Sciences, 362*(1480), 679–704.

De Jaegher, H., & Di Paolo, E. (2007). Participatory sense-making: An enactive approach to social cognition. *Phenomenology and the Cognitive Sciences, 6*, 485–507.

DeLancey, C. (2001). *Passionate Engines: What Emotions Reveal about the Mind and Artificial Intelligence*. Oxford University Press.

Di Paolo, E. A. (2003). Organismically-inspired robotics: Homeostatic adaptation and teleology beyond the closed sensorimotor loop. In *Dynamical Systems Approach to Embodiment and Sociality* (19–42). Available at: https://users.sussex.ac.uk/~ezequiel/dp-erasmus.pdf.

Dreyfus, H. L. (2007). Why Heideggerian AI failed and how fixing it would require making it more Heideggerian. *Philosophical Psychology, 20*(2), 247–268.

Dumouchel, P. (1999). *Émotions*. Institut Synthélabo.

Dumouchel, P. (2008). Social emotions. In L. Cañamero & R. Aylett (Eds.), *Animating Expressive Characters for Social Interaction* (pp. 1–19). John Benjamin Publishing Company.

Dumouchel, P. (2019a). Embodiment: The ecology of mind. *Philosophies, 4*(2), 12.

Dumouchel, P. (2019b). Intelligence, artificial and otherwise. *Forum Philosophicum, 24*(2), 241–258.

Dumouchel, P. (2022a). Making faces. *Topoi, 41*(4), 631–639.

Dumouchel, P. (2022b). Ethics & robotics, embodiment and vulnerability. *International Journal of Social Robotics*, 1–13.

Dumouchel, P., & Damiano, L. (2017). *Living with Robots*. Harvard University Press.

Fong, T., Nourbakhsh, I., & Dautenhahn, K. (2003). A survey of socially interactive robots. *Robotics and Autonomous Systems, 42*(3–4), 143–166.

Froese, T. (2007). On the role of AI in the ongoing paradigm shift within the cognitive sciences. In M. Lungarella, F. Lida, J. Bongard, & R. Pfeifer (Eds.), *50 Years of Artificial Intelligence*. Lecture Notes in Computer Science, *Vol. 4850* (pp. 63–75). Springer.

Gallese, V. (2005). Being like me: Self–other identity, mirror neurons, and empathy. In S. Hurley & N. Chater (Eds.), *Perspectives on Imitation: From Neuroscience to Social Science: Vol. 1. Mechanisms of Imitation and Imitation in Animals* (pp. 101–118). MIT Press.

Heerink, M., Kröse, B., Evers, V., & Wielinga, B. (2008). The influence of social presence on acceptance of a companion robot by older people. *Journal of Physical Agents, 2*, 33–40.

Jones, R. A. (2017). What makes a robot 'social'? *Social Studies of Science, 47*(4), 556–579.

Lehmann, H. (2020). *Social Robots for Enactive Didactics*. Franco Angeli.

Lehmann, H., Iacono, I., Dautenhahn, K., Marti, P., & Robins, B. (2014). Robot companions for children with Down Syndrome. *Interaction Studies*, 15(1), 99–112.

Malinowska, J. K. (2021). What does it mean to empathise with a robot? *Minds & Machines, 31*, 361–376.

Malinowska, J. K. (2022). Can I feel your pain? The Biological and socio-cognitive factors shaping people's empathy with social robots. *International Journal of Social Robotics, 14*, 341–355.

Matarić M. J., & Scassellati, B. (2016). Socially assistive robotics. In B. Siciliano & O. Khatib (Eds.), *Springer Handbook of Robotics* (pp. 1973–1993). Springer.

Maturana, H. R., & Varela, F. J. (1980). *Autopoiesis and Cognition: The Realization of the Living* (1st Ed.). Reidel Publishing Company.

Mead, G. H. (1934). *Mind, Self and Society*. Chicago University Press.

Meister, M. (2014). When is a robot really social? An outline of the robot sociologicus science. *Technology & Innovation Studies, 10*, 1, 107–134.

Montebelli, A., Lowe, R., & Ziemke, T. (2009). The cognitive body: From dynamic modulation to anticipation. In G. Pezzulo, M. V. Butz, O. Sigaud, & G. Baldassarre (Eds.), *Anticipatory Behavior in Adaptive Learning Systems: From Psychological Theories to Artificial Cognitive Systems* (pp. 132–151). Springer.

Nussbaum, M. (2004). *Upheavals in Thought*. Cambridge University Press.

Nyholm, S., & Smids, J. (2020). Can a robot be a good colleague? *Science and Engineering Ethics*, *26*, 2169–2188.

Ocnarescu, I., & Sciamma, D. (2021). Homes through the design shift in the digital age. In A. Argandoña, J. Malala, & R. Peatfield (Eds.), *The Home in the Digital Age* (pp. 103–119). Routledge.

Pfeifer, R., & Bongard, J. (2006). *How the Body Shapes the Way We Think*. MIT Press.

Pfeifer, R., & Scheier, C. (1999). *Understanding Intelligence*. MIT Press.

Rajaonah, B., & Zio, E. (2022). Social robotics and synthetic ethics: A methodological proposal for research. *International Journal of Social Robotics*, *15*, 2075–2085.

Searle, J. R. (1980). Minds, brains, and programs. *Behavioral and Brain Sciences*, *3*(3), 417–424.

Shapiro, L. (2007). The embodied cognition research program. *Philosophy Compass*, *2*(2), 338–346.

Sharkey, N., & Sharkey, A. (2010). The crying shame of robot nannies: An ethical appraisal. *Interaction Studies*, *11*, 161–190.

Steels, L. (1995). When are robots intelligent autonomous agents?. *Robotics and Autonomous systems*, *15*(1–2), 3–9.

Steels, L., & Brooks, R. (Eds.) (1995). *The Artificial Life Route to Artificial Intelligence*. Routledge.

Thompson, E. (2007) *Mind and Life*. Harvard University Press.

Thompson, E., & Varela, F. (2001). Radical embodiment. *Trends in Cognitive Science*, *5*(10), 418–425.

Tisseron, S. (2015). *Le Jour où mon robot m'aimera: Vers l'empathie artificielle*. Albin Michel.

Turkle, S. (2010). In good company? In Y. Wilks (Ed.), *Close Engagements with Artificial Companions: Key Social, Psychological, Ethical and Design Issues* (pp. 3–10). Benjamins.

Turkle, S. (2011). *Alone Together*. Basic Books.

Varela, F., Thompson, E., & Rosch, E. (1991). *The Embodied Mind*. MIT Press.

Van Wynsberghe, A. (2013). Designing robots for care: Care centered value-sensitive design. *Science and Engineering Ethics*, *19*, 407–433.

Ziemke, T. (2001, September). Are robots embodied? In *First International Workshop on Epigenetic Robotics: Modeling Cognitive Development in Robotic Systems* (85, pp. 701–746).

Ziemke, T. (2003). What's that thing called embodiment? In *Proceedings of The Annual Meeting of the Cognitive Science Society*, *25*. https://escholarship.org/uc/item/60w6v9jz.

Ziemke, T. (2007). What's life got to do with it? In A. Chella & R. Manzotti (Eds.), *Artificial Consciousness* (pp. 48–66). Imprint Academic.

Ziemke, T. (2016). The body of knowledge: On the role of the living body in grounding embodied cognition. *Biosystems*, *148*, 4–11.

Złotowski, J., Proudfoot, D., Yogeeswaran, K., & Bartneck, C. (2015). Anthropomorphism: Opportunities and challenges in human–robot interaction. *International Journal of Social Robotics*, *7*, 347–360.

David J. Gunkel

Persons or Things: The Role of Robots in Society

Abstract: In the face of increasingly capable and socially interactive robots, we are confronted with a question: Are robots things, technological objects that we can use or even abuse as we decide and see fit? Or is it the case that robots can or even should be something like a person, that is, another subject who would need to be recognized as a kind of socially significant other? The chapter investigates and seeks to respond to this question. It begins by demonstrating how we typically resolve such inquiry by way of a moral and legal ontology that has deep roots. It then details how robots and other seemingly intelligent artifacts effectively interrupt or resist efforts at both reification and personification. In response to these documented difficulties, the final part of the chapter evaluates three possible modes of response that can be taken up and operationalized as we struggle to integrate these technologies into human society and culture.

Keywords: Legal person, natural person, robot, slavery, thing

5.1 Introduction

Robots are a curious sort of thing. On the one hand, they are designed and manufactured technological artifacts. They are things. And, like any of the other things that we encounter and use each and every day, they are typically considered to be objects with instrumental value. Yet, and on the other hand, these things are not quite like other things. They seem to have social presence, they are able to talk and even converse with us, and many are designed to mimic or simulate the capabilities and behaviors that are commonly associated with other socially significant subjects.

So are robots things, technological objects that we can use or even abuse as we decide and see fit? Or is it the case that robots can or even should be something like a person, that is, another subject who would need to be recognized as a kind of socially significant other with some legitimate claim on us? These questions concern not only how we understand the ontology of robots (i.e., what they are) but also how we decide—as individuals and communities—to situate robots in social reality.

The following investigates and seeks to respond to these questions. It begins by demonstrating how we typically resolve such inquiry by way of a moral and legal ontology that has deep roots. It then details how robots and other seemingly intelligent artifacts effectively interrupt or resist efforts at both reification and personification, rending robots and similarly situated interactive artifacts something other, that is,

https://doi.org/10.1515/9783110792270-006

something that does not quite fit or easily accommodate the existing ways of thinking. In response to these documented difficulties, the final part of the chapter evaluates three possible modes of response that can be taken up and operationalized as we struggle to integrate these technologies into human society and culture.

5.2 Persons and Things

In confronting and dealing with others we inevitably make a decision that has the effect of dividing things into one of two types, arranging an exclusive distinction that separates those others who we are obligated to respect from what remain mere things. As the eighteenth-century German philosopher Immanuel Kant described it:

> Beings whose existence rests not indeed on our will but on nature, if they are non-rational beings, still have only a relative worth, as means, and are therefore called *things*, whereas rational beings are called *persons*, because their nature already marks them out as ends in themselves, i.e., as something that may not be used merely as means, and hence to that extent limits all choice (and is an object of respect). (Kant, 1786/2012, p. 40)

But even if you've never heard of Kant, this just sounds intuitively correct. We go out into the world and deal with others, knowing there's a difference between other persons to whom we owe respect as ends in themselves and those things that are mere objects with instrumental value as a means to an end. As Roberto Esposito, who arguably wrote the book on this subject, explains: "If there is one assumption that seems to have organized human experience from its very beginnings it is that of a division between persons and things. No other principle is so deeply rooted in our perception and in our moral conscience as the conviction that we are not things—because things are the opposite of persons" (2015, p. 1). This is not just good reasoning, it's the law. In fact, the binary distinction separating who is a person from what is a thing has been the ruling conceptual opposition in both moral philosophy and jurisprudence for close to 2000 years. When the Roman jurist Gaius (130–180 CE), in a treatise he titled *Institutes*, explained that law involved two kinds of entities, either persons or things, he instituted a fundamental ontological division that has been definitive of Western (but not just Western) moral and legal systems.

Consequently, in the face of others—another human being, a non-human animal, a tree, an extraterrestrial, a robot, and so on—the first and perhaps most important question that must be addressed and resolved is 'What is it?' Is it another *person* similar to myself, to whom I would be obligated? Or is it just a *thing* that can be taken up, possessed, and used without any further consideration or concern? Sounds easy enough. But, when it comes to robots and other seemingly intelligent and socially interactive artifacts, this determination turns out to be much easier said than done, as robots appear to resist efforts at both reification and personification.

5.3 Things

We all kind of know what things are, such that asking a question like 'What is a thing?' seems to be impertinent and immaterial. But, as the German philosopher Martin Heidegger pointed out, this is precisely the problem. Because things are already familiar—perhaps too familiar—we have little or no critical distance from them as things. This is because, as Heidegger (1927/1962, pp. 96–97) explains, things are not typically disclosed to us as things but encountered as objects. In other words, things are not experienced as mere entities laying around out there in the world. They are always pragmatically situated and characterized in terms of our involvements and interactions with the world in which we live. For this reason, things are first and foremost made available to us or revealed as objects for a subject. They are objectified.

The term *object* is crucial in this context. Things are always and already objectified by being situated in opposition to and placed in the service of a human subject. This mode of existence—what Graham Harman (2002) calls "tool being"—applies not just to artifacts, like hammers, pens, and smartphones. It also describes the basic existential condition of natural objects, which are, as Heidegger (1962, /1927p. 100) explains, discovered in the process of being put to use: "The wood is a forest of timber, the mountain a quarry of rock, the river is water-power, the wind is wind 'in the sails.'" Everything therefore exists and becomes what it is insofar as it is useful for some humanly defined purpose. Things are not just out there in a kind of raw and naked state but come to be what they are in terms of how they have been objectified and put to work or used by us as equipment for living.

This objectivist understanding of things can be correlated with what scholars of media and technology call the 'instrumental theory of technology,' which Heidegger subsequently addressed in *The Question Concerning Technology*. "We ask the question concerning technology," Heidegger (1977, pp. 4–5) writes, "when we ask what it is. Everyone knows the two statements that answer our question. One says: Technology is a means to an end. The other says: Technology is a human activity. The two definitions of technology belong together. For to posit ends and procure and utilize the means to them is a human activity." According to this explanation, the presumed role and function of any kind of technological object, whether it be a simple mechanical device, like a corkscrew or bicycle, or a digital computer or algorithm, is that it is a means employed by human users for specific ends. As Andrew Feenberg (1991, p. 5) has neatly summarized it, "the instrumentalist theory offers the most widely accepted view of technology. It is based on the common sense idea that technologies are 'tools' standing ready to serve the purposes of users."

Robots, however, do not seem to fit easily within this theoretical framework. The thing with robots, AI, and similar artifacts is that they are, or often appear to be, something other than mere objects or instruments. Consider three examples.

5.3.1 Things that talk

There are things that talk and use language in ways that appear to be intelligible, like chatbots, digital assistants, and large language models. In fact, the emulation of human conversational behaviors has been the defining condition of machine intelligence since Alan Turing's agenda-setting paper from 1950, "Computing machinery and intelligence." Users of these systems often find it difficult to distinguish between human communication mediated through the instrumentality of a computer and conversational behaviors generated by the machine itself. So much so that there have been a number of legislative efforts stipulating that any and all implementations of what Simone Natale (2021, p. 11) calls "communicative AI" disclose their true artificial nature prior to interacting with human users—a kind of 'Warning: I am AI' label. Unlike artificial general intelligence, which would presumably occupy a subject position reasonably close to that of another human person, these actually existing and ostensibly mindless but very loquacious things simply muddy the water (which is probably worse) by complicating and leaving undecided questions regarding who or what is actually doing the talking.

5.3.2 Things that think for themselves

There are things that are deliberately designed to exceed (or at least seem to exceed) the comprehension, oversight, and control of their human developers—things like DeepMind's AlphaGo and its successor AlphaZero, self-driving automobiles and lethal autonomous weapons systems, and many other machine-learning implementations and algorithms. Although the extent to which one might reasonably assign 'autonomy' to these mechanisms remains a contested issue, what is not debated is the fact that the rules of the game seem to be in flux. As Andreas Matthias (2004, p. 177) points out, summarizing his survey of learning automata: "Presently there are machines in development or already in use which are able to decide on a course of action and to act without human intervention. The rules by which they act are not fixed during the production process, but can be changed during the operation of the machine, by the machine itself. This is what we call machine learning." The instrumental theory, which effectively tethers machine decision and action to human persons, no longer adequately applies to or fully explains mechanisms that have been deliberately designed to operate and exhibit some form, no matter how rudimentary, of independent action or autonomous conduct. Contrary to the usual instrumentalist way of thinking, we now have things that have (or at least seem to possess) a mind of their own.

5.3.3 Things that are more than things

There are things that are designed to be and promoted as being something that is more than a mere thing, like robot companions and social robots. In 2014, for instance, social robotics pioneer Cynthia Breazeal introduced the Jibo robot with the following explanation: "This is your car. This is your house. This is your toothbrush. These are your things. But these [and the camera zooms into a family photograph] are the things that matter. And somewhere in between is this guy. Introducing Jibo, the world's first family robot." According to this ontological classification, Jibo is not just another instrument or tool, like the automobile or toothbrush. But he/she/it (and the choice of pronoun is not unimportant) is also not quite another member of the family pictured in the photograph. The social robot inhabits a place in between the two categories of thing and person, becoming what Katrin Etzrodt and Sven Engesser (2021) have called—using a curious mashup of existing terminology—"personified things." This is, it should be noted, not necessarily unprecedented. We are already familiar with other entities that occupy a similar ambivalent social position, like the family dog.

In fact, as Kate Darling (2021) has effectively argued, animals and pets, in particular, provide a good template for understanding the changing nature of things. "Looking at state of the art technology," Darling (2012, p. 1) writes in one of her early essays on the subject, "our robots are nowhere close to the intelligence and complexity of humans or animals, nor will they reach this stage in the near future. And yet, while it seems far-fetched for a robot's legal status to differ from that of a toaster, there is already a notable difference in how we interact with certain types of robotic objects." This occurs, Darling explains, because of our tendencies to anthropomorphize things by projecting into them cognitive capabilities, emotions, and motivations that do not necessarily exist in the mechanism *per se*. But it is this inescapable emotional reaction that necessitates new forms of response and responsibility in the face of these things. "It's important," Darling (2021, p. 154) concludes, "to understand that we will treat robots like living things. Our tendency runs deep, and as much as we could decry it and argue against it, it's not going away. When it comes to interacting with machines, even more so than with animals, we know that we're projecting something that's not there, and we do it anyway."

All these 'things' (though that word might not be entirely suitable in this context, hence the scare quotes) seem to resist efforts at reification and/or impede classification as an instrument and object. And they do so not only because they have been designed to be or to appear to be (and the difference between 'being' and 'appearing to be' will not be insignificant) intelligent, autonomous, or social; but also because of our innate (and rather mindless) human tendency to anthropomorphize all kinds of things that are situated in such a way that they seem to have social presence, as documented and explained across several decades of research with what Byron Reeves and Clifford Nass (1996) have called "the media equation."

5.4 Persons

So, if robots are not easily accommodated by the category of 'thing,' does this mean that they are or should be persons? This question again seems to be direct and immediately understandable. But, as was the case with things, the concept of 'person' also turns out to be far more interesting and complicated than one might initially anticipate, making answers to this inquiry equally difficult to resolve.

Typically, when we use the word 'person,' we are referring to another human being. But this seemingly natural and everyday understanding is not entirely accurate. The word 'person' has a long and intricate history. It is originally derived from the Latin noun *persona*, which referred to the mask worn by actors in a stage play. As a result of this, 'person' denotes not just an individual human being but more accurately names the role that one plays or is assigned within the context of a social situation or performance. This can be seen especially in law, where who is considered a person has not been limited to human individuals but can also be extended to other kinds of non-human entities that occupy or play roles that are important for our shared social reality—corporations, organizations, ships, animals, and the natural environment. In order to attend to this important difference, 'person' is often divided into two types: *natural person* and *legal person*.

5.4.1 Natural person

Whether something is or is not a natural person is usually decided on the basis of individual properties and capabilities—for example rationality, consciousness, sentience, and so on. So there is a kind of litmus test for achieving recognition as a person. We (and who is interpellated by this first-person-plural pronoun will not be insignificant) first define the criteria for what makes an entity a person. In other words, we devise a standard by deciding what we believe are the necessary person-making properties or capabilities. We then use this standard to test and evaluate whether some entity qualifies as a natural person or not. So the question—whether robots or AI could ever become natural persons—is a question that we get to evaluate based on criteria that we get to define.

Answers to this question typically divide into a simple and mutually exclusive yes/no set of responses. On the side of 'no,' there have been efforts to limit who can be included in the category 'person.' According to this way of thinking, no matter how sophisticated robots and AI may become or seem to be, they will never actually achieve any of the person-making properties, like consciousness, rationality, or sentience. They might appear to do so through various programmed behaviors that we take as signs of an inner life (e.g., consciousness or sentience), but they do not and cannot really possess the capability that they have been designed to simulate.[1] On the other side, you have those who assert the exact opposite, arguing that artifacts, like AI, robots, and other seemingly autonomous systems, either are or will be able to achieve the benchmarks of 'person.' If and when—and it is more often than not a matter of 'when'—this happens, then withholding the title of person from these things would be unethical and unjust.

Perhaps the best example of this tension and its significance can be seen with recent events surrounding former Google engineer Blake Lamoine and the LaMDA large language model. In June of 2022, Lemoine claimed that the LaMDA system was conscious and therefore was a person deserving of moral respect and legal consideration. Google—the corporation that funded the research and built LaMDA—took the opposite position, not only arguing that LaMDA, like any computer application, was not conscious but also suspending and then eventually firing Lemoine. Both sides in this debate asserted and sought to justify their positions by mobilizing what Mark Coeckelbergh (2012, p. 14) has called the 'properties approach' to deciding moral status. And each side struggled with the same set of problems—accurately defining the qualifying ontological or psychological properties and then proving (or disproving) that these properties were actually present in the algorithm. In fact, the entire debate was defined by and circulated around the inability to resolve these fundamental issues.

5.4.2 Legal person

Unlike natural person, which is grounded in the ontological conditions or essential properties of the individual entity, legal person is a socially constructed and conferred recognition. In other words, the title to person has little or nothing to do with what something *is*. Instead, it concerns how something comes to be situated in the context of socially constructed situations and relationships. It is, to return to the etymology of the word 'person,' a matter of the role that an individual entity occupies or performs within social reality. Consequently, the criterion for decision, as Chopra and White (2011, p. 155) explain, is not a matter of "physical makeup, internal constitution, or other ineffable attributes of the entity." It is a practical matter that concerns the integrity and functioning of existing legal systems. "The recognition of legal personality by legislatures or courts," they continue, "takes place in response to legal, political, or moral pressure. The legal system, in so doing, seeks to ensure its internal functional coherence. Legal entities are recognized as such in order to facilitate the working of the law in consonance with social realities."

---

1 This difference—something philosophers call 'ontological difference'—is rooted in Platonism and can be found in, for example, "The Allegory of the Cave" situated in Bk. VII of *The Republic* (Plato, 1987).

If the paradigmatic natural person is the human being in possession of a set of natural capabilities that make one a person, the paradigmatic legal person is the corporation, which is a person not because of its intrinsic nature but due to the fact that it is recognized and situated within the law as a subject of the law. Thus, the question confronted in the face or the faceplate of the robot or other seemingly intelligent and/or socially interactive artifact is whether it would make sense to extend the category of legal person to these other kinds of entities—not, it should be pointed out, because of what they need and deserve as 'natural persons,' but because of the role we want or need them to play in *our* social reality.

Responses to this question also tend to divide into two opposed and seemingly irreconcilable positions. Those who are critical of the idea argue that recognizing technological artifacts as persons, although clearly possible and entirely legal, is wrong and should not be allowed to happen. Extending the title of person to robots and AI, they argue, would only exacerbate existing problems with corporate personhood concerning responsibility gaps, shell companies, and liability shields (Bryson et al., 2017). Those who have advocated for it take an entirely different position, arguing that extending legal personhood or personality to robots and other intelligent (or at least seemingly intelligent) artifacts will be necessary for integrating these technologies into our social reality and moral/legal systems. As Visa A. J. Kurki (2020, p. 189) neatly summarizes it:

> endowing an AI with the incidents of legal personhood that enable it to function as an independent commercial actor does not bespeak any acceptance of the notion that AIs are endowed with ultimate value. The legal personhood of an AI can rather serve various purposes that might have nothing to do with the AI itself, such as economic efficiency or risk allocation.

The terms of this debate became evident in May of 2016 when the Committee on Legal Affairs of the European Parliament—the legislative branch of the European Union—proposed that 'sophisticated autonomous robots' be considered 'electronic persons' with specific rights and obligations for the purposes of contending with the challenges of technological unemployment, tax policy, and legal liability (Delvaux, 2016). The proposal, which did not pass as originally formulated, was strongly opposed by over 250 experts in the field of AI and robotics, who signed an open letter addressed to EU lawmakers. According to the opinion expressed in this document, the very idea of "creating a legal personality for a robot is inappropriate whatever the legal status model" (Robotics Open Letter, 2017).

5.5 Disruptions and Available Solutions

The person/thing dichotomy has been an undeniably useful and influential ordering principle, one that not only has the weight of history behind it but has also been codified in language and law. For this reason, the principal challenge that can now be seen taking shape in the face or the faceplate of robots and AI concerns how *we* decide to fit these artifacts into this often unquestioned and seemingly unassailable social order. And the problem is that accommodating these technological innovations to one category or the other has been debatable, inconclusive, and seemingly irresolvable. “In the dichotomous model that has long opposed the world of things to the world of persons,” Esposito (2015, p. 3) writes, “a crack appears to be showing.” And ‘robot’ designates that fissure. So now what? How can or should we respond to this challenge or opportunity?

5.5.1 Status quo

We can decide to reassert and defend the existing moral and legal ontology, forcing robots, AI, and other artifacts to fit into one or the other designation. As Esposito (2015, pp. 1–2) has explained:

> If there is one assumption that seems to have organized human experience from its very beginnings it is that of a division between persons and things. . . . Since Roman times, this distinction has been reproduced in all modern codifications, becoming the presupposition that serves as the implicit ground for all other types of thought—for legal but also philosophical, economic, political, and ethical reasoning.

Proceeding according to this standard way of organizing things (actually not just things but things and persons) seems reasonable, obvious, and natural, but that is *the* problem, and for three reasons.

First, binary oppositions, although useful for categorizing things, restrict what is possible to know and to say about the world and our own experiences. This is because conceptual opposites push things towards mutually exclusive options. In this either/or mode, any phenomenon is assumed to be reducible to x or its opposite, not-x. Though this kind of exclusivity has a certain functionality and logical attraction it is often criticized for not being entirely in touch with the complexity and exigency of facts on the ground. And the person/thing dichotomy, as Anna Beckers and Gunther Teubner (2021, p. 13) point out in the context of AI and robot law, provides an almost perfect illustration of the problem, insofar as existing technology already seems to resist both reification, where the AI or robot is ‘just a tool’ of human action, and personsification, that is, promoting the artifact to a moral or legal position that would be similar to that of a natural human person. In the face of these technological innovations, the efficacy of the person/thing binary appears to be, if not already broken, then at least straining against its own limitations to the point of breaking.

Second, the existing debates about how robots, AI, and other seemingly intelligent artifacts fit into the existing ontological categories of person or thing has produced a lot of activity, but the effort seems to be caught in what nineteenth-century German philosopher G. W. F. Hegel (2010/1816, p. 202) called a "bad infinity"—a seemingly endlessly repetition of the same arguments and disputes that do not make much progress on the issue. After several decades of work and robust discursive activity, we haven't got very far and find ourselves in a kind of stalemate or *cul de sac*. There are seemingly good reasons to conclude that robots and AI cannot and should not be regarded as mere things. But there are just as many good arguments and evidence that say extending the status of person to these machines would be just as bad, if not worse. So at this juncture, we find ourselves in something of an impasse. Each side in the debate—the critics and the advocates—continues to heap up arguments and evidence to prove or substantiate their position. But neither side appears to have won the contest or is even showing signs of making progress on it.

Finally, the seemingly natural opposition between person and thing proceeds from a distinctly Western way of thinking, "genetically composed of the confluence between Greek philosophy, Roman law, and the Christian conception" (Esposito, 2015, p. 3). This way of organizing things—this way of dividing up all of existence into one of two mutually exclusive types—has not only been exported around the world through colonial conquest and religious conversion but has, for better or worse, dominated the entire domain. Other cultures and traditions, distributed across time and space, do not divide up and make sense of the diversity of beings in this arguably binary fashion. They perform decisive cuts separating the *who* from the *what* according to other ways of seeing, valuing, and acting. As Archer Pechawis explains in his contribution to the essay "Making Kin with the Machines":

> *nēhiyawēwin* (the Plains Cree language) divides everything into two primary categories: animate and inanimate. One is not 'better' than the other, they are merely different states of being. These categories are flexible: certain toys are inanimate until a child is playing with them, during which time they are animate. A record player is considered animate while a record, radio, or television set is inanimate. But animate or inanimate, all things have a place in our circle of kinship or *wahkohtowin*. (Lewis et al., 2018, n. p.)

This alternative formulation—which is just one among many—runs counter to the dominant ways of proceeding, seeing the boundary between what Western philosophy and law call 'person' and 'thing' as being endlessly flexible, permeable, and more of a continuum than an exclusive, binary opposition.

5.5.1.1 Both/and alternatives

One understandable frustration with conceptual oppositions, like that which has distinguished person from thing, is that they have "compressed and continue to compress human experience into the confines of this exclusionary binary equation" (Esposito, 2015, p. 4), such that between person and thing "there appears to be nothing" (Esposito, 2015, p. 16). Though binary oppositions have a certain functionality and logical attraction, they often seem to be unable to represent or to capture the rich experiences of actual existing empirical reality, which always seems to complicate simple reduction into one of two options. It is for this reason that we are generally critical of 'false dichotomies'—the parsing of complex experience into a simple and irreducible either/or distinction.

One method for resolving this problem is to formulate a third alternative that is neither one thing nor the other but a kind of combination or synthesis of the one and the other. Alternatives like this sound liberating and hold considerable promise, precisely because they appear to interrupt the structural limitations imposed by either/or logic and arrange for a more nuanced representation and understanding that is both/and. And such an alternative is already available to us with the legal institution of slavery.[2]

Already in Roman times, slaves were regarded as something more than a mere thing but not quite a full person. They occupied a position that was situated in between the one and the other, being both thing and person. And there has been, in both the legal and philosophical literature, a surprising number of serious proposals arguing for instituting what can only be called Slavery 2.0 (Gunkel, 2018).

Clearly use of the term 'slave' is provocative and morally charged, and it would be impetuous to presume that the various proposals for repurposing the paradigm of slavery to deal with robots and AI would be the same or even substantially similar to what had occurred (and is still unfortunately occurring) with human bondage. But, and by the same token, we also should not dismiss or fail to take into account the documented evidence and historical data concerning slave-owning societies and how institutionalized forms of slavery affected both individuals and human communities. The corrupting influence of socially sanctioned, institutionalized bondage concerns not just the enslaved population but also those who would occupy the position of mastery. "The problem," as Lantz Fleming Miller (2020, p. 301) concludes, "is the term 'slave.' If slavery is, as most of the world now concurs, not morally good, it is reasonable to deduce that not only should no one be anyone's slave, but also no one should be anyone's master. There is something about the relationship that is wrong." Consequently, even if one would be inclined to agree that "robots should be slaves" (Bryson, 2010, p. 63) or that we can, in the words of Steve Petersen (2011, p. 283), "design people to serve" us (also see Musiał, 2017), we still need to ask ourselves whether we would ever want to risk becoming masters.

Additionally (and adding yet another layer of ethnocentric complexity to the problem), the extension of the seemingly paradigmatic master–slave relationship to robots, AI, and other things is culturally specific and distinctly Western. This is some-

2 The word 'robot' was first introduced and initially used in Karel Čapek's 1921 stage-play *Rossumovi Univerzální Roboti* (Rossum's Universal Robots). In Czech, as in several other Slavic languages the word *robota* (or some variation thereof) denotes 'servitude or labor,' and *robot* was the word that Čapek employed to name a class of manufactured, artificial servants.

thing that is identified and explained by Raya Jones in her examination of the work of Masahiro Mori, the Japanese robotics engineer who first formulated the uncanny valley hypothesis back in 1970. In a statement that directly contravenes the largely Western robot-as-slave model, Mori (quoted in Jones, 2016, p. 154) offers the following counterpoint: "There is no master–slave relationship between human beings and machines. The two are fused together in an interlocking entity." As Jones (2016, p. 154) explains, Mori's statement "connotes two ways that the concepts of 'human' and 'robot' can relate to each other. The 'master–slave' viewpoint that Mori eschews accords with individualism and the conventional understanding of technology in terms of its instrumentality. The viewpoint that Mori prompts is based in the Buddhist view of the interconnectedness of all things."

5.6 Duty Now and for the Future

There is no doubt that robots, AI, and other technological artifacts are things, but they are not necessarily the kind of things that are (or should be) situated and conceptualized as the polar opposite of persons. They are not limited to being merely objects for a subject, instrumental means as opposed to an end, or *res* in distinction to *persona*. Instead, it is with these things that "the bivalent logic of modernity is opening up to other paradigms"—other ways of thinking and responding in the face of "things that are no longer merely objects, and of subjects who are increasingly difficult to confine inside the dispositif of the person" (Esposito, 2015, p. 132). Robots and AI are a kind of glitch or anomaly that interrupts and deconstructs the existing binary logic that differentiates person from thing. They destabilize the very terms and conditions of the debate, and in doing so they invite us to think differently and otherwise. It is, therefore, in the face or the faceplate of the robot that we are called to respond to and take responsibility for beings that are other—that do not fit or accommodate the existing moral and legal ontology.

Consequently, the task before us—a task that begins to stand out in relief in the face of technological things that challenge us and our "misunderstood anthropocentrism" (Esposito's term)—is to learn to question our privilege and develop critical perspectives on our own sense of exceptionalism. We—we human beings who have been bestowed with or granted to ourselves the power to decide all things—need to learn how to take responsibility for our privilege, for the damage and harm that it has perpetrated across the centuries, and for the myriad of exclusions and marginalizations that it has instituted and justified.

Questioning privilege from the position of privilege is never an easy undertaking, precisely because it is often rendered transparent by the assumption that it is just normal, somehow part of the natural order of things, and true for all times and places. But these ways of thinking and acting come from somewhere, and they are supported by a particular ideological formation, one that has effectively concealed itself from view by adopting the guise of universality. Yet even here (with this critical self-reflection), all things are not equal, as this privilege is something that has not been evenly distributed. Fortunately, it is in the face of the robot and other things—things that resist, for one reason or another, both reification and personification—that we are able to catch a glimpse of our own unique position and status such that we are invited (or compelled) to investigate and take responsibility for its lineage, its current formations, and its possible futures. It is only by challenging ourselves to engage in this critical self-reflection that we can begin to assemble a moral and legal ontology that can respond to and take responsibility in the face of things.

Ultimately, then, this is not really about robots, AI, and other artifacts. It is about us.[3] It is about the moral and legal institutions that we have fabricated to make sense of all that is. And it is with the robot—who plays the role or occupies the place of a kind of spokesperson for these things—that we are now called to take responsibility for this privileged situation and circumstance. What is needed in response is not some forceful reassertion of the usual ways of thinking but a significantly reformulated moral and legal ontology that can scale to the opportunities and the challenges of the twenty-first century and beyond. What this new framework will look like and how it will be developed is the task for thinking and acting from this point forward. Confronting this will be as terrifying and exhilarating as any of the robot uprisings that have been imagined in science fiction because getting this right will require nothing less than a thorough rethinking of everything we thought was right, natural, and beyond question.

References

Beckers, A., & Teubner, G. (2021). *Three Liability Regimes for Artificial Intelligence: Algorithmic Actants, Hybrids, Crowds*. Hart Publishing.

Bryson, J. (2010). Robots should be slaves. In Y. Wilks (Ed.), *Close Engagements with Artificial Companions: Key Social, Psychological, Ethical and Design Issues* (pp. 63–74). John Benjamins.

Bryson, J., Diamantis, M. E., & Grant, T. D. (2017). Of, for, and by the people: The legal lacuna of synthetic persons. *Artificial Intelligence and Law, 25*, 273–291. https://doi.org/10.1007/s10506-017-9214-9.

Chopra, S., & White, L. F. (2011). *A Legal Theory for Autonomous Artificial Agents*. University of Michigan Press.

Coeckelbergh, M. (2012). *Growing Moral Relations: Critique of Moral Status Ascription*. Palgrave Macmillan.

Darling, K. (2012, September 10). Extending legal protection to social robots. *IEEE Spectrum*. http://spectrum.ieee.org/automaton/robotics/artificial-intelligence/extending-legal-protection-to-social-robots.

3 On the way that robots provide the occasion for and motivate this kind of self-reflection, see Fortunati (2013).

Darling, K. (2021). *The New Breed: What Our History with Animals Reveals about Our Future with Robots*. Henry Holt & Company.

Delvaux, M. (2016). Draft report, with recommendations to the Commission on Civil Law Rules on Robotics, 2015/2103(INL). Committee on Legal Affairs. European Parliament. https://www.europarl.europa.eu/doceo/document/JURI-PR-582443_EN.pdf?redirect.

Esposito, R. (2015). *Persons and Things*. Trans. Z. Hanafi. Polity.

Etzrodt, K., & Engesser, S. (2021). Voice-based agents as personified things: Assimilation and accommodation as equilibration of doubt. *Human–Machine Communication*, *2*, 57–79. https://doi.org/10.30658/hmc.2.3.

Feenberg, A. (1991). *Critical Theory of Technology*. Oxford University Press.

Fortunati, L. (2013). Afterword: Robot conceptualizations between continuity and innovation. *Intervalla: Platform for Intellectual Exchange*, *1*, 116–129.

Gunkel, D. J. (2018). *Robot Rights*. MIT Press.

Harman, G. (2002). *Tool Being: Heidegger and the Metaphysics of Objects*. Open Court Publishing.

Hegel, G. W. F. (1816/2010). *The Science of Logic*. Trans. G. Di Giovanni. Cambridge University Press.

Heidegger, M. (1927/1962). *Being and Time*. Trans. J. Macquarrie and E. Robinson. Harper & Row.

Heidegger, M. (1962/1977). *The Question Concerning Technology and Other Essays*. Trans. W. Lovitt. Harper & Row.

Jibo (2014). JIBO: The world's first social robot for the home. https://www.youtube.com/watch?v=H0h20jRA5M0.

Jones, R. (2016). *Personhood and Social Robotics: A Psychological Consideration*. Routledge.

Kant, I. (1786/2012). *Groundwork of the Metaphysics of Morals*. Trans. Mary Gregor and Jens Timmermann. Cambridge University Press.

Kurki, V. A. J. (2019). *A Theory of Legal Personhood*. Oxford University Press.

Lewis, J. E., Arista, N., Pechawis, A., & Kite, S. (2018, July 16). Making kin with the machines. *Journal of Design and Science*. https://doi.org/10.21428/bfafd97b.

Matthias, A. (2004). The responsibility gap: Ascribing responsibility for the actions of learning automata. *Ethics and Information Technology*, *6*, 175–183. https://doi.org/10.1007/s10676-004-3422-1.

Miller, L. F. (2020). Responsible research for the construction of maximally humanlike automata: The paradox of unattainable informed consent. *Ethics and Information Technology*, *22*, 297–305. https://doi.org/10.1007/s10676-017-9427-3.

Musiał, M. (2017). Designing (artificial) people to serve—The other side of the coin. *Journal of Experimental & Theoretical Artificial Intelligence*, *29*(5), 1087–1097. https://doi.org/10.1080/0952813X.2017.1309691.

Natale, S. (2021). *Deceitful Media: Artificial Intelligence and Social Life after the Turing Test*. Oxford University Press.

Petersen, S. (2011). Designing people to serve. In P. Lin, K. Abney, & G.A. Bekey (Eds.), *Robot Ethics: The Ethical and Social Implications of Robotics* (pp. 283–298). MIT Press.

Plato (1987). *Plato V & VI: Republic*. Trans. P. Shorey. Harvard University Press.

Reeves, B., & Nass, C. (1996). *The Media Equation: How People Treat Computers, Television, and New Media Like Real People and Places*. Cambridge University Press.

Robotics Open Letter (2017). Open letter to the European Commission Artificial Intelligence and Robotics. http://www.robotics-openletter.eu/.

Turing, A. M. (1950). Computing machinery and intelligence. *Mind*, *LIX*(236), 433–460. https://doi.org/10.1093/mind/LIX.236.433. Reprinted in P. A. Mayer (Ed.) (1994). *Computer Media and Communication* (pp. 37–58). Oxford University Press.

Robert W. Gehl and Sean Lawson

Automated Masspersonal Social Engineering

Abstract: Manipulative communication, including disinformation, propaganda, con artistry, and fraud, is not only being aided by artificial intelligence, automation, and machine learning. It is going to be dominated by such practices in the near future. In this chapter, we will draw on the concept of masspersonal social engineering as developed in our recent book (Gehl & Lawson, 2022) and show how this form of manipulative communication is and will continue to be automated. Masspersonal social engineering comprises distinct stages—trashing, pretexting, and bullshitting. We'll show how those stages are automated, and argue that automation does not necessarily make any stage more effective—rather, automation can intensify each stage, or act as a prosthesis to human communication. We'll conclude by considering how automation can help with the ultimate goal of mass personal social engineering: penetration, arguing that communication-as-penetration and automation share many assumptions about the role of communication in society.

Keywords: Masspersonal social engineering, deepfakes, ChatGPT, bullshitting, pretexting, communication-as-domination

6.1 Introduction

Manipulative communication, including disinformation, propaganda, con artistry, and fraud, is not only being aided by artificial intelligence (AI), automation, and machine learning. It is going to be dominated by such practices in the near future.

If you don't believe us, here is what one interested party had to say about automated manipulative communication:

> AI can be used for manipulative communication in a few ways:
> – *Social media bots*: AI-powered bots can be programmed to mimic human behavior and spread false or misleading information to manipulate public opinion.
> – *Deepfake technology*: AI algorithms can be used to generate fake audio or video content that appears to be real, making it easier to spread false information or manipulate individuals.
> – *Personalized advertising*: AI algorithms can analyze data about a person's behavior, preferences, and relationships to create targeted advertisements that manipulate their emotions and influence their purchasing decisions.

https://doi.org/10.1515/9783110792270-007

This is the response one of the authors received when he asked ChatGPT to "explain how AI could be used for manipulative communication."[1] ChatGPT is an artificial intelligence chatbot, capable of parsing text input and quickly providing text responses. At the time of this interview with ChatGPT, that project was garnering a great deal of media attention—arguably, a panic was happening: could ChatGPT replace all of us?

Whether or not ChatGPT will replace all of us remains to be seen, but at the very least it got a lot of details correct when the author talked to it. After all, ChatGPT is right: scholars, journalists, and media watchdogs have already documented the use of socialbots to manipulate public discourse (Beskow & Carley, 2019; Myers et al., 2022) and the use of deepfakes to muddy distinctions between true and false (Boyd, 2019; Vincent, 2023); and of course, personalized advertising is one of the pillars of the dominant political economy of the internet, surveillance capitalism (Gehl, 2014; Zuboff, 2019). Improvements in AI technologies—such as the development of content creation systems like ChatGPT—will no doubt intensify these practices.

Could ChatGPT *itself* be used for manipulative communication? Again, one of the authors asked it this very question. Its response:

> As an AI language model, ChatGPT itself is not inherently manipulative. However, like any technology, it can be used for manipulative purposes if the user intends to do so. For example, someone could use ChatGPT to generate false or misleading information that could be used to manipulate others. Additionally, if ChatGPT is trained on biased data, it may generate biased responses, which could also be used for manipulative purposes.

To be fair to ChatGPT, it also repeatedly said such uses would be unethical. However, given the already long history of manipulative communication in the digital age, the brute fact is that tools such as ChatGPT will be used for manipulative communication. They already have been. And they will continue to be.

In this chapter, we will draw on the central concept from our recent book (Gehl & Lawson, 2022), *Social Engineering*, and show how this form of manipulative communication is and will continue to be automated. Masspersonal social engineering comprises distinct stages—trashing, pretexting, and bullshitting. We'll show how those stages are automated, and argue that automation does not necessarily make any stage more effective—rather, automation can intensify each stage, or act as a prosthesis to human communication. We'll conclude by considering how automation can help with the ultimate goal of masspersonal social engineering: penetration, arguing that communication-as-penetration and automation share many assumptions about the role of communication in society.

1 Conversation happened February 3, 2023 on the January 30, 2023 iteration of the ChatGPT beta.

6.2 On 'Masspersonal Social Engineering'

The media panic over ChatGPT often includes poorly informed 'hot takes' meant to garner likes and shares in social media. When it comes to pressing issues related to AI—including the uses of AI in disinformation and manipulative communication—it's apparent we need clear analysis.

But analyzing automated manipulation is challenging because the domain is incredibly complex. Even from the simple answers ChatGPT provided, we see a great deal of this complexity. Its answers included:
- socialbots mimicking humans;
- the spread of false information;
- manipulation at an individual scale;
- manipulation at a societal scale; and
- personalization of manipulative information.

The range of manipulative techniques identified (in a matter of seconds!) by ChatGPT is staggering, as is the range of targets—both individuals and publics can be affected. How do we make sense of this bewildering set of phenomena? What unites it all?

Elsewhere, we have argued that the full range of manipulative communication can fall under the umbrella of 'social engineering' (Gehl & Lawson, 2022). As we argue in our latest book, social engineering has had two distinct meanings over the past century and a half. Today, 'social engineering' typically is used in a cybersecurity context, referring to a type of attack on an organization (Conheady, 2014; Hadnagy, 2018). Instead of attempting to break into a computer system, a social engineer will instead target the humans who use that system. This tends to be focused on one individual at a time—it's an interpersonal form of manipulative communication.

However, the earliest uses of the phrase 'social engineering' actually go back to the late nineteenth and early twentieth centuries, obviously pre-dating computer hacking (e.g., Earp, 1911; van Marken, 1894). The context here is progressive-era social reform movements, which sought to apply new social sciences (e.g., sociology, psychology, and economics) to society, much in the same manner as a civil engineer applies physics to the landscape by building bridges or dams. As we show in our book, this particular concept of social engineering, which functions on a societal scale, was eagerly adopted by propagandists and early public relations practitioners. Most notable among these were Doris Fleischman and Edward Bernays, a married couple who developed ideas about 'the engineering of consent' in the 1930s and 1940s (e.g., Bernays, 1947; Fleischman, 1935). Their social engineering was an exercise in using mass media (among other channels) to shape opinion on behalf of their clients, including tobacco companies and politicians.

So, we have two meanings of 'social engineering': an interpersonal act of con artistry, and a mass communication practice of public relations. The interpersonal form of social engineering has its epistemological roots in the turn to individualism (e.g., neocapitalism) found in the 1970s, while the mass form has its epistemological roots in early twentieth-century modernism and theories of mass society (Gehl & Lawson, 2022).

But despite these differences, the two forms of social engineering are not unrelated (Hatfield, 2018). Both the mass and interpersonal forms of social engineering take as their object humans, which is to say that they consider society to be a complex of social and technical relations and see 'human nature', 'human sociality', or 'human psychology' as the most malleable elements in the system. Whether it is breaking into a computer database, or propagandizing on behalf of a state, both forms of social engineering hold that 'the human is the weakest link' (to use a phrase common among hacker social engineers).

And now, thanks in part to digital media, the two forms are merging even closer. Drawing on theories of 'masspersonal communication' (French & Bazarova, 2017; O'Sullivan & Carr, 2018), which breaks down the distinction between interpersonal and mass communication, we would suggest that contemporary social engineering combines interpersonal hacking with mass-scale propaganda into a new form we call *masspersonal social engineering* that can deftly move from the interpersonal to the mass scale and back again. We see this most especially in contemporary social media, which has further blurred the artificial distinction between mass and interpersonal communication.

We define masspersonal social engineering as

> an emerging form of manipulative communication enabled by the unique affordances of the internet and social media platforms. It brings together the respective tools and techniques of hackers and propagandists, interpersonal and mass communication, in an attempt to shape the perceptions and actions of audiences. To manipulate, masspersonal social engineers gather data on their targets; create fake personas to share messages; mix deception, accuracy, and friendliness as they engage with targets; and penetrate communication systems. Manipulation, in this case, can involve a range of goals, which might include attempts to change actions and beliefs. But it could also include discouraging action (e.g., voting) and amplifying or intensifying preexisting beliefs (e.g., racism, sexism, or other social divisions) when doing so is in the perceived interests of the masspersonal social engineers or their clients. (Gehl & Lawson, 2022, p. 4)

We suggest masspersonal social engineering encompasses the wide range of manipulative communication practices ChatGPT so ably identified. Whether the target is an individual or a public, whether the specific technique is to use socialbots, deepfakes, or personalized advertising, all these activities can be understood as masspersonal social engineering, so long as the goal is to penetrate the target.

While we have detailed the hybridization of interpersonal and mass social engineering explicitly, we're not the only ones who are beginning to see the merging of the two. For example, in its March 2021 warning about the potential for malicious actors to use AI-generated 'synthetic content' (which would include the text ChatGPT can produce) the US Federal Bureau of Investigation conflated the interpersonal and mass manipulation when it warned of synthetic content's use in both mass-scale influence campaigns and more focused cyberattacks. The conflation extends as well to the FBI's recommended responses, which included a mix of techniques related to critical thinking and media literacy, on one hand, and care when clicking on email attachments and the use of multi-factor authentication, on the other. In other words, the manipulation can happen at mass or individual scales and the solutions thus require mass and individual mitigation.

As should be clear, the digital age is marked by masspersonal communication and thus masspersonal social engineering. Socialbots can shape online political discussions (Gehl & Bakardjieva, 2017), while spearphishing attacks can target individuals, and the two practices can be run by the same organization. A deepfake can be used in a targeted romance scam, or it could be used to muddy political discourse. Advertisements can be targeted at individuals, but the advertiser's goal will often to be to affect societal-level thinking about its advertised product, service, or politician.

6.3 How Masspersonal Social Engineering Can Be Automated

To get a grip on masspersonal social engineering—including how it has been and will continue to be automated—we break the process of masspersonal social engineering down into three stages: *trashing*, *pretexting*, and *bullshitting*. These stages are meant to support the overall goal: communicative *penetration* of either individuals, organizations, or even large segments of society. In the following sections, we'll define these terms further and demonstrate how each of these stages is being automated right now. We'll then turn to the automation of penetration and what that implies for communication.

6.3.1 Trashing

'Trashing' is the information-gathering stage of masspersonal social engineering. This is where the social engineer gathers information on potential targets. We use the term 'trashing' instead of other terms (e.g., 'information gathering,' 'OSINT,' 'research') because it hearkens back to the mid-twentieth-century social engineers—the phone phreaks—who would literally go through corporate trash in order to find passwords and restricted information (Slatalla & Quittner, 1995). (For a Hollywood depiction of this in practice, see the 1993 movie *Sneakers*.) We argue that the phone phreak trashing practice was made possible by the ontology of garbage: it is meant to be tossed aside and forgotten. This can include informational objects: credit card receipts, old technical manuals, or small pieces of paper with passwords written on them (Long, 2007).

We extend that ontological conception forward to the digital age: think of all the online accounts you haven't used and haven't thought about in years, the various digital storage devices you have thrown out that might contain personal information, or the old emails that are sitting on a server somewhere. Like a sticky note with a password sitting in a dumpster, this forgotten digital information is incredibly valuable despite the fact that you may have forgotten about it. Contemporary masspersonal social engineers scour this information in order to develop profiles of their targets.

For example, trashing was apparent in the Russian hack of John Podesta's emails in 2016. Podesta, who was in charge of Hillary Clinton's 2016 presidential campaign, fell victim to a phishing attack (Franceschi-Bicchierai, 2016). However, he wasn't the first target. The Russian hackers worked their way through several email addresses used by Clinton staffers—including staffers who were no longer part of her campaign and who had abandoned their email accounts years before (Satter, 2017). In other words, they had to work their way through the digital detritus of old and abandoned emails until they were able to find active ones—including Podesta's—and *then* they could send a phishing email.

Poring through trash—digital or otherwise—is time-consuming work, ripe for automation. Automated trashing was conducted by a team of European security researchers as early as 2009, which create a automatic data-gathering Facebook bots to gather personal data on potential targets (Huber et al., 2009). Such techniques were commercialized by people like Alexandr Kogan, who used a Facebook app to gather data on millions of Facebook users. Kogan's data would eventually become part of Cambridge Analytica's massive Facebook dataset, used in political influence campaigns around the world (Briant, 2018; Cadwalladr, 2020; Cadwalladr & Graham-Harrison, 2018).

With automation, trashing can take on a very large scale—in fact, a more polite term for it might be 'Big Data analysis.' Indeed, the European security researchers and later Cambridge Analytica cases both occurred during the initial hype over the possibilities of Big Data analysis—coinciding, of course, with the rise of corporate social media. Facebook/Meta, Twitter, and especially LinkedIn have been ripe targets for the automated gathering of data on potential targets (Bossetta, 2018; Hadnagy, 2018). Beyond social media, websites—particularly those that have organizational structures, names of employees, and email addresses—are targets for scraping (Gray, 2020).

As with non-automated trashing, this stage is about finding targets. And again, as the 'masspersonal' aspect reminds us, these targets can be individuals, or swathes of society, such as a particular demographic or (to use a favorite term from Cambridge Analytica) 'psychographic.' Based on our research, we would argue that the danger of automation is not that targets will be identified with more precision when compared with, say, hand-sorting trash. Instead, as social engineer Joe Gray notes, data gathering is an iterative process (Gray, 2020). Moreover, social engineering is an exercise in persistence: if at first a social engineer does not succeed, they will try, try again, moving from target to target until they get a result. Thus, the danger of automated trashing will be that the iterations will happen faster, giving social engineers more opportunities in less time.

6.3.2 Pretexting

A 'pretext' is the role a masspersonal social engineer plays during an engagement. To take a simple example, when you receive a phishing email, the phisher is playing a role as a representative of your bank, Amazon, or a government. Pretexts can be quite complex and conducted in person, as well—common in-person pretexts include playing the role of janitor, serviceperson, or caterer (Conheady, 2018). The most effective pretexts are developed based on the information gathered in trashing—the more information a social engineer has on their target, the more likely the social engineer can construct a compelling role to play.

One of the intersections between AI, machine learning, and social engineering to receive the most attention so far is the use of this technology for the creation of pretexts. For example, in the aftermath of the 2016 US presidential election, journalists, academics, and state officials spent years examining how Russians used automated Twitter accounts in their attempt to sway American opinion away from Clinton towards Trump. These techniques were replicated in subsequent elections around the world (Beskow & Carley, 2019).

In the past, such automated social media profiles simply copied and pasted profile pictures from the internet. These deceptive pretexts could be uncovered relatively easily through reverse image searches. Today, thanks to machine learning and AI-generated images, social engineers can simply create realistic pictures of people (Federal Bureau of Investigation, 2021, p. 2). This relatively cheaply and easily produced 'synthetic content' (as the US FBI calls it) can be quite convincing to both mass audiences and individual people. Evidence for their power extends beyond social media: the FBI is also reporting that social engineers are using machine learning-generated images (along with stolen personal information) to apply for remote IT jobs (Gatlan, 2022b), hoping to gain access to computer networks through the pretext of hired employees.

But of course, static media, such as still images, are not the only synthetic content to watch out for. For several years, disinformation scholars have been warning about deepfake video and audio being used to interfere with democratic deliberation. Even something as seemingly trivial as using voice-cloning software to imitate celebrities' voices can muddy distinctions between real and fake and stoke division, but of course the danger increases when politicians are imitated. A recent example was a deepfake video of Ukrainian president Zelenskyy appearing in the early days of the 2022 Russian invasion. The synthetic Zelenskyy urged Ukrainians to surrender (Gatlan, 2022a).

While imitations of celebrities and politicians is one thing, imitating a close friend or colleague may be even more effective and dangerous. In fact, a case of fraud involving an AI-based imitation of a CEO's voice has already occurred (Stupp, 2019). This is an AI-aided update on an old deceit—a technology-enhanced version of Kevin Mitnick's relatively clumsy imitation of a Japanese businessman in the 1990s (Gehl & Lawson, 2022, p. 108; Mitnick & Simon, 2012). As security researcher and psychologist Matthew Canham argues, technology-aided impersonations of people we are familiar with, such as colleagues, are "immensely powerful because they encourage the victim to suspend disbelief." Any anomalous artifacts associated with a familiar voice or even visage, such as those that inevitably occur with synthetic imitations, are explained away: "Humans are so accustomed to believing what we see and hear that when we believe we are interacting with familiar [sic] associate or family member, we are more likely to cognitively justify behavioral anomalies that otherwise might alert us if we knew we were interacting with a stranger" (Canham, 2021, p. 3).

We may cast doubts on Canham's claims here—how could any of us be fooled by a fake version of a friend?—but keep in mind that social engineering is about persistence and iteration. We might not be fooled all of the time, but when we are exhausted, anxious, or distracted, we are more susceptible. And, as social engineer Jenny Radcliffe points out, the past few pandemic years have been a time of exhaustion, anxiety, and distraction (Tanium, 2020).

Finally, as we argue in our book, corporate social media sites themselves are a kind of pretext. Ostensibly, they are designed to allow us to socialize with friends and family, but in practice, of course, they gather data on our interactions in order to sell our attention to marketers. Similarly, critical researchers are already exploring the ways in which social robots (e.g., caretaker robots, retail robots, or humanoid assistants), coupled with invasive terms of service agreements, represent a new and potentially deeply invasive form of privacy violation that extends the practices of corporate social media (Lutz et al., 2019).

Indeed, much as in automated trashing, automated pretexting is dangerous for precisely these reasons: rather than being concerned about fidelity with reality, when it comes to AI-generated content, we should be concerned about its volume and its utility for iterative attempts at manipulation. Political campaigns, advertising campaigns, or infrastructural security have one thing in common: small numbers actually matter a great deal. Political campaigns can be decided by a tiny percentage of an electorate, ad campaigns are successful if they can boost sales by fractions of a percent, and it only takes one breach in security to threaten massive swathes of an infrastructure. Likewise, automated or synthetic pretexts can be successful if they sway tiny percentages of people.

6.3.3 Bullshitting

Our third and final stage, 'bullshitting,' is the actual engagement with the target. In an in-person or synchronous online interaction, this is the moment when the masspersonal social engineer interacts with the person they are targeting. All the trashing and development of a pretext leads to this point: trashing gives the social engineer background information, and the pretext a role-playing framework, but in any back-and-forth engagement with the target, the social engineer engages in improvisation.

We call this 'bullshitting' purposely. Drawing on moral philosophy and sociology (Frankfurt, 2005; Mukerji, 1978), we define bullshitting as "a peculiar, truth-indifferent mix of deception, friendliness, and accuracy" (Gehl & Lawson, 2022, p. 116). Bullshitting is not lying—often, the social engineer gets many details exactly right. Obviously, the pretext is most likely deceptive, but even then it has to be accurate enough to be believable. Bullshitting is often friendly and affable, too, since research shows that friendly people are more likely to be successful in deception (Levine et al., 2011). Overall, the social engineer does not care whether or not what they say is true or false, so long as it has the desired effect on the target. All in all, bullshitting is, as the moral philosopher Harry Frankfurt noted, a realm of creativity and artistry (Frankfurt, 2005).

As for mass social engineering, AI could produce bullshit content that floods a media system at a large scale. So-called 'deepfakes' are implicated once again (Hoque et al., 2021). The primary concern here is the production of seemingly real content that is meant to deceive or muddy debate. This is a concerning development, to be sure, but there are lots of other ways that AI can potentially aid in the production of mass-scale bullshit. We are already seeing interest among so-called content marketers, for example, in the use of AI to help them crank out more content for their blogs. Tools like ChatGPT could be effective here—imagine a content-marketing blog tapping into such a tool via an application programming interface (API) and generating a new blog post every hour. Even if no one piece is particularly effective, the flood of such content online will further add to what RAND termed the "firehose of falsehood" when talking about Russian disinformation tactics (Paul & Matthews, 2016). In our case, we might call it the 'firehose of bullshit,' which could have the effect of further muddying the waters of online discourse and eroding our sense of what is true, false, and authentic online.

At an interpersonal level, AI can aid in the production of bullshit content at the center of the social engineer's engagement with the target. As with manually created bullshit, AI-created bullshit would be a mix of accuracy, deception, friendliness and sociability. The use of AI could make the production of bullshit more efficient and potentially more effective by allowing it to be more targeted and produced at scale. For example, text generators like ChatGPT could be used to create scripts a social engineer could use for over-the-phone engagements. Or, for chat-based, interpersonal social engineering, AI such as ChatGPT could plausibly handle the interaction entirely on its own.

It's not hard to imagine synthetic video chat systems, complete with a realistic-looking person appearing on the screen and using voice synthesis, bullshitting a target.

However, of all the areas of masspersonal social engineering, bullshit is probably the most difficult to automate. Whether it is con artistry or mass oratory, the actual performance of manipulation is a highly skilled, and likely very human, act. Here, we would suggest automation would likely be a supplement to human engagement. Indeed, many masspersonal social engineering engagements are a mix of human and machinic communication with targets—a practice dating back to the early 2010s, when the US Air Force inadvertently revealed its program of 'persona management' disinformation infrastructure (Webster, 2011). A few years later, the Russian interference campaigns mixed human-controlled social media accounts in with their more famous use of bots, successfully convincing their targets to take real-world actions (Jankowicz, 2020). While something like ChatGPT could generate scripts—which have been shown to be effective in situations such as romance scams (Foster, 2019)—it is likely human beings will have to make most of the in-the-moment decision making when it comes to bullshitting their targets. It's difficult to imagine AI completely replacing humans when it comes to our ability to bullshit.

6.4 Implications

There is an obvious implication to the automation of social engineering: social engineering will occur at more expansive scale and at faster speed than before. Trashing (or information gathering) can be done at a faster pace and a larger scale, social engineers can try more and more pretexts through the aid of automation, and bullshitting can be enhanced through AI-generated scripts and media. That implication is obvious—it's clearly stated across most of the literature. Books like Woolley's (2023) *Manufacturing Consensus* or Howard's (2020) *Lie Machines* make this case.

Another implication, however, is less obvious: automation of social engineering is further evidence that the act of communication has been reduced to an instrumentalized act of dominance. This is what we mean when we say the goal of masspersonal social engineering is *penetration*. 'Penetration'—another term we borrow from the hacker lexicon—is the marker of success. The targeted system is breached, or the targeted human takes an action the social engineer wants. It is a vision of communication where achieving dominance, rather than collectively making meaning, is what matters most.

Drawing on the work of scholars such as Simpson (1996), Williams (2000), and Yang (2020), we have argued in our book that masspersonal social engineering is deeply elitist. Whether the social engineer is manipulating at a mass scale or interpersonally, their practice stems from the belief that they are master communicators who are able to subtly shape others' perceptions. This elitism can be found in the thinking of Edward Bernays, who in the early twentieth century saw democracy as a process of elites engineering public opinion (Ewen, 1996). And it can be found in today's hacker social engineers—especially those hired by corporations to test security—who "manipulate" in order to "educate" people (Conheady, 2014, p. 50).

In this vision of society, the goal of communication is to effect results. Such results are entirely up to the master communicators, and we have to hope that they are acting ethically—and we have little recourse if they are not. When masspersonal social engineers are pressed on this point, they will retreat to the argument that these manipulative tools are neutral; they could be used for good or ill. Hacker social engineers use this defense, as do states (we have to propagandize because *they* are propagandizing). This neutrality defense is eerily similar to the defense of AI offered by ChatGPT. "The way AI technology is used depends largely on the intentions and motivations of the people using it," ChatGPT told one of the authors of this chapter. "Some people may use it to generate accurate information and improve people's lives, while others may use it to spread false or misleading information for personal gain or to manipulate public opinion." Again, we have to trust that our (algorithmic) betters have our best interests in mind when they communicate.

Put another way, the neutrality defense of communication-as-penetration is that the ends justify the means. This includes the exploitation of existing biases and inequalities in society. Propagandists recognized the power of stereotypes in the early twentieth century (Jansen, 2013), and today's hacker social engineers do, as well (Hadnagy, 2018, pp. 4–6). As critical scholars Noble (2018) and O'Neill (2017) remind us, machine learning has to be trained on existing datasets, and the existing datasets reflect longstanding social inequalities. This has a capillary effect on everything AI touches, from facial recognition to the construction of images and texts: racial, gender, sexual, ethnic, and class stereotypes are suffused in big datasets and inform the AI based upon them.

But rather than decrying such biases, a masspersonal social engineer will exploit them, because the goal is to communicatively penetrate the other, and anything that serves that goal is welcome. When a masspersonal social engineer seeks to stoke confusion or play on base prejudices, existing social stereotypes—including those that are embedded in machines—are tools ready-to-hand for the task. In this sense, social engineers are our most reliable participants in communication-as-domination: whatever trick, mental shortcut, or bigotry does the job is acceptable so long as the target is penetrated. The fact that this field has so readily adopted artificial intelligence and machine learning confirms the compatibility of those tools with communication-as-domination.

References

Bernays, E. L. (1947). *The Engineering of Consent*. The University of Oklahoma Press.

Beskow, D. M., & Carley, K. M. (2019, April). Social cybersecurity: An emerging national security requirement. *Military Review: The Professional Journal of the U.S. Army*, 117–127.

Bossetta, M. (2018). The weaponization of social media: Spear phishing and cyberattacks on democracy. *Journal of International Affairs*, *71*(1.5), 97–106.

Boyd, C. (2019, November 20). Deepfakes and LinkedIn: Malign interference campaigns. *Malwarebytes Labs*. https://blog.malwarebytes.com/social-engineering/2019/11/deepfakes-and-linkedin-malign-interference-campaigns/.

Briant, E. L. (2018, April 17). Cambridge Analytica and SCL—How I peered inside the propaganda machine. *The Conversation*. http://theconversation.com/cambridge-analytica-and-scl-how-i-peered-inside-the-propaganda-machine-94867.

Cadwalladr, C. (2020, January 4). Fresh Cambridge Analytica leak "shows global manipulation is out of control." *The Guardian*. http://www.theguardian.com/uk-news/2020/jan/04/cambridge-analytica-data-leak-global-election-manipulation.

Cadwalladr, C., & Graham-Harrison, E. (2018, March 17). Revealed: 50 million Facebook profiles harvested for Cambridge Analytica in major data breach. *The Guardian*. https://www.theguardian.com/news/2018/mar/17/cambridge-analytica-facebook-influence-us-election.

Canham, M. (2021, April). Deepfake social engineering: Creating a framework and implementing countermeasures. *Black Hat, Las Vegas*. https://i.blackhat.com/USA21/Wednesday-Handouts/us-21-Deepfake-Social-Engineering-Creating-A-Framework-For-Synthetic-Media-Social-Engineering.pdf

Conheady, S. (2014). *Social Engineering in IT Security: Tools, Tactics, and Techniques*. McGraw-Hill Education.

Conheady, S. (2018, October 3). Social engineering for penetration testers. *Brucon*. https://www.youtube.com/watch?v=oAJ1pNJnJHQ.

Earp, E. L. (1911). *The Social Engineer*. Jennings & Graham. http://archive.org/details/cu31924014043370.

Ewen, S. (1996). *PR! A Social History of Spin*. Basic Books.

Federal Bureau of Investigation (2021). Private industry notification (No. 210310–001; p. 6). https://www.ic3.gov/Media/News/2021/210310-2.pdf.

Fleischman, D. E. (1935, October 30). *Public Relations and the Consumer*. Fashion Group.

Foster, A. (2019, April 3). Secrets in scammer's 'how-to' handbook. *News.Com.Au*. https://www.news.com.au/technology/online/security/creepy-handbook-reveals-how-scammers-trick-their-victims/news-story/6b04f43c52fc8906cc02b1ffd3f5967f.

Franceschi-Bicchierai, L. (2016, October 20). How hackers broke into John Podesta and Colin Powell's Gmail accounts. *Vice*. https://www.vice.com/en/article/mg7xjb/how-hackers-broke-into-john-podesta-and-colin-powells-gmail-accounts.

Frankfurt, H. G. (2005). *On Bullshit*. Princeton University Press.

French, M., & Bazarova, N. N. (2017). Is anybody out there? Understanding masspersonal communication through expectations for response across social media platforms. *Journal of Computer-Mediated Communication*, *22*(6), 303–319.

Gatlan, S. (2022a, March 6). Facebook removes deepfake of Ukrainian President Zelenskyy. *BleepingComputer*. https://www.bleepingcomputer.com/news/technology/facebook-removes-deepfake-of-ukrainian-president-zelenskyy/.

Gatlan, S. (2022b, June 28). FBI: Stolen PII and deepfakes used to apply for remote tech jobs. *BleepingComputer*. https://www.bleepingcomputer.com/news/security/fbi-stolen-pii-and-deepfakes-used-to-apply-for-remote-tech-jobs/.

Gehl, R. W. (2014). *Reverse Engineering Social Media: Software, Culture, and Political Economy in New Media Capitalism*. Temple University Press.

Gehl, R. W., & Bakardjieva, M. (2017). *Socialbots and their Friends: Digital Media and the Automation of Sociality*. Routledge.

Gehl, R. W., & Lawson, S. T. (2022). *Social Engineering: How Crowdmasters, Phreaks, Hackers, and Trolls Create a New Form of Manipulative Communication*. MIT Press.

Gray, J. (2020, July 28). OSINT primer [Workshop]. Hope 2020. https://scheduler.hope.net/hope2020/talk/TRTLRT/.

Hadnagy, C. (2018). *Social Engineering: The Science of Human Hacking* (2nd ed.). Wiley.

Hatfield, J. M. (2018). Social engineering in cybersecurity: The evolution of a concept. *Computers & Security, 73*, 102–113. https://doi.org/10.1016/j.cose.2017.10.008.

Hoque, M. A., Ferdous, M. S., Khan, M., & Tarkoma, S. (2021). Real, forged or deep fake? Enabling the ground truth on the internet. *IEEE Access, 9*, 160471–160484. https://doi.org/10.1109/ACCESS.2021.3131517.

Howard, P. (2020). *Lie Machines: How to Save Democracy from Troll Armies, Deceitful Robots, Junk News Operations, and Political Operatives*. Yale University Press.

Huber, M., Kowalski, S., Nohlberg, M., & Tjoa, S. (2009). Towards automating social engineering using social networking sites. In *Computational Science and Engineering, 2009. CSE'09. International Conference On Computational Science and Engineering*, 3, (pp. 117–124).

Jankowicz, N. (2020, July 5). How an anti-Trump flash mob found itself in the middle of Russian meddling. *Politico*. https://www.politico.com/news/magazine/2020/07/05/how-an-anti-trump-flash-mob-found-itself-in-the-middle-of-russian-meddling-348729.

Jansen, S. C. (2013). Semantic tyranny: How Edward L. Bernays stole Walter Lippmann's mojo and got away with it and why it still matters. *International Journal of Communication, 7*. Available at: https://ijoc.org/index.php/ijoc/article/view/1955.

Levine, T. R., Serota, K. B., Shulman, H., Clare, D. D., Hee Sun, P., Shaw, A. S., Jae Chul, S., & Jung Hyon, L. (2011). Sender demeanor: Individual differences in sender believability have a powerful impact on deception detection judgments. *Human Communication Research, 37*(3), 377–403.

Long, J. (2007). *No Tech Hacking: A Guide to Social Engineering, Dumpster Diving, and Shoulder Surfing*. Syngress.

Lutz, C., Schöttler, M., & Hoffmann, C. P. (2019). The privacy implications of social robots: Scoping review and expert interviews. *Mobile Media & Communication, 7*(3), 412–434. https://doi.org/10.1177/2050157919843961.

Mitnick, K. D., & Simon, W. L. (2012). *Ghost in the Wires: My Adventures as the World's Most Wanted Hacker*. Little, Brown.

Mukerji, C. (1978). Bullshitting: Road lore among hitchhikers. *Social Problems, 25*(3), 241–252. https://doi.org/10.2307/800062.

Myers, S. L., Mozur, P., & Kao, J. (2022, February 18). Bots and fake accounts push China's vision of Winter Olympic wonderland. *New York Times*. https://www.nytimes.com/2022/02/18/technology/china-olympics-propaganda.html.

Noble, S. U. (2018). *Algorithms of Oppression: Data Discrimination in the Age of Google*. New York Press.

O'Neil, C. (2017). *Weapons of Math Destruction*. Penguin Books.

O'Sullivan, P. B., & Carr, C. T. (2018). Masspersonal communication: A model bridging the mass–interpersonal divide. *New Media & Society, 20*(3), 1161–1180. https://doi.org/10.1177/1461444816686104.

Paul, C., & Matthews, M. (2016). *The Russian 'Firehose of Falsehood' Propaganda Model: Why It Might Work and Options to Counter It* (PE-198-OSD). Rand Corporation. https://doi.org/10.7249/PE198.

Satter, R. (2017, November 4). Inside story: How Russians hacked the Democrats' emails. *AP News*. https://apnews.com/dea73efc01594839957c3c9a6c962b8a.

Simpson, C. (1996). *Science of Coercion: Communication Research and Psychological Warfare, 1945–1960* (1st ed.). Oxford University Press.

Slatalla, M., & Quittner, J. (1995). *Masters of Deception: The Gang that Ruled Cyberspace*. HarperCollins Publishers. http://archive.org/details/mastersofdecepti00slat.
Stupp, C. (2019, August 30). Fraudsters used AI to mimic CEO's voice in unusual cybercrime case. *Wall Street Journal*. https://www.wsj.com/articles/fraudsters-use-ai-to-mimic-ceos-voice-in-unusual-cybercrime-case-11567157402.
Tanium (Director) (2020, September 16). A 'people hacker' explains how the pandemic opened human-centered vulnerabilities. https://www.youtube.com/watch?v=0wxkig_yiIE.
Van Marken, J. C. (1894). *Sociale ingenieurs*. Delftsche Corps-Almanack.
Vincent, J. (2023, January 31). 4chan users embrace AI voice clone tool to generate celebrity hatespeech. *The Verge*. https://www.theverge.com/2023/1/31/23579289/ai-voice-clone-deepfake-abuse-4chan-elevenlabs.
Webster, S. C. (2011, February 22). Military's 'persona' software cost millions, used for 'classified social media activities.' *The Raw Story*. http://www.rawstory.com/rs/2011/02/22/exclusive-militarys-persona-software-cost-millions-used-for-classified-social-media-activities/.
Williams, R. (2000). Advertising: The magic system. *Advertising & Society Review*, *1*(1), 419–423. https://doi.org/10.1353/asr.2000.0016.
Woolley, S. (2023). *Manufacturing Consensus: Understanding Propaganda in the Era of Automation and Anonymity*. Yale University Press.
Yang, G. (2020). Communication as translation: Notes toward a new conceptualization of communication. In M. Powers & A. Russell (Eds.), *Rethinking Media Research for Changing Societies* (1st ed., pp. 184–194). Cambridge University Press. https://doi.org/10.1017/9781108886260.013.
Zuboff, S. (2019). The Age of Surveillance Capitalism: The Fight for a Human Future at the New Frontier of Power (1st ed.). PublicAffairs.

Section 2: **Humanistic and Social Scientific Perspectives**

Linguistics

Naomi S. Baron

AI and Human Writing: Collaboration or Appropriation?

Abstract: Since the earliest days of artificial intelligence (AI) research, getting computers to mirror human linguistic abilities has been of keen interest. Recent developments in natural language processing have made possible not only impressive machine translation and speech-based personal assistants but generation of human-like written text.

This chapter probes how AI writing tools are increasingly pervading traditional human writing space. We begin by considering humans as writers: how writing affects our minds and brains, and potential motivations for writing. We then turn to the roles AI can play in our writing lives. Key issues here include shortcomings of today's language generation models, challenges in distinguishing between human- and machine-produced text, and implications of AI as author, both for students and for writing-intensive professions such as journalism or translation. The chapter then turns to human–AI collaboration, including its benefits and potential pitfalls. We close by returning to human motivations for writing, asking which motivations invite AI collaboration (or appropriation) and which remain uniquely human.

Keywords: Artificial intelligence, ChatGPT, GPT-3, machine translation, natural language processing, writing

7.1 Introduction

Language has always been central to the artificial intelligence (AI) enterprise. As far back as 1948, Alan Turing identified translation and the learning of languages as two tasks an intelligent machine might tackle (Turing, 1948). In 1955, a quartet of mathematicians, computer scientists, and engineers requested funding from the Rockefeller Foundation for a conference to be held at Dartmouth College the next summer. They proposed studying 'artificial intelligence'—the first time the term was used to talk about intelligent machines. One area they intended to explore was language: "An attempt will be made to find out how to make machines use language, form abstractions and concepts, solve problems now reserved for humans, and improve themselves" (McCarthy et al., 1955).

Now, more than half a century later, these speculations have substantially become reality. Thanks to powerful computers, machine learning, deep neural networks, massive datasets, and new programming models, machines can simulate many linguistic skills, including speech recognition, speech synthesis (producing voice from

https://doi.org/10.1515/9783110792270-008

text), natural language understanding (making linguistic sense out of written text or speech), natural language generation (creating new written text or speech), and translation.

As of the early 2020s, the most successful model for handling language generation and translation has been transformers, an approach pioneered by Google researchers (Vaswani et al., 2017). The most widely known of the transformer models is GPT-3 (Generative Pretrained Transformer), built by the company OpenAI and debuting in 2020. GPT-3 was designed to spin out new text: feed it an opening sentence or two, and the program continues writing by drawing on a huge dataset (collected from the internet, Wikipedia, and books) to predict the most likely next word and sentence. Today's transformers are coupled with large language models, which contain the massive language datasets on which they draw. It is important that we remain aware of what language models can—and cannot—purport to 'know.' The data upon which they draw may be out of date, incomplete, or flat out wrong. And of course, the models themselves don't actually have knowledge. They simply draw upon text available online to generate new text.

In 2021, OpenAI used the same principle—predict what comes next—to construct two new tools: DALL-E (for generating illustrations from written descriptions) and Codex (for producing computer code from ordinary language instructions). These tools rapidly evolved into DALL-E2 for art and Copilot for coding, respectively. November 30, 2022 marked the public launch of ChatGPT (GPT-3.5), a chatbot that is eerily good at responding to queries with impeccable prose, though not always accurate information (Roose, 2022). Then, in early 2023, OpenAI did a limited launch of GPT-4. Other tech companies, most notably Google, have followed with their own large language models and chatbots.

But it is GPT-3 that has most notably made its way into commercial products for text generation and editing. The biggest player to date is Microsoft, which has been incorporating GPT-3 (and its successors) into multiple functions, including Microsoft Editor and Microsoft Translate (both now components of Word). Companies like Copysmith and Sudowrite have licensed GPT-3 to build programs enabling users to automatically write text such as advertising copy, blogs, and short amounts of fiction, for a monthly fee. As of spring 2023, Grammarly was in the process of incorporating GPT-3 for some of its writing-assistance functions. Similar tools are available from companies utilizing alternative AI platforms.

While this chapter focuses on the use of AI to generate written language, essentially the same core technology (with speech recognition and speech synthesis added) can generate and 'understand' spoken language. Such technology has been evidenced for some time in older systems like Siri and Alexa. With the coming of language models like GPT-3, GPT-4, and similar architectures, the prospects for more sophisticated conversations between human and machine continue to grow. Those machines might also be embodied, as with social robots like Ameca, a highly realistic humanoid robot built by Engineering Arts. Recent integration of GPT-3 (and sometimes GPT-4) into

Ameca adds a linguistic dimension to her earlier prowess with facial expressions and gestures (Video "AMECA Humanoid Robot's GPT-3 & GPT-4 Upgrade Shocks the World (Approaching SINGULARITY)" by user AI Revolution (@airevolutionx): https://www.youtube.com/watch?v=8kwWLDUwEPY).

The question this chapter wrestles with is what place AI-generated writing holds —or should hold—in the lives of human writers. We start by considering writing itself, including its impact on people's minds and brains, and our motivations for writing.

7.2 Humans as Writers

Our ability to speak probably only emerged once in human history. Writing is different. Multiple writing systems developed over the millennia, ranging from cuneiforms in Sumeria to hieroglyphs in Egypt to characters in China. The functions of writing have varied across civilizations. In Mycenean Greece, Linear B was used for running the palace bureaucracy. In ancient Israel, a consonantal alphabet recorded religious texts. In Mesoamerica, Mixtec glyphs recounted historical events.

But writing can do much more than provide durable accounts of what people say, do, or believe. It can also impact what happens in our heads.

7.2.1 Writing's connection with mind and brain

Modern arguments linking writing with mental activity trace back at least to work of the classicist Eric Havelock (1963). Havelock argued that the coming of literacy in late archaic and early classical Greece made possible the flowering of Greek philosophy. The thesis is not without its critics (Halverson, 1992). However, what's clear is that writing provides cognitive benefits to both writer and reader. When we write, we can pause, reflect, and rewrite. When reading, we can also pause, think, and reread. Psychologist David Olson has argued that the transition from the oral Homeric tradition to the classical Greek use of literacy reconfigured the way people thought: "Writing . . . gives rise to the idea of an idea and the mind becomes the storehouse of those ideas" (Olson, 1994, p. 242).

Psychologists and reading specialists also ask whether literacy affects our brains. While we previously believed that brain functioning was set early in life, we now know the brain is 'plastic,' meaning it can reorganize and lay down new pathways. Perhaps the most familiar example is London black cab taxi drivers, who must acquire 'the Knowledge,' which entails memorizing thousands of routes, streets, and landmarks. These drivers have larger posterior hippocampi—the area of the brain linked to physical navigation—than control groups (Maguire et al., 2000).

What about reading? Using fMRI scanners, researchers have compared brain activity in adults who never learned to read with people who became literate as children or only as adults. Those who became literate (as children or adults) show more neural activity in areas associated with vision and speech than non-literate adults (Dehaene et al., 2010). The neurological impact of literacy has been documented in Colombia, Portugal, Brazil, and India.

Becoming literate physically changes us. Literacy also provides a versatile tool for human expression.

7.2.2 Motivations for writing

Why do people write? By 'people,' I mean not just professional writers, poets, or novelists, but also everyday writers. Collectively, we have a host of potential motivations, some of which can overlap. In the list below, the first two categories focus on everyday writers. The next two are largely about writers who publish their works, while the final three categories target writing that is more psychologically fueled.

- *Quotidian acts.* Much of the writing we do is of the everyday, self-initiated sort: making grocery lists, keeping diaries, writing letters, doing online blog postings, or sending emails and text messages. In the public arena, we might write laws, contracts, or project proposals.
- *Writing at others' bidding.* We write because others ask us to. If we're students, the context may be an essay assignment for school. If we're journalists, it could be a story we're assigned to cover. As citizens of literate societies, we are obligated to fill out forms and respond to questionnaires.
- *Tangible gains.* Many professionals (not just journalists but grant writers, editors, or translators) write for a living. Others—like academics—are expected to publish in order to be promoted, which may be intertwined with making a living.
- *Sharing expertise.* Those with expertise often write to share it. Think of magazine articles or books designed to inform, reveal, or advise. Often writers are paid, but such writing can also fill personal, non-monetary needs.
- *Looking outward.* Many write to help readers make sense of their world, including expressing hope or offering advice on coping with realities. Literature tends to fall in this category. In the words of the Spanish novelist Carlos Ruiz Zafrón, "It is my way of bringing something to the table, contributing what I believe is the best thing I have to offer for others to enjoy" (Zafrón, n. d.).
- *Looking inward.* An especially important motivation for writing is self-discovery. A surprising number of writers have explained they write to clarify their thinking. For instance, in the words of Joan Didion, "I write entirely to find out what I am thinking, what I'm looking at, what I see and what it means. What I want and what I fear" (Didion, 1976).

- *Personal release.* We also sometimes write to vent our emotions. It might be a blistering diary entry, an angry letter to an employer, or graffiti spray-painted on the side of a building.

Articulating human reasons for writing helps set the stage for talking about the place of artificial intelligence as a writer. Today's AI programming has become incredibly sophisticated, both in editing what we humans have written and in producing written material on its own. However, the natural language processing tools used to generate that text are not sentient. They have no motivations, at least of the recognizable human kind. A critical issue is therefore whether using these tools undermines human motivation to write and, what is equally important, to think.

Choosing between human or AI as editor or writer need not be a zero-sum game. An alternative is collaboration. The issue becomes how much effort is put forth by each side, and the consequences if humans relinquish too much control to the machine. We'll turn first to AI as editor, since that is the capacity in which most people encounter AI's writing abilities. From there we move to AI as solo author, AI as translator, and then to collaboration.

7.3 AI as Editor

Today's AI is skilled at amending or augmenting text that humans have written. The tools run the gamut from spellcheck to autocomplete to predictive texting to grammar and style check programs. Our question is whether availability of these AI capacities undermines human motivation to know how to spell, master grammar rules, proofread what we've written, and make our own stylistic choices.

7.3.1 Overview of tools

One of the oldest editing programs is spellcheck. Early versions were designed for large computers in the 1970s and 1980s. With the coming of personal computers and a variety of word-processing programs, spellcheck became available to everyday writers. By 1993, the spellcheck built into Microsoft Word autocorrected misspellings rather than writers needing to authorize proposed corrections.

A host of autocomplete or autogeneration writing tools followed over the years. These days, when doing online searches, if we type a few words into Google like "What famous," the search engine completes the request with a string of options such as "people live in New York" or "died on the Titanic." Autocomplete also shows up in email platforms. I type "I'm ver" and Microsoft Outlook suggests I add "y sorry to hear." When texting on smartphones, predictive texting second-guesses how we might want to finish a word: I type "She qu" and my iPhone proposes three choices: "gu," "quickly," and "quit". In a similar mold, in 2015 Google introduced Gmail Smart Reply, which autogenerates three responses from which to select. Gmail Smart Compose followed in 2018, proposing sentence completions.

A different type of AI language tool is grammar and style checkers. Again, these date back to the 1970s, running on mini-computers of the time. By 1992, Microsoft Word had incorporated a grammar checker. Over the decades, the scope and sophistication of its program increased enormously. Other developers have created their own grammar and style programs, including Grammarly (a commercial program used widely in the US) and the Educational Testing Service's (ETS's) program Criterion (which incorporates a version of e-rater, ETS's natural language processing tool for evaluating essay components of standardized tests). These programs make recommendations on everything from grammar and punctuation to style (e.g., be more concise) and potential bias in word choice (e.g., to replace *chairman* with *chair* or *chairperson*).

7.3.2 Programming challenges

As with any evolving computer technology, programming attempts at language editing haven't always been successful. Take spellcheck. Earlier versions of the autocorrect program relied on a dictionary against which to compare what the user typed. If the intended word wasn't in that dictionary, the program offered its nearest approximation. Some results made no sense, as when Microsoft "autocorrected" *Barak* to *Boatman* or *Barracks*, and *Obama* to *Osama* (Wilson, 2008). Thanks to today's deep neural networks and large language models, this sort of problem is largely a thing of the past.

Every language has standard spellings for each word, though standards might vary across language varieties (e.g., American English *theater* vs. British English *theatre*). However, grammar, punctuation, and style are different stories. Depending upon what choices the software programmers have made, the writer's words might pass muster or might be flagged. For instance, do you need a comma after the word *however*? Can you begin a sentence with the word *and*?

More intrusive are recommendations involving word choice. Microsoft Editor (available in many versions of Word) relentlessly advises making writing more concise. When I write "be able to," I'm counseled to replace the phrase with *can*. "In the near future" should be shortened to *soon*. There is a place for concision, but not when it obliterates the individual writer's style or even renders a sentence ungrammatical. When I typed in the opening line of the US Declaration of Independence ("When in the course of human events"), Word suggested I replace "in the course of" with *during*. However, "during human events" is grammatically off. Even if the phrase passed grammatical muster, *during* is too stylistically abrupt for a document of this solemnity.

Even more challenging are the new tools Microsoft has developed to help writers counteract bias and promote inclusion. Word balked when I wrote about the need to "master grammar rules," suggesting I should use a gender-neutral term like *expert*, *head*, or *primary*. The software failed to understand that *master* was a verb here, not a noun. Similarly, when I wrote about an AI program called AlphaGo winning "against a Go master," again I was urged to make a substitution. Here, *master* is a noun, but "a Go primary" makes no sense. Writers who are not confident in their skills, including many English-language learners, could be misled into following such recommendations, with unfortunate results.

7.3.3 What do users say?

AI editing tools such as these are used by millions of people worldwide. It's relevant to ask what everyday writers feel about these software programs. Are they always perceived as helpful? Do users wonder if their own skills or writing voice is being affected? Do they sense they are thinking less when they write with AI at their side than when writing on their own, perhaps by hand?

There's experimental evidence that when writers use predictive texting, their vocabulary becomes more constrained (more succinct—and less interesting) than when generating their own words (Arnold et al., 2020). The philosopher Evan Selinger suggests that "[B]y encouraging us not to think too deeply about our words, predictive technology may subtly change how we interact with each other. As communication becomes less of an intentional act, we give others more algorithm and less of ourselves. . . . [A]utomation . . . can stop us thinking" (Selinger, 2015).

Grammar and style programs also come in for criticism. Daniel de Beer, a high school student in London, described how a tool like Grammarly can undermine the writer's individuality: "Grammarly can remove students' artistic voice when writing. Rather than using their own unique style when writing, Grammarly can strip that away from students by suggesting severe changes to their work" (de Beer, 2020).

AI grammar and style programs have other potential effects as well. One of these is to make the writer feel that by accepting the program's recommendation, the piece of writing (say, a student essay) has actually been effectively revised. Anne Herrington and Charles Moran, both professors of English composition and rhetoric, analyzed ETS's grammar and style program Criterion. They argued that the program doesn't look at text holistically but only at the word and sentence level: "it presents the revising task to come as one of moving through categories and eliminating errors, not of rethinking the argument or adapting to a particular rhetorical situation" (Herrington and Moran, 2012, p. 226).

Another issue is whether students using these programs are becoming better writers as a result. On its website, Grammarly reports that, among those using the premium version of its software, 85% say the program has made them better writers. Perhaps. A 2011–2012 company study of university students reported that 99% of those surveyed felt that Grammarly had improved their grades. But grades and learning are not the same. Daniel de Beer observed that "Quickly correcting the red line [through which Grammarly indicates an error] without even acknowledging why the word is incorrect does not make its users better writers; instead, it will make them rely too much on the app." He wondered if "Grammarly gives users false security . . . by simply accepting suggestions" and whether users aren't actually improving but are "simply being tricked into thinking they are better writers.").

Given my interest in user perceptions of AI writing tools, I surveyed around 200 young adults (in Europe and the US), probing their attitudes towards such editing tools as spellcheck, predictive texting, and grammar and style programs (Baron, 2023). While there were many positive comments about the usefulness of these tools, respondents also expressed concerns about effects upon their skills and especially on their writing voice. Here are highlights from my findings.

7.3.3.1 Spelling

While one out of five participants felt spellcheck made them worse at spelling, two out of five believed spellcheck made them better able to spell words correctly on their own. (I would have expected the reverse distribution.) In some cases, respondents indicated that although they felt their spelling skills were reasonably good, they looked to spellcheck to catch mistakes. However, more than a quarter of participants didn't always trust spellcheck's corrections.

7.3.3.2 Predictive texting

When asked to judge whether they felt predictive texting affected the messages they sent, 11% said it resulted in longer or more complex messages, while more than 20% concluded it led to simpler or shorter messages. One person commented that predictive texting "makes me a little too repetitive."

There was also frustration when predictive texting came up with words they hadn't intended. Almost 30% of the Europeans and nearly 60% of the US respondents said they would rather compose text messages entirely themselves—as one person wrote, "[Predictive texting] wastes time because of the amount of time I have to erase the generated response."

Perhaps more significantly, several people were troubled that predictive texting undermined their personal writing voice. As one wrote, "I feel like the message I sent is not mine."

Comments such as this reinforce Evan Selinger's point about giving others "more algorithm and less of ourselves."

7.3.3.3 Grammar and style programs

Study participants also had mixed feelings about AI software that offered grammatical suggestions and corrections. Fifteen percent indicated they were happy to let technology do the editing and proofreading for them. At the same time, there was recognition that tools like Grammarly could make them lazy. One person commented, “Grammarly’s effects on me as a writer are mixed, as its corrections help, but they also make it so I rely on some crutches in writing.”

As with spellcheck, users didn’t always trust the program’s recommendations: “I don’t trust Grammarly but I do need to see its suggestions.”

An important lesson from this research is that many users of personal editing software are aware of programming shortcomings as well as potential effects on their own writing ability and individual writing voice. As the technologies become yet more sophisticated and accurate, it remains to be seen to what extent everyday writers will be tempted to surrender control to AI.

We move now from editing issues to full-blown AI composition.

7.4 AI as Author

The idea of computers writing text on their own is hardly new. In 1951—even before the field of artificial intelligence was so named by John McCarthy and his colleagues —British computer scientist Christopher Strachey was fashioning a program that cranked out primitive love letters (Roberts, 2017). His inspiration was conjectures by Alan Turing, his colleague, that it would be possible to build a machine that could imitate human intelligence.

Over the years, as both computers and programming models grew more powerful, the kinds of writing that computers could do became incrementally impressive. In 1984, William Chamberlain and Thomas Etter’s novel *The Policeman’s Beard is Half-Constructed* appeared, with its cover proclaiming this to be “The First Book Ever Written by a Computer” (Henrickson, 2021). In 2019, the publisher Springer Nature published *Lithium-Ion Batteries*, the first computer-generated textbook (Vincent, 2019).

Meanwhile, AI researchers and those who enjoyed tinkering with natural language generation were using the technology to produce both informational prose and literary works. These productions now increasingly rely on deep neural networks, a technology pioneered in the 2010s by Geoffrey Hinton, Yann LeCun, and Yoshua Bengio. Deep neural networks underlie the transformer model that appeared in 2017.

When OpenAI released GPT-2 and then GPT-3, the level of text generation took a quantum leap forward. Particularly with the newest refinements to the GPT-3 model (along with parallel developments in large language models by Google, DeepMind, and Meta), the language that is produced is increasingly indistinguishable from what a human might have written (Clark et al., 2021; Jakesch et al., 2023). The AI-generated language appears to be authentic.

But note the qualifier "appears." As of 2023, there are subtle but critical differences. Some involve content and others, writing style.

Three key content issues are bias, accuracy, and what's known as hallucination, that is, making things up that have no factual basis. Regarding bias, all large language models suffer from reflecting prejudices existing in the datasets from which they draw. AI generation is notorious for producing text that is prejudicial against females, people of color, and Muslims (Abid et al., 2021). As for accuracy, the best advice is to verify with trusted sources. Inputting the same prompt twice can produce opposite answers. When I asked ChatGPT, "Should it be possible for short stories written by GPT to be copyrighted?", the first response was yes (since the words are original) and the second was no (which is factually true in American law).

Then there's the content problem of hallucination. A much-cited example involved Galactica, a large language model Meta trained for use in science. However, when Galactica produced an article on bears in space, Meta retracted the tool after only three days (Heaven, 2022). Unless one is writing science fiction, it's unlikely scientists would mistake this text for human generated.

Writing style can also help reveal whether text was written by a human or AI. The longer the text, the more likely it is for AI to repeat itself. Similarly, with longer texts, it's easier to spot that high-frequency words in the language will be more prevalent in generative AI, while less common words will be less often produced, reflecting the dataset upon which AI draws. Another clue is that AI-generated text tends to have consistent, bland, evenhanded wording. Don't expect variation in writing style (variety is more typical from humans). And as with Wikipedia, tools like ChatGPT are designed to be objective, reporting information but not expressing a point of view.

Another clue is that text produced by AI looks too perfect: essentially no misspellings, no punctuation issues, everything grammatically correct. Even when using traditional software tools like spellcheck and grammar check programs, humans typically make errors. In the words of Daphne Ippolito, senior research scientist at Google Brain, "A typo in the text is actually a good indication that it was human written" (quoted in Heikkilä, 2022).

Finally, AI researchers are using technology to help distinguish between text written by humans and AI. One idea is to insert digital 'watermarks' into AI-generated text (Wiggers, 2022). By early 2023, tools like GPTZero were being developed for spotting when ChatGPT was writing text that students were submitting as their own (Bowman, 2023; https://gptzero.me).

7.4.1 Educational use: the ethics question

Even before ChatGPT burst on the scene in late 2022, educators were worrying they couldn't distinguish whether students had done the work themselves or AI had done the writing (D'Agostino, 2022; Peritz, 2022; Schatten, 2022). The issue of originality in the assignments students turn in is hardly new.

Ever since English composition classes were introduced at Harvard in the late nineteenth century, students have found ways of submitting written work that's not their own. Schemes now range from paying someone to write the paper for you to copying another person's essay to pulling material from books or the internet without attribution. To detect plagiarism, schools and colleges increasingly turn to software like Turnitin to compare student submissions against texts housed in the company's huge dataset. However, the challenge with both contract cheating (paying someone to ghostwrite for you) and text generated by tools like GPT-3 is that the text is new and therefore not discoverable in these datasets.

Most teachers who read students' assignments have neither the time nor technical expertise for determining whether their student or a machine did the writing. Moreover, determined students figure out that by tweaking what AI produces—adding a couple of typographical errors or an ungrammatical sentence, replacing common words with less frequent vocabulary, incorporating an opinion or local information that wouldn't be in the machine's dataset—they can make AI-generated text sound as if they did the work themselves.

In the years ahead, educators will need to find ways to formulate meaningful writing assignments they can be assured were actually produced by their students. Writing essays by hand, especially in class, sounds like a logical solution, though not one that's generally practical. Substituting oral for written assignments is another option, but then students forego the benefits of both learning good writing and using writing as a tool for thinking. As teachers of language and literature in Norway have argued, writing is not only about the final product but about the process (Eriksen, 2022; Valand, 2022). A productive writing process might entail searching for information, trying out ideas, formulating arguments, rethinking, and revising.

Education is the journey, not just the destination.

7.5 AI as Translator

AI as author goes beyond writing novels or student essays. AI authorship has the potential to reshape professions by supplanting human writing in fields ranging from advertising to journalism to law to translation. We focus here on human versus machine translation, since its history traces back to the early days of AI, when the Cold

War build-up after World War II fueled American interest in rapid access to scientific research written in Russian.

The first American proposal to use computers for translation came in 1947 from Warren Weaver, director of the National Science Division of the Rockefeller Foundation (Weaver, 1949). By the early 1950s, the idea caught fire. Considerable government funding was devoted to the endeavor. It was generally agreed that a 'mixed machine translation model' would be used, where humans would do some pre-editing and then post-edit text translated by the computer (Bar-Hillel, 1951). However, in 1966 a report issued by a National Academy of Sciences task force concluded not only that the quality of machine translation was poor but also that it was cheaper to hire human translators, who were in ample supply (Hutchins, 1996).

Yet work on machine translation persisted in a number of quarters. One system was TAUM-METEO, created at the University of Montreal for automatic translation between French and English of Canadian weather forecasts. Another was SYSTRAN, a project seeded at Georgetown University in the early 1960s but later developed into a program used by the US Department of Defense and the European Community. SYSTRAN also became the basis for Babel Fish, the first free online translator, made available in 1997 by the early search engine Altavista.

The next two decades saw dramatic increase in both availability and then quality of AI translation. Google Translate was launched in 2006. The biggest next step came with new programming techniques, shifting from statistical algorithms to deep neural networks (Cho et al., 2014). When Google Translate made this transition in 2016, its translation error rate dropped overnight by 60% (Lewis-Kraus, 2016).

Today's AI translation is hardly perfect. A stark illustration comes from a study of how hospital emergency rooms prepared patient discharge orders (Wetsman, 2021). Since the orders needed to be written and not all patients knew (in this case) English, it seemed reasonable to enlist Google Translate to convert the English instructions written by the medical staff into a language the patient could read. However, since programs like Google Translate have much larger datasets to draw upon for some languages than for others, translation accuracy declines when the dataset isn't robust. Two glaring examples from the study:

– English "You can take over the counter ibuprofen as needed for pain" was rendered in Armenian as "You may take anti-tank missile as much as you need for pain."
– English "Your Coumadin level was too high today" was translated into Chinese as "Your soybean level was too high today."

And there are other challenges. Translation programs struggle with such issues as gender bias, with translations of words like *doctor* defaulting to male and *nurse* to female (Savoldi et al., 2021). Moreover, if you don't know the language you are translating from or translating into, you have no assurance the translation is accurate.

Equally importantly, machine translation rarely takes into account the cultural understanding and stylistic nuance that skilled human translators take years to develop.

Reliance on automatic translation continues growing, both by everyday users of free online tools (like Google Translate, Microsoft Translate, and DeepL Translator) and by business. Thousands of companies offer paid translation services. Among the biggest are TransPerfect, RWS, Language Solutions, Keyword Studies, and Lionbridge (The world's top translation companies, n. d.).

What are the dynamics between automatic translation and the (human) translation profession? According to the US Department of Labor Statistics, employment projections for translators (of written text) and interpreters (who do simultaneous translation of spoken or signed language) are highly encouraging. A 24% growth rate between 2020 and 2030 is anticipated in the US, compared with an overall national job increase of 8% (Job Outlook, n. d.). The issue is what kind of jobs these will be.

In talking about the impact of AI on the labor market, social scientists distinguish between enough jobs remaining for humans and enough 'good' jobs (Tyson & Zysman, 2022). Continuing expansion of international business, along with requests for translations of everything from text messages to YouTube videos, will ensure the need for translators. However, since so much of the basic work is being handed off to computers, the jobs left for humans are increasingly as post-editors for cleaning up machine infelicities.

Many professional translators don't see post-editing as a 'good' job. In an international survey of more than 7,000 translators and interpreters, 89% preferred doing traditional (human) translation. While 35% engaged in some post-editing of machine translation, only 3% preferred it. The remaining 8% favored editing human translations—the original form of post-editing (Pielmeier & O'Mara, 2020). In a study of professional literary translators, participants voiced such comments as "[Post-editing] makes you a bit lazy. . . . You don't feel like changing too many things" (quoted in Moorkens et al., 2018). In yet another study, one translator said, "In the post-editing phase, I felt that my creativity was limited, and I found it more difficult to think outside the box when I already had a translation provided" (quoted in Guerberof Arenas & Toral, 2020, p. 261).

Our discussion of AI as author has thus far focused on AI doing all, or at least the majority, of the writing. An alternative model is to think of AI as a writing collaborator.

7.6 AI as Collaborator

Since at least the 1980s, computer scientists have reminded us that humans and computers need to work in tandem, to communicate with each other. The concept was initially known as human–computer interaction (Card et al., 1983; Shneiderman, 1987). That interaction typically referred to the way interfaces and feedback loops were designed. At the same time, work on artificial intelligence continued to probe what cognitive tasks computers might be able to take over on their own (the automation model).

The mid to late 2010s saw increased emphasis on the role of humans in the interaction. The phrase 'human in the loop' (or, as I prefer to say, 'humans in the loop,' since multiple people are often involved) became increasingly prevalent, as did the notion of 'human-centered AI.' Instead of asking what AI could accomplish on its own, the question became, What can AI do in the service of humans? New research centers began springing up, including Stanford University's Institute for Human-Centered Artificial Intelligence (Stanford HAI) and the University of California at Berkeley's Center for Human-Compatible Artificial Intelligence.

7.6.1 Augment, don't automate

For a growing number of AI researchers, the aim is to discover ways in which AI can augment, not automate, human activity (Brynjolfsson, 2022). Accordingly, Stanford HAI suggests speaking of AI being in the loop, rather than humans, since the collaboration is controlled by people, not computers (https://hai.stanford.edu/events/2022-hai-fall-conference-ai-loop-humans-charge).

Across the AI spectrum, we find examples of AI programs devised to augment rather than replace human skills. Such human-managed collaborations run the spectrum from reading mammograms to co-creating art or music to generating computer code. They also appear with writing.

7.6.2 AI as writing collaborator

Many of the commercial writing programs driven by large language models emphasize their collaborative potential. For example, Copysmith, which is designed to draft text like advertising copy or blog posts, uses the website tag line "Content crafted by AI, perfected by humans." Here, the human part involves setting the parameters of what is to be written (topic, tone, length, and so on) and then reviewing the output.

AI is also being harnessed for collaboration in creative writing endeavors. Examples include interactive storytelling and poetry writing (Branch et al., 2021; Rezwana & Maher, 2022). Sometimes collaboration takes the form of interleaving which party is doing the writing, such as K Allado-McDowell's *Pharmako-AI* (2021). Other times, the human does the writing and then AI evaluates. For instance, with the AI 'self-editing tool' called Marlowe, would-be authors input their draft and then Marlowe produces an extensive report analyzing components such as narrative arc, word choice, and character personality traits (https://authors.ai/marlowe/).

The most typical human–AI collaboration is using an AI text-generating program to jumpstart your thinking or provide suggestions for, say, alternative wording, next paragraphs, or new plot directions. Sudowrite promises to "bust writer's block" (https://www.sudowrite.com/). But, as users discover, it can easily become more invasive. Jennifer Lepp, a self-published author who produces a novel roughly every nine weeks, turned to Sudowrite to generate descriptions of scenes or objects. However, as she increasingly relied on Sudowrite, she discovered herself becoming alienated from her own plot and characters: "It didn't feel like mine anymore. It was very uncomfortable to look back over what I wrote and not really feel connected to the words or ideas" (quoted in Dzieza, 2022).

Reading Lepp's words, I kept hearing in my head the comment about predictive texting from a student in my research: "I feel like the message I sent is not mine."

7.7 What If the Dream Came True?

Use of artificial intelligence to edit or generate text is still a work in progress. But we need to think ahead and imagine what would happen if the dream of AI as infallible editor and author came true.

By the time you read this chapter, the sophistication of AI writing tools will surely have surpassed what I've described here. I can envision that as an editor, AI might be more flexible on stylistic choices, not insisting that I change "in the near future" to *soon.* I suspect that tools like Microsoft Word will have access to historical documents like the US Declaration of Independence, therefore not instructing me to write "When during human events." And I predict that today's well intended but awkward attempts at urging inclusive language will be softened, no longer advising me to write "a Go primary" rather than "a Go master."

Similarly, newer versions of GPT-3 and similar large language models are already becoming increasingly skilled at nuance. Through the kinds of machine-learning initiatives now ongoing, AI will become less likely to hallucinate. They also make it possible to train an AI program on the writings (and therefore writing styles) of particular authors, enabling the program to generate new text in that style. The author might be Shakespeare or J. D. Salinger, or it might be yourself. As a result, discerning whether a human or computer wrote a text becomes yet more difficult.

Such developments will reinforce patterns already underway—in everyday writers' use of AI editing, in students turning to programs like ChatGPT to draft their essays, in the impact of generative AI on writing-intensive professions, in possibilities for collaborative use of AI. As AI text generation becomes increasingly refined and powerful, the stakes for humans as writers also rise.

Let's take each of these contexts in turn.

7.7.1 AI editing and human writing skills

A vital question for us all is whether availability of AI editing tools undermines personal motivation to know how to spell, master grammar rules, proofread what we have written, or make our own stylistic choices. My research with young adults yielded mixed results: some felt tools like spellcheck and grammar programs helped them learn these writing mechanics, while others voiced concern that AI undermined their own abilities.

As with use of calculators, individuals will need to make their own decisions about how much to rely on AI for core literacy skills. More worrisome to me is potential effects of AI style programs on our writing voice. The decision to use multiple words rather than one is often a conscious choice, not a case of verbosity. When predictive programs for email or texting automatically offer up wording for us, our motivation diminishes for taking the time to craft what we, ourselves, want to say.

7.7.2 Student use of programs like ChatGPT

By early 2023, with millions of students having access to ChatGPT, the teaching profession was divided on how to respond. One group was primarily concerned with how to detect if students were using the bot to cheat. The other camp was intent on identifying ways of harnessing the technology (Roose, 2023). Suggestions for the latter included having ChatGPT generate an essay and asking students to critique it, or directing students to write an essay and then getting ChatGPT to assess it.

It's too early to know what a productive balance might be for classroom use of tools like ChatGPT. For certain, though, one of the most important potential effects should be encouraging teachers in secondary schools and colleges to rethink the purpose of writing assignments, and to find ways of focusing at least as much on the writing process as on the final product.

7.7.3 Impact of AI on writing-intensive professions

Again, it's too soon to accurately predict the eventual impact of generative AI on writing-intensive professions. Consider journalism, where a growing number of stories are being written by AI. For more than a decade, there has been talk of AI freeing journalists to focus on things like investigative stories or in-depth analyses (Colford, 2014). However, the more skilled AI as author becomes, the more it is capable of producing nuanced and analytical writing. Moreover, a perfect storm is brewing between the heightened abilities of AI and ongoing shrinkage of positions in the news business. Between 2004 and 2020, the number of US jobs in the industry plummeted 57% (Newspapers fact sheet, 2021).

Similar uncertainty applies to the legal profession (Susskind, 2017). And while the number of jobs for translators may actually rise, a lot of the work will be for post-editing, not 'good' assignments for doing full human translation.

7.7.4 Collaboration with AI

Much optimism is being expressed about the prospects of human–AI writing collaboration. Fascinating writing partnerships are already proliferating for everything from essays to novels to songs (e.g., Vara, 2021; Wilk, 2021; Huang et al., 2020).

The challenge will be that collaboration can be a slippery slope. Jennifer Lepp discovered the problem when using Sudowrite. Many students and everyday writers sense how easy it is to cede control of their writing to AI editors or text generators. We need to worry at what point the written product is no longer our own.

7.8 Motivations for Writing: Which are Uniquely Human?

Finally, we return to a central question posed early in the chapter: Why do people write? To this query we now add another: Of the human purposes for writing, which could be assumed in part or whole by AI and which remain unique to people?

7.8.1 Quotidian acts

For everyday writing produced on digital technologies like computers or mobile phones, it's nearly inevitable these days that we draw on AI's editing prowess. Spellcheck and autocomplete show up by default. Regarding AI's recommendations for grammar, punctuation, or word choice, it's tempting simply to agree rather than pause to weigh the options. What's more, as generative AI becomes increasingly incorporated into functions like email, bots are at the ready to take over much of our everyday business and personal communication (Rogers, 2023).

Of course, in our everyday lives, we still have the option of using earlier technologies—pen and paper—to commit words to a durable medium. When doing so, we lose out on digital affordances like editing tools and ease of transmission, but may gain in the convenience of not needing to power up a computer or locate our mobile phone.

7.8.2 Writing at others' bidding

In professional contexts, those making the business decisions understandably enlist AI to do greater amounts of the writing. Whether the task is composing advertising copy or a news story, drafting a legal brief or doing a translation, AI text generation is likely less expensive. The jury is currently out among labor economists and those in the professions as to how far the balance will tip towards AI as author (Handel, 2022).

When it comes to students, many already grapple with the temptation to let a tool like ChatGPT do writing assignments for them (Cu, 2023). Yet rather than assume that most students are looking to cheat, we need to do a better job of explaining to students why learning to write—and doing their own writing—matters.

7.8.3 Tangible gains

Jennifer Lepp's collaboration with Sudowrite sped up her writing production, with the derivative benefit of financial gain. Of course, not all gains from assistance in writing are monetary. They might also be in professional standing, especially in academia. In the scientific world, researchers are increasingly calling upon AI text generation to contribute to their journal articles. However, there is debate over whether AI should be listed as a co-author, especially given that a computer program can't be held responsible for what it has written (Stokel-Walker, 2023).

Current developments in generative AI suggest that collaboration will prove a likely direction when writing for tangible gain. Humans will presumably come up with the initial concept, plot, or research problem, and then draw on AI to produce background information, alternative approaches, analysis, and some of the sentences and paragraphs.

7.8.4 Sharing expertise

Especially where the text involves a lot of information, AI is already capable of authoring articles and books. *Lithium Ion Batteries* is one such example. So are the more than 200,000 volumes produced by Philip M. Parker, a marketing professor in France, who created a patented system that incorporates a template, databases, and internet searches to crank out books automatically (Cohen, 2008).

There is also ample room here for collaboration between humans and AI. As with writing for tangible gain (which often overlaps with sharing expertise), humans will likely set the agenda, with AI contributing research and text.

With all four of these motivations for writing, AI already plays or will play a substantial role. However, it is the next three categories of motivation that are unique to humans. Given this uniqueness, I combine my comments about them.

7.8.5 Looking outward, looking inward, personal release

One reason many of us write is to reach out to others in a communal human attempt to make sense of our world and perhaps improve upon it. Computers, not being human or even animate, are incapable of such motivation. They do our bidding but have no aspirations of their own.

In much the same way, literate humans use writing to formulate what's going on in their minds. To reprise Joan Didion, "I write entirely to find out what I am thinking." While AI programs are increasingly capable of simulating human intelligence as expressed in text they generate, these programs are incapable of actually thinking.

As for using writing for personal release, AI has no emotions. It can't write a diary based on personal experiences, and it's not driven to write angry letters or spray-paint buildings.

7.9 Concluding Remarks

Today's AI offers abundant opportunities for assisting us as we write—as editor or collaborator—but also for appropriating the writing process. As with many technologies, individuals have some latitude as to what role they choose for AI to play in their writing lives.

Human writers can decide how much it matters to retain skills in spelling, grammar, and punctuation. Though it may require additional effort, it behooves us to stop AI from overriding our writing voice or lulling us into believing we have written a piece when AI produced the lion's share. And as human writers, we need to push back against economic arguments that if AI is capable of generating a text instead of people writing it, we should regularly acquiesce. Given how incredibly proficient AI has already become at generating text, it's easy to lapse into thinking there's little point in expending the thought and energy needed to write. We need to remind ourselves that writing is a precious human tool for articulating our own thoughts and for sharing our thinking with others. For unlike computers, we are sentient and social beings.

References

Abid, A., Farooqi, M., & Zou, J. (2021). Large language models associate Muslims with violence. *Nature Machine Intelligence, 3*, 461–463. https://www.nature.com/articles/s42256-021-00359-2.

Allado-McDowell, K (2021). *Pharmako-AI*. Ignoto Books.

Arnold, K. C., Chauncey, K., & Gajos, K. Z. (2020). Predictive text encourages predictable writing. *IUI '20. Intelligent User Interfaces. Association for Computing Machinery*. https://www.eecs.harvard.edu/~kgajos/papers/2020/arnold20predictive.pdf.

Bar-Hillel, Y. (1951). The present state of research on mechanical translation. *American Documentation, 2*(4), 229–237.

Baron, N. (2023). *Who Wrote This? How AI and the Lure of Efficiency Threaten Human Writing*. Stanford University Press.

Bowman, E. (2023, January 9). A college student created an app that can tell whether AI wrote an essay. *NPR*. https://www.npr.org/2023/01/09/1147549845/gptzero-ai-chatgpt-edward-tian-plagiarism.

Branch, B., Mirowski, P., & Mathewson, K. (2021). Collaborative storytelling with human actors and AI narrators. In *Proceedings of the 12th International Conference on Computational Creativity* (pp. 97–101). https://arxiv.org/abs/2109.14728.

Brynjolfsson, E. (2022). The Turing trap: The promise & peril of human-like artificial intelligence. *Daedalus, 151*(2), 272–287.

Card, S. K., Moran, T. P., & Newell, A. (1983). *The Psychology of Human–Computer Interaction*. Lawrence Erlbaum Associates.

Cho, K., van Merrienboer, B., Bahdanau, D., & Bengio, Y. (2014). On the properties of neural translation: Encoder–decoder approaches. In *Eighth Workshop on Syntax, Semantics and Structure in Statistical Translation. Association for Computational Linguistics* (pp. 103–111). https://arxiv.org/abs/1409.1259.

Clark, E., August, T., Serrano, S., Haduong, N., Gururangan, S., & Smith, N. A. (2021). All that's 'human' is not gold: Evaluating human evaluation of generated text. In *Proceedings of the 59th Annual Meeting of the Association for Computational Linguistics and the 11th International Joint Conference on Natural Language Processing* (pp. 7282–7296). https://aclanthology.org/2021.acl-long.565/.

Cohen, N. (2008, April 14). He wrote 200,000 books (but computers did some of the work). *New York Times*. https://www.nytimes.com/2008/04/14/business/media/14link.html.

Colford, P. (2014, June 30). A leap forward in quarterly earnings stories. *AP: The Definitive Source Blog*. https://blog.ap.org/announcements/a-leap-forward-in-quarterly-earnings-stories.

Cu, M. A. (2023, January 22). Scores of Stanford students used ChatGPT on final exams, survey suggests. *Stanford Daily*. https://stanforddaily.com/2023/01/22/scores-of-stanford-students-used-chatgpt-on-final-exams-survey-suggests/.

D'Agostino, S. (2022, October 26). Machines can craft essays: How should writing be taught now? *Inside HigherEd*. https://www.insidehighered.com/news/2022/10/26/machines-can-craft-essays-how-should-writing-be-taught-now.

de Beer, D. (2020, November 3). Grammarly both helps, hinders students. *The Standard*. The American School in London. https://standard.asl.org/16178/opinions/does-grammarly-help-or-hinder-students/.

Dehaene, S., Pegado, F., Braga, L.W., Ventura, P., Nunes Filho, G., Jobert, A., Dehaene-Lambertz, G., Kolinsky, R., Morais, J., & Cohen, L. (2010). How learning to read changes the cortical networks for vision and language. *Science, 330* (December), 1359–1364.

Didion, J. (1976, December 5). Why I write. *New York Times*. https://www.nytimes.com/1976/12/05/archives/why-i-write-why-i-write.html.

Dzieza, J. (2022, July 20). The great fiction of AI. *The Verge*. https://www.theverge.com/c/23194235/ai-fiction-writing-amazon-kindle-sudowrite-jasper.

Eriksen, D. (2022, December 7). Teachers despair over new artificial intelligence. *NRK Culture*. Original Norwegian available at https://www.nrk.no/kultur/laerere-fortvilet-over-ny-kunstig-intelligens-1.16210580. English translation through Microsoft Translate.

Guerberof Arenas, A., & Toral, A. (2020). The impact of post-editing and machine translation on creativity and reading experience. *Translation Spaces, 9*(2), 255–282.
Halverson, J. (1992). Goody and the implosion of the literacy thesis. *Man (N.S.), 27*, 301–317.
Handel, M. (2022, July). Growth trends for selected occupations considered at risk from automation. *Monthly Labor Review*, US Bureau of Labor Statistics. https://www.bls.gov/opub/mlr/2022/article/growth-trends-for-selected-occupations-considered-at-risk-from-automation.htm.
Havelock, E. (1963). *Preface to Plato*. Harvard University Press.
Heaven, W. (2022, November 18). Why Meta's latest large language model survived only three days online. *MIT Technology Review*. https://www.technologyreview.com/2022/11/18/1063487/meta-large-language-model-ai-only-survived-three-days-gpt-3-science/.
Heikkilä, M. (2022, December 19). How to spot AI-generated text. *MIT Technology Review*. https://www.technologyreview.com/2022/12/19/1065596/how-to-spot-ai-generated-text/.
Henrickson, L. (2021, April 4). Constructing the other half of the policeman's beard. *Electronic Book Review*. https://electronicbookreview.com/essay/constructing-the-other-half-of-the-policemans-beard/.
Herrington, A., & Moran, C. (2012). Writing to a machine is not writing at all. In N. Elliot & L. Perelman (Eds.), *Writing Assessment in the 21st Century* (pp. 219–232). Hampton Press.
Huang, C.-Z. A., Vincent Koops, H., Newton-Rex, E., Dinculescu, M., & Cai, C. J. (2020). AI song contest: Human–AI co-creation in songwriting. In *21st International Society for Music Information Retrieval Conference*. https://arxiv.org/abs/2010.05388.
Hutchins, J. (1996). ALPAC: The (in)famous report. *MT News International, 14*, 9–12. https://aclanthology.org/www.mt-archive.info/90/MTNI-1996-Hutchins.pdf.
Jakesch, M., Hancock, J. T., & Naaman, M. (2023). Human heuristics for AI-generated language are flawed. *PNAS, 120*(11). https://www.pnas.org/doi/10.1073/pnas.2208839120#:~:text=A%20computational%20analysis%20of%20language,topics%20with%20human%2Dwritten%20language.
Job Outlook (n. d.) *Occupational Outlook Handbook*. Interpreters and translators. US Bureau of Labor Statistics. https://www.bls.gov/ooh/media-and-communication/interpreters-and-translators.htm#tab-6.
Lewis-Kraus, G. (2016, December 14). The great A.I. awakening. *New York Times Magazine*. https://www.nytimes.com/2016/12/14/magazine/the-great-ai-awakening.html.
Maguire, E. A., Gadian, D. G., Johnsrude, I. S., Good, C. D., Ashburner,J., Frackowiak, R. S., & Frith, C. D. (2000). Navigation-related structural change in the hippocampi of taxi drivers. *Proceedings of the National Academy of Sciences, 97*(8), 4398–4403.
McCarthy, J., Minksy, M. L., Rochester, N., & Shannon, C. E. (1955, August 31). A proposal for the Dartmouth Summer Research Project on Artificial Intelligence. Proposal to the Rockefeller Foundation. http://www-formal.stanford.edu/jmc/history/dartmouth/dartmouth.html.
Moorkens, J., Toral, A., Castilho, S., & Way, A. (2018). Translators' perceptions of literary post-editing using statistical and neural machine translation. *Translation Spaces, 7*(2), 240–262.
Newspapers fact sheet (2021, June 29). Pew Research Center. https://www.pewresearch.org/journalism/fact-sheet/newspapers/.
Olson, D. (1994). *The World on Paper: The Conceptual and Cognitive Implications of Writing and Reading*. Cambridge University Press.
Peritz, A. (2022, September 6). A.I. is making it easier than ever for students to cheat. *Slate*. https://slate.com/technology/2022/09/ai-students-writing-cheating-sudowrite.html.
Pielmeier, H., & O'Mara, P. (2020, January). The state of the linguistic supply chain: Translators and interpreters in 2020. *CSA Research*. https://cdn2.hubspot.net/hubfs/4041721/Newsletter/The%20State%20of%20the%20Linguist%20Supply%20Chain%202020.pdf.
Rezwana, J., & Maher, M. L. (2022). Designing creative AI partners with COFI: A framework for modeling interaction in human–AI co-creative systems. *ACM Transactions on Computer–Human Interaction*. https://dl.acm.org/doi/pdf/10.1145/3519026.

Roberts, S. (2017, February 14). Christopher Strachey's nineteen-fifties love machine. *New Yorker*. https://www.newyorker.com/tech/annals-of-technology/christopher-stracheys-nineteen-fifties-love-machine.
Rogers, R. (2023, March 1). How AI could transform email. *Wired*. https://www.wired.com/story/how-to-use-ai-for-email-artificial-intelligence-tips-advice/.
Roose, K. (2022, December 5). The brilliance and weirdness of ChatGPT. *New York Times*. https://www.nytimes.com/2022/12/05/technology/chatgpt-ai-twitter.html.
Roose, K. (2023, January 12). Don't ban ChatGPT in schools. Teach with it. *New York Times*. https://www.nytimes.com/2023/01/12/technology/chatgpt-schools-teachers.html.
Savoldi, B., Gaido,M., Bentivogli,L., Negri,M., & Turchi, M. (2021). Gender bias in machine translation. *Transactions of the Association for Computational Linguistics*, *9*, 845–874. https://arxiv.org/abs/2104.06001.
Schatten, J. (2022, September 14). Will artificial intelligence kill college writing? *The Review*. https://www.chronicle.com/article/will-artificial-intelligence-kill-college-writing.
Selinger, E. (2015, January 15). Will autocomplete make you too predictable? *BBC Future*. https://www.bbc.com/future/article/20150115-is-autocorrect-making-you-boring.
Shneiderman, B. (1987). *Designing the User Interface: Strategies for Effective Human–Computer Interaction*. Addison-Wesley.
Stokel-Walker, C. (2023, January 18). ChatGPT listed as author on research papers: Many scientists disapprove. *Nature*. https://www.nature.com/articles/d41586-023-00107-z.
Susskind, R. (2017). *Tomorrow's Lawyers* (2nd ed.). Oxford University Press.
The world's top translation companies by revenue (2022) (n. d.). *LanguageWire*. https://www.languagewire.com/en/blog/top-translation-companies.
Turing, A. (1948). Intelligent machinery. National Physical Laboratory. Typescript. https://www.npl.co.uk/getattachment/about-us/History/Famous-faces/Alan-Turing/80916595-Intelligent-Machinery.pdf?lang=en-GB. Also in B. J. Copeland (Ed.), (2004). *The Essential Turing* (pp. 410–432). Oxford University Press.
Tyson, L. D., & Zysman, J. (2022). Automation, AI & work. *Daedalus*, *151*(2), 256–271.
Valand, S. S. (2022, December 19). Is it now that the living writing dies? *Klassekampen*. Original Norwegian available at https://klassekampen.no/utgave/2022-12-19/debatt-er-det-na-den-levende-skriften-dor1/. English translation through Microsoft Translate.
Vara, V. (2021, August 9). Ghosts. *The Believer Magazine*. https://believermag.com/ghosts/.
Vaswani, A., Shazeer, N., Parmar, N., Uszkoreit, J., Jones, L., Gomez, A. N., Kaiser, L., & Polosukhinet, I. (2017). Attention is all you need. In *31st Conference on Neural Information Processing Systems (NIPS 2017)*. https://arxiv.org/abs/1706.03762.
Vincent, J. (2019, April 10). The first AI-generated textbook shows what robot writers are actually good at. *The Verge*. https://www.theverge.com/2019/4/10/18304558/ai-writing-academic-research-book-springer-nature-artificial-intelligence.
Weaver, W. (1949). Translation.Rockefeller Foundation. Memorandum. http://gunkelweb.com/coms493/texts/weaver_translation.pdf.
Wetsman, N. (2021, March 9). Google Translate isn't good enough for medical instructions. *The Verge*. https://www.theverge.com/2021/3/9/22319225/google-translate-medical-instructions-unreliable.
Wiggers, K. (2022, December 10). OpenAI's attempts to watermark AI text hit limits. *TechCrunch*. https://techcrunch.com/2022/12/10/openais-attempts-to-watermark-ai-text-hit-limits/.
Wilk, E. (2021, March 28). What AI can teach us about the myth of human genius. *The Atlantic*. https://www.theatlantic.com/culture/archive/2021/03/pharmako-ai-possibilities-machine-creativity/618435/.
Wilson, C. (2008, November 6). Introducing President 'Barracks Boatman'—Updated. *Slate*. https://slate.com/news-and-politics/2008/11/introducing-president-barracks-boatman-updated.html.
Zafrón, C. R. (n. d.). Why I write. https://www.carlosruizzafon.co.uk/landing-page/carlos-ruiz-zafon/carlos-ruiz-zafon-why-i-write/.

Law

Caja Thimm and Laura Thimm-Braun

Policies, Regulation, and Legal Perspectives on Social Robots

Abstract: The chapter gives an overview of general principles for policy and lawmaking and introduces measures that have been put into place in selected countries. Taking the example of Europe, policy measures like the AI Act and other preliminary regulation measures like the 'robot law' are introduced. The chapter looks at social robots from a contextual point of view, with the main argument being that all social situations with social robots call for some basic policies and principles, although situational parameters apply. Apart from this human user perspective, the role of the robot as a 'person' is introduced in order to follow discourses on data ownership, intellectual property rights, and machine ethics.

Technological advances like social robots need a global perspective, although lawmaking and policy are, for the most part, under national authority. Legal and regulatory action for social robots poses a particular challenge to international law, such as human rights, national regulation, and, for example, industrial use. It is argued that for further policy and regulatory actions, it is necessary to understand social robots beyond their physical embodiment and put regulations in place that protect human rights and control for data justice and data dignity.

Keywords: Law, international policy, governance, social robots, AI Act

8.1 Introduction

Robots are not only technological innovations that aim to simplify life and make it more efficient—they also evoke very mixed reactions and attitudes. Initially, for example, Europeans in particular showed very positive attitudes towards robots. According to an EU-barometer survey of 2012 (European Commission, 2012), more than two-thirds of EU citizens (70%) had a positive view of robots; the majority agreed that robots "are necessary as they can do jobs that are too hard or too dangerous for people" (88%) and that "they are a good thing for society because they help people" (76%). The survey found that people who have some personal experience of robots are more likely to have a positive view (88%) than the wide majority who lack this experience (68%). EU citizens also had clear views about the areas where robots should operate: they should work in areas too difficult or too dangerous for humans, like space exploration (52% consider this a priority), manufacturing (50%), military and security uses (41%), and search and rescue tasks (41%) (see Taipale et al., 2015). More recently, however, attitudes not only in Europe have changed. People have become more skeptical

https://doi.org/10.1515/9783110792270-009

and less trusting towards this technology (Naneva et al., 2020). For the US, studies show a fear of the loss of jobs (Smith & Anderson, 2017) and concerns about losing control over data and personal autonomy (Anderson & Rainie, 2023). Other studies bring up worries about data ownership, privacy, and the non-democratic use of information gathered by technologically advanced social robots (Gnambs & Appel, 2019). These changes in attitudes not only concern ordinary users, but also include specialists and, in particular, policymakers (Stanford University HAI Research, 2023). The more social robots are becoming 'social' and the more they are equipped with AI-based technologies, the more, so it seems, they evoke feelings of threat and risk.

As robots are now increasingly used in a multitude of ways related to social interaction with humans, from teaching children (Van Ewijk et al., 2020) to protecting family homes or distributing medication in hospitals (Henschel et al., 2021), assisting in restaurants (Pieska et al., 2013), or serving as personal companions (Stroessner, 2020), ethical and legal concerns have become more pronounced (Subramanian, 2017). These concerns are based on the technological development itself (Maroto-Gómez et al., 2023), and concentrate on those social situations that are particularly sensitive to human–machine interactions, such as care contexts (Sætra, 2020). But not just specific social situations or the equipping of the robots with, for example, sensor technologies, feed into such more negative assessments; it is the combination of advanced artificial intelligence (AI), data collection, and data usage that worries more and more users and, consequently, lawmakers as well (Leenes et al., 2017). Neither should economic perspectives be forgotten, as they play heavily into policymaking. Social robots represent an important market; the European Union alone, in collaboration with industrial partners, is investing EUR 2.6 billion in ten years of research (2021–2030) in AI, data, and robotics. Future-of-work concerns seem well founded when regarding these major investments (Stanford University HAI Research, 2023). But it is not just the future of jobs that worries the public, but also the possibilities for the control of public space (Oruma et al., 2022; Fortunati et al., 2020), a decline in social well-being, and the loss of autonomy, such as non-consensual control from corporations and governments (Formosa, 2021). From both perspectives, the personal-human and the economic side of social robots, it can be noted that the identification of ethical issues for robotics and AI, and the debate on ethical frameworks, were the first steps to creating a regulatory environment for these technologies (Langman et al., 2021).

These preliminary remarks not only point to a need for a wider public debate on the matter but also demonstrate the demand for political and legal action. As technology does not stop at national borders, this need for regulation is a global one, although lawmaking operates for most cases under national authority. It can hence be claimed that legal and regulatory perspectives for social robots pose a particular challenge for international law, such as human rights, but also for national regulation, for example industrial use. We will highlight these perspectives in the following section, with our main argument being that all social situations with social robots call for some basic policies and principles. As regulation policies and legal frameworks refer to the role, functionalities, and type of use in relation to humans, it is necessary to define and differentiate between types of social robots.

8.2 Social Robots in Society—Agency, Embodiment, Relationships

The deciding issue for policies and regulation from a legal perspective is to define and differentiate social robots along categories like agency, tasks, relationships, and types of embodiment (Edwards, 2018). Taking into account the many definitions and descriptions that exist to define social robots as a class of technology, it becomes clear that, due to the variety and complexity of tasks and behaviors that robots with social intentions can already and will furthermore perform, there is no consensus on how to define social robots along their various types (Dautenhahn, 2007; Hegel et al., 2009). An important aspect was put forward by Darling (2016, p. 4), who defined a social robot as a "physically embodied, autonomous agent that communicates and interacts with humans on an emotional level."The noteworthy description as 'emotional' points to assumed psycho-social susceptibility, which equates the robot with another human.

In general, social robots are defined as robots that have the ability to perceive, reason about, and interact with humans in a social context. They are human-made technologies that can take physical or digital forms. This can include activities like holding conversations, providing information, recognizing emotions, and performing tasks. A robot is not a social robot *per se*, and it needs specific communicative capabilities. First, it implies the robot to behave (function) socially within a context and, second, it implies the robot to have an appearance (form) that explicitly expresses it to be social in a specific respect to any user. From this point of view, a social robot can be regarded as a social interface (McGinn, 2020) that can resemble humans. Such humanoid social robots represent a specific category, as they evoke anthropomorphism and influence source credibility (Finkel & Krämer, 2022). A social interface encloses all the designed features by which a user judges the robot as having social qualities. Fox and Gambino (2021, p. 294) underscore that "a common assumption is that social robots can and should mimic humans, such that human–robot interaction (HRI) closely resembles human (i.e., interpersonal) interaction." Consequently, it is argued that they should be responsible for their actions (Bigman et al., 2019).

Regarding the increasing integration of more and more advanced AI systems, it is necessary to widen the scope of defining robots beyond their physical appearance. Bendel, for example, argues that in addition to hardware robots, certain software robots must be included in the typology, such as (socially designed) chatbots or voice bots (voice assistants) and social bots (Bendel, 2021, p. 150). As social robots undergo rapid changes in their physical as well as their non-physical appearance, human–robot relations change, and new legal challenges come up (Tonkin et al., 2017).

If embodied, robots are typically equipped with sensors, cameras, actuators, and other technologies that allow them to perceive and respond to their environment, as well as natural language processing and other forms of artificial intelligence that allow them to understand and respond to human language and gestures. Social robots are often used in settings such as homes, hospitals, and businesses, where they can assist with tasks and provide companionship. Situational usage of social robots includes, for example, personal assistant robots, educational robots, and therapeutic robots. Henschel et al. (2021, p. 11) conclude, "the largest clusters in social robotics research can be summarized as 'robots as social partners' and 'human factors and ergonomics in human–robot interaction.'" To be able to naturally interact with humans in this role of a social partner, more and more social robots collect large amounts of data, both about users themselves but also about their environment. To enable this activity, the equipment of robots has become more elaborated and refined. Adaptive technologies and sensors are part of the development of adaptive behavioral models, the design of cognitive architectures, and the ability to establish empathy with the user (Nocentini et al., 2019). In addition, current social robots are wireless and connected to computers or the internet, and multi-robot interactions have become more frequent. And particularly when resembling humans, social robots are attributed with human-like qualities.

When looking at policies and legal frameworks for social robots, the described contextual and functional differences on the level of agency have to be considered in order to assess the diverse abilities and functionalities of the robots.

8.3 General Frameworks for the Governance of Social Robots

As social robots are becoming more integrated into our society and daily life, the need to establish governance frameworks and regulations to ensure that they are used ethically and responsibly has become an internationally acknowledged challenge (Fosch-Villaronga et al., 2020). The field of 'machine ethics' has become a wide and productive field in ethics and has influenced policymaking decisively (overview in Tolmeijer et al., 2020). Some of these ethical reflections are currently the basis for political decision making in national and supranational parliaments, such as the 'Ethics Committee Recommendations on AI/Deutsche Ethikkommission' in Germany (German Ethics Council, 2023) or the Directorate-General for Research and Innovation & European Group on Ethics in Science and New Technologies (2018) in its *Statement on Artificial Intelligence, Robotics and 'Autonomous Systems.'*Many general principles are based on the famous 'robot laws' by Isaac Asimov and work with Asimov's three laws as a basic rule for robots in society (for a discussion, see Clarke, 2011). Pasquale (2020) added four new principles to Asimov's original three in order to frame policy concepts:

1. Robotic systems and AI should complement professionals, not replace them.
2. Robotic systems and AI should not counterfeit humanity.
3. Robotic systems and AI should not intensify zero-sum arms races.
4. Robotic systems and AI must always indicate the identity of their creator(s), controller(s), and owner(s).

Pasquale points out that these general principles can be broken down into selected social situations, which will, however, call for specific policies and legal solutions. These policies need to be based on existing principles and legal frameworks.

8.3.1 Types of social robots, governance, and regulation

Different social robots present different types and levels of risk to users and society, depending on their tasks and autonomy. In 2017, the European Parliament adopted a Resolution on *Civil Law Rules in Robotics* (2015/2103(INL)). The resolution underlines the importance of establishing a common definition of intelligent autonomous robots, taking note of features such as:
- the ability to learn, thanks to experience and interaction, and the ability to acquire autonomy through the exchange of data with the environment and its analysis;
- the physical support of the robot;
- the ability to adapt behavior and actions to the surrounding environment.

The report suggests that these laws bind developers and users of robots and apply to robots themselves. It discusses various perspectives regarding robots, which lawmakers will need to address. Laws already exist that specify how, where, and when people can use machines. But these laws do not deal with machines with 'brains,' and as robots can make decisions, react to situations, and move on their own, new regulations are needed. Policymakers have, with reference to the aforementioned robot laws of the EU in 2017, particularly looked at specific issues like the increase of robot vs. human autonomy and the growing risk of dependence (Thimm & Bächle, 2018; Formosa, 2021). Additionally, some industries, such as healthcare and finance, have put specific regulations in place regarding the use of robots (Fosch-Villaronga, 2020). However, with the increasing complexity of functions and application contexts, new legal frameworks will have to be developed to address these issues.

An overriding concern in this respect is connected to governance in general. Some of the challenges in policymaking and law related to social robots include:
1. *Liability*. Determining who is responsible for any harm caused by a social robot, such as accidents or malfunctions, can be difficult.
2. *Privacy and data protection*. Social robots often collect and process personal data, and there are concerns about how this data is used, stored, and protected.

3. *Discrimination.* Social robots may perpetuate or reinforce biases and discrimination, and there may be legal questions about how to address this.
4. *Autonomy.* As social robots become more autonomous, there may be legal questions about their decision making, accountability, and moral status.
5. *Intellectual property.* Social robots may create or use protected works, such as images or software, and there may be legal questions about who owns and controls these works.
6. *Ethical considerations.* Social robots raise a number of ethical questions, such as how to ensure their actions align with human values and how to balance the benefits and risks of their use.

Like other AI-based technologies, social robots can be classified and thus governed on different levels depending on the likelihood and severity of risk or situational factors. Considerations can include conducting human rights impact assessments, implementing transparency and accountability measures, and ensuring that there are appropriate protections in place for individuals whose rights may be affected. It does, for example, make a decisive difference where the activities take place. As Fortunati (2018, p.2674) points out, one of the most noteworthy changes is "the pivotal role of the machines from factories and services into the domestic sphere" (see also Oruma et al., 2022). This goes hand in hand with a closer relationship of humans with these machines in private contexts like the home. Consequently, contextualized approaches to assess the opportunities and risks have been called for (Heuer et al., 2019). These contextualized assessments of social robots are based on agency, location, and control in the relations between the robot and humans (Boch et al., 2021). Some examples include:

- *General healthcare.* Social robots can be used to assist patients with physical, cognitive, and emotional deficits or impairments, providing companionship and helping with tasks such as medication reminders and exercise prompts.
- *Education.* Social robots can be used to teach children and assist with language learning and other educational activities.
- *Retail and customer service.* Social robots can be used to greet customers, provide product information, and assist with checkouts and other tasks in retail settings, airports, or administrative settings.
- *Eldercare.* Social robots can be used to provide companionship and assistance to older adults living in assisted living facilities or at home.
- *Intimacy.* Sex robots as simulations of or as substitutions for human companionship (or, to make a parallel with the others: social robots can be used to simulate or substitute for human companionship and sex).
- *Hospitals and long-term care facilities.* Social robots can be used to provide patients with information and assistance.
- *Entertainment and leisure.* Social robots can be used to entertain people in public places like museums, aquariums, and theme parks.

- *Personal assistance.* Social robots can also be used in homes to provide assistance with tasks such as cleaning, cooking, and monitoring the safety of occupants.

But other contexts emerge, and many institutions will be affected by robot development. For example, AI systems can be used to automate decision-making processes in the criminal justice system, such as in the form of 'robot judges' (Zekos, 2022), which could lead to bias and discrimination if the systems are not properly designed and tested. Depending on the AI technology embedded in social robots, other justice-related issues may emerge, such as for law firms or non-governmental organizations. For instance, the use of face recognition algorithms which enable detection and surveillance on a larger scale has been an increasing concern.

There are also concerns about the use of AI in warfare, as it could potentially lead to more indiscriminate attacks and greater civilian casualties. A particular case of social robots is drones. Drones, also known as unmanned aerial vehicles, are aircraft that can be operated without a human pilot on board. They are typically controlled remotely or through pre-programmed flight plans. Drones can be used for a variety of tasks, such as aerial photography, delivery, surveillance, and search and rescue, but they also function as weapons of deadly force (Amoroso, 2020). This 'dual use' perspective is one of the complex challenges in the regulation of robots, notwithstanding their social functions, as discussed in the case of 'drone laws.' Which moral principles social robots in a war situation should follow has not been addressed widely. Some general frameworks, however, are under discussion and have, in some instances, already been put into legal measures (for an international comparison, see Lee et al., 2022).

8.3.2 Policy frameworks: international perspectives

In approaches to the law and regulation of AI and social robots, two dominant perspectives have been addressed so far. The first is legalistic and is based on the respective legal system. It proceeds by drawing lists of legal fields or issues affected by AI and robots, such as liability, privacy, cyber security, and so on. The second concerns practical applications in diverse settings, such as driverless cars, social robots, exoskeletons, and so on (Palmerini et al., 2016).

One of the important reflections from a legal point of view concerns the linkage between social robots and AI. As more and more social robots will be based on AI, this perspective overrides many other legal concerns. As Langman et al. (2021) point out, there are first steps to creating a regulatory environment for these technologies, particularly in Europe and the US.

For Europe, the Directorate-General for Research and Innovation & European Group on Ethics in Science and New Technologies (2018) listed criteria that align with the current EU Treaties and the EU Charter of Fundamental Rights. These principles are summarized below and can be adapted to social robots, even if they are not (yet) entirely autonomous:

- *Human dignity*. Autonomous technologies must not violate the inherent human right to be respected.
- *Autonomy*. Humans are free to live by their own standards, and humans are responsible for exerting control over autonomous technologies. Autonomous technologies must not impair human freedom, responsibility, or control.
- *Responsibility*. The development and use of autonomous technologies must benefit society and the environment on a global scale. Such benefits must be defined by democratic means.
- *Justice, equity, and solidarity*. Regulators and practitioners must prevent or neutralize discriminatory datasets from training AI systems, working in the interest of global justice and equality. All humans should benefit from autonomous technologies.
- *Democracy*. The regulation of autonomous technologies must result from democratic, public debate and engagement.
- *Rule of law and accountability*. Regulation of autonomous technologies must uphold all human rights standards, such as protections for safety and privacy. These protections rely on the rule of law, access to justice, the right to redress, and the right to a fair trial.
- *Security, safety, and bodily and mental integrity*. Safe autonomous systems promote external, internal, and emotional safety. External safety protects environments and users. Internal safety ensures consistent performance and protects against hacking. Emotional safety protects users from exploitation and abuse when interacting with autonomous machines.
- *Data protection and privacy*. Digital communication technologies employ autonomous technologies to amass and store vast quantities of users' personal data. Therefore, autonomous technologies challenge protections for personal information and privacy.
- *Sustainability*. Autonomous technologies must align with our human responsibility to protect our planet's ability to support life, preserve the continued quality of the environment, and maintain the prosperity of our species.

As robots can collect data from the users as well as from their social environment, there are data-related challenges. It must be determined how data will be collected and processed, to what extent this data should be stored or uploaded to the cloud, how to inform and get the users' consent to do so, and how to prevent unauthorized external actors from accessing personal detailed information. While the EU's General Data Protection Regulation (GDPR), for instance, offers a few layers of protection, which could be applied to the above principles, the GDPR also has its limits (European Parliament & Council of the European Union, 2016). Other initiatives like the SIENNA Project, an EU initiative aimed at understanding ethical and human rights challenges posed by new technologies, published a report on EU legislation about key legal issues in robotics, such as safety, liability, privacy, and equity. The report highlights the different regulatory approaches inside and outside of Europe (Rodrigues et al., 2019).

In the US, there is currently no specific legal framework for the control and regulation of social robots. Some strategies have been published, such as a plan by the National Science and Technology Council, which would support effective AI innovation and ethical implementation, and the US Congress has also passed the National Artificial Intelligence Initiative Act of 2020. It is worth noting that there is no specific federal law that regulates the use or development of social robots in the United States, but several federal agencies have issued guidance or regulations that may apply to social robots, such as regulations related to consumer protection, privacy, and intellectual property. Other agencies involved are the Federal Trade Commission, which has issued guidance for manufacturers and developers of social robots, emphasizing the need for transparency, security, and consumer control in the collection and use of personal data, and the Food and Drug Administration, which has suggested the regulation of certain types of social robots, such as those used in healthcare settings or for medical purposes. The majority of specific ethical principles and frameworks have been released by individual departments, such as the Department of Defense (US Department of Defense, 2020), the Department of Transportation, and the Department of Health and Human Services with a list of five qualities of ethical AI, which are aimed to supplement existing US military ethical frameworks (https://www.ai.gov) (for details see Langman et al., 2021).

In Germany, the regulation of social robots is primarily handled by the Federal Ministry for Economic Affairs and Climate Action (BMWK), the Federal Network Agency, and the Federal Office for Information Security. The BMWK is responsible for the development and promotion of the German robotics industry, including social robots. It has established a national robotics strategy and a robotics innovation center to promote research and development in the field of robotics. This is, like many policies on a national level, a framework for private–public partnerships and publicly funded joint projects. The German Robotics Innovation Center designs and realizes intelligent robots for a variety of applications such as underwater; space; search and rescue and security robotics; logistics, production, and consumer robotics; cognitive robotics; e-mobility; and rehabilitation robotics. The focus lies on the rapid transfer of results of basic research into real-world applications (https://www.dfki.de). Another important state agency is the Federal Office for Information Security (https://www.bsi.bund.de), which is responsible for cybersecurity, data safety, and regulation of robots, particularly for industrial use. It has issued guidelines for the secure development and use of social robots. This includes recommendations for the protection of personal data and the secure handling of sensitive information.

On the level of supranational organizations, it is the Organization for Economic Co-operation and Development (OECD) that proposed a policy paper. In 2019, a paper entitled *Artificial Intelligence in Society* highlighted the potential benefits and risks of

AI and the need for international cooperation to maximize the benefits and minimize the risks (OECD, 2019). The paper states that AI has the potential to improve productivity and economic growth but also poses risks to privacy, security, and human rights. It also emphasizes the importance of trust in AI systems and the need for transparency and accountability in their development and deployment. Overall, the paper emphasizes the importance of international cooperation in addressing the challenges and opportunities posed by AI to ensure that its benefits are widely shared and its risks are minimized. These Principles have been endorsed by, among others, the US and six non-member states (Argentina, Brazil, Colombia, Costa Rica, Peru, and Romania). In June 2019, ministers from the Group of 20 major economies agreed on a set of guiding principles for using AI, which is derived from the OECD Principles but is also characterized as non-binding, and in 2023 the BRICS states started a discussion on tightening regulation on AI regulation (Cyman et al., 2021).

Although many open questions remain (see Subramanian, 2017, p. 81), there is widespread agreement on some of the main issues (Fosch-Villaronga et al., 2020). These issues concern security and privacy, legal uncertainty, autonomy and agency, employment, and the possible replacement of human interactions by social robots. In light of these regulatory challenges the proposed AI Act of the EU Parliament takes a big step forward towards a more comprehensive and global approach in public policy, and will be explained in more detail below.

8.3.3 AI Act and new robot laws

While many countries are hesitant to draw up legal frameworks, the European Union proposed the first-ever legal framework on AI, which addresses the risks of AI and positions Europe to play a leading role globally. The regulatory proposal aims to provide AI developers, deployers, and users with clear requirements and obligations regarding specific uses of AI and, hence, social robots as well. It focuses primarily on strengthening rules around data quality, transparency, human oversight, and accountability for AI, but also for embodiments and applications like social robots (Księżak & Wojtczak, 2020). It also aims to address ethical questions and implementation challenges in various sectors ranging from healthcare and education to finance and energy (Charisi et al., 2021).

In the spring of 2023, the European Parliament agreed on a position paper, which was published in 2020 as a first draft (European Commission, 2020), as a comprehensive approach to the development and deployment of AI in Europe, underlining its benefits and technological advances in different areas, including medicine, security, and farming, as well as identifying its potential risks: opaque decision making, gender inequality, discrimination, and lack of privacy, just to name a few. Accompanied by input from other EU Agencies like the European Union Agency for Fundamental Rights (2020), a wide set of rules and regulations was suggested. The core element of the legal framework is a risk-based approach that includes various requirements and prohibitions based on potential capabilities and hazards. These regulations focus on the development, placement on the market, and use of AI systems. The regulation will, if adopted in 2024, apply throughout the EU. In addition, however, there is also likely to be the so-called 'Brussels effect.' This refers to the effect that other markets and regions will have to respond to the requirements of the European Union if they want to continue to offer corresponding services in the EU. This often involves de-facto convergence to the legal standards of the EU (see GDPR). In more detail, the paper lays out several key principles for the development and deployment of AI in Europe, including the need for human oversight, transparency, and explainability as well as the importance of data protection and privacy.

The law assigns applications of AI to three risk categories: (1) applications and systems that create an unacceptable risk, such as government-run social scoring of the type used in China, are banned; (2) high-risk applications, such as a CV-scanning tool that ranks job applicants, are subject to specific legal requirements; and (3) 'limited risk' refers to AI systems with specific transparency obligations. When using AI systems such as chatbots, users should be aware that they are interacting with a machine so they can make an informed decision to continue or step back, whereas (4) 'minimal' or 'no risk' includes applications such as AI-enabled video games or spam filters.

In his critical reflection on the proposed law, Hacker (2021) draws attention, among other points, to data quality and data ownership. To mitigate data power risks the future law will demand that the data behind such AI systems should be "sufficiently relevant, representative, error-free, and complete." But the word "representative," argues Hacker, raises the question of what exactly an AI system's data is supposed to represent. After all, AI systems often reflect the very dark sides of a sexist and racist structured reality and threaten to further entrench it. With new generative AI systems like ChatGPT, the AI Act will have to be adapted accordingly and will have to find ways to regulate these powerful tools (Helberger & Diakopoulos, 2023).

Despite many other critical voices on the EU AI Act, the hope is that, like the EU's General Data Protection Regulation in 2018, it could become a global standard, determining to what extent AI and various forms of embodiment such as social robots have a positive rather than negative effect on people's lives and society as a whole.

8.3.4 Robots as partners? Legal rights for machines

The policy approaches discussed above looked at regulation from a societal and human perspective. But the legal status of the robot itself has not gone unquestioned either. Although a separate legal status for robots seems unlikely, the introduction of a so-called 'electronic person' has been discussed already. A resolution on the development of a civil law framework for robots, which was formulated as a recommendation to the European Commission, reads like an instruction for action on this path (European Parliament, 2017, para. 59):

> Creating a specific legal status for robots in the long run, so that at least the most sophisticated autonomous robots could be established as having the status of electronic persons responsible for making good any damage they may cause, and possibly applying electronic personality to cases where robots make autonomous decisions or otherwise interact with third parties independently.

This view of robots as persons poses a question for both ethics and lawmakers about the fundamental status assignment of machines—if they were assigned comprehensive rights and duties, they would be placed on an equal legal level with humans. In their study, De Graaf et al. (2022) found that people are more willing to grant basic rights such as access to energy and the right to update to robots than socio-political rights such as voting rights and the right to own property. Attitudes towards granting rights to robots depend on the cognitive and affective capacities people believe robots possess or will possess in the future. Along the same line, Bertolini and Episcopo (2022) looked at the issue of robots as legal subjects, and Mamak (2021) questioned the ethical and legal limits of protections for robots when asking whether to save a robot or a human.

It is not surprising that there is massive criticism of the EU position paper measure on legal rights for robots and that major disagreement exists between those who find that question important and those who do not.

8.4 Conclusions and Outlook

We have highlighted some of the main points of the debate on policies and regulations on social robots, but it must be kept in mind that with the rapid technological advances we witness, the political and legal debate will evolve accordingly. The European Parliament's AI Act in particular could have a major impact on national policies and lead to more detailed measures on national levels.

As already explained, machines are no longer present only as visible and physically tangible objects—we also encounter them in the form of invisible program codes, algorithms and artificial intelligence. We want to argue that the differentiation between embodied, physical robots and robots as (software) machines overlooks an important development and an influential role of robots for society. In this respect, general principles need to be agreed on, which can then be developed in national and international law.

For many legal measures, ethical considerations have begun to play an important role and can lead to practical frameworks. As pointed out by Floridi et al. (2018) in their proposal for "AI4People," fundamental principles should be the base of political decision making. They formulate concrete recommendations to assess, develop, incentivize, and support regulation that may be "undertaken directly by national or supranational policy makers" (p. 689). They suggest the adoption of the premises laid down by Asimov and call them "beneficence" and "non-maleficence." Whereas the first is characterized as promoting well-being, preserving dignity, and sustaining the planet, the second is concerned with the prevention of infringements on personal privacy, which is listed as part of the human rights principles. Their suggestions are based on an ethical foundation and could be a starting point for future political strategies.

Many issues concern the possible impact of implementing social robotics in diverse sectors of society. Perhaps one of the most important concerns is 'data dignity.' Data dignity not only refers to the idea that people take ownership of their data and that training data for AI and social roots be fair and non-discriminatory, but also to the aim of preventing concentrated data power. And as social robots will sooner or later turn into AI-based, data-driven helpers, partners, or competitors, data issues will remain a core issue for a democratic and just usage of social robots.

But there are new concerns about the role of AI for society after the introduction of generative AI systems. Large generative AI models, such as ChatGPT, GPT-4, or Stable Diffusion, and others, are rapidly transforming the way we communicate, illustrate, and create. Challenges include harmful and biased answers and hallucinated outputs in which the AI invents content, which can then perpetuate harmful stereotypes or misinformation. There are also possible impacts on decision making, as AI systems are increasingly used to inform critical decisions in fields such as finance, healthcare, and law. The legal implications of these developments are profound and will also profoundly influence the future of the relationship between humans and robots. With this type of AI, regulation not only concerns more traditional regulatory aspects but will have to include the role of the robot as a social partner and perhaps even friend to humans, with many more implications than we know of today (Thimm & Bächle, 2019).

References

Amoroso, D. (2020). *Autonomous Weapons Systems and International Law: A Study on Human–Machine Interactions in Ethically and Legally Sensitive Domains*. Edizioni Scientifiche Italiane/Nomos. https://doi.org/10.5771/9783748909538.

Anderson, J., & Rainie, L. (2023). The future of human agency. *Pew Research Center*. https://www.pewresearch.org/internet/wp-content/uploads/sites/9/2023/02/PI_2023.02.24_The-Future-of-Human-Agency_FINAL.pdf.

Bendel, O. (Ed.) (2021). *Soziale Roboter: Technikwissenschaftliche, wirtschaftswissenschaftliche, philosophische, psychologische und soziologische Grundlagen*. Springer Gabler. https://doi.org/10.1007/978-3-658-31114-8.

Bertolini, A., & Episcopo, F. (2022). Robots and AI as legal subjects? Disentangling the ontological and functional perspective. *Frontiers in Robotics and AI*, *9*, 1–15. https://doi.org/10.3389/frobt.2022.842213.

Bigman, Y. E., Waytz, A., Alterovitz, R., & Gray, K. (2019). Holding robots responsible: The elements of machine morality. *Trends in Cognitive Sciences*, *23*(5), 365–368. https://doi.org/10.1016/j.tics.2019.02.008.

Boch, A., Lucaj, L., & Corrigan, C. (2021). *A Robotic New Hope: Opportunities, Challenges, and Ethical Considerations of Social Robots. Research Brief*. Technical University of Munich. https://ieai.mcts.tum.de/wp-content/uploads/2021/05/ResearchBrief_April2021_SocialRobots_FinalV2.pdf.

Charisi, V., Šabanović, S., Cangelosi, A., & Gomez, R. (2021). Designing and developing better robots for children: A fundamental human rights perspective. In C. Bethel, A. Paiva, E. Broadbent, D. Feil-Seifer, & D. Szafir (Eds.), *Companion of the 2021 ACM/IEEE International Conference on Human–Robot Interaction (HRI '21 Companion)* (pp. 712–714). Association for Computing Machinery. https://doi.org/10.1145/3434074.3444875.

Clarke, R. (2011). Asimov's laws of robotics. In M. Anderson & S. L. Anderson (Eds.), *Machine Ethics* (pp. 254–284). Cambridge University Press. https://doi.org/10.1017/CBO9780511978036.020.

Cyman, D., Gromova, E., & Juchnevicius, E. (2021). Regulation of artificial intelligence in BRICS and the European Union. BRICS *Law Journal*, *8*(1), 86–115. https://doi.org/10.21684/2412-2343-2021-8-1-86-115.

Darling, K. (2016). Extending legal protection to social robots: The effects of anthropomorphism, empathy, and violent behavior towards robotic objects. In R. Calo, A. M. Froomkin, & I. Ker (Eds.), *Robot Law* (pp. 213 –232). Edward Elgar. https://doi.org/10.4337/9781783476732.00017.

Dautenhahn, K. (2007). Socially intelligent robots: Dimensions of human–robot interaction. *Philosophical Transactions of the Royal Society B*, *362*, 679–704. https://doi.org/10.1098/rstb.2006.2004.

De Graaf, M. M. A., Hindriks, F. A., & Hindriks, K. V. (2022). Who wants to grant robots rights? *Frontiers in Robotics and AI*, *8*, 781985, 1–13. In C. Bethel, A. Paiva, E. Broadbent, D. Feil-Seifer, & D. Szafir (Eds.), *Companion of the 2021 ACM/IEEE International Conference on Human–Robot Interaction (HRI '21 Companion)* (pp. 38–46). Association for Computing Machinery. https://doi.org/10.1145/3434074.3446911.

Directorate-General for Research and Innovation & European Group on Ethics in Science and New Technologies (2018). *Statement on Artificial Intelligence, Robotics and 'Autonomous Systems'*. Publications Office of the EU. https://doi.org/10.2777/531856.

Edwards, A. (2018). Animals, humans, and machines: Interactive implications of ontological classification. In A. L. Guzman (Ed.), *Human–Machine Communication: Rethinking Communication, Technology, and Ourselves* (pp. 29–49). Peter Lang. https://doi.org/10.3726/b14399.

European Commission (2012). *Public Attitudes Towards Roots: Special Eurobarometer 382*. European Commission. https://europa.eu/eurobarometer/api/deliverable/download/file?deliverableId=40114.

European Commission (2020). *White Paper on Artificial Intelligence: A European approach to Excellence and Trus*t. European Commission. https://commission.europa.eu/document/download/d2ec4039-c5be-423a-81ef-b9e44e79825b_en?filename=commission-white-paper-artificial-intelligence-feb2020_en.pdf.

European Parliament (2017). *Civil Law Rules on Robotics*. European Parliament. https://www.europarl.europa.eu/doceo/document/TA-8-2017-0051_EN.pdf.

European Parliament & Council of the European Union (2016). General Data Protection Regulation. *Official Journal of the European Union*, *L 119*, 1–88. https://eur-lex.europa.eu/legal-content/EN/TXT/PDF/?uri=CELEX:32016R0679.

European Union Agency for Fundamental Rights (2020). *Getting the future Right: Artificial Intelligence and Fundamental Rights. Report*. Publications Office of the European Union. https://fra.europa.eu/sites/default/files/fra_uploads/fra-2020-artificial-intelligence_en.pdf.

Finkel, M., & Krämer, N. C. (2022). Humanoid robots—Artificial. Human-like. Credible? Empirical comparisons of source credibility attributions between humans, humanoid robots, and non-human-like devices. *International Journal of Social Robotics*, *14*, 1397–1411. https://doi.org/10.1007/s12369-022-00879-w.

Floridi, L., Cowls, J., Beltrametti, M., Chatila, R., Chazerand, P., Dignum, V., Luetge, C., Madelin, R., Pagallo, U., Rossi, F., Schafer, B., Valcke, P., & Vayena, E. (2018). AI4People—An ethical framework for a good AI society: Opportunities, risks, principles, and recommendations. *Minds and Machines: Journal for Artificial Intelligence, Philosophy and Cognitive Science, 28*, 689–707. https://doi.org/10.1007/s11023-018-9482-5.

Formosa, P. (2021). Robot autonomy vs. human autonomy: Social robots, Artificial Intelligence (AI), and the nature of autonomy. *Minds and Machines: Journal for Artificial Intelligence, Philosophy and Cognitive Science, 31*, 595–616. https://doi.org/10.1007/s11023-021-09579-2.

Fortunati, L. (2018). Robotization and the domestic sphere. *New Media & Society, 20*(8), 2673–2690. https://doi.org/10.1177/1461444817729366.

Fortunati, L., Cavallo, F., & Sarrica, M. (2020). Multiple communication roles in human–robot interactions in public space. *International Journal of Social Robotics, 12*, 931–944. https://doi.org/10.1007/s12369-018-0509-0.

Fosch-Villaronga, E. (Ed.) (2020). *Robots, Healthcare, and the Law: Regulating Automation in Personal Care* Routledge. https://doi.org/10.4324/9780429021930.

Fosch-Villaronga, E., Lutz, C., & Tamò-Larrieux, A. (2020). Gathering expert opinions for social robots' ethical, legal, and societal concerns: Findings from four international workshops. *International Journal of Social Robotics, 12*, 441–458. https://doi.org/10.1007/s12369-019-00605-z.

Fox, J., & Gambino, A. (2021). Relationship development with humanoid social robots: Applying interpersonal theories to human–robot interaction. *Cyberpsychology, Behavior, and Social Networking, 24*(5), 294–299. https://doi.org/10.1089/cyber.2020.0181.

German Ethics Council (2023). *Humans and Machines—Challenges of Artificial Intelligence*. German Ethics Council. https://www.ethikrat.org/fileadmin/Publikationen/Stellungnahmen/deutsch/stellungnahme-mensch-und-maschine.pdf.

Gnambs, T., & Appel, M. (2019). Are robots becoming unpopular? Changes in attitudes towards autonomous robotic systems in Europe. *Computers in Human Behavior, 93*, 53–61. https://doi.org/10.1016/j.chb.2018.11.045.

Gunkel, D. J (2023). *Person, Thing, Robot: A Moral and Legal Ontology for the 21st Century and Beyond*. MIT Press.

Hegel, F., Muhl, C., Wrede, B., Hielscher-Fastabend, M., & Sagerer, G. (2009). Understanding social robots. In *2009 Second International Conferences on Advances in Computer–Human Interactions, ACHI 2009* (pp. 169–174). IEEE. https://doi.org/10.1109/achi.2009.51.

Helberger, N., & Diakopoulos, N. (2023). ChatGPT and the AI Act. *Internet Policy Review, 12*(1). https://doi.org/10.14763/2023.1.1682.

Henschel, A., Laban, G., & Cross, E. S. (2021). What makes a robot social? A review of social robots from science fiction to a home or hospital near you. *Current Robotics Reports, 2*, 9–19. https://doi.org/10.1007/s43154-020-00035-0.

Heuer, T., Schiering, I., & Gerndt, R. (2019). Privacy-centered design for social robots. *Interaction Studies, 20*(3), 509–529. https://doi.org/10.1075/is.18063.heu.

Księżak, P., & Wojtczak, S. (2020). AI versus robot: In search of a domain for the new European civil law. *Law, Innovation and Technology, 12*(2), 297–317. https://doi.org/10.1080/17579961.2020.1815404.

Langman, S., Capicotto, N., Maddahi, Y., & Zareinia, K. (2021). Roboethics principles and policies in Europe and North America. *SN Applied Sciences, 3*, 857. https://doi.org/10.1007/s42452-021-04853-5.

Lee, D., Hess, D., & Heldeweg, M. (2022). Safety and privacy regulations for unmanned aerial vehicles: A multiple comparative analysis. *Technology in Society, 71*, 102079. https://doi.org/10.1016/j.techsoc.2022.102079.

Leenes, R., Palmerini, E., Koops, B.-J., Bertolini, A., Salvini, P., & Lucivero, F. (2017). Regulatory challenges of robotics: Some guidelines for addressing legal and ethical issues. *Law, Innovation and Technology, 9*(1), 1–44. https://doi.org/10.1080/17579961.2017.1304921.

Mamak, K. (2021). Whether to save a robot or a human: On the ethical and legal limits of protections for robots. *Frontiers in Robotics and AI, 8,* 712427. https://doi.org/10.3389/frobt.2021.712427.

Maroto-Gómez, M., Alonso-Martín, F., Malfaz, M., Castro-González, Á., Castillo, J. C., & Salichs, M. Á. (2023). A systematic literature review of decision-making and control systems for autonomous and social robots. *International Journal of Social Robotics, 15,* 745–789. https://doi.org/10.1007/s12369-023-00977-3.

McGinn, C. (2020). Why do robots need a head? The role of social interfaces on service robots. *International Journal of Social Robotics, 12,* 281–295. https://doi.org/10.1007/s12369-019-00564-5.

Naneva, S., Sarda Gou, M., Webb, T. L., & Prescott, T. J. (2020). A systematic review of attitudes, anxiety, acceptance, and trust towards social robots. *International Journal of Social Robotics, 12,* 1179–1201. https://doi.org/10.1007/s12369-020-00659-4.

Nocentini, O., Fiorini, L., Acerbi, G., Sorrentino, A., Mancioppi, G., & Cavallo, P. (2019). A survey of behavioral models for social robots. *Robotics, 8*(3), 54. https://doi.org/10.3390/robotics8030054.

OECD (2019). *Artificial Intelligence in Society.* OECD Publishing. https://doi.org/10.1787/eedfee77-en.

Oruma, S. O., Sánchez-Gordón, M., Colomo-Palacios, R., Gkioulos, V., & Hansen, J. K. (2022). A systematic review on social robots in public spaces: Threat landscape and attack surface. *Computers, 11*(12), 181. https://doi.org/10.3390/computers11120181.

Palmerini, E., Bertolini, A., Battaglia, F., Koops, B.-J., Carnevale, A., & Salvini, P. (2016). RoboLaw: Towards a European framework for robotics regulation. *Robotics and Autonomous Systems, 86,* 78–85. https://doi.org/10.1016/j.robot.2016.08.026.

Pasquale, F. (2020). *New Laws of Robotics: Defending Human Expertise in the Age of AI.* Harvard University Press. https://doi.org/10.4159/9780674250062.

Pieska, S., Luimula, M., Jauhiainen, J., & Spiz, V. (2013). Social service robots in wellness and restaurant applications. *Journal of Communication and Computer, 10*(1), 116–123.

Rodrigues, R., Siemaszko, K., & Warso, Z. (2019). *D4.2: Analysis of the Legal and Human Rights Requirements for AI and Robotics In and Outside the EU.* SIENNA. https://doi.org/10.5281/zenodo.4066811.

Sætra, H. S. (2020). The foundations of a policy for the use of social robots in care. *Technology in Society, 63,* 101383. https://doi.org/10.1016/j.techsoc.2020.101383.

Smith, A., & Anderson, M. (2017). Automation in everyday life. *Pew Research Center.* https://www.pewresearch.org/internet/wp-content/uploads/sites/9/2017/10/PI_2017.10.04_Automation_FINAL.pdf.

Stanford University HAI Research (2023). *Artificial Intelligence Index Report 2023.* Stanford University. https://aiindex.stanford.edu/wp-content/uploads/2023/04/HAI_AI-Index-Report_2023.pdf.

Stroessner, S. J. (2020). On the social perception of robots: Measurement, moderation, and implications. In R. Pak, E. J. de Visser, & E. Rovira (Eds.), *Living with Robots: Emerging Issues on the Psychological and Social Implications of Robotics* (pp. 21–47). Elsevier Academic Press. https://doi.org/10.1016/B978-0-12-815367-3.00002-5.

Subramanian, R. (2017). Emergent AI, social robots and the law: Security, privacy and policy issues. *Journal of International, Technology and Information Management, 26*(3), 81–105. https://doi.org/10.58729/1941-6679.1327.

Taipale, S., de Luca, F., Sarrica, M., & Fortunati, L. (2015). Robot shift from industrial production to social reproduction. In J. Vincent, S. Taipale, B. Sapio, G. Lugano, & L. Fortunati (Eds.), *Social Robots from a Human Perspective* (pp.11–24). Springer. https://doi.org/10.1007/978-3-319-15672-9_2.

Thimm, C., & Bächle, T. C. (2018). Autonomie der Technologie und autonome Systeme als ethische Herausforderung. In M. Rath, F. Krotz, & M. Karmasin (Eds.), *Maschinenethik: Normative Grenzen autonomer Systeme* (pp. 73–87). Springer. https://doi.org/10.1007/978-3-658-21083-0_5.

Thimm, C., & Bächle, T. C. (Eds.) (2019). *Die Maschine: Freund oder Feind? Mensch und Technologie im digitalen Zeitalter.* Springer. https://doi.org/10.1007/978-3-658-22954-2.

Tolmeijer, S., Kneer, M., Sarasusa, C., Christen, M., & Bernstein, A. (2020). Implementations in machine ethics: A survey. *ACM Computing Surveys, 53*(6), 132. https://doi.org/10.1145/3419633.

Tonkin, M., Vitale, J., Ojha, S., Clark, J., Pfeiffer, S., Judge, W., Wang, X., & Williams, M.-A. (2017). Embodiment, privacy and social robots: May I remember you? In *Social Robotics: 9th International Conference, ICSR 2017, Tsukuba, Japan, November 22–24, 2017, Proceedings* (pp. 506–515). Springer. https://doi.org/10.1007/978-3-319-70022-9_50.

US Department of Defense (2020, February 24). DOD adopts ethical principles for artificial intelligence. *U.S. Department of Defense*. https://www.defense.gov/News/Releases/Release/Article/2091996/dod-adopts-ethical-principles-for-artificial-intelligence/.

Van Ewijk, G., Smakman, M., & Konijn, E. A. (2020). Teachers' perspectives on social robots in education: An exploratory case study. In E. Rubegni, A. Vasalou, N. Parés, & N. Sawhney (Eds.), *IDC '20: Proceedings of the Interaction Design and Children Conference* (pp. 273–280). Association for Computing Machinery. https://doi.org/10.1145/3392063.3394397.

Zekos, G. I. (2022). *Advanced Artificial Intelligence and Robo-justice*. Springer. https://doi.org/10.1007/978-3-030-98206-5.

Christoph Lutz, Aurelia Tamò-Larrieux, and Eduard Fosch-Villaronga

How Social Robots Affect Privacy: Navigating the Landscape

Abstract: Social robots are increasingly used in different domains, such as healthcare, education, and retail. Researchers across disciplines have shown how social robots differ from other technologies, for example in terms of ethical, legal, and social (ELS) aspects. Privacy is a particularly important and relevant ELS aspect of social robots, given their heightened autonomy, data-processing capabilities, and physical mobility. This chapter provides an overview of key privacy implications in relation to social robots. It reviews useful privacy theories and discusses recent studies on privacy and social robots, showing how such research has become more empirical over time but still prioritizes data protection and data-related aspects of privacy over other dimensions, such as physical, psychological, and social ones. The chapter also combines a social science and legal lens, showing how the law addresses relevant social and ethical implications, particularly in Europe. It concludes with a future research agenda on how to investigate the topic.

Keywords: Privacy, social robots, data protection, human–machine communication, legal implications

9.1 Introduction

In recent years, embodied and interactive technologies such as social robots, smart speakers, and smart toys have become increasingly prevalent. These technologies are used as communicative agents with whom people interact in complex ways and direct social relationships (Fong et al., 2003). In contrast to earlier, less sophisticated technologies such as the telephone or the escalator, individuals often develop elaborate relationships with these machines and attach stronger meanings, going beyond a mere functional understanding (Rabb et al., 2022). This tendency can be explained by the 'computers are social actors' paradigm, which Nass and his colleagues established in the 1990s (e.g., Nass et al. 1994; Reeves & Nass, 1996). In particular, more 'intelligent' and adaptive technologies are often anthropomorphized (i.e., treated similarly to humans) (Fussel et al., 2008) and can be used as companions for emotional support, for example in therapy and education (Fosch-Villaronga & Albo-Canals, 2019; Fosch-Villaronga et al., 2020). Human–machine

Acknowledgments: We would like to thank Natalie Benou, who put together the initial table, for her help with the literature search. Christoph Lutz was partly funded by the Research Council of Norway (grants 275347 *Future Ways of Working in the Digital Economy* and 299178 *Algorithmic Accountability: Designing Governance for Responsible Digital Transformations*) while writing this chapter.

https://doi.org/10.1515/9783110792270-010

communication (HMC), as a field within media and communication research interested in the “creation of meaning among humans and machines” (Guzman, 2018, p. 1), has enhanced this understanding of such technologies (Fortunati & Edwards, 2020; Gambino et al., 2020; Guzman, 2020; Rodríguez-Hidalgo, 2020). At the same time, the ethics of technology literature has sharpened our understanding of the ethical aspects of embodied and interactive technologies, especially of social robots (Fosch-Villaronga et al., 2020; Coeckelbergh, 2021; Sætra, 2020) but also of adjacent technologies such as self-driving cars (Himmelreich, 2018; Othman, 2021). However, these two bodies of literature—the fledgling area of HMC research and the more established but rapidly growing area of technology ethics—stand relatively separate and would benefit from integration (Guzman & Lewis, 2020). One topic where such an integration would be particularly fruitful, as it constitutes an essential subject of interest to both communities, is privacy.

Privacy is a fundamental right and an ethical and social challenge when it comes to embodied technologies. For example, privacy and security issues of smart speakers have received ample attention in the media, and social scientists have become increasingly interested in the topic as well (e.g., Brause & Blank, 2020; Liao et al., 2019; Lutz & Newlands, 2021; Maccario & Naldi, 2022), especially as users clearly perceive these devices as “intrusive, invasive, always listening . . . and even spying” (Fortunati et al., 2022b, p. 83). Other AI-based technologies, such as facial recognition applications, have resulted in concerns about dataveillance and infringements of citizens’ civic rights (Stark et al., 2020), particularly in the vein of the COVID-19 pandemic (Newlands et al., 2020; Vitak & Zimmer, 2020) during which privacy-invasive technology was employed in the name of public safety (e.g., control of mask wearing in public spaces through monitoring via mobile robots equipped with 360-degree cameras; Barrett, 2021). In the following, our contribution offers a comprehensive overview of the privacy implications of social robots. We approach the topic from a multidisciplinary perspective, combining a social science (especially rooted in communication and HMC), ethical, and legal lens. Doing so shows how approaching the topic from different angles is synergetic and yields interesting insights. As such, the chapter will hopefully provide value for different communities, from more technical scholars to sociologically and psychologically oriented researchers to legal experts.

The chapter is divided into four sections. After this introduction, we will briefly discuss privacy in general, offering a definition of how we understand it here. This section contains an overview of key privacy theories used in the context of emerging technologies (i.e., not only social robots) and discusses the synergies between social theories and legal requirements. In the third section, we will look at privacy and social robots more specifically. We will summarize relevant literature on the privacy implications of social robots and provide an overview of recent research on the topic. In the final section, we will conclude the chapter by synthesizing the main points and outlining a forward-looking research agenda, showing how both science and the law might want to approach the topic going forward.

9.2 Privacy: From Societal Values to Legal Protection

Privacy has become a central topic in both the technical and social sciences, given the development of digital technologies that rely on personal (and non-personal) data on a large scale, such as artificial intelligence (AI), robotics, social media, e-commerce, and mobile health. However, the interest in privacy as a concept goes further back (arguably since the beginning of civilization, Acquisti et al., 2022) and spans fields such as sociology, communication, psychology, law, information systems, and computer science (Altman, 1975; Warren & Brandeis, 1890; Westin, 1967). Nowadays, privacy is a highly interdisciplinary topic (Bräunlich et al., 2021). However, the multitude of perspectives complicates a common understanding of the notion of privacy. Privacy has been understood in different ways, including control over personal data (Westin, 1967), protection of one's reputation, freedom of thought, bodily autonomy, and solitude at home (Solove, 2008). Privacy scholarship considers both 'freedom to' aspects that focus on autonomy and 'freedom from' aspects that capture if somebody is let alone (Koops et al., 2017). Privacy should also be construed in relational, zonal, and dimensional ways (Aeschlimann et al., 2015; Koops et al., 2017; Marwick & boyd, 2014). More recently, however, informational privacy has emerged as the critical type, especially given the growing importance of ICTs and legal frameworks such as the European General Data Protection Regulation (GDPR) that approach privacy from a data protection perspective (Smith et al., 2011). Such an informational understanding presents itself in the widely adopted definition by Westin (1967, p. 7) of privacy as "the claim of individuals, groups, or institutions to determine for themselves when, how, and to what extent information about them is communicated to others."

The use of smart and interactive technologies impacts users in various ways (Fosch-Villaronga et al., 2020), and the need to consider privacy multidimensionally, beyond just informational privacy (Aeschlimann et al., 2015), has increased. Burgoon's (1982) classification of privacy into informational, social, psychological, and physical categories is helpful in this regard, mainly because it covers a broader spectrum of related impacts that other approaches do not consider (Lutz et al., 2019). *Physical privacy* refers to the degree to which a person has the right to be left alone (Warren & Brandeis, 1890) and is physically inaccessible to others, including spatial considerations. *Social privacy* is related to the communicative aspects of privacy and involves managing interpersonal boundaries, including intimacy, and protecting the individual from control by others and the state ~~and falls~~. Social privacy thus falls into the category of 'freedom from.' *Psychological privacy* pertains to control over cognitive and affective inputs and thus falls into the category of 'freedom to' (Koops et al., 2017). Finally, *informational privacy* concerns the protection of a person's information in relation to other actors, especially companies that collect, process, and exploit our data. Therefore, it is essential to understand privacy in terms of its physical, social, psychological, and informational dimensions, which often connect to other important aspects, such as user safety (Martinetti et al., 2021).

Privacy has been studied using different theories, including contextual integrity (Nissenbaum, 2010), the privacy calculus (Dinev & Hart, 2006), the privacy paradox (see Baruh et al., 2017 and Kokolakis, 2017 for overviews), protection motivation (Youn, 2009), digital inequality (Park, 2021; Lutz & Hoffmann, 2021), digital and privacy literacy (Trepte et al., 2015; Park, 2013; Masur, 2020), and communication management (Petronio, 2002). Some of these theories, especially those discussed in more detail next, share a relational view of privacy that challenges individualistic, control-based perspectives (Marwick & boyd, 2014).

While covering the breadth of such theories in full is outside the scope of this chapter (see Bazarova & Zhao, 2023 and Trepte, 2023 for recent overviews), two important ones are worth mentioning within the context of social robotics. First is the theory of contextual integrity (Nissenbaum, 2010), which describes privacy as safeguarding context-specific information norms. Contexts are social and institutional spheres such as politics, the law, and family, and five parameters make up context-specific information norms: sender, subject, recipient, information type, and transmission principle. For example, a social robot employed at home should not share the information obtained in a private context with external parties (e.g., when guests are invited to the house and when it shares the information related to the family, including children, with companies that have sold the robot or any other actors; Fosch-Villaronga et al., 2023), as this would go against the principle of contextual integrity. Different authors have studied and used that theory in practice (e.g., Apthorpe et al., 2018; Lutz & Newlands, 2021; Martens et al., 2021; Zimmer, 2018). These studies reveal that the context where these technologies are inserted is very complex, including many parties such as users, family members, device manufacturers, application developers, cloud service providers, and third parties such as moderation contractors, data brokers, and sub-providers of all of them (Fosch-Villaronga & Millard, 2019; Lutz & Newlands, 2021; Lutz & Tamò, 2018). They also reveal that users' privacy perceptions depend on actor typology, for example being more concerned about institutional actors (such as manufacturers) than social ones (such as household members) (Apthorpe et al., 2018; Lutz & Newlands, 2021). In the context of social robotics, these concerns are exacerbated by the robot's appearance, which often distracts users from knowing that there are complex information flows behind those 'eyes' and 'ears,' which are cameras and microphones (Kaminski et al., 2016; Fosch-Villaronga et al., 2018b).

A second prominent theory is communication privacy management (Petronio, 2002). The baseline here is that people balance disclosure and withdrawal in a dialectic process: they set up privacy boundaries, describing which information is considered private or public. Boundary management includes boundary rule formation, boundary coordination, and boundary turbulence. These boundaries are shaped, and even blurred, with the use of digital technologies, for example when workers use platforms like Facebook to engage with other co-workers (Frampton & Child, 2013) or in light of sharenting, where children often lack the power to control what content is posted by their parents online (Ranzini et al., 2020; Walrave et al., 2022). To retake the example from above, a social robot should respect boundaries within a home (e.g., bedroom, bathrooms) and understand such physical and informational boundaries, which will then, inevitably, be shaped by technological affordances, relationships, and context of use.

The regulatory realm has unsurprisingly tried to catch up with new technological developments and proposed ways to address corresponding privacy challenges. Such developments are old (Acquisti et al., 2022; Warren & Brandeis, 1890) but still today, seemingly unproblematic technologies ~~(such as the Kodak film)~~ that helped capture reality have impacted how we interact with each other and seek privacy protection in many environments (Letwin, 1982). However, technology itself can never be seen as acting in a vacuum. For instance, the calls for a 'right to privacy' and 'right to be let alone' in 1890 in the US (Warren & Brandeis, 1890) can only be understood in light of the broader picture: the increase in urbanization (and thus people living close to each other), the rise in literacy, and the developments of the tabloid press all influenced the need to ensure that seclusion from the public sphere remained possible. Likewise, the increase in digital communication and information sharing led to calls for enabling greater informational self-determination (Hornung & Schnabel, 2009). From a regulatory perspective, we see thus an evolution of new ways of thinking about informational privacy in particular and the movement towards regulating how personal data is processed (Mayer-Schönberger, 1997), which is in continuous evolution as technology evolves in different shapes and forms, such as AI, generative AI, or social robotics (Guzman & Lewis, 2020; Fosch-Villaronga et al., 2020).

In Europe, data protection laws emerged first on a national level (throughout the 1970–1990s) and later in a harmonized way, initially in the Data Protection Directive (1995) and later in the GDPR (2018). New technological developments have propagated data protection law in Europe to become the central regulatory tool (even referred to as the 'Law of Everything' due to its broad scope; Purtova, 2018) to address the multitude of challenges raised by data-processing services and products (such as social robots; Fosch-Villaronga et al., 2018a). At its core, data protection law sets out fundamental principles of how personal data (i.e., data relating to an identified or identifiable person) must be collected, used, stored, and deleted.

These fundamental principles include the principle of lawfulness, which requires, among others, that the data processing is based on a legal ground (such as the performance of a contract, a legal obligation, or consent); the principle of transparency, which requires informing the individuals about the data processing that occurs; the principle of data minimization and purpose limitation, which require that the processed data is kept to a necessary minimum and only for the purpose set out (e.g., recommend certain content, interact with the user); the principle of accuracy, which requires the data being processed to be accurate and kept up to date; the principle of storage limitation, which requires the data to be stored only as long as necessary to achieve a predetermined and communicated purpose; and the principle of data security, which ensures the confidentiality, integrity, and availability of data (see Art. 5

GDPR). Importantly, these principles must be implemented within the organizational setup of the service or product provider and its technical design (Langheinrich, 2001; Tamò-Larrieux, 2018). The codification of the principle of privacy by design (Cavoukian, 2009) into the GDPR (now called data protection by design and default) (Tamò-Larrieux, 2018; Bygrave, 2017) is one of the efforts that aim to bring the goals of such a 'technologically neutral' piece of legislation closer to specific types of technology. Such a principle has led to research in the field of how to transpose the legal norm into the design of different types of technology, including social robots (Heuer et al., 2019) or active assisted living technologies, among many others (Ake-Kob et al., 2021).

9.3 Social Robots and Their Privacy Implications

Social robots are a sub-type of robots that interact with humans and are enabled with high-level dialogue skills (Fong et al., 2003). Unlike industrial robots, social robots are employed in more private settings, such as at home, school, or healthcare facilities. As a wide range of social robots exist and are currently employed (Leite et al., 2013; Belpaeme et al., 2018), such as in the education of non-neurotypical children or older adult care, researchers have been increasingly focused on understanding the ethical, legal, and social (ELS) implications of human–robot interactions (HRI) (Peter & Kühne, 2018; Fosch-Villaronga et al., 2020; Søraa et al., 2021).

Based on a series of workshops at leading robotics conferences, Fosch-Villaronga and colleagues (2020) identified five salient aspects of ELS discussion and corresponding recommendations: privacy and security, legal uncertainty, autonomy and agency, employment effects, and replacement of human interactions. We will return to the first point of privacy and security, which was seen as a critical issue, below, but briefly go into the other four aspects here. *Legal uncertainty* refers to whether and how current regulations apply to social robots in a given context. With it comes the question and disputes of how (moral) responsibility should be addressed. The topic of liability, especially, has led to many policymaking debates, which resulted even in the establishment of the proposed AI Liability Directive in the EU (2022). *Autonomy and agency* are linked to the question of responsibility as it is triggered by the uncertainty of attribution of agency to social robots: How do autonomous robots make decisions? How much transparency is there over the decision of a social robot? How much does the appearance of a social robot alter the interaction and perception of agency? The fear of losing human autonomy in a world driven by (social) robots culminates in fear of being replaced by robots in the workforce (*employment effects*) and having social interactions replaced by social robots (*replacement of human interactions*).

However, privacy and data protection were the most recurrent and worrying topics raised during the workshop ELS discussions. Scholars, practitioners, and developers showed increasing interest in the conundrum of balance between the richness of information to support robot-mediated tasks and respect for users' personal data. In this respect, the academic community has approached the topic from different angles. In a pivotal contribution to the field, Calo (2010) claimed that three key privacy implications of robots could be distinguished: direct surveillance, increased access, and social meaning. The first privacy risk of robots is the potential for direct surveillance through their sophisticated technology, including processors and sensors that enable extensive monitoring, such as in law enforcement and military settings. The second privacy issue, increased access, refers to robots' ability to enter protected areas such as bedrooms and bathrooms. Finally, social meaning corresponds to robot design, which can facilitate forming relationships and disclosing sensitive information. The privacy implications of social meaning are particularly relevant to the study of HRI and HMC, where the emergence of increasingly anthropomorphic machines using conversational AI capabilities generates meaningful interactions between humans and machines, affecting how we perceive, interact with, and live with these new agents (Guzman, 2018; Guzman & Lewis, 2020).

Since Calo's (2010) influential contribution, considerable research has furthered our understanding of privacy in the context of social robots. Lutz et al. (2019) conducted a systematic review of research on social robots and privacy, considering publications indexed up to 2017. They draw on Burgoon's (1982) privacy typology and note increased research interest over time. The authors show how the research on privacy and social robots is interdisciplinary, with the law being the discipline that has contributed the most articles considered for the review, followed by computer science and the medical sciences. Most of the articles were conceptual in nature, and the authors note that in the empirical works, small sample sizes and specific interaction contexts make it difficult to draw generalizable conclusions. Additionally, the review identifies privacy as contextual integrity and fair information practices as important conceptual frameworks. Finally, informational privacy appears to be the dominant type, rather than physical, social, and psychological privacy—a finding that mirrors Krupp et al.'s (2017) qualitative study.

Since the publication of Lutz et al.'s (2019) systematic review, more than three years have passed, and much research on the topic has been done. We will discuss empirical studies investigating privacy in the context of social robots between 2017 and 2022. The literature search for this update was done in late November 2022 with the help of a research assistant (see Acknowledgments), who compiled a table of search results and meta-information based on a comprehensive Google Scholar search, using the keywords *privacy* AND *robot* within the titles of each publication. We used the compiled table to download, read, or skim all available full-text documents from the search. While not fully systematic, this process gave us a relatively comprehensive overview of recent research on social robots and privacy. In this process, and for reasons of parsimony, we focused on social sciences and ethical and legal texts. We did not further engage with a few very technical texts (e.g., Tang, 2020) that develop privacy-preserving systems via machine learning or advancements in cybersecurity. Of the first 46 selected papers, Table 9.1 compiles some of the findings from 24[1] relevant texts, demonstrating the breadth and evolution of research and academic discussion on privacy and social robots in recent years.

Table 9.1: Overview of research on privacy and social robots published 2017–2023.

Article	Summary
Rueben et al., 2017a	– Privacy-sensitive robotics constitutes an emerging area of HRI research. – The impact of contextual frames on users' privacy judgments in a home telepresence setting is studied. – Based on four studies, a large effect of manipulating the robot operator's identity between a stranger and a close confidante is found.
Tonkin et al., 2017	– Anthropomorphism has an effect on privacy perceptions. – A new privacy measure is used to analyze disclosure to an embodied vs. disembodied robotic system. – Robot embodiment may increase users' risk tolerance and reduce privacy concerns. – An embodied social robot may be able to gather consented private information from users more easily than disembodied systems.
Van Cleynenbreugel, 2017	– The right to erasure of one's personal data is a key aspect of data protection regulation. – This is also the case for HRI, where this right matters. – The paper maps the relevance of linking the right to be forgotten and privacy-sensitive HRI literature.
Fosch-Villaronga et al., 2018a	– Many privacy aspects of healthcare robots in the European context are underexplored: 1. The role of subconsciously incentivised disclosure in HRI. 2. The complexity of consent requirements. 3. Privacy implications of conversational approaches to privacy management. 4. Data portability requirements when someone has to rely substantially on a robot. 5. The privacy risks related to robot-based data collection in the workplace. 6. The need to go beyond simpler Privacy by Design approaches.
Schulz & Herstad, 2018	– Palen and Dourish's privacy framework can be applied to living with moving robots at home. – Sensor placement and the role of proxemics, or motion technologies, for the negotiation of privacy is investigated, zooming in to concrete privacy issues of networked robot technologies at home.

1 In the revision process of the chapter in spring 2023, three additional texts were added to the table, resulting in a total number of 27 entries in the table.

Table 9.1 (continued)

Article	Summary
Vitale et al., 2018	– A face enrolment system on a humanoid robot is compared with a screen-based kiosk system. – Adding to earlier work, the effect of transparency on users' privacy considerations is investigated. – Within the same system (robot or kiosk) transparency does not affect users' privacy considerations. – However, transparency leads to a significantly better user experience for both systems.
Balali et al., 2019	– People are most concerned about hiding identifiable personal or financial information and valuables from a household robot, preferring stronger filters to hide such items. – There seem to be correlations between a person's familiarity with technology, sociability, and trust and their privacy concerns.
Henkel et al., 2019	– Ethical issues of social robots in sensitive contexts are discussed, especially among children. – A user design study and two controlled experiments with adults and children are conducted. – Adult participants felt a therapeutic robotic dog would be most useful for children compared with other age groups. – Children perceive social robots as providing more socially protective factors than adults.
Heuer et al., 2019	– Given the increasing human–robot interaction, and to prevent pivacy risks of social robots, there should be privacy-centered design for social robots. – This framework should include *seven types of privacy*: 1) Privacy of the Person; 2) Privacy of Action and Behavior; 3) Privacy of Communication; 4) Privacy of Data and Image; 5) Privacy of Thoughts and Feelings; 6) Privacy of Location and Space; and 7) Privacy of Association.
Kerr, 2019	– Epistemic privacy offers a useful analytic framework for understanding the kind of cognizance that gives rise to diminished privacy. – Legal theory and doctrine will have to expand their understanding of privacy relationships to include robots and AIs that meet these epistemic conditions.

Table 9.1 (continued)

Article	Summary
Tonkin et al., 2019	– There is a shortage of research covering user challenges with respect to robots in-the-wild. – A user registration system for a social robot using facial recognition systems in an innovation center presents new privacy risks for users. – Compared to the use of a QR code, people prefer facial recognition technology when registering personal data with the robot to access services although that option is more privacy demanding. – Addressing privacy considerations at the beginning of designing social robot applications leads to more inclusive systems. – Since transparency impacts user trust, designing for privacy with inclusiveness and transparency upfront leads to more responsible and trustable social robot applications.
Abbas et al., 2020	– Privacy perceptions around crowd-operated robots are investigated. – Based on an experiment of an audio+video vs. audio-only manipulation, no significant differences in social intelligence and user experience were found. – Less privacy was perceived by the group with an audio-only broadcast feed compared with the audio-video feed (however, a very small sample of only 14).
Horstmann et al., 2020	– A companion app that informs users what data a robot processes and why they are using it in their private households can give users greater control over their personal data. – A companion privacy app could mitigate privacy concerns, enhancing privacy compliance. – A prototype of a companion privacy app, tested on a small sample, proves promising, showing how the app helps maintain privacy and provide control over personal data.
Lutz & Tamó-Larrieux, 2020	– Based on a survey of non-experts, differentiates three distinct types of privacy concerns around social robots: institutional, social, and physical. – Tests how privacy concerns and antecedents from adjacent scholarship (e.g., trust) affect robot use intention. – Respondents are most concerned about institutional privacy, followed by social and physical privacy. – A privacy paradox is found, where the perceived benefits of social robots override privacy concerns.
Pearson, 2020	– There are many opportunities and use cases of robots interacting with children. – Some risks need to be considered, including privacy risk, robot abuse, and deception. – Alongside the potential harms, however, the collection and use of information have the potential to protect children's welfare.

Table 9.1 (continued)

Article	Summary
Su & Yuan, 2021	– Social robots lack suitable design strategies for privacy notices that announce camera status to users. – Embodiment and notice style significantly impact the user's privacy consideration. – Physically embodied robots are favored over virtual ones.
Chatterjee et al., 2021	– There is not much research at the intersection of privacy, security and legal concerns about robots for household purposes. – Privacy and security concerns affect robot use but few people are aware of legal implications. – Unless legal frameworks cover households' safety, security, and privacy concerns more precisely and concretely, households may not use robots.
Chatzimichali et al., 2021	– The role of privacy policies and terms of conditions is studied in consumer robotics. – It is then assessed how eight robot manufacturers comply with the eight OECD Privacy Guidelines principles (collection limitation, data quality, purpose specification, use limitation, security safeguards, openness, individual participation, and accountability). – Dyson has the best score, followed by ANKI and Wonder Workshop. Ecovacs has the worst score.
Garcia et al., 2021	– Studies how the GDPR principles can be mapped to relevant aspects of toy robots, based on the example of GoPiGo3. – The data flows and data-processing requirements of the toy robot are investigated and an automatic consent mechanism developed that aligns with the GDPR and parental consent requirements of EU member states.
Heuer et al., 2021	– A privacy framework is proposed and put in the context of a user-centered design approach. – Feasible methodologies to simplify the risk assessment and meet the privacy principles are developed. – Two different use case scenarios with a robot as an interaction partner show the importance of context awareness.
Lutz & Tamò-Larrieux, 2021	– Following up on Lutz & Tamò-Larrieux (2020) and based on an experimental vignette study. – Unlike in the survey, privacy concerns affect user intention significantly and negatively. – Potential robot users take into account privacy as consideration for future behavior when it is made salient.

Table 9.1 (continued)

Article	Summary
Dietrich & Weisswange, 2022	– Privacy challenges of social robots that serve as mediators in human–human interactions within domestic assistance scenarios are studied. – The literature on the related technology of smart home and smart speaker devices serves as the basis. – Do common user concerns and data-sharing requirements also apply for a robotic mediator? – Privacy implications of shared devices and telepresence robots are discussed too. – Social privacy emerges as a particular challenge, pointing to future research directions.
Fortunati et al. 2022	– Users usually perceive smart speakers as intrusive, invasive, always listening, and even spying, provoking a surveilling 'Big Brother' effect. – A cross-cultural comparison revealed differences in smart speaker (positive) acceptance, with US respondents being more positive but more concerned with privacy than Italian respondents.
Xie & Lei, 2022	– The nonlinear relationship between anthropomorphism and privacy concerns about service robots is investigated. – Three empirical studies showed that anthropomorphism has a nonlinear effect on customers' privacy concerns, which in turn negatively affects subsequent usage intention. – The negative impacts of anthropomorphism are attenuated when (1) information sensitivity is low, and (2) service personalization is high.
Yang et al., 2022	– Robot designers can help users understand how the robot protects their privacy, for instance through the robot gaze, the distance between the robot and user, and the clarity in expressing the intent of the robot. – Users feel more comfortable seeing the robot respecting users' privacy, and performing actions like changing clothes. In this context, turning away and showing the robot's back is the most effective behavior. – As the robot combines more privacy-protective behaviors, the concern of users tends to decrease.

Table 9.1 (continued)

Article	Summary
Fosch-Villaronga et al., 2023	– Socially assistive robots interact with users and entail monitoring in terms of behavior and physiological environment. – The human–robot interaction happens within a complex ecosystem that involves the humans involved in the interaction, the robot, environmental sensors, wearable technology, and cloud services. – Collecting behavioral and physiological data is critical for the active development, deployment, and implementation of healthcare robots and their capability to support users adequately. However, this raises privacy and security issues and concerns, and causes a loss of autonomy. – Roboticists need instruments to help them translate privacy ideals (from the EU GDPR): privacy and data protection impact assessments applied to robot technology in consonance with the Robot Impact Assessment and the Assessment List of Trustworthy AI; Privacy by Design mechanisms.
Schrum & Gombolay, 2023	– An individual's personality and prior experiences may influence their trust in an autonomous robot, and thus it is essential first to understand these characteristics that affect the shareability of personal information. – Robots should learn and adapt to end users' uniqueness to optimize the human–robot relationship while respecting end-user privacy rights. – Researchers hypothesize that users may be more comfortable giving specific information to a robot if it aligns with their expectations in the context of applications. – Explainability may increase understanding, comfort, and intention to share personal information with a robot.

Taken together, our literature review shows the breadth and complexity of the privacy implications of social robots. The publication landscape is diverse, with relevant contributions made mainly as conference/workshop proceedings papers (13 out of 27, especially ACM and IEEE robotics conferences such as *HRI, RO-MAN, ARSO*, and *IROS*) or as journal articles (12 out of 27; especially robotics journals that welcome non-technical contributions such as *Paladyn* and the *International Journal of Social Robotics*). However, the fact that some publications come from outlets of other disciplines, such as communication (Fortunati et al., 2022b; Lutz & Tamò-Larrieux, 2020), hospitality management (Xie & Lei, 2022), and law (Kerr, 2019), shows that there is ample interest in the topic beyond robotics and HRI. In contrast to before 2017 (Lutz et al., 2019), there is a much higher proportion of empirical studies showing the field's maturation.

9.4 Conclusion

In this contribution, we discussed the privacy implications of social robots. Through an interdisciplinary literature review and putting special emphasis on recent studies, we showed how privacy constitutes an important consideration in the context of social robots that has received increasing attention. The publication landscape is interdisciplinary, with contributions from the social sciences, engineering, ethics, law, and policy. The emerging fields of privacy-sensitive robotics (Rueben et al., 2018) and human–machine communication (Guzman, 2018) represent promising lenses to approach privacy in the context of social robots. Both spotlight the intertwinement of social and technological aspects. Social robots differ from other smart technologies in important ways, given their interactive capabilities, physical embodiment, mobility, ability to record conversations and images, and their use in areas such as healthcare and education. Their characteristics, especially how humans anthropomorphize these devices, have relevant consequences for privacy research. Calo's (2010) aspect of social meaning stands out as a main privacy implication, also for future research, primarily because if social robots become increasingly normalized, adopted, and used, then human scripts and theories of privacy, such as thosementioned above, might need to be re-evaluated. Interpersonal privacy theories, in particular, such as communication privacy management (Petronio, 2002), the privacy calculus (Dinev & Hart, 2006), and protection motivation theory (Youn, 2009), might need to be reoriented, although their key concepts (such as privacy rules, boundary management, privacy turbulence, risk/benefit trade-off, threat appraisal, coping appraisal) could still be applied or adapted. Relational and situated theories such as privacy as contextual integrity (Nissenbaum, 2010) and actor-network theory (Lutz & Tamò, 2018) are especially useful given the autonomy of social robots and the complex ecosystem in which they are embedded.

We propose the following directions for future research on privacy and social robots. First, such research should be (more) interdisciplinary, fostering collaboration between privacy scholars from different disciplines (Lutz & Tamò, 2015). Studies could, for example, combine legal, technical, and social expertise by investigating how particular principles in data protection regulation (such as those discussed above) can be implemented into the technological design and how different design solutions are in turn evaluated by the users, for example in terms of trust and satisfaction.

Second, future research should study privacy and social robots in real-life settings in addition to lab settings (Tonkin et al., 2019). Observational studies that accompany the introduction of robots in specific contexts such as hospitals, care homes, shopping centers, or schools would deepen our understanding of privacy processes (Calleja et al., 2022). For example, do users' privacy preferences change over time as they get to know the robot as an interlocutor? Can robots' data processing be tailored to specific, personalized privacy preferences? How do users develop privacy literacy around the robot, individually and in conversations with others?

Third, future research should look at privacy and social robots from a comparative perspective. Cross-cultural survey-based research shows substantial differences in privacy perceptions between countries and regions (Trepte et al., 2017; Fortunati et al., 2022a, 2022b). Such research could, for example, explore whether privacy plays less of a role in robot adoption and user interactions in countries where social robots are more accepted (e.g., Japan), in contrast to countries where they are less accepted (e.g., Germany). The comparative lens could also feature theory comparisons, where extant privacy theories are tested against each other, or technology comparisons, where privacy around social robots is contrasted with privacy around adjacent technologies such as (disembodied) chatbots or (less autonomous/mobile) smart speakers. Particular emphasis should be put on emerging economies, and Non-WEIRD (Western, educated, industrialized, rich, democratic) countries as contexts outside of Europe, North America, and East Asia are severely under-represented in robotics research, privacy research, and AI research and development more broadly (Eke et al., 2023; Gupta & Heath, 2020; Masur et al., 2021). Such studies should be mindful of indigenous knowledge and conceptions of privacy that differ from mainstream Western ones (e.g., Rennie et al., 2018).

Fourth, future research on privacy and social robots should look into emerging privacy topics and discourses, such as those on privacy cynicism and digital resignation (Draper & Turow, 2019; Lutz et al., 2020), the potential of personalization to tailor privacy settings to specific individuals (Sadeh et al., 2021), or the interrelation of privacy and other important ELS implications such as transparency and trust (Čartolovni et al., 2022).

Fifth, there should be more practical research that establishes mechanisms that help roboticists translate privacy and data protection ideals into concrete actions (Lutz & Tamò, 2015). This could be done, for instance, via the use of privacy or transparency by design principles as an ex-ante consideration (Tamò-Larrieux, 2018; Felzmann et al., 2020; Garcia et al., 2021) and as an ex-post consideration through data protection impact assessments combined with a technology (robot) impact assessment (Fosch-Villaronga, 2015; Fosch-Villaronga et al., 2023).

Sixth and finally, advancements in large language models (LLM) have led to recent breakthrough applications such as GPT-4, which substantially affect work, education, and leisure (e.g., Lund et al., 2023). The focus of the discussion has been on text-based interaction modalities through user prompts, with multimodal capacities (e.g., images, videos) emerging most recently (Liu et al., 2023). Robotics firms are starting to integrate LLM into embodied technologies. For example, Engineered Arts uses GPT-3 for its Ameca robot (Tangermann, 2022). Embedding state-of-the-art LLM into a new generation of social robots will raise new ethical and privacy implications, not least because embodied agents capable of mimicking natural languages will increase the capabilities of interacting and connecting with people on an emotional and personal level. While we are currently at the very beginning of this journey, we have already witnessed some potential drawbacks of conversational chatbots based on LLM, like the persuasive and manipulative tone of Microsoft Bing's Chatbot (from OpenAI) with a New York Times journalist (Roose, 2023) and the persuasion of a Google engineer that their LaMDA chatbot is sentient (Tiku, 2022). These examples show an increased blurring of boundaries between human–human interaction and human–robot interaction. The changes brought forward by the increased interactive capabilities of social robots and abilities of being more pervasive and persuasive than disembodied chatbots will require interdisciplinary research on the impact of these new generative social robots.

Bibliography

Abbas, T., Corpaccioli, G., Khan, V. J., Gadiraju, U., Barakova, E., & Markopoulos, P. (2020). How do people perceive privacy and interaction quality while chatting with a crowd-operated robot? In *Companion of the 2020 ACM/IEEE International Conference on Human–Robot Interaction* (pp. 84–86). ACM.

Acquisti, A., Brandimarte, L., & Hancock, J. (2022). How privacy's past may shape its future. *Science, 375*(6578), 270–272.

Aeschlimann, L., Harasgama, R., Kehr, F., Lutz, C., Milanova, V., Müller, S., Strathoff, P., & Tamò, A. (2015). Re-setting the stage for privacy: A multi-layered privacy interaction framework and its application. In S. Brändli, R. Harasgama, R. Schister, & A. Tamò (Eds.), *Mensch und Maschine—Symbiose oder Parasitismus?* (pp. 1–41). Stämpfli.

Ake-Kob, A., Blazeviciene, A., Colonna, L., Cartolovni, A., Dantas, C., Fedosov, A., Florez-Revuelta, F., Fosch-Villaronga, E. . . . & Colantonio, S. (2021). State of the art on ethical, legal, and social issues linked to audio- and video-based AAL solutions. *Cost Action Good Brother*, 1–56. https://goodbrother.eu/wp-content/uploads/2021/12/GoodBrother-State-of-the-art-on-ethical-legal-and-social-issues-linked-to-audio-and-video-based-AAL-solutions.pdf.

Altman, I. (1975). *The Environment and Social Behavior: Privacy, Personal Space, Territory, Crowding*. Wadsworth Publishing Company.

Apthorpe, N., Shvartzshnaider, Y., Mathur, A., Reisman, D., & Feamster, N. (2018). Discovering smart home internet of things privacy norms using contextual integrity. *Proceedings of the ACM on Interactive, Mobile, Wearable and Ubiquitous Technologies, 2*(2), 1–23.

Balali, S., Sowell, R. T., Smart, W. D., & Grimm, C. M. (2019). Privacy concerns in robot teleoperation: Does personality influence what should be hidden? In *International Conference on Social Robotics* (pp. 719–729). Springer.

Barrett, C. (2021). "Exterminate bad habits": Meet Xavier, Singapore's COVID-compliance robot. *Sydney Morning Herald* (September 9). https://www.smh.com.au/world/asia/exterminate-bad-habits-meet-xavier-singapore-s-covid-compliance-robot-20210908-p58q1c.html.

Baruh, L., Secinti, E., & Cemalcilar, Z. (2017). Online privacy concerns and privacy management: A meta-analytical review. *Journal of Communication, 67*(1), 26–53.

Bazarova, N. N., & Zhao, P. (2023). Individualistic privacy theories. In S. Trepte & P. K. Masur (Eds.), *The Routledge Handbook of Privacy and Social Media* (pp. 16–24). Routledge.

Belpaeme, T., Kennedy, J., Ramachandran, A., Scassellati, B., & Tanaka, F. (2018). Social robots for education: A review. *Science Robotics, 3*(21), eaat5954.

Bräunlich, K., Dienlin, T., Eichenhofer, J., Helm, P., Trepte, S., Grimm, R., Seubert, S., & Gusy, C. (2021). Linking loose ends: An interdisciplinary privacy and communication model. *New Media & Society, 23*(6), 1443–1464.

Brause, S. R., & Blank, G. (2020). Externalized domestication: Smart speaker assistants, networks and domestication theory. *Information, Communication & Society*, *23*(5), 751–763.

Burgoon, J. K. (1982). Privacy and communication. *Annals of the International Communication Association*, *6*(1), 206–249.

Bygrave, L. A. (2017). Hardwiring privacy. In R. Brownsword, E. Scotford, & K. Yeung (Eds.), *The Oxford Handbook of Law, Regulation and Technology* (pp. 754–775). Oxford University Press.

Calleja, C., Drukarch, H., & Fosch-Villaronga, E. (2022). Harnessing robot experimentation to optimize the regulatory framing of emerging robot technologies. *Data & Policy*, *4*, 1–15, e20.

Calo, R. (2010). Robots and privacy. In P. Lin, K. Abney, & G. A. Bekey (Eds.), *Robot Ethics: The Ethical and Social Implications of Robotics* (pp. 187–201). MIT Press.

Čartolovni, A., Fedosov, A., Fosch-Villaronga, E., Lutz, C., & Tamò-Larrieux, A. (2022). *Proceedings of the GoodBrother International Conference on Privacy-Friendly and Trustworthy Technology for Society 2022*. COST. https://www.zora.uzh.ch/id/eprint/220632/1/Complete_Proceedings_GoodBrother_Conf_2022.pdf.

Cavoukian, A. (2009). *Privacy by Design: The 7 Foundational Principles*. Information and Privacy Commissioner of Ontario, Canada, 1–5.

Chatterjee, S., Chaudhuri, R., & Vrontis, D. (2021). Usage intention of social robots for domestic purpose: From security, privacy, and legal perspectives. *Information Systems Frontiers*, 1–16. https://doi.org/10.1007/s10796-021-10197-7.

Chatzimichali, A., Harrison, R., & Chrysostomou, D. (2021). Toward privacy-sensitive human–robot interaction: Privacy terms and human–data interaction in the personal robot era. *Paladyn: Journal of Behavioral Robotics*, *12*(1), 160–174.

Coeckelbergh, M. (2021). How to use virtue ethics for thinking about the moral standing of social robots: A relational interpretation in terms of practices, habits, and performance. *International Journal of Social Robotics*, *13*(1), 31–40.

Dietrich, M., & Weisswange, T. (2022). Exploring privacy implications for domestic robot mediators. In *IEEE/RSJ International Conference on Intelligent Robots and Systems (IROS 2022)* (pp. 1–6). Workshop on Robot Trust for Symbiotic Societies (RTSS 2022).

Dinev, T., & Hart, P. (2006). An extended privacy calculus model for e-commerce transactions. *Information Systems Research*, *17*(1), 61–80.

Draper, N. A., & Turow, J. (2019). The corporate cultivation of digital resignation. *New Media & Society*, *21*(8), 1824–1839.

Edwards, C., Edwards, A., Kim, J., Spence, P. R., de Graaf, M., Nah, S., & Rosenthal-von der Pütten, A. (2019). Human–machine communication: What does/could communication science contribute to HRI? In *2019 14th ACM/IEEE International Conference on Human–Robot Interaction* (pp. 673–674). IEEE.

Edwards, A., Edwards, C., Westerman, D., & Spence, P. R. (2019). Initial expectations, interactions, and beyond with social robots. *Computers in Human Behavior*, *90*, 308–314.

Eke, D. O., Wakunuma, K., & Akintoye, S. (2023). *Responsible AI in Africa: Challenges and Opportunities*. Palgrave Macmillan.

Felzmann, H., Fosch-Villaronga, E., Lutz, C., & Tamò-Larrieux, A. (2020). Towards transparency by design for artificial intelligence. *Science and Engineering Ethics*, *26*(6), 3333–3361.

Fong, T., Nourbakhsh, I., & Dautenhahn, K. (2003). A survey of socially interactive robots. *Robotics and Autonomous Systems*, *42*(3–4), 143–166.

Fortunati, L., & Edwards, A. P. (2020). Opening space for theoretical, methodological, and empirical issues in human–machine communication. *Human–Machine Communication*, *1*(1), 7–18.

Fortunati, L., Edwards, A., Edwards, C., Manganelli, A. M., & de Luca, F. (2022a). Is Alexa female, male, or neutral? A cross-national and cross-gender comparison of perceptions of Alexa's gender and status as a communicator. *Computers in Human Behavior*, *137*, 107426.

Fortunati, L., Edwards, A. P., Manganelli, A. M., Edwards, C., & De Luca, F. (2022b). Do people perceive Alexa as gendered? A cross-cultural study of people's perceptions, expectations, and desires of Alexa. *Human–Machine Communication*, *5*(1), 75–97.

Fosch-Villaronga, E. (2015). Creation of a care robot impact assessment. *International Journal of Social, Behavioral, Educational, Economic, Business and Industrial Engineering*, *9*, 1913–1917.

Fosch-Villaronga, E., & Albo-Canals, J. (2019). "I'll take care of you," said the robot. *Paladyn: Journal of Behavioral Robotics*, *10*(1), 77–93.

Fosch-Villaronga, E., Drukarch, H., & Custers, B. (forthcoming). Privacy considerations for socially assistive robots. In W. Barfield, Y.-H. Weng, & U. Pagallo (Eds.), *Cambridge Handbook on Law, Policy, and Regulations on Human–Robot Interaction*.

Fosch-Villaronga, E., Felzmann, H., Pierce, R., De Conca, S., De Groot, A., Robins, S., & Del Castillo, A. P. (2018a). Nothing comes between my robot and me: Privacy and human–robot interaction in robotised healthcare. In R. Leenes & R. Van Brakel (Eds.), *Data Protection and Privacy: The Internet of Bodies* (pp. 135–170). Hart Publishing.

Fosch-Villaronga, E., Felzmann, H., Ramos-Montero, M., & Mahler, T. (2018b). Cloud services for robotic nurses? Assessing legal and ethical issues in the use of cloud services for healthcare robots. In *2018 IEEE/RSJ International Conference on Intelligent Robots and Systems (IROS)* (pp. 290–296).

Fosch-Villaronga, E., Lutz, C., & Tamò-Larrieux, A. (2020). Gathering expert opinions for social robots' ethical, legal, and societal concerns: Findings from four international workshops. *International Journal of Social Robotics*, *12*(2), 441–458.

Fosch-Villaronga, E., & Millard, C. (2019). Cloud robotics law and regulation: Challenges in the governance of complex and dynamic cyber-physical ecosystems. *Robotics and Autonomous Systems*, *119*, 77–91.

Fosch-Villaronga, E., Van der Hof, S., Lutz, C., & Tamò-Larrieux, A. (2023). Toy story or children story? Putting children and their rights at the forefront of the artificial intelligence revolution. *AI & Society*, *38*(1), 133–152.

Frampton, B. D., & Child, J. T. (2013). Friend or not to friend: Coworker Facebook friend requests as an application of communication privacy management theory. *Computers in Human Behavior*, *29*(6), 2257–2264.

Fussell, S. R., Kiesler, S., Setlock, L. D., & Yew, V. (2008). How people anthropomorphize robots. In *2008 3rd ACM/IEEE International Conference on Human–Robot Interaction* (pp. 145–152).

Gambino, A., Fox, J., & Ratan, R. A. (2020). Building a stronger CASA: Extending the computers are social actors paradigm. *Human–Machine Communication*, *1*, 71–85.

Garcia, K., Zihlmann, Z., Mayer, S., Tamo-Larrieux, A., & Hooss, J. (2021). Towards privacy-friendly smart products. In *2021 18th International Conference on Privacy, Security and Trust* (pp. 1–7). IEEE.

Gupta, A., & Heath, V. (2020, September 14). AI ethics groups are repeating one of society's classic mistakes. *MIT Technology Review*. https://www.technologyreview.com/2020/09/14/1008323/ai-ethics-representation-artificial-intelligence-opinion/.

Guzman, A. L. (2018). Introduction: What is human–machine communication, anyway? In A. Guzman (Ed.), *Human–Machine Communication: Rethinking Communication, Technology, and Ourselves* (pp. 1–28). Peter Lang.

Guzman, A. L. (2020). Ontological boundaries between humans and computers and the implications for human–machine communication. *Human–Machine Communication*, *1*, 37–54.

Guzman, A. L., & Lewis, S. C. (2020). Artificial intelligence and communication: A human–machine communication research agenda. *New Media & Society*, *22*(1), 70–86.

Harkous, H., Fawaz, K., Shin, K. G., & Aberer, K. (2016). Pribots: Conversational privacy with chatbots. In *Twelfth Symposium on Usable Privacy and Security* (SOUPS, 2016). USENIX.

Henkel, Z., Baugus, K., Bethel, C. L., & May, D. C. (2019). User expectations of privacy in robot assisted therapy. *Paladyn: Journal of Behavioral Robotics*, *10*(1), 140–159.

Heuer, T., Schiering, I., & Gerndt, R. (2019). Privacy-centered design for social robots. *Interaction Studies* *20*(3), 509–529.

Heuer, T., Schiering, I., & Gerndt, R. (2021). Privacy framework for context-aware robot development. *Paladyn: Journal of Behavioral Robotics*, *12*(1), 468–480.

Himmelreich, J. (2018). Never mind the trolley: The ethics of autonomous vehicles in mundane situations. *Ethical Theory and Moral Practice*, *21*(3), 669–684.

Hornung, G., & Schnabel, C. (2009). Data protection in Germany I: The population census decision and the right to informational self-determination. *Computer Law & Security Review*, *25*(1), 84–88.

Horstmann, B., Diekmann, N., Buschmeier, H., & Hassan, T. (2020). Towards designing privacy-compliant social robots for use in private households: A use case based identification of privacy implications and potential technical measures for mitigation. In *2020 29th IEEE International Conference on Robot and Human Interactive Communication* (pp. 869–876). IEEE.

Kaminski, M. E., Rueben, M., Smart, W. D., & Grimm, C. M. (2016). Averting robot eyes. *Maryland Law Review*, *76*, 983.

Kerr, I. (2019). Schrödinger's robot: Privacy in uncertain states. *Theoretical Inquiries in Law*, *20*(1), 123–154.

Kokolakis, S. (2017). Privacy attitudes and privacy behaviour: A review of current research on the privacy paradox phenomenon. *Computers & Security*, *64*, 122–134.

Koops, B. J., Newell, B. C., Timan, T., Skorvanek, I., Chokrevski, T., & Galic, M. (2017). A typology of privacy. *University of Pennsylvania Journal of International Law Review*, *38*, 483–575.

Krupp, M. M., Rueben, M., Grimm, C. M., & Smart, W. D. (2017, August). A focus group study of privacy concerns about telepresence robots. In *2017 26th IEEE International Symposium on Robot and Human Interactive Communication* (pp. 1451–1458). IEEE.

Langheinrich, M. (2001). Privacy by design—Principles of privacy-aware ubiquitous systems. In *International Conference on Ubiquitous Computing* (pp. 273–291). Springer.

Leite, I., Martinho, C., & Paiva, A. (2013). Social robots for long-term interaction: A survey. *International Journal of Social Robotics*, *5*(2), 291–308.

Letwin, A. Z. (1982). Open doors, open minds: The privacy issue. *Update on Law-Related Education*, *6*(2), 10–15.

Liao, Y., Vitak, J., Kumar, P., Zimmer, M., & Kritikos, K. (2019). Understanding the role of privacy and trust in intelligent personal assistant adoption. In *International Conference on Information* (pp. 102–113). Springer.

Lintvedt, N. (2022). Thermal imaging in robotics as a privacy enhancing or privacy invasive measure? The necessity of a holistic approach to privacy in human–robot interaction. In *Proceedings of the International Conference on Privacy-Friendly and Trustworthy Technology for Society* (paper 12). COST.

Liu, Y., Han, T., Ma, S., Zhang, J., Yang, Y., Tian, J., . . . & Ge, B. (2023). Summary of chatgpt/gpt-4 research and perspective towards the future of large language models. *arXiv preprint arXiv:2304.01852*. https://arxiv.org/abs/2304.01852.

Lund, B. D., Wang, T., Mannuru, N. R., Nie, B., Shimray, S., & Wang, Z. (2023). ChatGPT and a new academic reality: Artificial Intelligence-written research papers and the ethics of the large language models in scholarly publishing. *Journal of the Association for Information Science and Technology*, *74*(5), 570–581.

Lutz, C., & Hoffmann, C. P. (2021). Inequalities in online political participation: The role of privacy concerns. In E. Hargittai (Ed.), *Handbook of Digital Inequality*. Edward Elgar.

Lutz, C., Hoffmann, C. P., & Ranzini, G. (2020). Data capitalism and the user: An exploration of privacy cynicism in Germany. *New Media & Society*, *22*(7), 1168–1187.

Lutz, C., & Newlands, C. (2021). Privacy and smart speakers: A multi-dimensional approach. *Information Society*, *37*(3), 147–162.

Lutz, C., Schöttler, M., & Hoffmann, C. P. (2019). The privacy implications of social robots: Scoping review and expert interviews. *Mobile Media & Communication*, *7*(3), 412–434.

Lutz, C., & Tamò, A. (2015). RoboCode-Ethicists: Privacy-friendly robots, an ethical responsibility of engineers? In *Proceedings of the ACM Web Science Conference* (pp. 1–12).
Lutz, C., & Tamò, A. (2018). Communicating with robots: ANTalyzing the interaction between healthcare robots and humans with regards to privacy. In A. Guzman (Ed.), *Human–Machine Communication: Rethinking Communication, Technology, and Ourselves* (pp. 145–165). Peter Lang.
Lutz, C., & Tamò-Larrieux, A. (2020). The robot privacy paradox: Understanding how privacy concerns shape intentions to use social robots. *Human–Machine Communication, 1*, 87–111.
Lutz, C., & Tamò-Larrieux, A. (2021). Do privacy concerns about social robots affect use intentions? Evidence from an experimental vignette study. *Frontiers in Robotics and AI, 8*, 63.
Maccario, G., & Naldi, M. (2022). Privacy in smart speakers: A systematic literature review. *Security and Privacy*, 6(1):e274. doi:10.1002/spy2.274
Martens, M., De Wolf, R., Vadendriessche, K., Evens, T., & De Marez, L. (2021). Applying contextual integrity to digital contact tracing and automated triage for hospitals during COVID-19. *Technology in Society, 67*, 101748.
Martinetti, A., Chemweno, P. K., Nizamis, K., & Fosch-Villaronga, E. (2021). Redefining safety in light of human–robot interaction: A critical review of current standards and regulations. *Frontiers in Chemical Engineering, 3*, 1–12, 666237.
Marwick, A. E., & boyd, d. (2014). Networked privacy: How teenagers negotiate context in social media. *New Media & Society, 16*(7), 1051–1067.
Masur, P. K. (2020). How online privacy literacy supports self-data protection and self-determination in the age of information. *Media and Communication, 8*(2), 258–269.
Masur, P. K., Epstein, D., Quinn, K., Wilhelm, C., Baruh, L., & Lutz, C. (2021). A comparative privacy research framework. *OSF Preprint*. https://osf.io/preprints/socarxiv/fjqhs/.
Mayer-Schönberger, V. (1997). Generational development of data protection in Europe. In P. Agre & M. Rotenberg (Eds.), *Technology and Privacy: The New Landscape* (pp. 219–236). MIT Press.
Nass, C. L., Steuer, J., & Tauber, E. (1994). Computers are social actors. In *Proceedings of the CHI Conference*.
Newlands, G., Lutz, C., Tamò-Larrieux, A., Villaronga, E. F., Harasgama, R., & Scheitlin, G. (2020). Innovation under pressure: Implications for data privacy during the Covid-19 pandemic. *Big Data & Society, 7*(2), 1–14.
Nguyen, T. (n. d.). A deep look into privacy and security of vacuum robot. https://www.cpp.edu/cyberfair/poster-information/documents/vacuum-robot-abstract.pdf.
Nissenbaum, H. (2010). *Privacy in Context: Technology, Policy and the Integrity of Social Life*. Stanford University Press.
Othman, K. (2021). Public acceptance and perception of autonomous vehicles: A comprehensive review. *AI and Ethics, 1*(3), 355–387.
Park, Y. J. (2013). Digital literacy and privacy behavior online. *Communication Research, 40*(2), 215–236.
Park, Y. J. (2021). Why privacy matters to digital inequality. In E. Hargittai (Ed.), *Handbook of Digital Inequality*. Edward Elgar Publishing, 284–295.
Pearson, Y. (2020). Child–robot interaction: What concerns about privacy and well-being arise when children play with, use, and learn from robots? *American Scientist, 108*(1), 16–22.
Peter, J., & Kühne, R. (2018). The new frontier in communication research: Why we should study social robots. *Media and Communication, 6*(3), 73–76.
Petronio, S. (2002). *Boundaries of Privacy: Dialectics of Disclosure*. State University of New York Press.
Proposal for a Directive of the European Parliament and of the Council on adapting non-contractual civil liability rules to artificial intelligence (AI Liability Directive) (2020). COM/2022/496 final, retrieved from https://eur-lex.europa.eu/legal-content/EN/TXT/?uri=CELEX%3A52022PC0496.
Purtova, N. (2018). The law of everything: Broad concept of personal data and future of EU data protection law. *Law, Innovation and Technology, 10*(1), 40–81.

Rabb, N., Law, T., Chita-Tegmark, M., & Scheutz, M. (2022). An attachment framework for human–robot interaction. *International Journal of Social Robotics*, *14*(2), 539–559.
Ranzini, G., Newlands, G., & Lutz, C. (2020). Sharenting, peer influence, and privacy concerns: A study on the Instagram-sharing behaviors of parents in the United Kingdom. *Social Media + Society*, *6*(4), 1–13.
Reeves, B., & Nass, C. (1996). *The Media Equation: How People Treat Computers, Television, and New Media Like Real People*. Cambridge University Press.
Rennie, E., Yunkaporta, T., & Holcombe-James, I. (2018). Privacy versus relatedness: Managing device use in Australia's remote Aboriginal communities. *International Journal of Communication*, *12*, 1291–1309.
Rodríguez-Hidalgo, C. (2020). Me and my robot smiled at one another: The process of socially enacted communicative affordance in human–machine communication. *Human–Machine Communication*, *1*, 55–69.
Roose, K. (2023, February 17). Why a conversation with Bing's chatbot left me deeply unsettled. *New York Times*. https://www.nytimes.com/2023/02/16/technology/bing-chatbot-microsoft-chatgpt.html.
Rueben, M., Aroyo, A. M., Lutz, C., Schmölz, J., Van Cleynenbreugel, P., Corti, A., . . . & Smart, W. D. (2018). Themes and research directions in privacy-sensitive robotics. In *2018 IEEE Workshop on Advanced Robotics and its Social Impacts* (pp. 77–84).
Rueben, M., Bernieri, F. J., Grimm, C. M., & Smart, W. D. (2017a). Framing effects on privacy concerns about a home telepresence robot. In *Proceedings of the 2017 ACM/IEEE International Conference on Human–Robot Interaction* (pp. 435–444).
Rueben, M., Grimm, C. M., Bernieri, F. J., & Smart, W. D. (2017b). A taxonomy of privacy constructs for privacy-sensitive robotics. *arXiv*, 1701.00841.
Sadeh, N., Acquisti, A., Bauer, L., Fredrikson, M., Crano, L., Datta, A., & Tschantz, M. (2021). Personalized privacy assistants for big data and the internet of things. *Carnegie Mellon University*. https://apps.dtic.mil/sti/pdfs/AD1140163.pdf.
Sætra, H. S. (2020). The foundations of a policy for the use of social robots in care. *Technology in Society*, *63*, 101383.
Schrock, A. R. (2015). Communicative affordances of mobile media: Portability, availability, locatability, and multimediality. *International Journal of Communication*, *9*, 1229–1246.
Schrum, M. L., & Gombolay, M. C. (2023). Privacy and personalization: Transparency, acceptance, and the ethics of personalized robots. In *ACM Human Robot Interaction '23, March 13–16, Stockholm, Sweden*.
Schulz, T. W., & Herstad, J. (2018). Walking away from the robot: Negotiating privacy with a robot. In *Electronic Workshops in Computing (eWiC)*. British Computer Society.
Smith, H. J., Dinev, T., & Xu, H. (2011). Information privacy research: An interdisciplinary review. *MIS Quarterly*, *35*(4), 989–1015.
Solove, D. J. (2008). *Understanding Privacy*. Harvard University Press.
Søraa, R. A., Nyvoll, P., Tøndel, G., Fosch-Villaronga, E., & Serrano, J. A. (2021). The social dimension of domesticating technology: Interactions between older adults, caregivers, and robots in the home. *Technological Forecasting and Social Change*, *167*, 120678.
Stark, L., Stanhaus, A., & Anthony, D. L. (2020). "I don't want someone to watch me while I'm working": Gendered views of facial recognition technology in workplace surveillance. *Journal of the Association for Information Science and Technology*, *71*(9), 1074–1088.
Su, P., & Yuan, X. (2021). Are you watching me? A study on privacy notice design of social robot. In *International Conference on Applied Human Factors and Ergonomics* (pp. 339–344). Springer.
Tang, Y. (2020). Privacy-preserving pose estimation for human-robot interaction. Master's Thesis, Cornell University. https://ecommons.cornell.edu/server/api/core/bitstreams/1a3d47c2-5d31-4fe1-a7b3-bedb7d155c8c/content
Tangermann, V. (2022, September 14). They put GPT-3 into that robot with creepily realistic facial expressions and yikes. *Futurism*,. https://futurism.com/the-byte/gpt-3-ameca-robot-facial-expressions.

Tamò-Larrieux, A. (2018). *Designing for Privacy and its Legal Framework*. Springer.

Tiku, N. (2022, June 11). The Google engineer who thinks the company's AI has come to life. *The Washington Post*, https://www.washingtonpost.com/technology/2022/06/11/google-ai-lamda-blake-lemoine/.

Tonkin, M., Vitale, J., Herse, S., Raza, S. A., Madhisetty, S., Kang, L., . . . & Williams, M. A. (2019). Privacy first: Designing responsible and inclusive social robot applications for in the wild studies. In *2019 28th IEEE International Conference on Robot and Human Interactive Communication* (pp. 1–8). IEEE.

Tonkin, M., Vitale, J., Ojha, S., Clark, J., Pfeiffer, S., Judge, W., . . . & Williams, M. A. (2017). Embodiment, privacy and social robots: May I remember you? In *International Conference on Social Robotics* (pp. 506–515). Springer.

Trepte, S. (2023). Privacy theory—Social, networked, relational, collective. In S. Trepte & P. K. Masur (Eds.), *The Routledge Handbook of Privacy and Social Media* (pp. 25–36). Routledge.

Trepte, S., Reinecke, L., Ellison, N. B., Quiring, O., Yao, M. Z., & Ziegele, M. (2017). A cross-cultural perspective on the privacy calculus. *Social Media + Society*, *3*(1), 1–13.

Trepte, S., Teutsch, D., Masur, P. K., Eicher, C., Fischer, M., Hennhöfer, A., & Lind, F. (2015). Do people know about privacy and data protection strategies? Towards the 'Online Privacy Literacy Scale' (OPLIS). In S. Gutwirth, R. Leenes, & P. de Hert (Eds.), *Reforming European Data Protection Law* (pp. 333–365). Springer.

Van Cleynenbreugel, P. (2017). The EU right to be forgotten in a privacy-sensitive human–robot interaction context. In *HRI 2017—'Smart interaction'*.

Vitak, J., & Zimmer, M. (2020). More than just privacy: Using contextual integrity to evaluate the long-term risks from COVID-19 surveillance technologies. *Social Media + Society*, *6*(3), 1–4.

Vitale, J., Tonkin, M., Herse, S., Ojha, S., Clark, J., Williams, M. A., . . . & Judge, W. (2018). Be more transparent and users will like you: A robot privacy and user experience design experiment. In *2018 13th ACM/IEEE International Conference on Human—Robot Interaction (HRI)* (pp. 379–387). IEEE.

Walrave, M., Verswijvel, K., Ouvrein, G., Staes, L., Hallam, L., & Hardies, K. (2022). The limits of sharenting: Exploring parents' and adolescents' sharenting boundaries through the lens of communication privacy management theory. *Frontiers in Education*, *7*, 803393. https://doi.org/10.3389/feduc.2022.803393.

Warren, S. D., & Brandeis, L. D. (1890). The right to privacy. *Harvard Law Review*, *4*(5), 193–220.

West, S. M. (2019). Data capitalism: Redefining the logics of surveillance and privacy. *Business & Society*, *58*(1), 20–41.

Westin, A. (1967). *Privacy and Freedom*. Atheneum Press.

Xie, L., & Lei, S. (2022). The nonlinear effect of service robot anthropomorphism on customers' usage intention: A privacy calculus perspective. *International Journal of Hospitality Management*, *107*, 103312.

Yang, D., Chae, Y. J., Kim, D., Lim, Y., Kim, D. H., Kim, C., . . . & Nam, C. (2022). Effects of social behaviors of robots in privacy-sensitive situations. *International Journal of Social Robotics*, *14*(2), 589–602.

Youn, S. (2009). Determinants of online privacy concern and its influence on privacy protection behaviors among young adolescents. *Journal of Consumer Affairs*, *43*(3), 389–418.

Zimmer, M. (2018). Addressing conceptual gaps in big data research ethics: An application of contextual integrity. *Social Media + Society*, *4*(2), 1–11.

Sociology

Richard Ling

Artificial Intelligence, Robotics, and the Evolution of the Social Sciences

Abstract: In this chapter, I examine the co-evolution of method (along with the technologies/tools that enable various methods) and theory in the social sciences. This is done with an eye towards understanding the eventual role of artificial intelligence[1] (which, in the context of this chapter, I consider a part of robotics) in the future evolution of the social sciences.

The chapter examines several examples of how developments in technologies have changed the trajectory of how sciences are done. Examples include the development of tools in astronomy and computational statistics. In each case, the development of new technology and access to new forms of data played into the trajectory of the area of inquiry. This runs counter to the normal notion of the hypothetico-deductive method that is often seen as central. Given this history, the chapter then poses the question of how the eventual development of artificial intelligence will affect the practice of social science.

Keywords: Technical development, artificial intelligence, computational statistics, astronomy, scientific methods

10.1 Introduction

Innovations have nudged us to revise how we do academic work. As will be discussed below, Galileo's telescope changed our thinking about celestial mechanics, and computational statistics (e.g., SPSS and R) have changed the theoretical trajectory of the social sciences. This chapter considers how robotics (and, in particular, artificial intelligence) will affect the development of the social sciences. These technical developments are in essence, new tools that help us to better evaluate social phenomena. The chapter will look at historical examples of how new tools changed the trajectory of science since they allowed us new insight. In many cases, empirical insight precedes theoretical codi-

1 In this chapter, the term 'artificial intelligence' will be used. This is one of several terms used in this area. Others include 'machine learning,' 'neural networks,' 'deep learning,' 'symbolic reasoning,' and so on. While there are differences between all of these, they often describe the use of computing devices that iteratively 'learn' tasks without explicitly being programmed to do so. This 'learning process' can be guided by humans at steps of the way, or it can be unguided.

Note: This is an expanded version of a paper presented at the online conference *Online Communication and the New World of Scholarship* held on October 9, 2020 at the Hungarian Academy of Sciences.

https://doi.org/10.1515/9783110792270-011

fication. This is backward when seen in the context of the hypothetico-deductive approach.

With the new ability to gather empirical insight, it can be that we observe things that are beyond our ability to make sense of. There is no theoretical foundation with which to describe the phenomenon. We do not have a mental framework with which to clarify it. In these cases, the theoretical explanation often comes *ex post facto*. It is the assertion here that the application of artificial intelligence to science is such a situation. This chapter, then, is a cautionary tale that will look at how science has been challenged and developed as a result of new technologies.[2] After examining the interaction between artificial intelligence and robotics, I will look at a medieval example and then an example from the middle of the twentieth century. Finally, I will look at the contemporary situation and the eventual use of automated tools, specifically artificial intelligence.

10.2 Robotics and Artificial Intelligence

Robotics often assumes that the device or apparatus being considered is a type of physical machine that carries out some tasks. Indeed, often the term 'robotics' evokes images of a humanoid device (e.g., Okafuji et al., 2022) that includes artificial intelligence. Thus, we need to consider its social impact.

Social robotics describes the robotics designed as applied in social areas such as education, health, and entertainment. In general, robotics as applied in factories is outside this domain. At the point of departure, the devices are developed as tools we use to address certain tasks. They are intended to lighten our load. There is a long legacy of applying tools to various tasks. The printing press allowed the mass production of books. The steel plow allowed us to cultivate the soil more efficiently. The gasoline motor (along with the development of a roadway system) allowed us to travel at a speed and at a level of comfort that were previously unknown. The clock replicated the rotation of the Earth and notified us of the passage of the day/hour/minute/, and so on. The transistor allowed us to develop tools that calculate and analyze information quickly and efficiently. All of these are either tools, automata, or simultaneously tools and automata.

Moving on to contemporary robots and robotics, there is the extension of classic automata in that the devices often can respond to different stimuli and to change behavior accordingly. They may be a robotic 'dog' that is able to negotiate uneven terrain, stairs, and so on, they may be a self-driving car that needs to understand the

2 There is a significant literature on the interaction between science, instrumentation, and technology. This includes careful discussions regarding the distinctions between, for example, the idea of the mechanical arts vs. science and their interrelations (Hansson, 2015; Shapin & Schaffer, 2011).

difference between rain and snow, or they may be a chatbot that understands the evolution of slang terms.

In general, tools are intended to augment human activities. In the case of robots and artificial intelligence applications these are often used to release humans from dangerous (e.g., bomb detection), tiring (e.g., lifting heavy objects), or repetitive (e.g., assembly work) tasks. Artificial intelligence fits into this area and allows robotic devices powered by it to better understand and cooperate with humans (Rajan & Saffiotti, 2017). Both are being used for tasks where there are a set of issues to consider and evaluate and/or the repetitive nature of the task can lead to human error (e.g., examination of x-ray images or insurance adjustment).

Putting this into the context of social robotics, artificial intelligence can facilitate various human tasks. Rather than simply being able to carry out a set of instructions based on a human-authored program of eventual alternatives, artificial intelligence focuses on fulfilling a particular objective through the interpretation of available data. Artificial intelligence 'learns' how to complete the task without explicit programming. For example, a bank could write a program with which to filter good and bad loan risks. It would likely contain a set of criteria (e.g., whether the individual has paid their previous loans, has a job, has other loans, etc.). By contrast, an artificial intelligence application with the same task (and access to a wide range of data on the individual) may consider some of the same elements, but would also draw on other, perhaps unexpected, dimensions of the individual's background. The exact mixture of these background elements remains obscure in the case of artificial intelligence. The former case focuses more squarely on explanation while the latter focuses on the outcome.

Thus, artificial intelligence is squarely within the domain of social robotics. That said, there is not always a physical device associated with artificial intelligence. It can be that the artificial intelligence system powers a pre-existing device such as a mobile phone or a computer. However, there can also be artificial intelligence applications that are oriented towards decision processes (e.g., judging loan applications, examining x-rays, evaluating parole risk, etc.). When brought into the context of this chapter, artificial intelligence systems can be used in academic research to analyze, for example, complex datasets.

To better understand the dynamics of artificial intelligence and its effect on the sciences, it is useful to examine historical developments where new tools changed the trajectory of scientific inquiry.

The following two sections will examine how technical developments changed the trajectory of scientific inquiry. This will be followed by a discussion of how artificial intelligence can be a disruptive force in the social sciences.

10.3 Galileo and Saturn 'Swallowing his Children'

In 1610, Galileo Galilei put together a primitive telescope of about 1 meter in length with somewhat impure and imperfectly formed lenses. His device, however weak and imperfect, brought the heavens closer by many orders of magnitude. Indeed, it was an instrument of discovery (Boon, 2015). For the first time in history, he used it to examine the mountains, pits, and other geographical features of the Moon. He discovered the moons of the Earth's neighboring planets, Venus, and Jupiter. Indeed, his observation of moons orbiting other planets contradicted the religiously enforced belief that all heavenly bodies orbit the Earth. As we shall see, the clash between evidence and belief worked out poorly for Galileo. But there were other unaccountable discoveries that awaited.

When he looked at Saturn, which at that time was thought to be the most distant planet, Galileo saw what he thought were either 'ears' protruding from the main planet; or perhaps a combination of three fuzzy planets. When he again looked at Saturn two years later, he was even more surprised since there was only one object, the planet itself, instead of three. He noted, "I do not know what to say in a case so surprising, so unlooked for and so novel. . . . Has Saturn swallowed his children?"

Galileo was living at a time of ferment when it came to understanding the solar system. Only one year before he first looked at Saturn, Johannes Kepler, with the help of Tycho Brahe, had published *Astronomia nova* showing, among other things, that the Copernican model with the Earth orbiting the Sun was more accurate than the Ptolemaic model of the heavens with the Earth at the center. Based to some degree on his observations of the moons of Jupiter, three years later Galileo asserted that the Earth rotated around the Sun, and not the opposite. He did this in a letter to the monk Benedetto Castelli, an action that eventually led to his inquisition 20 years later. Thus, Galileo was in the midst of both scientific and cultural/religious ferment. He was lauded for his work on Saturn, and it also helped to confirm the Copernican heliocentric model (Boon, 2015). This work eventually landed him in trouble with the Inquisition, where he was judged "vehemently suspect of heresy," and he indeed spent the last part of his life under house arrest.

Galileo died in 1642. This was before Christiaan Huygens used a telescope with greater magnification and improved optics to determine that Galileo's observation of Saturn's 'three planets' were what we now know to be its rings. They were not 'eaten' by the planet, but as Saturn moved around its 29.5 (Earth years) orbit of the Sun, sunlight was variously reflected off the rings to produce Galileo's 'ears' or disappeared when the rings were viewed 'edge-on.'[3]

3 Interestingly, though we take Saturn's rings for granted, there is still not a clear sense of why they have formed. Indeed, they seem to contradict the laws of entropy.

The episode illuminates one messy model in the progression of science. A new tool is developed that gives rise to a new way to interrogate existing data. Unrecognizable phenomena are observed. New systems of understanding need to be developed (*post hoc*) to grasp the meaning. Insight and intuition are marshaled to clarify unexpected findings. In the meanwhile, the insight and its clarification ruffle the feathers of the established sense of order but, with time, it becomes the orthodoxy.

10.3.1 The messiness of science

There is a notion of orderly progression in the hypothetico-deductive tradition that we often use. Interestingly, it was Christiaan Huygens (1690), who we met above, who outlined this approach to science in his *Treatise on Light*. This line of development often frames the narration in our academic articles.

In broad terms, the hypothetico-deductive approach says that we become grounded in a particular orientation to the phenomenon at hand. We then seek out a gap or an opening in this structure and develop conjectures (or hypotheses) about how the interaction or phenomenon functions. We then draw on the toolkit of methods to address the conjecture. In this process, we work to eliminate alternative explanations and we seek to falsify our conjectures. If they stand up, then we have a contribution. This is, in essence, a deductive system. There is the sense that the stuff that is happening 'out there' is, in one way or another, explained by relating them to the wider principles and definitions of the broader theoretical perspective.

In many cases, this is indeed how we do our work. These small incremental developments in insight accrue to form the body of a discipline. This is how much of our work takes place. It is how we find out 'what is happening out there,' whatever that might be.

However, there is also the recognition that it is not always so well formulated. One is reminded of the quip attributed to John Godfrey Saxe who in 1869 said, "Laws, like sausages, cease to inspire respect in proportion as we know how they are made."[4] The ordered way of telling the story of our research is not always how it is done. In some cases, we chance upon new lenses with which to view phenomena. In other cases, new forms of empirical information let us view the existent theories in a new light. In yet other situations, we buck the constraint of theory and tradition dictated by a theoretical orthodoxy in order to chart new ground.

I posit that we are on the cusp of a period when the rise of a new methodological lens (artificial intelligence) and access to very different types of large databases (big data) will show that the process of social science is not so linear (Yarkoni & Westfall,

4 https://quoteinvestigator.com/2010/07/08/laws-sausages/.

2017). Indeed, these developments promise to add a fundamentally new approach to understanding 'what is happening out there.'

In the sections below, I want to consider several examples where the line of narration is different from the deductive approach. In each case, the approach does not really support the simple line of progression noted by Huygens, but rather is more like the work in the sausage factory.

10.4 Computational Statistics, SPSS, and the Quantitative Turn in Social Sciences

10.4.1 Computational statistics as a new lens

Another example of disruptive technology is the development of computational statistics. Statistics has been used for various purposes for millennia. It gained focus, for example, in the mid-1700s as a tool[5] with which to analyze demographic information. In addition, the development of probability theory in the 1600s (Grinstead & Snell, 2012), and the least squares model (Stigler, 1986; see also Pearson, 1978) provided a way to understand various phenomena. From its development until the early 1970s, most often the work of doing statistics relied on manually calculating the material, perhaps using mechanical adding machines, to derive the answer. The ability to determine a mean or a standard deviation for a survey question with several hundred respondents could be an onerous task that was vulnerable to error.

With time, mainframe computers were used for statistical analysis. According to Watnik (2011), the work on the development of the atomic bomb at Los Alamos during the Manhattan Project was an important turning point. He describes how the need to model nuclear cascades and other tasks meant that the use of the computer to derive statistics was a central tool in this work. H. L. Anderson (1986) noted that computation would also produce solutions that the scientists had not suspected were possible. Indeed, the whole sub-discipline of soliton theory (i.e., the study in physics of self-reinforcing waves) owes its insight to unexpected computational results.

When examining the behavior of an equilibrium, Enrico Fermi and some colleagues tested their notions using statistical computing and:

> As expected, the computations showed that the initial vibrational energy gradually transferred into neighboring modes and eventually achieved equilibrium [. . . however], the completely un-

---

5 It is perhaps reductive to think of statistics as a tool since it is indeed a fully developed discipline within mathematics. Indeed, there are people who devote their careers to the study of statistics. That said, there are many other academics who use these methods as tools in their own areas of inquiry. The point in this work is to consider the latter situation.

> expected happened one day when a typical problem was being computed. Owing to a very lively and distracting discussion, the computations continued beyond the usual cutoff. The results were so strange and mysterious that everyone around was quick to assume that the computer, the traditional whipping boy, had gone awry. The vibrational energy had returned to its initial state, within a few percent! The rest is history—nonlinear studies were shown to have some fascinating aspects. The ideas of soliton theory emerged and there was an outpouring of papers that became a minor industry. Today this classic work is known as the FPU problem. (Anderson, 1986, p. 743)

Here we see unexpected insight arising from the use of new tools. As with Galileo's first viewing of Saturn's rings, there was the development of soliton theory.[6]

10.4.2 Computational statistics in the social sciences

During the Manhattan Project, statistical computing was limited to only a small number of themes. However, with the development of the transistor in the late 1940s, and the integrated circuit a decade later, we see the democratization of computing that eventually led to the personal computer.

These developments meant that computational statistics could be applied to the study of social interaction. To be sure, statistical analysis has been a part of the social science landscape for many decades extending back to Galton (Clauser, 2007) and beyond. However, until the 1970s there was extremely limited access to computerized data analysis. This often limited analyses to simple descriptive statistics given the laborious nature of calculating more advanced statistics.

With access to computing, many applications became possible. Statistical analysis was one of them, and one of the first broadly available programs was SPSS. In the early 1970s, three students at Stanford, Norman H. Nie, C. Hadlai Hull, and Dale H. Bent developed the core of the statistical package. The initial version was developed for mainframe computers and then for so-called mini-computers such as the DEC. With the later development of personal computers, SPSS moved to that platform in the early 1980s.

It can be said that SPSS, and other PC-based statistical software such as SAS, led to a particular trajectory in the social sciences (Wellman, 1998). Following Wellman, with the coming of programs such as SPSS:

> We no longer have to queue and beg an expert who possesses the rare knowledge of how to get a user-unfriendly statistical package to work. Instead, we are now routinely able to do complex analyses from a number of perspectives instead of just talking about complex processes but measuring simple relationships. We can easily check a variety of alternative causes and correlates, and we can take endless alternative views of our data. We revel in our post-Pearsonian ability to

6 In other examples from physics, Lenhard (2015) has discussed the role of computation in the case of Schrödinger's equation in quantum chemistry and Markov chain Monte Carlo methods.

> try dozens of complex procedures, to view things in ten different ways, to obsessively clean and re-weight data, to transform the intractable, to hunt down pesky residuals, and to apply once-obscure statistical tests with exotic-sounding names. It is an orgy of do-it-yourself quantitative analysis. (Wellman, 1998, p. 72)

Wellman, in many ways, describes the sausage factory. There is little question, however, that this tool has dramatically changed the types of analysis that are being done.

A quick review of recently published articles shows the regular use of methods such as regression, ANOVA, structural equation models, multi-level random effects models, and mixed effects models. These are used in the elaboration of a wide variety of theoretical directions. Looking back over the recent history of communication studies, their theoretical approaches include, but are hardly limited to, areas such as moral foundations theory (Graham et al., 2013), third-person effects (Davison, 1983; Rucinski & Salmon, 1990), confirmation bias (Nickerson, 1998; Wason, 1960), para-social intimacy (Perse & Rubin, 1989), social information processing (Walther, 2008), hyperpersonal theory (Walther, 1996), and media effects theory (Scheufele, 1999). This is not to assert that SPSS was the inspiration for these approaches; nor is it to assert that these methods cannot be done in the absence of statistical software. It can be argued, however, that access to computational statistics has facilitated their development.

Following Wellman (1998), the ease with which these procedures can be done means that other approaches to studying society do not have the same academic vitality. Wellman notes that qualitative analysis is often seen as the poorer cousin of quantitative analysis. The difficulty of transcribing, coding, and analyzing qualitative material often means that in our pursuit of a job, tenure, or promotion, it is often more expedient to do a survey and perform statistical analysis.

Another dimension of this issue is that since much of the work in social science is based on survey data or perhaps experimental work, the unit of analysis is often the individual. Following Wellman, "The very assumption of statistical independence, which makes standard statistical analyses so powerful detaches individuals from social structures and forces analysts to treat them as parts of a disconnected mass" (1998, p. 75). The focus is less on social structures and social processes, and more on individuals. Dyadic and triadic analysis, information flows, the position of ties within social networks, contextual information, and the like, are not examined as a system, but at best, as an attribute of the individual.

In some cases, statistical analysis is the "orthodoxy" (Wellman, 1998, p. 74) of social science research. To be sure, there are increasingly sophisticated approaches to examining qualitative material. That said, easily adopted statistical programs such as SPSS, SAS, and more recently MATLAB, Statistica, R, and so on, have facilitated a particular trajectory in the social sciences.

Thus, the development of computational statistics and later their diffusion to PCs in the form of, for example, SPSS show that the development of a new scientific tool uncovered unexpected theoretical directions (soliton theory in physics) and it fostered a particular trajectory in the social sciences. The development of computational statistics has allowed us to frame statistically based questions while neglecting others. It has also promoted a Kuhnian scientific community (Kuhn, 2012) with its version of how 'normal science' is done. I will return to this theme below.

10.5 Social Science Theory and the Disruptive Effect of Artificial Intelligence

The previous example showed how a transition in our ability to access or examine information has affected the direction of scientific inquiry. Galileo's primitive telescope and the development of computational statistics have each facilitated a new and often unexpected dynamic in their respective domains.

As noted above, we are on the verge of another transition in the social sciences because of artificial intelligence. Indeed, we are already starting to see the use of this approach (Rudin, 2018). Social scientists are using newly accessible data sources and artificial intelligence to, for example, examine linguistic drift in society (Garg et al., 2018), interrogate economic behavior (Marwala & Hurwitz, 2017), and understand the dynamics of dyadic ties and their surrounding clusters of friends (Mattie et al., 2018). Thus, as with other disruptive events, it is time to discuss whether, like Galileo's telescope and its effect on astronomy, artificial intelligence is a new lens that will adjust, challenge, and refine our sense of how to do social science.

According to Yarkoni and Westfall, there is merit in this position. In the case of psychology, they note,

> Psychology has historically been concerned, first and foremost, with explaining the causal mechanisms that give rise to behavior. Randomized, tightly controlled experiments are enshrined as the gold standard of psychological research, and there are endless investigations of the various mediating and moderating variables that govern various behaviors. We argue that psychology's near-total focus on explaining the causes of behavior has led much of the field to be populated by research programs that provide intricate theories of psychological mechanism but that have little (or unknown) ability to predict future behaviors with any appreciable accuracy. (Yarkoni & Westfall, 2017, p. 1100)

They go on to say that the application of artificial intelligence "will reveal major holes in an otherwise elegant explanatory story, and many well-known findings—some with seemingly large effect sizes—will likely fail to survive rigorous cross-validated analysis in large samples" (p. 1118). They also suggest that the adoption of artificial intelligence can bring a different and more predictive focus.

10.5.1 Explanation vs. prediction

As noted above, Yarkoni and Westfall consider the distinction between explaining and predicting a phenomenon (Yarkoni & Westfall, 2017; see also Domingos, 2012). The goal of science should both explain and predict or model the outcome. Indeed, the two are intertwined since often explanation facilitates prediction. However, Yarkoni and Westfall as well as, for example, Hofman and colleagues (2017) posit that there is the possibility of focusing too closely on the explanatory issues while ignoring the predictive side. Hofman et al. note, "Rather than asking whether a given theory can predict some outcome of interest, the accepted practice in social science instead asks whether a particular coefficient in an idealized model is statistically significant and in the direction predicted by the theory" (2017, p. 486).

Hofman et al. (2017) go on to say that the traditional social science model, perhaps ironically founded on the notion of computational statistics described above, has also been criticized as being the search for statistical significance resulting in non-replicable findings. They note, however, that because of artificial intelligence there is also currently a growing interest in computational social science examining areas such as the dynamics of social networks, information diffusion, and social inequality. These have blossomed because of access to large databases and artificial intelligence. As is obvious, this line of research brings into question central issues in the ontological framing of social sciences.

10.5.2 Artificial intelligence and variables

While social science is often looking for explanations, artificial intelligence applications are teasing out predictions that include insights and interactions that are difficult for humans to understand. Just as the new tool of the telescope gave Galileo incomprehensible information (e.g., a planet with 'ears' that were subsequently 'eaten'), so is artificial intelligence. We are at a point where we understand that artificial intelligence can passively provide reliable findings (note: we are also finding that it can provide cockeyed findings). However, we are not able to understand how these findings were derived. In some cases, it seems that artificial intelligence is even conceptualizing phenomena using hidden state variables that are unknown to us (Chen et al., 2022). This is an extremely disruptive finding. Rather than simply relying on a recognized set of variables, the artificial intelligence system is finding new variables with which to model the systems. It is most often the case that we know which variables affect the outcome of a particular system. However, Chen et al. have applied artificial intelligence to the examination of dynamic systems with the intention of predicting the future state of the system. Their results show that artificial intelligence can do this, but that the very variables used to model the system are unknown.

10.5.3 Variables, theory, and artificial intelligence

A central issue associated with the use of artificial intelligence, and that is an extension of the explanation/prediction issue noted above, is the role of theory. What is the position of theory in the social sciences and how will it likely change in an eventual era of artificial intelligence? How will our current trove of theories help us to understand the role of digital technology in society? How will the development of technology itself change the way that we theorize? To start with, let us think about what constitutes a theory.

Theory in the social sciences can be seen, at one level, as a creation of the mind. In this view it is a conceptual framing that has been collectively developed to guide the study of various social phenomena. To be sure, a theory is nudged and prodded by empirical findings, but it is nonetheless a socio-cognitive construction.

Following Kuhn (2012), theory can be seen as a collectively held belief that directs the trajectory of scientific inquiry. This is not to say that it is correct, but rather that it has the power to direct and control activity.

As suggested above, the application of artificial intelligence to the questions of social science will, in all probability, mean that we need to reconsider portions of our theoretical legacy. Just as Galileo's conceptualization of Saturn needed revision as the empirical picture was refined, notions of cognition, social bonding, the flow of influence, and other social phenomena may be seen in a new light with the evolution of artificial intelligence.

The notion of theory can vary across different areas of academia. According to the US National Academies of Science,[7] a theory explains and importantly predicts phenomena. In the introduction to the fiftieth-anniversary edition of the *Structure of Scientific Revolutions*, Ian Hacking says that "Theories should be accurate in their predictions, consistent, broad in scope, present phenomena in an orderly and coherent way, and be fruitful in suggesting new phenomena or relationships between phenomena (Hacking & Kuhn, 2012). In this phase of its use, which is perhaps more aligned with the natural sciences' use of the concept, a theory is seen as a rather inviolable connected system of theorems that models a phenomenon. In its strongest form, this is seen as a natural law. In this strong form, the theoretical model always describes a phenomenon when the prescribed conditions are present. Examples of this include things like Kepler's laws, Boyle's Law, Maxwell's equations, the General Laws of Motion, Avogadro's Law, Charles's Law, and so on. Thus, by referring to Charles's Law we know that a gas increases in volume by a fixed amount for each increase in degree centigrade temperature. Boyle's Law states that the product of absolute pressure and volume of a gas is constant. In these cases, there are rarely, if ever, variations from the theory.

7 http://www.nas.edu/evolution/TheoryOrFact.html.

As discussed by Hofman et al. (2017) the application of theory in the social sciences has another dynamic. In this case, a theory does not have the absolute formulation as seen in the natural laws. Rather, theories are often our best attempt at describing the interactions. Take, for example, the economic 'law' of supply and demand.[8] This is a reasonable model of pricing in the marketplace. However, it does not have the same precision as, for example, Kepler's laws. To give supply and demand its due, there are many situations where, as supply increases, prices fall. Nonetheless, there are inconsistencies in the interaction between the elements that cannot be explained. There are many reasons for this, not the least of which is that we rarely behave completely in the spirit of John Stuart Mill's 'homo economicus.' We are only rarely rational and narrowly self-interested in our economic activities. The idea of supply and demand is a useful heuristic. Indeed, in a broad way, it is insightful. That said, it also has its obvious drawbacks. Coming back to the idea of artificial intelligence, Marwala and Hurwitz (2017) have suggested that economic concepts such as supply and demand, rational choice, bounded rationality, portfolio theory, and a variety of other economic perspectives will be challenged and modified by the use of artificial intelligence.

10.5.4 Spuriousness

Clearly one of the bugaboos associated with the coming black box of artificial intelligence is spuriousness. Assuming that we can, for example, find a large accommodating database and that we are interested in understanding the interaction, is it possible to simply enter the data into a machine and turn on the switch? There is obviously the immediate concern that the outcomes will simply be spurious relationships. Indeed, in the social sciences, one role of theory is to guard against exactly this situation. If there is no guidance provided by theory, there is the idea that the machine will likely spin out any number of odd, unfounded combinations.[9] Indeed there is a small industry of websites showing these kinds of relationships (e.g., the number of people who drowned by falling into a swimming pool as associated with the number of films Nicolas Cage appeared in, etc.). These findings suggest that analysis which is not guided by theory can report findings that are simply coincidences. This is clearly an issue, but there can be aspects of throwing out the baby with the bathwater.

There are many examples of what seem to be spurious findings that become the core of science. For example, Louis Pasteur noted in his speech upon becoming the Dean of the Faculty of Sciences at Lille, the experience of the Danish scientist Ørstad (Andel, 1994). Pasteur recounted Ørstad's surprise when he saw by chance that when current passes through a wire it deflects the needle of a compass. This finding, how-

8 It is rather an overstatement of the economists to call this a law.
9 https://hbr.org/2015/06/beware-spurious-correlations.

ever, unlocked the core of electromagnetics that again spawned our understanding of electricity and radio, and is the essential insight that facilitated modern computing (and artificial intelligence). Pasteur noted the brilliance of the serendipitous discovery by saying that "in the sciences . . . chance favors the prepared mind." Indeed as shown with Fermi's soliton theory, Garfinkel's adoption of the tape recorder, and Galileo/Huygens's analysis of Saturn, an openness to the unexpected can provide insight where others see only odd findings. Citing Bernard and his comments on innovative insight in psychology, Andel (1994, p. 634) notes that these are most commonly "born by chance."

10.5.5 Issues to consider

There are obviously issues to consider in the use of artificial intelligence. The following is an incomplete list of some of these.

- Consider whether the explanation of the mechanism or the prediction of an outcome is important. The two are clearly not mutually exclusive. However, there needs to be a consideration of this issue. It may be possible to, for example, examine the outcome from artificial intelligence-based analyses for different sets of inputs, thus helping to isolate the contribution of different inputs and thereby clarifying the mechanism. In this way, there can be an interaction between explanation and prediction (Hofman et al., 2017; Yarkoni & Westfall, 2017).
- What is the nature of the data being used to train the artificial intelligence application? This issue points to an ethical dimension of artificial intelligence. Systems that are trained on biased data (e.g., that contain cultural prejudices) will not produce unbiased predictions (Bostrom & Yudkowsky, 2011). The data used to train systems can reflect latent cultural biases. This can include tacit racism, and biases towards, for example, profit seeking, and so on (Char et al., 2018).
- Be open to and creative in thinking about alternative explanations. The development of artificial intelligence will likely uncover new and unexpected interactions. It will require greatness of thought (or Pasteur's 'prepared mind') to understand them and to consider them in concert with existing theoretical concepts. In some cases, the 'findings' will be wrong. In other cases, they can shake the cage of orthodoxy. What other theoretical or conceptual resources can be brought to bear on clarifying seemingly inconsistent findings? These new insights and unexpected findings can be the core of great new approaches.
- To the degree possible, there is a need to provide all insight into how the analyses are done. Further, there is the need to develop standards, and indeed a praxis for judging this type of work in, for example, the peer review process. This means that the authors of these studies will need to work to make their results as transparent and comprehensible as possible and the academic community will need to work out how to approach this material.

- As Kuhn (2012) suggests, there will likely be resistance to accepting the results of the computational analyses. The critiques perhaps need to be parsed in relation to whether they are a comment on the intrinsic analysis or arising from Kuhnian paradigmatic motivations.
- Not every issue can be examined using artificial intelligence. A prerequisite is often access to large databases. When these are not available, artificial intelligence is not feasible.

10.6 Conclusion: The Social Structure of Theoretical Communities

I have considered academic transitions associated with the development of new approaches to a particular domain. In each case, the transition caused comment. Galileo faced the Inquisition. Garfinkel and Goffman fell out over the different perspectives on social interaction. We are all familiar with the fault lines that can exist between the practitioners of qualitative and quantitative analysis. Indeed, in some cases, the method is more of an ideology than a tool.

According to Kuhn, the process of science necessarily includes the rise and the withering of paradigms. As a paradigm rises, it affords insight and organizes understanding in a way that was not possible given the previous academic regime. As it matures, it enters the period called 'normal science' (Kuhn, 2012). As the paradigm becomes central, it also, in a sense, enforces its position. The paradigm and its practitioners determine what counts as a contribution, what approaches are legitimate, and how the careers of novices practicing in the area are formed. In short, there is what Berger and Luckmann (1967) would call 'legitimation structures.'

Following Thomas Kuhn, academia has undeniable social inertia (2012). Kuhn described groupings and in some cases coalitions associated with a particular theory or perspective. His book focuses largely on the development, flowering, and eclipsing of paradigms in the natural sciences (Ewoldsen, 2017). At its core, Kuhn was interested in the social dynamics of academia and the social dynamics among scientists. For example, he notes:

> The more carefully they study, say, Aristotelian dynamics, phlogistic chemistry, or caloric thermodynamics, the more certain they feel that those once current views of nature were, as a whole, neither less scientific nor more the product of human idiosyncrasy than those current today. If these out-of-date beliefs are to be called myths, then myths can be produced by the same sorts of methods and held for the same sorts of reasons that now lead to scientific knowledge. . . . Rather than seeking the permanent contributions of an older science to our present vantage, they attempt to display the historical integrity of that science in its own time. They ask, for example, not about the relation of Galileo's views to those of modern science, but rather

> about the relationship between his views and those of his group, i.e., his teachers, contemporaries, and immediate successors in the sciences. (Kuhn, 2012)

As a paradigm shift approaches, there is sometimes the urge to, in effect, circle the wagons around particular academic pre-existing orthodoxy.

Kuhn focused his analysis on how scientific revolutions form and take place. Interestingly, he wrote only sparingly on transitions afforded by technical advancement and the disruptive development of new methods. Just as with the Kuhnian transitions that arise from the development of an alternative paradigm, it is possible to assert that these technical transitions have some of the same dynamics. The development of new ways of approaching the object of study brings to the surface all the social issues outlined by Kuhn.

Placing this into the context of artificial intelligence and contemporary theorizing, and research in the social sciences, I posit that we need to be prepared and open to surprise. All too often, I get the sense that we treat our theories and our methods as ideologies and not as tools. They are tools, not ideologies. It is better to seek out the contributions that this approach can provide than to condemn it.

Finally, we do not know how these issues will play out. The process, like making sausage, will probably be messy. The analyses done using artificial intelligence can perhaps confirm our pre-existing theoretical approaches. It may also be that they will uncover unknown or unfamiliar interactions. Thus, it is my sense that we need to approach theory and research with humility. We need to be open to the unexpected. It will be interesting, but it may also be fraught. Thus, I see a new (perhaps inductive) moment arising where our current way of looking at the field will be challenged.

References

Andel, P. van (1994). Anatomy of the unsought finding. Serendipity: Origin, history, domains, traditions, appearances, patterns and programmability. *British Journal for the Philosophy of Science, 45*(2), 631–648.

Anderson, H. L. (1986). Scientific uses of the MANIAC. *Journal of Statistical Physics, 43*(5–6), 731–748. https://doi.org/10.1007/BF02628301.

Berger, P., & Luckmann, T. (1967). *The Social Construction of Reality: A Treatise in the Sociology of Knowledge*. Anchor.

Boon, M. (2015). The scientific use of technological instruments. In S. O. Hansson (Ed.), *The Role of Technology in Science: Philosophical Perspectives* (pp. 55–79). Springer.

Bostrom, N., & Yudkowsky, E. (2011). The ethics of artificial intelligence. In W. Ramsey & K. Frankish (Eds.), *Cambridge Handbook of Artificial Intelligence* (pp. 316–334). Cambridge University Press.

Char, D. S., Shah, N. H., & Magnus, D. (2018). Implementing machine learning in health care—Addressing ethical challenges. *New England Journal of Medicine, 378*(11), 981–983. https://doi.org/10.1056/NEJMp1714229.

Chen, B., Huang, K., Raghupathi, S., Chandratreya, I., Du, Q., & Lipson, H. (2022). Automated discovery of fundamental variables hidden in experimental data. *Nature Computational Science*, *2*(7), 7. https://doi.org/10.1038/s43588-022-00281-6.

Clauser, B. E. (2007). The life and labors of Francis Galton: A review of four recent books about the father of behavioral statistics. *Journal of Educational and Behavioral Statistics*, *32*(4), 440–444. https://doi.org/10.3102/1076998607307449.

Davison, W. P. (1983). The third-person effects and the differential impact in negative political advertising. *Journalism Quarterly*, *68*, 680–688.

Domingos, P. (2012). A few useful things to know about machine learning. *Communications of the ACM*, *55*(10), 78–87.

Ewoldsen, D. R. (2017). Normal science and paradigm shift. In J. Matthes, C. S. Davis, & R. F. Potter (Eds.), *The International Encyclopedia of Communication Research Methods* (pp. 1–17). John Wiley & Sons, Inc.

Garg, N., Schiebinger, L., Jurafsky, D., & Zou, J. (2018). Word embeddings quantify 100 years of gender and ethnic stereotypes. *Proceedings of the National Academy of Sciences*, 201720347. https://doi.org/10.1073/pnas.1720347115.

Graham, J., Haidt, J., Koleva, S., Motyl, M., Iyer, R., Wojcik, S. P., & Ditto, P. H. (2013). Moral foundations theory: The pragmatic validity of moral pluralism. *Advances in Experimental Social Psychology*, *47*, 55–130.

Grinstead, C. M., & Snell, J. L. (2012). *Introduction to Probability: Second Revised Edition* (rev. ed.). American Mathematical Society.

Hansson, S. O. (2015). Science and technology: What they are and why their relation matters. In S. O. Hansson (Ed.), *The Role of Technology in Science: Philosophical Perspectives* (pp. 11–23). Springer.

Hofman, J. M., Sharma, A., & Watts, D. J. (2017). Prediction and explanation in social systems. *Science*, *355*(6324), 486–488.

Huygens, C. (1690). *Treatise on Light* (S. P. Thompson, trans.). Project Gutenberg. file:///C:/Users/rili/Google%20Drive/1a%20Active/1a%20Writing/1%20Korea%20keynote/Lit/pg14725-images.epub.

Lenhard, J. (2015). Iteration unleashed: Computer technology in science. In S.O. Hansson (Ed.), *The Role of Technology in Science: Philosophical Perspectives* (pp. 113–130). Springer.

Marwala, T., & Hurwitz, E. (2017). *Artificial Intelligence and Economic Theory: Skynet in the Market*. Springer.

Mattie, H., Engø-Monsen, K., Ling, R., & Onnela, J.-P. (2018). Understanding tie strength in social networks using a local 'bow tie' framework. *Scientific Reports*, *8*(1), 9349. https://doi.org/10.1038/s41598-018-27290-8.

Nickerson, R. S. (1998). Confirmation bias: A ubiquitous phenomenon in many guises. *Review of General Psychology*, *2*(2), 175.

Okafuji, Y., Ozaki, Y., Baba, J., Nakanishi, J., Ogawa, K., Yoshikawa, Y., & Ishiguro, H. (2022). Behavioral assessment of a humanoid robot when attracting pedestrians in a mall. *International Journal of Social Robotics*. https://doi.org/10.1007/s12369-022-00907-9.

Pearson, K. (1978). *The History of statistics in the 17th and 18th Centuries against the Changing Background of Intellectual, Scientific and Religious Thought; Lectures by Karl Pearson Given at University College London during the Academic Sessions 1921–1933*. Cambridge University Press.

Perse, E. M., & Rubin, R. B. (1989). Attribution in social and parasocial relationships. *Communication Research*, *16*(1), 59–77.

Rajan, K., & Saffiotti, A. (2017). Towards a science of integrated AI and robotics. *Artificial Intelligence*, *247*, 1–9. https://doi.org/10.1016/j.artint.2017.03.003.

Rucinski, D., & Salmon, C. T. (1990). The 'other' as the vulnerable voter: A study of the third-person effect in the 1988 US presidential campaign. *International Journal of Public Opinion Research*, *2*(4), 345–368.

Rudin, C. (2018). *Can Machine Learning Be Useful for Social Science?* The Cities Papers. http://citiespapers.ssrc.org/can-machine-learning-be-useful-for-social-science/.

Scheufele, D. A. (1999). Framing as a theory of media effects. *Journal of Communication*, *49*(1), 103–122.

Shapin, S., & Schaffer, S. (2011). *Leviathan and the Air-Pump: Hobbes, Boyle, and the Experimental Life*. Princeton University Press.

Stigler, S. M. (1986). *The History of Statistics: The Measurement of Uncertainty before 1900*. Harvard University Press.

Walther, J. B. (1996). Computer-mediated communication: Impersonal, interpersonal, and hyperpersonal interaction. *Communication Research, 23*(1), 3–43.

Walther, J. B. (2008). Social information processing theory. In L.A. Baxter & O. Dawn Braithwaite (Eds.), *Engaging Theories in Interpersonal Communication: Multiple Perspectives* (pp. 391–404). Sage.

Wason, P. C. (1960). On the failure to eliminate hypotheses in a conceptual task. *Quarterly Journal of Experimental Psychology, 12*(3), 129–140. https://doi.org/10.1080/17470216008416717.

Watnik, M. (2011). Early computational statistics. *Journal of Computational and Graphical Statistics, 20*(4), 811–817.

Wellman, B. (1998). Doing it ourselves: The SPSS manual as sociology's most influential recent book. In D. Clauson (Ed.), *Required Reading: Sociology's Most Influential Book* (pp. 71–78). University of Massachusetts Press.

Yarkoni, T., & Westfall, J. (2017). Choosing prediction over explanation in psychology: Lessons from machine learning. *Perspectives on Psychological Science, 12*(6), 1100–1122.

Astrid Weiss

Human Interactions With (Embodied) AI: The Future of Authenticity in Human–AI Relation(ship)s

Abstract: Approximately 20 years ago, the research field of human–robot interaction (HRI) emerged with the objective of designing robots that laypeople could interact with in an 'intuitive' manner, among other goals. It was argued that HRI differs from human–human interaction and human–computer interaction due to the materiality of robots and the apparent agency associated with them. Over the past decade, an increasing amount of empirical evidence has been accumulated regarding the similarities and differences in how laypeople perceive and interact with robots. However, studying human–robot interactions presents unique challenges due to the 'wicked' nature of the problems involved and the diverse range of robot embodiments. Presently, we are witnessing significant growth in non-embodied artificial intelligence (AI), surpassing the advancements in physically embodied AI. In this chapter, I argue that the forthcoming evaluation criterion to evaluate interactions with both embodied and non-embodied AI systematically is authenticity, drawing upon the insights gained from HRI research as an ideal foundation to address this criterion. What will authenticity even mean in the future when AI generates dialogues, voices, movements, and so on, and humans become more and more immersed in these interactions?

Keywords: Human–robot interaction, human–agent interaction, embodiment, AI, parasociality, authenticity

11.1 Introduction

There was a time, approximately 10–15 years ago, when the 'rise of the robots' was widely discussed, particularly in the media. However, despite the buzz, our everyday lives have not witnessed significant advancements beyond the presence of autonomous vacuum cleaners, lawnmowers, and delivery robots. As the research field of human–robot interaction (HRI) progressed with the aim (among others) of designing robots in a way that enabled laypeople to 'intuitively' interact with them, Kiesler and Hinds (2004, p. 3), in the preface to the special issue on HRI in the *Human–Computer Interaction* journal noted that "people seem to perceive autonomous robots differently than they do most other computer technologies." Similarly, Dautenhahn (2018, p. 2) argued that "human–robot interaction is very different from human–human interaction, human–computer interaction, and human–animal interaction."

https://doi.org/10.1515/9783110792270-012

What was once largely hypothetical has now been supported by empirical evidence. A considerable amount of research in HRI has sought to determine the specific factors that contribute to the differences observed: Is it the autonomy of the robots, the expectations, and narratives surrounding them, or the apparent agency[1] that is supported by humans' tendency to anthropomorphize robots and the robot's appearance, task, and interaction context (Takayama, 2012)? Of particular interest in HRI and social robotics is the topic of agent embodiment, which has witnessed a surge in studies. In a comprehensive review of 57 experiments that examined different aspects of embodiment in HRI, Deng et al. (2019) identified six different notions of embodiment, where, according to the authors, physical and social embodiment were the most relevant in the context of socially interactive robots. Physical and social embodiment provide a possibility to leverage multimodal communication and enhance the perception of an agent as more trustworthy, helpful, and engaging than disembodied agents. Importantly, empirical evidence suggests that robot embodiment has a measurable effect on performance and perception of social interaction. In their own words, various embodiments "in physically embodied socially interactive robots . . . are used as tools for communication, acceptance, and engagement" (Deng et al., 2019, p. 10). The researchers identified 43 studies that demonstrated the superiority of physically embodied agents in enhancing how individuals perceive and interact with these technologies.

In our own research, we have observed similar effects. For example, we compared anthropomorphic language use in online forums about the embodied agents Q. bo One, Vector, and Echo Show (see Figure 11.1). Initially, we expected to find the highest degree of anthropomorphic language use for Q.bo One due to its humanoid shape; however, "findings suggest that the life-likeness of the artifact is not pre-dominantly linked to the appearance, but to its interactivity and attributed agency and gender" (Weiss et al., 2020, p. 1).

Similarly, other research suggests that decreasing human likeness in robots does not necessarily correspond to a decrease in perceived cognitive and affective capabilities (Fortunati et al., 2023). However, anthropomorphizing consumer robots can increase their perceived warmth but decrease liking due to the 'uncanny valley' effect (Kim et al., 2019). This preliminary investigation already provides valuable insights into the complexity of anthropomorphism in connection to different materializations/embodiments of voice assistants. There is a high impact of apparent agency and voice design, rather than the physical embodiment we initially expected to be influential.

Related research has provided similar findings, including a study that examined the perceptions of university students towards the voice assistant Alexa. Through an

1 Apparent agency refers to the phenomenon on where individuals tend to attribute intentions and beliefs to robots, and treat them as if they possess genuine agency—see more details in Hannibal (2023).

Figure 11.1: Three differently embodied agents with similar functionality: Q.bo One, Vector, and Echo Show (from left to right). (Image source: ©SalzburgResearch).

online survey, participants regarded Alexa as a distinct entity, attributing both technical and social characteristics to the system (Fortunati et al., 2022a). Similarly, a YouTube video analysis of the Sophia robot showed that its highly human-like design did not significantly influence the participants' perception of the robot as human-like. Participants attributed only certain mental functions to Sophia and did not associate her with any emotions. Additionally, the use of Sophia was found to be associated with gender stereotypes typically associated with women's professions and occupations, although not entirely (Fortunati et al., 2022b).

In this chapter, I argue that the distinction in perception arises not solely from either the human's preferences or specific robot attributes but rather from the process of interaction itself. My particular interest lies in investigating whether similar mechanisms come into play when interacting with robots or embodied artificial intelligence (AI) in comparison to other technologies or humans, and how people perceive and experience these interactions. Given the rapid advancements in non-embodied AI, I deem it crucial to focus on the aspect of authenticity in human–AI relation(ship)s. However, I exercise caution in using the terminology of human–AI relation(ship)s, as our current understanding falls short in determining which aspects of authentic human–human or human–AI relation(ship)s can be replicated or substituted by technology.

The Merriam-Webster dictionary defines authentic as both "worthy of acceptance or belief as conforming to or based on fact" and "conforming to an original so as to reproduce essential features." While topics such as acceptance, trust, fairness, reliability, and understandability are commonly associated with responsible AI and HRI, I believe that the concept of authenticity in AI and people's relation(ship)s with it deserves more focused attention. Understanding the interplay between these topics is crucial for advanced AI applications, particularly in the case of social robots that interact in close proximity with humans. My research in recent years has primarily centered around human interaction with embodied AI. However, with the rise of generative AI applications, the significance of authenticity in these interactions has become increasingly apparent. Questions about what is real, what is fake, and how authenticity will be defined in the future are of paramount importance (especially in relation to virtual and hybrid reality solutions like the metaverse). We need to enhance our understanding of how AI affects human–human and human–technology relation(ship)s in order to investigate how we can design responsible AI that fosters a feeling of authenticity and supports humans in understanding the contribution of AI in their relations. The lessons we have learned from human-centered HRI research on embodied AI serve as an ideal foundation to embark upon this endeavor.

11.2 Embodied vs. Non-embodied AI

The key distinction between embodied and non-embodied AI lies in the technological embodiment and the ability of robots to navigate and operate within our physical reality. In his book *Robot Futures*, Nourbakhsh (2015) proposed the idea that artificially intelligent robots could serve as a bridge between the real and virtual worlds, leveraging their unique attributes. However, at present, we are witnessing a rise in non-embodied AI, exemplified by AI text and art generators, as well as the vision of the metaverse as a hybrid space where the boundaries between the real and virtual worlds are expected to merge.

The concept of the technological embodiment of robots encompasses far more than just their physical components like hardware, sensors, and actuators. It extends to their overall materiality, which includes the cognitive architecture that orchestrates their behavior in the world. This materiality plays a crucial role in defining and limiting the capabilities of embodied AI, particularly in the context of interaction with humans. For instance, the capacity for conveying personality in a social robot can be realized through specific design features. A social robot equipped with the ability to detect and respond to a person's gaze enables more human-like interactions compared with a robot with more limited interaction possibilities, such as a touchscreen interface (Willemse et al., 2018).

Additional essential elements of a robot's materiality, as understood in the research fields of HRI and social robotics, include its physical and morphological embodiment. Deng et al. (2019) conducted a comprehensive review of studies examining various aspects of robot embodiment and found that physical and morphological embodiment significantly contributes to communication, acceptance, and engagement in HRI. Empirical evidence indicates that the physical body of a robot impacts human interaction and the associated expectations regarding functionality and sociality (Grigore et al., 2016). The embodiment of a robot elicits a wide range of human responses due to its bodily presence and design features (Williams et al., 2021). For example, Hwang et al. (2013) ran a study that used 27 different shapes of robots (including visual representations and physical prototypes) to explore whether any of them aroused affective responses in humans. The study concluded that certain robot bodies elicited particular emotions and that these and the Big Five personality traits were perceived more strongly through present physical prototypes than through images of the same robots. Similarly, Dennler et al. (2022) developed an open-source database of 165 robot embodiments and showed through three crowd-sourced studies the effect of robot design on user expectations of robot capabilities.

The presence of a wide range of robot embodiments poses a unique challenge for HRI research and contributes to the current replication crisis experienced in the field (Weiss, 2022). Increasingly, findings from experimental user studies are failing to be replicated. This replication crisis is not exclusive to HRI and social robotics but extends to other disciplines such as human–computer interaction, psychology, medicine, and all fields involving human participants in research. The HRI research community, in its origins, aimed at analytic generalization for claims, such as: 'a trustworthy robot looks like x, behaves like y, and so on,' considering quality criteria for the interaction with robots as either a property of the robot or the human (see Hannibal and Weiss (2022) for more details on the argument to consider trust as an event between a human and a robot). Instead, I propose considering a human–robot interaction setting as a "technology probe" (Hutchinson et al., 2003), which provides opportunities to explore the relationship between people, data, and technology, surpassing the focus solely on the system itself (Weiss, 2022).

In the following, I present three different approaches of how to consider the human and their relation towards a robot in human-centered HRI research, namely: (1) human translated, (2) stakeholder involvement, and (3) co-shaping. I will briefly outline the differences between these approaches and then shift focus towards the insights they provide for future research on embodied AI. This perspective will pave the way for reflecting on the significance of authenticity in gaining a deeper understanding of potential parasocial relationships between humans and AI in the future.

11.3 Three Different Approaches to Human-Centered HRI Research

The distinguishing factor of HRI research is the human component, which sets it apart from other technical disciplines such as robotics, computer vision, and AI. While the field focuses on the development of (semi-) autonomous robotic systems capable of interacting effectively with laypeople, user studies play a significant role in contributing to the research field alongside technical, design, methodological, and theoretical considerations. Contributions in HRI research can be categorized into robot-centered, cognition-centered, or human-centered perspectives, as outlined by Dautenhahn (2007). The categorized approaches discussed in the following primarily aim to investigate human responses towards robots. A similar categorization is proposed by H. R. Lee et al. (2022).

11.3.1 The human-translated approach

In my perspective, HRI research that aims to replicate human interaction styles in robots can be described as a human-translated approach (Weiss, 2022). Such approaches frequently involve studying interaction paradigms through observations of human–human interactions and then translating the findings into the design of robot behaviors to the fullest extent possible. The rationale behind this approach stems from the fact that people are already acquainted with human behavioral strategies (Breazeal, 2003) and can intuitively relate to them (Okuno et al., 2009).

HRI research that follows a human-translated approach is grounded in the belief that human–human interaction serves as the 'gold standard for intuitiveness.' For instance, in the context of a study for developing a humanoid robotic head, we used the recognition of human facial expressions as a benchmark to explore the recognition rate of facial expressions (Mirnig et al., 2015). Similarly, in another study on the perception of congruent and incongruent bodily emotional expressions of the humanoid robot Pepper, we aimed to closely mimic the bodily expressions observed in humans (Tsiourti et al., 2019). It is important to note that human-translated studies conducted in a laboratory setting often involve participants and research environments that may not represent the intended users or the real-world deployment settings for the robot.

When taken to an extreme, a human-translated approach in HRI research can assume that humans exist in a vacuum, where they are interchangeable and always exhibit the same reactions to identical stimuli (H. R. Lee et al., 2022). However, research on Joint Action demonstrates that certain human behaviors naturally emerge during repetitive actions, such as movement synchronization. This phenomenon has primarily been studied in undirected tasks, like when two individuals rock in chairs together (Richardson et al., 2007) or walk side by side (van Ulzen et al., 2008). However, it has also been observed in goal-directed movements, such as pick-and-place tasks in HRI (Lorenz et al., 2011). The integration of insights derived from neurocognitive and neurophysical rehabilitation into the behavior design could, therefore, prove particularly advantageous for social companion robots in the care domain (Lorenz et al., 2016).

Within the human-translated approach, laypeople primarily act as informants and may have limited opportunities to share their ideas and thoughts on shaping human–robot interaction. As the emphasis is not on participants' identities (such as gender, age, or professional background), many studies within this approach tend to involve convenience samples from university populations, which has been criticized for its lack of diversity (Lee et al., 2022).

11.3.2 The stakeholder approach

The stakeholder approach in HRI research takes into account the identities of primary users, including factors such as age, gender, cultural background, and more, as well as their surrounding environments, in order to develop robotic solutions that fit their actual context of use. The first projects in which I focused on involving primary users in the development were in two very different domains: (1) human–robot collaboration for the factory context (Weiss et al., 2016); and (2) care robot development for enabling independent living at home for older adults (Bajones et al., 2019). In both domains, it became apparent and necessary to involve the target users (factory operators and older adults) at every stage of a user-centered design approach. Additionally, conducting HRI research in the actual context of use, whether it be in an industrial setting or individuals' private homes, was essential to ensure the relevance and effectiveness of the developed solutions.

Through this stakeholder approach, I gained a profound understanding of the inherently complex nature of interacting with embodied AI, often described as 'wicked' problems.[2] It became evident that every prototype design for human–robot interactions is intricately intertwined with its unique context of use, the involved stakeholders, technology readiness, and various other factors. In contrast to the human-translated approach, attempting to generalize from a single solution to encompass 'all robots in the industry' or 'all care robots for older adults' loses its significance, as the problems and solutions in each domain are highly interdependent, subjective, and constantly evolving.

Kiesler and Hinds (2004, p. 4) have already pointed towards the fact that "designing these robots appropriately will require a deep understanding of the context of use and of the ethical and social considerations surrounding this context.' Similarly, Dautenhahn (2004, p. 21) argued that the "exploring of the design space . . . is likely never to be completed." The stakeholder approach in HRI research has brought to light the realization that developing a comprehensive understanding of how various design choices impact the broad range of human responses towards embodied AI—ranging from cognitive and affective to emotional and relational—is a bigger challenge than originally anticipated by the research community. It turns out that a more constructivist perspective, which views people, social systems, and technological development as interrelated, is more fruitful to this approach than the post-positivist stance of the human-translated approach (see, for example, Lee et al., 2022; Šabanović et al., 2014; Weiss, 2022).

However, the stakeholder approach often overlooks the crucial aspect of designing for the future and the evolving use of (embodied) AI. It is imperative to explore strate-

---

2 The terminology of 'wicked problems' was apparently first coined by Rittel and Webber (1973) and describes problems that resist complete definition and resolution.

gies that account for future usage and the ability to shape interactions through the involvement of individuals impacted by the technology. One such approach is enabling customization and penalization of robots, allowing for adaptability and responsiveness to individual needs. It is crucial to acknowledge that human–robot interaction can never be designed to its end, as it is an ongoing and dynamic process that requires continuous adaptation and improvement.

11.3.3 The co-shaping approach

In the co-shaping approach, human–robot interaction is no longer understood as a dyadic interaction between one robot and one human but as a complex socio-technical arrangement of human and non-human agents such as robots, sensors, programs, and other devices embedded in their social context, whereby the social context not only relates to stakeholders, domain, and cultural backdrop, but also to the views, values, intuitions, and expertise of researchers (Weiss, 2022). In such an approach, humans are no longer just considered as informants but as lay experts who collaboratively engage with researchers in the shaping of human–robot interaction. As early as 2010, Šabanović stressed that in this process, society (the context of use and stakeholders involved) is an active shaper rather than a 'passive receptor' of robots (Šabanović, 2010). In a co-shaping approach, we therefore must study how people engage with robots over time. Long-term field trials, which use this approach, are still an exception and are primarily conducted with commercially available robots (see, e.g., De Graaf et al., 2017; Weiss et al., 2021).

However, co-shaping approaches will need to go even beyond long-term studies of people using commercially available robots in the field. The proper goal is to develop means for co-shaping in developing highly capable service robots. Clearly, involving lay experts in the active decision making of HRI development projects poses significant challenges. As my experiences in the factory context as well as in the care context have taught me, traditional stakeholder involvement in HRI research does not allow for gradually equalizing the input of researchers and other participants in robot design across sessions. Therefore, participants do not develop competency in interacting with it and subsequently do not feel confident about their ability to contribute to design. In turn, researchers are not becoming more conscious of different perspectives on robot capabilities, usefulness, and challenges and instead consider their expert knowledge as superior (Weiss & Spiel, 2022).

The co-shaping approach will become inadvertently more prominent with the fast development of non-embodied generative AI, and this will also impact how people relate to these technologies in the future. We have learned from HRI research so far that embodiment does play a role in how people relate, but it is only one relevant aspect. The more people get involved and co-shape their interactions, the more the perception of authenticity will matter. How will people try to keep their interactions authentic and personal instead of perceiving them as automated? As researchers, we will have to ask ourselves in the development of (embodied) AI whether and when we should sacrifice efficiency in order to increase authenticity, and which contributions and values people assign to the technology and to themselves.

11.4 The Future of Human Interactions With (Embodied) AI

Research in HRI and social robotics has revealed valuable insights into why people tend to anthropomorphize and sociomorphize robots. Anthropomorphism refers to the tendency to attribute human characteristics to non-human agents (Damholdt et al., 2020b; Spatola & Wykowska, 2021), and sociomorphism to the tendency to perceive actual non-human capacities as social (Seibt et al., 2020). In simpler terms, I would describe this phenomenon as a form of 'pretend play.' Despite being aware of the limited social and cognitive abilities of robots, individuals can engage in interactive relationships with them, even if these relationships are founded on false perceptions. This willingness to interact with robots as if they were fellow social agents may serve practical or emotional purposes, establishing relationships that are economically or emotionally beneficial (Sullins, 2020, p. 320).

As noted by Seibt et al. (2020), there is a need to explore in more detail the conceptual and methodological aspects of why individuals attribute social capabilities to (embodied) AI. The descriptive Ontology of Asymmetric Interactions (OASIS) framework serves as an initial step in providing theoretical foundations for empirical evaluation of sociomorphizing and experienced sociality in HRI, incorporating conceptual resources as starting points. Research within the HRI field increasingly demonstrates that robotic embodiment represents a minimal contextual factor that modulates the impact of the robot on a dispositional level (Spatola & Wykowska, 2021). For instance, when a robot's morphology and movement patterns closely resemble those of a human, it is more likely to activate the human action–perception system, leading people to attribute human-like traits to it (Lorenz et al., 2016; Wykowska, 2020). However, the subsequent challenge lies in identifying other relevant factors, beyond morphology and movement, that influence sociality and perceptions of authenticity.

11.4.1 How will future human–AI relation(ship)s look?

Embodied AI, in the form of physical objects and technological artifacts, is often defined by its function, leading to descriptions of human–artifact relation(ship)s in terms of usability and functionality (Verbeek, 2015). Similar to Verbeek (2015), I believe that usability and functionality alone are insufficient when accounting for the complex dynamics of human–technology relations(ship)s. Turkle et al. (2006) refer to social robots as "relational artifacts," intentionally designed to facilitate social and affective dimensions of interactions. Turkle (2007) also raises ethical concerns regarding the development of robots that elicit emotions without genuine authenticity.

The modes of relating to technology are emergent and can exceed those intended by designers and engineers. As mentioned earlier, we observe the intentional sociomorphization of robots even when design cues are absent (Dobrosovestnova et al., 2022). Hence, I approach the terminology of human–AI relation(ship)s cautiously, as our understanding of which aspects of authentic human–human relationships can be replicated or replaced by technology remains limited. A recent theoretical framework proposed by Huang and Jung (2022) aims to explain how various virtual characters, regardless of their ontological agency, can be perceived as authentic by human interactants without necessarily being perceived as humans themselves.

Another useful concept to address this complexity is that of parasociality and parasocial interactions. Coined by Donald Horton and Richard Wohl (1956), the concept has been widely studied across various scientific disciplines. Recent research, for instance, demonstrates that parasocial interaction mediates the relationship between authenticity attributes and buying behavior, excluding expertise and uniqueness attributes (Agnihotri et al., 2023). In the context of HRI, parasocial interaction can be defined as the tendency to engage in a two-way conversation with a robot, involving cognitive and affective components from the human side for both interactants. HRI and social robotics studies have provided ample evidence that robots with limited social behavior, coupled with a human-like embodiment, elicit perceptions of social robots as social agents (e.g., Beran et al., 2011; Severson & Carlson,2010; You & Robert Jr., 2018). Humans' innate tendency to anthropomorphize (Duffy, 2003) and sociomorphize (Seibt et al., 2020) artifacts fosters their willingness to form parasocial relation(ship)s with such robots. Perceptions of robots as social agents also mediate attitudes and overall acceptance of the technology (e.g., Lee et al., 2011).

Traditionally, the parasocial dimension of human–robot relations has been studied mainly from the perspective of human responses to robots, measuring variables such as acceptance (e.g., Meerbeek et al., 2006; Sehili et al., 2014), affect (e.g., Lee et al., 2006; Niculescu et al., 2013; Yamashita et al., 2016), and trust (e.g., Gockley & Mataric, 2006; Leuwerink, 2012). While these studies are valuable, they predominantly reflect one aspect of parasociality and fail to account for the intra- and intersubjective variability in how individuals perceive and relate to robots. Shifting the emphasis from the design of things (i.e., a robot with a specific embodiment, with a fixed set of design features and behaviors) to the design of meaningful human–technology–world relations requires adopting appropriate concepts that aid in the investigation of the structure and dynamics of such relations. Therefore, I propose embracing an interactionist and process-oriented perspective that places less emphasis on acceptance and trustworthiness as inherent characteristics of AI and instead emphasizes the co-construction and enactment of authenticity within diverse configurations of humans, technology, and other non-human factors, including physical space.

11.4.2 Authenticity as a boundary object

Artificial intelligence presents both remarkable opportunities and significant threats to human–technology relation(ship)s. AI holds the promise of enhancing our efficiency, aiding decision making, facilitating connections with others across time and space, and fostering new avenues for creativity. However, it also introduces notable risks, such as the manipulation of public opinion, responsibility gaps, and potential challenges to the value of authentic relationships. In my opinion, the current body of academic research on the societal implications of AI often remains fragmented and confined to specific disciplines, thereby risking both technical solutionism and unrealistic conceptualizations. Overcoming this challenge necessitates identifying the key elements of responsibility in practice, encompassing institutional cultures, individual agency, and necessary skills, and providing concrete guidance on achieving responsible innovation in research and development (Fisher et al., 2019).

To address this, I propose a multidisciplinary approach to explore the future of human relation(ship)s with embodied AI, utilizing authenticity as a boundary object that encourages the involvement of various disciplines. "Boundary objects are entities that enhance the capacity of an idea, theory or practice to translate across culturally defined boundaries, for example, between communities of knowledge or practice" (Fox, 2011, p. 1). Boundary objects, therefore, have the capacity to enhance the adoption, dissemination, and advancement of research findings, technology, and other intellectual property within the realms of social policy, organization, and management, as well as commercial and public services. As a first high-level roadmap for authenticity research, I suggest focusing on the following aspects:

- Understanding how existing evaluation criteria from HRI and AI research, such as acceptance, trust, fairness, reliability, and explainability, manifest in practical applications and how they relate to the concept of authenticity.
- Expanding the scope of studying human–technology relation(ship)s to incorporate considerations of how, when, and why laypeople consider interactions as authentic and how this perception changes over time.
- Organizing stakeholder engagement and citizen participation when exploring authenticity in order to ensure responsible and safe development of (embodied) AI.

By delving into these areas, we can advance our understanding of the practical dimensions of responsible AI, explore its intersection with authenticity, and establish mechanisms for inclusive stakeholder involvement that promote responsible and secure development in the field of (embodied) AI.

11.4.3 The need for epistemological diversity

As mentioned earlier, to comprehensively explore the diverse and novel forms of human–AI relation(ship)s, we must embrace epistemological diversity, accommodating various approaches such as positivist, relativist, constructivist, and interpretivist paradigms (Weiss, 2022). Given the rapid advancements in (embodied) AI, we will encounter a multitude of phenomena that may not be effectively investigated solely through controlled empirical studies. Relying solely on criteria such as reliability, replicability, and representativeness to evaluate research quality may overlook valuable insights generated outside of post-positivist paradigms. I propose adopting rigor as a high-level criterion that acknowledges the distinct qualities stemming from relativist, interpretivist, and critical perspectives in addition to traditional criteria. This approach will foster a more inclusive and comprehensive examination of the complexities arising from human–AI relation(ship)s.

My proposal for a set of quality criteria is built upon the six criteria for rigor developed for visualization design studies (Meyer & Dykes, 2019), with an additional component of ethical reflection. A comprehensive explanation of these criteria and their relevance to human-centered HRI research can be found in Weiss (2022). For more in-depth guidance on how to apply these criteria throughout the research process, researchers can refer to the original paper on visualization design studies (Meyer & Dykes, 2019). In Table 11.1, I present an overview of these criteria, along with best practice examples drawn from HRI studies. This table can serve as a valu-

Table 11.1: Description of the seven criteria for rigor.

Criteria	Description	Best Practice
Informed	Existing knowledge informs design and facilitates new interpretations	Azenkot et al., 2016; Fink et al., 2013
Reflexive	Positionality is adequately presented and reflected	Frennert et al., 2013; Leeet al., 2017
Abundant	Complexity is accounted for by involving many different perspectives	Lehmann et al., 2013; Sauppe & Mutlu, 2015
Plausible	Knowledge claims are evidenced, appropriate, and persuasive	Hanheide et al., 2017; Jeong et al., 2018
Resonant	The research inspires understanding and invites action	De Graaf et al., 2017; Sirkin et al., 2015
Transparent	The reporting invites scrutiny	Karreman et al., 2015; Weiss et al., 2015
Ethical	The developed design follows ethical conduct	Fischer et al., 2020; Kahn et al., 2012

able starting point for fellow researchers interested in investigating the concept of authenticity in future human–AI relation(ship)s.

11.5 Conclusion

Over the past 10–20 years, a considerable amount of research in HRI and social robotics has focused on understanding how robot embodiment influences people's responses to them. Researchers have been particularly interested in investigating factors that contribute to people's acceptance and trust of robots, the roles they assign to them, and the reasons behind anthropomorphizing or sociomorphizing these technologies. Deng et al. (2019) revealed in their systematic review a positive impact of robot embodiment on performance and the attribution of sociality. However, the diverse array of robot embodiments has made it challenging to conduct a structured analysis of effects, contributing to a replication crisis within the HRI research community. This issue is particularly problematic for human-centered HRI research that follows a human-translated approach but also presents challenges for the stakeholder involvement and co-shaping approaches.

Based on the insights from previous HRI research and the "rise of non-embodied generative AI," I propose a shift in how we study future human–AI relation(ship)s, which I intentionally refer to that way as we cannot predict how they will evolve. I advocate for considering authenticity as a boundary object (Fox, 2011) that needs to be studied in multidisciplinary approaches in order to understand the interaction between humans and (embodied) AI. To support this perspective, the concepts of parasocial interactions (Horton & Wohl, 1956) and sociomorphism (Damholdt et al., 2020a) can serve as theoretical underpinnings. Furthermore, I emphasize the importance of allowing for epistemological diversity in future research endeavors within the HRI and social robotics communities. This inclusivity should encompass positivist, relativist, constructivist, and interpretivist approaches, among others. Considering these, I propose quality criteria for conducting research in this domain, accompanied by best practice examples specifically tailored for HRI research. By adopting this approach, we can enrich our understanding of human–AI relation(ship)s and pave the way for more nuanced and comprehensive investigations in all related research fields.

References

Agnihotri, D., Chaturvedi, P., Kulshreshtha, K., & Tripathi, V. (2023). Investigating the impact of authenticity of social media influencers on followers' purchase behavior: Mediating analysis of parasocial interaction on Instagram. *Asia Pacific Journal of Marketing and Logistics*, *35*(3–4). https://doi.org/10.1108/apjml-07-2022-0598.

Azenkot, S., Feng, C., & Cakmak, M. (2016). Enabling building service robots to guide blind people: A participatory design approach. In *2016 11th ACM/IEEE International Conference on Human–Robot Interaction (HRI)* (pp. 3–10).

Bajones, M., Fischinger, D., Weiss, A., Puente, P. D. L., Wolf, D., Vincze, M., Kortner, T., Weninger, M., Papoutsakis, K., Michel, D., Qammaz, A., Panteleris, P., Foukarakis, M., Adami, I., Ioannidi, D., Leonidis, A., Antona, M., Argyros, A., Mayer, P., Panek, P., Eftring, H., & Frennert, S. (2019). Results of field trials with a mobile service robot for older adults in 16 private households. *Journal of Human–Robot Interaction*, *9*(2), 10. https://doi.org/10.1145/3368554.

Beran, T. N., Ramirez-Serrano, A., Kuzyk, R., Fior, M., & Nugent, S. (2011). Understanding how children understand robots: Perceived animism in child–robot interaction. *International Journal of Human–Computer Studies*, *69*(7–8), 539–550.

Breazeal, C. (2003). Toward sociable robots. *Robotics and Autonomous Systems*, *42*(3–4), 167–175.

Damholdt, M. F., Vestergaard, C., Nørskov, M., Hakli, R., Larsen, S., & Seibt, J. (2020a). Towards a new scale for assessing attitudes towards social robots: The attitudes towards social robots scale (ASOR). *Interaction Studies*, *21*(1), 24–56. https://doi.org/10.1075/is. 18055.fle.

Damholdt, M. F., Vestergaard, C., & Seibt, J. (2020b). Testing for 'Anthropomorphization': A case for mixed methods in human–robot interaction. In C. Jost, B. Le Pevedic, T. Belpaeme, C. Bethel, D. Chrysostomou, N. Crook, M. Grandgeorge, & N. Mirnig (Eds.), *Human–Robot Interaction: Evaluation Methods and Their Standardization* (pp. 203–227). Springer International Publishing. https://doi.org/10.1007/978-3-030-42307-08.

Dautenhahn, K. (2004). Robots we like to live with?! A developmental perspective on a personalized, life-long robot companion. In *RO-MAN2004, 13th International Conference on Robot and Human Interactive Communication* (pp. 17–22).

Dautenhahn, K. (2007). Socially intelligent robots: Dimensions of human–robot interaction. *Philosophical Transactions of the Royal Society B: Biological Sciences*, *362*(1480), 679–704.

Dautenhahn, K. (2018). Some brief thoughts on the past and future of human–robot interaction. *Journal of Human–Robot Interaction*, *7*(1), 4. https://doi.org/10.1145/3209769.

De Graaf, M., Ben Allouch, S., & van Dijk, J. (2017). Why do they refuse to use my robot? Reasons for non-use derived from a long-term home study. In *Proceedings of the 2017 ACM/IEEE International Conference on Human–Robot Interaction* (pp. 224–233). https://doi.org/10.1145/2909824. 3020236.

Deng, E., Mutlu, B., & Mataric, M. J. (2019). Embodiment in socially interactive robots. *Foundations and Trends® in Robotics*, *7*(4), 251–356.

Dennler, N., Ruan, C., Hadiwijoyo, J., Chen, B., Nikolaidis, S., & Mataric, M. (2022). Using design metaphors to understand user expectations of socially interactive robot embodiments. *arXiv preprint arXiv:2201.10671*.

Dobrosovestnova, A., Schwaninger, I., & Weiss, A. (2022). With a little help of humans: An exploratory study of delivery robots stuck in snow. In *2022 31st IEEE International Conference on Robot and Human Interactive Communication* (pp. 1023–1029).

Duffy, B. R. (2003). Anthropomorphism and the social robot. *Robotics and Autonomous Systems*, *42*(3–4), 177–190.

Fink, J., Bauwens, V., Kaplan, F., & Dillenbourg, P. (2013). Living with a vacuum cleaning robot. *International Journal of Social Robotics*, *5*(3), 389–408.

Fischer, K., Seibt, J., Rodogno, R., Rasmussen, M. K., Weiss, A., Bodenhagen, L., Juel, W. K., & Krüger, N. (2020). Integrative social robotics hands-on. *Interaction Studies*, *21*(1), 145–185. https://doi.org/ https://doi.org/10.1075/is.18058.fis.

Fisher, M., List, C., Slavkovik, M., & Weiss, A. (2019, January 1). *Ethics and Trust: Principles, Verification and Validation (Dagstuhl Seminar 19171)* (M. Fisher, C. List, M. Slavkovik, & A. Weiss, Eds.; tech. rep. No. 4). Dagstuhl, Germany, Schloss Dagstuhl-Leibniz-Zentrum fuer Informatik. https://doi.org/10.4230/DagRep.9.4.59.

Fortunati, L., Edwards, A., Manganelli, A. M., Edwards, C., & de Luca, F. (2022a). Do people perceive Alexa as gendered? A cross-cultural study of people's perceptions, expectations, and desires of Alexa. *Human–Machine Communication, 5*, 75–97.

Fortunati, L., Manganelli, A. M., Höflich, J., & Ferrin, G. (2022b). How the social robot Sophia is mediated by a Youtube video. *New Media & Society*, online first, https://doi.org/10.1177/14614448221103176.

Fortunati, L., Manganelli, A. M., Höflich, J., & Ferrin, G. (2023). Exploring the perceptions of cognitive and affective capabilities of four, real, physical robots with a decreasing degree of morphological human likeness. *International Journal of Social Robotics, 15*(3), 547–561.

Fox, N. J. (2011). Boundary objects, social meanings and the success of new technologies. *Sociology, 45*(1), 70–85.

Frennert, S., Eftring, H., & Östlund, B. (2013). Older people's involvement in the development of a social assistive robot. In *International Conference on Social Robotics* (pp. 8–18).

Gockley, R., & Mataric, M. J. (2006). Encouraging physical therapy compliance with a hands-off mobile robot. In *Proceedings of the 1st ACM SIGCHI/SIGART Conference on Human–Robot Interaction* (pp. 150–155).

Grigore, E. C., Pereira, A., Yang, J. J., Zhou, I., Wang, D., & Scassellati, B. (2016). Comparing ways to trigger migration between a robot and a virtually embodied character. In *Proceedings of the 8th International Conference on Social Robotics, Kansas City, USA* (pp. 839–849). Springer.

Hanheide, M., Hebesberger, D., & Krajnık, T. (2017). The when, where, and how: An adaptive robotic info-terminal for care home residents. In *Proceedings of the 2017 ACM/IEEE International Conference on Human–Robot Interaction* (pp. 341–349).

Hannibal, G. (2023). The trust–vulnerability relation—A theory-driven and multidisciplinary approach to the study of interpersonal trust in human–robot interaction. Doctoral dissertation, Technische Universitat Wien.

Hannibal, G., & Weiss, A. (2022). Exploring the situated vulnerabilities of robots for interpersonal trust in human–robot interaction. In S. T. Koeszegi & M. Vincze (Eds.), *Trust in Robots* (pp. 33–56). TU Wien Academic Press.

Horton, D., & Wohl, R. (1956). Mass communication and para-social interaction: Observations on intimacy at a distance. *Psychiatry, 19*(3), 215–229.

Huang, J., & Jung, Y. (2022). Perceived authenticity of virtual characters makes the difference. *Frontiers in Virtual Reality, 3*, 1033709. https://doi.org/10.3389/frvir.2022.1033709.

Hutchinson, H., Mackay, W., Westerlund, B., Bederson, B. B., Druin, A., Plaisant, C., Beaudouin-Lafon, M., Conversy, S., Evans, H., Hansen, H., Roussel, N.,Eiderbäck, B., Lindquist, S., & Sundblad, Y. (2003). Technology probes: Inspiring design for and with families. In *Proceedings of the SIGCHI Conference on Human Factors in Computing Systems* (pp. 17–24).

Hwang, J., Park, T., & Hwang, W. (2013). The effects of overall robot shape on the emotions invoked in users and the perceived personalities of robot. *Applied Ergonomics, 44*(3), 459–471.

Jeong, K., Sung, J., Lee, H.-S., Kim, A., Kim, H., Park, C., Jeong, Y., Lee, J., & Kim, J. (2018). Fribo: A social networking robot for increasing social connectedness through sharing daily home activities from living noise data. In *Proceedings of the 2018 ACM/IEEE International Conference on Human–Robot Interaction* (pp. 114–122).

Kahn, P. H., Jr., Kanda, T., Ishiguro, H., Gill, B. T., Ruckert, J. H., Shen, S., Gary, H. E., Reichert, A. L., Freier, N. G., & Severson, R. L. (2012). Do people hold a humanoid robot morally accountable for the harm it causes? In *Proceedings of the Seventh Annual ACM/IEEE International Conference on Human–Robot Interaction* (pp. 33–40). https://doi.org/10.1145/2157689. 2157696.

Karreman, D., Ludden, G., & Evers, V. (2015). Visiting cultural heritage with a tour guide robot: A user evaluation study in-the-wild. In *International Conference on Social Robotics* (pp. 317–326).

Kiesler, S., & Hinds, P. (2004). Introduction to this special issue on human–robot interaction. *Human–Computer Interaction, 19*(1–2), 1–8.

Kim, S. Y., Schmitt, B. H., & Thalmann, N. M. (2019). Eliza in the uncanny valley: Anthropomorphizing consumer robots increases their perceived warmth but decreases liking. *Marketing Letters*, *30*, 1–12.

Lee, H. R., Cheon, E., Lim, C., & Fischer, K. (2022). Configuring humans: What roles humans play in HRI research. In *2022 17th ACM/IEEE International Conference on Human–Robot Interaction* (pp. 478–492).

Lee, H. R., Šabanović, S., Chang, W.-L., Nagata, S., Piatt, J., Bennett, C., & Hakken, D. (2017). Steps toward participatory design of social robots: Mutual learning with older adults with depression. In *Proceedings of the 2017 ACM/IEEE International Conference on Human–Robot Interaction* (pp. 244–253).

Lee, K. M., Peng, W., Jin, S.-A., & Yan, C. (2006). Can robots manifest personality? An empirical test of personality recognition, social responses, and social presence in human–robot interaction. *Journal of Communication*, *56*(4), 754–772.

Lee, N., Shin, H., & Sundar, S. S. (2011). Utilitarian vs. hedonic robots: Role of parasocial tendency and anthropomorphism in shaping user attitudes. *Proceedings of the 6th International Conference on Human–Robot Interaction*, 183–184.

Lehmann, H., Walters, M. L., Dumitriu, A., May, A., Koay, K. L., Saez-Pons, J., Syrdal, D. S., Wood, L., Saunders, J., Burke, N., Duque-Garcia, I., Christianson, B., & Dautenhahn, K. (2013). Artists as HRI pioneers: A creative approach to developing novel interactions for living with robots. In G. Herrmann, M. J. Pearson, A. Lenz, P. Bremner, A. Spiers, & U. Leonards (Eds.), *Social Robotics* (pp. 402–411). ICSR 2013. Lecture Notes in Computer Science vol. 8239. Springer. https://doi.org/10.1007/978-3-319-02675-6_40.

Leuwerink, K. (2012). A robot with personality: Interacting with a group of humans. In *Proceedings of the 16th Twente Student Conference on IT, 4.*

Lorenz, T., Mortl, A., Vlaskamp, B., Schubö, A., & Hirche, S. (2011). Synchronization in a goal-directed task: Human movement coordination with each other and robotic partners. In *2011 RO-MAN 20th International Conference on Robot and Human Interactive Communication* (pp. 198–203).

Lorenz, T., Weiss, A., & Hirche, S. (2016). Synchrony and reciprocity: Key mechanisms for social companion robots in therapy and care. *International Journal of Social Robotics*, *8*(1), 125–143. https://doi.org/10.1007/s12369-015-0325-8.

Meerbeek, B., Hoonhout, J., Bingley, P., & Terken, J. (2006). Investigating the relationship between the personality of a robotic tv assistant and the level of user control. In *ROMAN 2006 15th IEEE International Symposium on Robot and Human Interactive Communication* (pp. 404–410).

Meyer, M., & Dykes, J. (2019). Criteria for rigor in visualization design study. *IEEE Transactions on Visualization and Computer Graphics*, *26*(1), 87–97.

Mirnig, N., Strasser, E., Weiss, A., Kühnlenz, B., Wollherr, D., & Tscheligi, M. (2015). Can you read my face? *International Journal of Social Robotics*, *7*(1), 63–76. https://doi.org/10.1007/s12369-014-0261-z.

Niculescu, A., van Dijk, B., Nijholt, A., Li, H., & See, S. L. (2013). Making social robots more attractive: The effects of voice pitch, humor and empathy. *International Journal of Social Robotics*, *5*(2), 171–191.

Nourbakhsh, I. R. (2015). *Robot Futures*. MIT Press.

Okuno, Y., Kanda, T., Imai, M., Ishiguro, H., & Hagita, N. (2009). Providing route directions: Design of robot's utterance, gesture, and timing. In *Proceedings of the 4th ACM/IEEE International Conference on Human Robot Interaction* (pp. 53–60).

Richardson, M. J., Marsh, K. L., Isenhower, R. W., Goodman, J. R., & Schmidt, R. C. (2007). Rocking together: Dynamics of intentional and unintentional interpersonal coordination. *Human Movement Science*, *26*(6), 867–891.

Rittel, H. W., & Webber, M. M. (1973). Dilemmas in a general theory of planning. *Policy Sciences*, *4*(2), 155–169.

Šabanović, S. (2010). Robots in society, society in robots. *International Journal of Social Robotics*, *2*(4), 439–450.

Šabanović, S., Reeder, S. M., & Kechavarzi, B. (2014). Designing robots in the wild: In situ prototype evaluation for a break management robot. *Journal of Human–Robot Interaction*, *3*(1), 70–88.

Sauppe, A., & Mutlu, B. (2015). The social impact of a robot co-worker in industrial settings. In *Proceedings of the 33rd Annual ACM Conference on Human Factors in Computing Systems* (pp. 3613–3622).

Sehili, M., Yang, F., Leynaert, V., & Devillers, L. (2014). A corpus of social interaction between NAO and elderly people. In *5th International Workshop on Emotion, Social Signals, Sentiment & Linked Open Data (ES3LOD2014). LREC* .

Seibt, J., Vestergaard, C., & Damholdt, M. F. (2020). Sociomorphing, not anthropomorphizing: Towards a typology of experienced sociality. In *Culturally Sustainable Social Robotics—Proceedings of Robophilosophy*, (pp. 51–67).

Severson, R. L., & Carlson, S. M. (2010). Behaving as or behaving as if? Children's conceptions of personified robots and the emergence of a new ontological category. *Neural Networks*, *23*(8–9), 1099–1103.

Sirkin, D., Mok, B., Yang, S., & Ju, W. (2015). Mechanical ottoman: How robotic furniture offers and withdraws support. In *Proceedings of the Tenth Annual ACM/IEEE International Conference on Human–Robot Interaction* (pp. 11–18).

Spatola, N., & Wykowska, A. (2021). The personality of anthropomorphism: How the need for cognition and the need for closure define attitudes and anthropomorphic attributions toward robots. *Computers in Human Behavior*, *122*, 106841. https://doi.org/https://doi.org/10.1016/j.chb. 2021.106841.

Sullins, J. P. (2020). Trust in robots. In J. Simon (Ed.), *The Routledge Handbook of Trust and Philosophy* (pp. 313–325). Routledge.

Takayama, L. (2012). Perspectives on agency interacting with and through personal robots. In M. Zacarias & J. V. de Oliveira (Eds.), *Human–Computer Interaction: The Agency Perspective. Studies in Computational Intelligence* (pp. 195–214), vol. 396. Springer. https://doi.org/10.1007/978-3-642-25691-28._

Tsiourti, C., Weiss, A., Wac, K., & Vincze, M. (2019). Multimodal integration of emotional signals from voice, body, and context: Effects of (in)congruence on emotion recognition and attitudes towards robots. *International Journal of Social Robotics*, *11*(4), 555–573. https://doi.org/10.1007/s12369-01900524-z.

Turkle, S. (2007). Authenticity in the age of digital companions. *Interaction Studies*, *8*(3), 501–517. https://doi.org/10.1075/is.8.3.11tur.

Turkle, S., Taggart, W., Kidd, C. D., & Daste, O. (2006). Relational artifacts with children and elders: The complexities of cyber companionship. *Connection Science*, *18*(4), 347–361.

van Ulzen, N. R., Lamoth, C. J., Daffertshofer, A., Semin, G. R., & Beek, P. J. (2008). Characteristics of instructed and uninstructed interpersonal coordination while walking side-by-side. *Neuroscience Letters*, *432*(2), 88–93.

Verbeek, P.-P. (2015). Cover story beyond interaction. *Interactions*, *22*(3), 26–31.

Weiss, A. (2022, January 24). One cannot know everything—On the need of epistemological diversity in human-centered HRI research. Habilitation Treatise, Universitat Salzburg.

Weiss, A., Huber, A., Minichberger, J., & Ikeda, M. (2016). First application of robot teaching in an existing industry 4.0 environment: Does it really work? *Societies*, *6*(3), 20.

Weiss, A., Mirnig, N., Bruckenberger, U., Strasser, E., Tscheligi, M., Kühnlenz, B., Wollherr, D., & Stanczyk, B. (2015). The interactive urban robot: User-centered development and final field trial of a direction requesting robot. *Paladyn: Journal of Behavioral Robotics*, *6*(1), 000010151520150005. https://doi.org/10.1515/pjbr-2015-0005, https://doi.org/10.1515/pjbr-2015-0005.

Weiss, A., Pillinger, A., Spiel, K., & Zauchner-Studnicka, S. (2020). Inconsequential appearances: An analysis of anthropomorphic language in voice assistant forums. In *Extended Abstracts of the 2020 CHI Conference on Human Factors in Computing Systems* (pp. 1–7). https://doi.org/10.1145/3334480. 3382793.

Weiss, A., Pillinger, A., & Tsiourti, C. (2021). Merely a conventional 'diffusion' problem? On the adoption process of anki vector. In *2021 30th IEEE International Conference on Robot & Human Interactive Communication* (pp. 712–719).

Weiss, A., & Spiel, K. (2022). Robots beyond science fiction: Mutual learning in human–robot interaction on the way to participatory approaches. *AI & Society*, *37*(2), 501–515.
Willemse, C., Marchesi, S., & Wykowska, A. (2018). Robot faces that follow gaze facilitate attentional engagement and increase their likeability. *Frontiers in Psychology*, *9*, 70. https://doi.org/10.3389/fpsyg.2018.00070.
Williams, T., Ayers, D., Kaufman, C., Serrano, J., & Roy, S. (2021). Deconstructed trustee theory: Disentangling trust in body and identity in multi-robot distributed systems. In *Proceedings of the 2021 ACM/IEEE International Conference on Human–Robot Interaction* (pp. 262–271).
Wykowska, A. (2020). Social robots to test flexibility of human social cognition. *International Journal of Social Robotics*, *12*(6), 1203–1211.
Yamashita, Y., Ishihara, H., Ikeda, T., & Asada, M. (2016). Path analysis for the halo effect of touch sensations of robots on their personality impressions. In *Social Robotics: 8th International Conference, ICSR 2016, Kansas City, MO, USA, November 1–3, 2016 Proceedings 8* (pp. 502–512).
You, S., & Robert Jr., L. P. (2018). Human–robot similarity and willingness to work with a robotic co-worker. In *Proceedings of the 2018 ACM/IEEE International Conference on Human–Robot Interaction* (pp. 251–260).

Psychology and Neuroscience

Monica Pivetti and Silvia Di Battista

Mind Perception During and After Interacting with Artificial Agents

Abstract: As social scientists, we are interested in people's attributions of mind to social robots, as it has long been assumed and experimentally shown that when people observe robots, they intuitively ascribe mind and intentionality to them, just as they do to other humans and animals. In literature, different terms have been used to address mind perception, such as theory of mind, anthropomorphism, and so on. In this chapter, we review the most recent literature on mind perception when interacting with robots by taking into consideration the different terms used, the line of research, and the main results. Then, we describe some empirical research dealing with mind perception in the lifespan. The interaction between mind perception and the social acceptability of robots is discussed in light of the uncanny valley phenomenon. Finally, we identify the main criticisms of the literature on the topic. Future developments are also envisioned.

Keywords: Mind perception, social perception, uncanny valley, lifespan

12.1 Introduction

Human–human interaction, as well as human–robot interaction (HRI), depends on the possibility that either party is able to understand the other. A pivotal role is played by the ability to correctly recognize and attribute intentions, goals, and beliefs to themselves and to other people (Wellman, 1992). In turn, other humans may use their understanding to predict (and facilitate) social interaction. In ascribing desires, thoughts, and beliefs to other individuals, one infers states that are not directly observable and that can differ from one's own, and can use these states anticipatorily to predict the behavior of others (Premack & Woodruff, 1978). This set of meta-representational abilities, which have been called a 'theory of mind' (ToM) or the ability to 'mentalize,' are a critical step in a child's development and are also considered to be instrumental to human interaction, as everyday social interaction relies on the recognition and understanding of others' points of view, and the decoding of non-verbal signs and emotional states.

As social scientists, we are interested in people's attribution of mind to social robots, even if there are several works in engineering dealing with the humanoid robot possessing a ToM (e.g., Scassellati, 2002).

In scientific literature, different terms have been used to refer to mental state attribution to robots (for a review, see Thellman et al., 2022; Sarrica et al., 2019). It is important to enlist the different terms here at the beginning of the chapter in order to

https://doi.org/10.1515/9783110792270-013

facilitate the subsequent reading. According to the number of publications where each term has been spotted, the terms most frequently used are: 'mental state attribution/ascription,' 'anthropomorphism,' 'mind perception,' 'theory of mind,' 'mentalizing,' and, to a lesser extent 'intentional stance,' 'folk psychology,' and 'mind reading.' While those terms may have overlapping meanings, they have been used in different lines of research or fields of study.

The term 'mental state/mind attribution' (or 'ascription', which is frequently used as a synonym) is by and large the most frequent term found in the literature. It is generally described as the process by which people tend to use information on others' mental states to explain the causes of events (Fiske & Taylor, 1991). Those terms are common in psychological and philosophical literature and refer to the attribution of social behaviors to underlying, sometimes hidden or internal, mental events.

In the case of robots, 'anthropomorphism' commonly refers to the attribution of both mental (such as motivations, emotions, and intentions) and non-mental (such as human-like appearance) states to robots (Epley et al., 2007). For example, the robot Pepper, in some cases, is perceived as a small child due to its size and is frequently considered appropriate to be used in Educational Robotics with children (Tanaka et al., 2015).

In many publications, 'mind perception' is closely associated with the work of Gray et al. (2007), who used self-report data to study two 'dimensions' of mind perception, that is, perception of mind agency and of mind experience. The focus is on the perceptual process and on the quantification of how much agency and experience is perceived for each non-human target (e.g., animals, robots, etc.). It is possible to argue that the terms 'mind perception' and 'mind attribution' do not differ significantly from a theoretical point of view, as both imply the ascription of mental state to non-human targets.

Developmental and experimental psychologists generally prefer to use the term 'theory of mind' to describe the capacity to attribute mental states to others, growing from early childhood to adolescence (Premack & Woodruff, 1978). It is usually said that a child who lacks the ability to understand others' viewpoints does not possess a theory of mind.

In neuroscientific research, the term 'mentalizing' is frequently used in accordance with the 'mentalizing system/network,' which comprises brain regions associated with mental state attribution (Van Overwalle & Baetens, 2009). Functional magnetic resonance imaging (fMRI) has been used to study how people make judgments about other people's behaviors, such as about their goals and intentions, desires, beliefs, and traits. This modern neuroscientific technique has allowed exploration of which brain areas are responsible for understanding the actions of others and the goals and motives underlying them.

Some less frequently used terms are 'intentional stance,' 'folk psychology,' and 'mind reading.' The first is the central construct in a broader philosophical theory about the nature of the mind called intentional systems theory (Dennett, 1989). This theory views the mental state attribution as one of the multiple 'stances' that people can adopt to interpret and interact with objects and others in their environment. The second term, 'folk psychology,' is also named naïve or common-sense psychology and refers to people's everyday non-scientific understanding of their own and others' minds. It includes views about others' beliefs, desires, and intentions, as well as traits, dispositions, and representation. The last one, 'mind reading,' is used in philosophical literature and refers to the ability to 'read someone's mind.' It points to the process by which people infer others' mental states and capacities (Nichols & Stich, 2003).

In this chapter, we review the most recent literature on mind perception when interacting with social robots by taking into consideration some of the most frequent terms used, their line of research, and their main results. Later, we will describe some empirical research dealing with mind perception in the lifespan. Some researchers have studied when children start to attribute a mind to robots and also whether there are differences in mind perceptions between young and elderly people. Next, we will address the 'uncanny valley' phenomenon, which has been widely investigated in the literature. In this regard, we will also explore the social acceptability of social robots and how mind perception could intervene in these relations. We will also try to identify the main limits and criticisms of the literature on the topic. Finally, we will put forward some recent contributions to the topic, and we will attempt to envision some future developments.

12.2 Mind Perception

In this section, we will present the work by Gray et al. (2007) on mind perception with robots. In HRI, it has long been assumed and experimentally shown that when people observe robots, they intuitively ascribe a mind and intentionality to them, just as they do to other humans and animals (De Graaf & Malle, 2019; Jacobs et al., 2022). Banks (2020) experimentally proved that people do implicitly and automatically mentalize robots as long as the cues or generalized behaviors mirror those of humans. However, there was no support for any correlation between implicit ToM tests and explicit mind ascription. It is possible that even if in some cases humans engage in implicit and automatic mentalizing of robots, the explicit mind ascription may evoke disruptive and unpleasant mental models for robots' ontological status, positioning the robot too close to the human essence.

While some philosophical studies have considered mind reading as a one-dimensional phenomenon, that is, whether a person has a mind or not (Dennett, 1996), some psychology researchers put forward the multidimensional nature of this process. Via a large-scale survey study exploring the mental capacities of different persons, animals, and robots, Gray et al. (2007) found that mind perception is divided into perception of mind experience and of mind agency. The dimension of mind experience can be summarised as a perceived ability for 'feeling,' such as the capacity to feel hunger, fear, pain, pleasure, rage, desire, consciousness, pride, embarrassment, and joy. Mind agency can be summarised as a perceived ability for 'doing,' such as the capacity for self-control, morality, memory, recognition, planning, communication, and thought (Gray & Wegner, 2012). In their study, an adult man and an adult woman were attributed with full experience and agency, whereas the robot was attributed with the lowest score for experience and with a moderate score for agency. Specifically, the robot scored lower than a dead man in experience and scored higher than a dog, a frog, or a baby in agency.

Generally, computers are attributed less mental state than robots, which in turn are attributed less mental state than human beings (Thellman et al., 2022). People's tendency to attribute mental states to robots has been reported across all demographic categories (e.g., age, gender and sex, cultural and socioeconomic background, various mental disorders).

Attribution of mental state is also determined by robot presentation, behavior, and appearance. For instance, Li (2015) identified the physical presence of the robot (i.e., the robot is present in the same room) rather than its physical embodiment (i.e., telepresence robot or virtual agent) as the stronger predictor of positive outcomes in interactions with artificial agents. Moreover, there is evidence of a stronger tendency to ascribe mental states to robots with more human-like physical appearance as compared with robots with less human-like appearance. Finally, some studies reported a stronger tendency to attribute mental states to robots when they display socially interactive behavior (Thellman et al., 2022).

To summarize, as conceptualized by Gray et al. (2007), mind perception in the case of robots points to robots as being able to 'do' things, and at the same time not being capable of 'feeling' things, as compared with human beings and other animals.

12.3 Anthropomorphism and Theory of Mind

This section discusses first the concept of anthropomorphism and, second, the relations between ToM and anthropomorphism in the perception of robots. Then we will present Haslam's model on dehumanization.

Anthropomorphism could be understood in terms of the appearance of the robot or in terms of the interaction situation with the robot (for a review, see Dubois-Sage et al., 2023). According to the mere appearance hypothesis, it is the appearance of the robot itself that activates processes usually used in human–human interactions. In this vein, the perspective taking of a robot by a human is only done if the robot has an appearance that resembles the human. According to the sociality, effectance, and elicited agent knowledge theory, which better explains the empirical results, individuals rely on their previous knowledge of human–human interaction to explain the behavior of a robot. They build a representation based on their past experience, which becomes the default model used when interacting with robots. According to this approach, individuals are motivated to interact with and make sense of their environment, and this motivation prompts individuals to anthropomorphize an object. In a longitudinal study, participants with more experience in interacting with robots showed a decreased propensity to anthropomorphism, thus supporting the idea that anthropomorphism is no longer activated once the individual is able to make his/her own representation of the robot (Kim et al., 2020).

While some studies consider the appearance of the robot, others also take into account the interaction situation in which the robot is involved. As for the appearance of the robot, studies have focused on its design, voice, verbal or non-verbal behavior, and the quality of its movements. As for the interaction situation of the robot, characteristics of the interaction, such as the way the robot is presented to individuals (the 'anthropomorphic framing'), the role of the robot, the frequency of the interaction, and the perceived degree of autonomy of the robot could change the level of anthropomorphism. For instance, as for the frequency of interaction with robots, negative attitudes and feelings decrease over time, regardless of the robot type (De Graaf et al., 2016; Złotowski et al., 2015) and regardless of the age of participants (Sinnema & Alimardani, 2019).

In literature, the ToM has been related to anthropomorphism, or the tendency to readily attribute human characteristics to non-human creatures and objects such as robots. If, on the one hand, humans are able to anthropomorphize any creature/object, on the other, research in robotics has focused on making robots communicate and behave more like humans to facilitate their use. This could be particularly important for social robots so as to create meaningful social interactions (Breazeal, 2003; Kühne & Peter, 2023).

However, in the literature, the relationship between ToM and anthropomorphism is unclear. Anthropomorphism has been defined as the attribution of human features to non-human entities (Epley et al., 2007; Fussel et al., 2008), and some scholars have pointed out that it involves possessing a ToM, which is a form of social cognition dealing with the attribution and recognition of thought, emotions, and beliefs in others (Baron-Cohen, 1999). In this sense, humanizing non-human entities and recognizing the point of view of others are closely connected (Atherton & Cross, 2018).

Some suggest that the two constructs, that is, anthropomorphism and ToM, are overlapping constructs since anthropomorphism is not just seeing human features (e.g., eyes or hands) in non-human agents but also involves actively attributing human mental states (e.g., emotions, intentionality) to non-human agents (e.g., Tahiroglu & Taylor, 2019). The analogy between the two constructs has been studied via behavioral methods and neuroimaging, leading to inconclusive results. Some studies have shown that when individuals engage in anthropomorphism, regions considered to be part of the ToM network are activated (Hortensius & Cross, 2018). Other studies using a multi-method approach, including magnetic resonance imaging, found no evidence for a clear relationship between dispositional measures of anthropomorphism and ToM. Altogether, those results provided unclear evidence as to whether dispositional anthropomorphism modulates activity in brain regions of the ToM network (Hortensius et al., 2021).

Moreover, some scholars have successfully used Haslam's dehumanization model (Haslam, 2006; Haslam & Loughnan, 2014) to study anthropomorphism when people interact with social robots (e.g., Eyssel et al., 2011). This model describes criteria that individuals may use when evaluating people's (and robots') human-likeness. Haslam (2006) distinguishes human uniqueness (HU), referring to civility, refinement, moral sensibility, rationality and logic, and maturity, from human nature (HN), referring to emotional responsiveness, interpersonal warmth, cognitive openness, agency and individuality, and depth. When HU is denied to others, they are seen implicitly or explicitly as animal-like (animalistic dehumanization). When HN is denied, they will be perceived as inert and cold, similar to robots (mechanistic dehumanization). Ferrari et al. (2016) found that android robots (more similar to a human body) were judged to possess more HN traits than mechanical and (biped) humanoid robots (being both less similar to a human body than androids). Eyssel et al. (2011) confirmed that anticipating an interaction with robots characterized by low predictability increased anthropomorphic inferences about them in terms of both increased HU and HN attribution, in line with Waytz et al. (2010). As unpredictability is uncomfortable for the human being, this would lead to increased attention for and motivation to understand the agent's goals and intentions, which in turn would lead to the attribution of humanness.

Finally, scholars have traced a conceptual overlap between Haslam's framework on dehumanization and Gray et al.'s frameworks on mind perception, with HU characteristics broadly tapping into agency and HN characteristics into experience (Hortensius & Cross, 2018).

In summary, anthropomorphism could be understood either in terms of the appearance/design of the robot, or in terms of the interaction situation with the robot. The latter approach posits that individuals rely on their previous knowledge of human–human interaction to explain the behaviour of a robot. Individuals are motivated to interact with and make sense of their environment, and this motivation prompts individuals to anthropomorphize an object. This approach better explains the empirical results of the study.

Some suggest that anthropomorphism, which is the tendency to readily attribute human characteristics to non-human creatures, objects, and robots, partially overlaps with other dimensions such as ToM, that is, the tendency to understand the point of view of others. However, neuroscientific studies provided unclear evidence as to whether dispositional anthropomorphism activates the same brain regions of the ToM network. Furthermore, studies using explicit measures such as questionnaires confirmed the attribution of anthropomorphic characteristics in terms of human nature (HN) to android robots, which are similar to a human body.

12.4 Mind Perception and Lifespan

In this section, we will present some studies investigating the mind perception in different age groups. People of all ages appear to attribute mental states to robots, even if to a different extent. The attribution of mental states to robots has been detected in young children and could evolve throughout the lifespan (Marchetti et al., 2018), with a stronger tendency in younger children (Thellman et al., 2022). For instance, children aged 24–35 months showed some signals of mind attribution to robots. Children this age tended to imitate goal-directed actions where the goal was not achieved by the robot only if the robot's action was preceded by eye contact between the robot and one adult on the scene (Itakura et al., 2008). According to the authors, this indirect evidence shows that it is not human-like morphology but rather human-like behavior, through eye contact, that plays an important role in inducing children to imitate the goal-directed behavior of a non-human agent. In this line, other studies showed that a mental simulation mechanism was activated in a group of three-year-old children only when the children imitated human actions (Moriguchi et al., 2010). Via studies that explicitly analyzed mind attribution, Kahn et al. (2012) found an effect of age, where 9- and 12-year-old children perceived the robotic partner as having a better mental state and being more social, and partially moral entity than did the 15-year-olds.

As for research on human-like features in robots, Okumura et al. (2013a, 2013b) confirmed that the presence of eyes in a robot is crucial for a 12-month-old child to direct their attention to the movement of a robot, as compared with an eyeless robot. However, Manzi et al. (2020) explored the role of faces in anthropomorphic robots as compared with human faces and found that seven- and nine-year-old children ascribed more mental state (e.g., sensory, perceptive, emotive, will, imaginative, and epistemic attributes) to the human face than to a robotic face. Also, in a simulated healthcare environment, Broadbent et al. (2013) used three robots: one with no face, one with a silver face, and one with a human face, to study how the facial characteristics of a robot may influence recognition of the robot's mind, personality, and eeriness. In a sample of young adults (mean age = 22 years old), they found a clear preference for the robot with the human-like face, which was rated as having the most mind and personality and the lowest eeriness.

Also, the tendency to anthropomorphize can decrease with age and experience with robots. According to the interaction situation approach, as anthropomorphism serves to fill a partial representation of the robots, once the child gains experience interacting with robots with age, their representation of robots becomes more accurate, and there is no more need for anthropomorphization (Kahn et al., 2012). This tendency even extends to adults, who are less likely to view a humanoid robot as having a free choice as compared with 5–7-year-old children, who assign more free will to them (Flanagan et al., 2021). It is likely that these age-related differences could be due to increased experience with technology. As young adults tend to have more experience with robots in daily life than small children, the former could have a more accurate representation of the robots' real capabilities than their younger counterparts.

As for the elderly, in a supermarket, Iwamura et al. (2011) found that an assistive robot that also engaged in non-shopping-oriented conversation was preferred to a robot capable only of functionally oriented conversation. On the contrary, in a retirement village, Stafford et al. (2014) analyzed attribution of mind agency and acceptability of social robots and found that older people preferred to interact with robots to which they attributed less mind agency. As these findings are counterintuitive, it is possible that older adults were frightened by the robots because of an overestimation of the robot's cognitive capacities.

To summarize, studies on mind perception at different age stages have found that young children aged 24–35 months showed some signals of mind attribution to robots. Specifically, the presence of eyes in a robot is crucial to direct children's attention to the movement of a robot. Robots with a human face are generally preferred to robots with a silver or no face.

Even if mind attribution is still persistent in adolescent children, there is some preliminary evidence showing that this declines at around 15 years old. According to the interaction situation approach, as anthropomorphism would serve to fill a partial representation of the robots, once the child gains experience interacting with robots, their representation of robots becomes more accurate, and there is no more need for anthropomorphism. Older people felt more comfortable interacting with robots characterised by low mind agency.

12.5 The Uncanny Valley Phenomenon

This section describes the so-called 'uncanny valley' phenomenon, describing the relation between robot anthropomorphism and positive/negative perception of robots. It has been suggested that interacting with extremely anthropomorphized robots could lead to appreciation up to a certain point, after which the resemblance would become unpleasant and unnerving for the human being. The upsetting nature of human-like robots was first suggested by Mori (1970), who suggested that robots can evoke strong feelings of eeriness if they approach a high level of realism. He called this dip in liking the 'uncanny valley. The hypothesized reasons behind this feeling are many. Starting from terror management theory (Greenberg et al., 1986), some scholars have suggested that human-like robots, by eliciting a subconscious fear of replacement and annihilation, could trigger an innate fear of death and, as a consequence, some coping mechanisms of unease and avoidance (MacDorman & Ishiguro, 2006). Others put forward the 'category conflict hypothesis,' suggesting that when human likeness is operationalized as a combination of human and non-human categories, negative affect is elicited due to some features being human and others being non-human, which results in competing stimulus interpretations (Burleigh et al., 2013). Also, some studies showed that perceptions of mind, in particular those of mind experience, are linked to feelings of unease. In other words, people might be happy with robots doing things, that is, high in mind agency, but not with robots feeling things, that is, high in mind experience (Gray & Wegner, 2012). Along these lines, Appel et al. (2020) compared agency and experience and found that a robot's capacity to feel (experience) elicits stronger feelings of eeriness than a robot's capacity to plan ahead and to exert self-control (agency), which in turn elicits more eeriness than a robot without mind, considered a tool.

Building from the literature on anthropomorphism and dehumanization, Wang et al. (2015) proposed that perceiving an anthropomorphized humanoid robot as lacking humanness, namely the process of dehumanization, is at work in the uncanny phenomenon. In other words, the more humans attribute human-like characteristics to a humanized robot (i.e., anthropomorphize), the likelier it is that humans will detect its mechanistic features, which in turn will activate the dehumanization process that would lead to eerie feelings.

Fortunati et al. (2023) proposed an empirical way to study the perception of cognitive and affective capabilities of four robots with a decreasing degree of morphological human likeness. They measured the attribution of perceptions, wishes, thoughts, and intentions to each of the four robots. They expected that the attribution (or absence of attribution) of mental and emotional capabilities to each of the four robots would be modulated according to their decreasing degrees of morphological and behavioral similarity to humans. Results showed a low attribution of mental and emotional capabilities for all four robots, and only the robot most similar to humans (i.e., InMoov) was attributed with more perceptions, wishes, thoughts, and intentions than the other three robots. However, even though InMoov is in general considered to have a medium–high level of similarity with humans, they found that it could generate the uncanny valley effect, too, as it evoked the emotion of pain (e.g., anxiety, disorientation, insecurity) among the respondents.

The same experimental procedure was applied to investigate the attribution of mental states and emotions to a highly human-like robot, Sophia, shown to participants in a YouTube video. Sofia expressed a high similarity with a woman and delivered a one-minute speech describing herself. They found that Sofia was attributed with more intentions than wishes, perceptions, and thoughts. Moreover, she was perceived as being incapable of feeling primary (e.g., anger) or secondary (e.g., pride) emotions (Fortunati et al., 2022).

By means of three experimental studies, Muller et al. (2021) confirmed that anthropomorphic appearance (but not mind perception) indeed mediated the relationship between the type of robot, human–machine distinctiveness, and perceived damage for humans and their identity. When people feel that the distinction between humans and robots is blurred, such as in the case of android and humanoid (and mechanical) robots, then intergroup distinctiveness is threatened. This could, in turn, lead to increased fear and perception of risk for humans (study 1). However, the results on the possible mediation of mind perception were inconsistent across two experimental studies, with one significant mediational pattern found out of six comparisons in study 2, and only the experience attribution mediating the effect on participants' perception of human/machine distinctiveness, for mechanical versus android robots only, in study 3. Taken together, these results support the idea that human-like appearance can lead to resistance, once again empirically confirming the uncanny valley phenomenon.

To sum up, the uncanny valley phenomenon refers to the idea that robots can evoke strong feelings of eeriness if they approach a high level of realism. Human-like robots are appreciated up to a certain point, then there is a dip in liking. Many hypotheses have been advanced: the terror management theory, the category conflict hypothesis, and the perception of mind experience. Some evidence points to human likeness playing a stronger role than mind perception in the uncanny valley phenomenon.

12.6 Mind Attribution and Acceptance of Social Robots

This section describes how mind attribution could affect the acceptance of social robots. Asking whether robots have a mind or not represents a factor capable of affecting the acceptance of social robots. Lay conceptions of robots could affect the way people approach robots and behave in their presence. In terms of motivational determinants of mind attribution, a stronger tendency to attribute mental states to robots was found in the literature when participants were motivated to predict robot behavior, when anticipating future interactions with a robot, when having high expectations of a robot's capabilities, when a robot was perceived as being mistreated or subjected to harm, and when the person was feeling lonely (Thellman et al., 2022).

Via self-report and behavioral data, Kont and Alimardani (2020) compared the engagement of young students and older people while interacting with the robot NAO. In line with the previous literature, they found that the young students reported a significantly higher engagement score compared with the older people. A composite score on mind perception was computed from mind agency and mind experience scores. For the older individuals, the engagement score was positively correlated with mind perception; that is, the higher the mind perception, the better the older person experienced engagement with the robot and *vice versa*.

By investigating implicit theories of the human mind, that is, people's beliefs regarding the malleability of psychological states, Dang and Liu (2022) found that sharing a malleable (vs. fixed) theory of the human mind was positively associated with mastery goals, which in turn positively predicted cooperative behaviors with robots in the UK and China. In other words, individuals who consider human characteristics to be changeable with learning tend to focus more on mastery goals and therefore express greater intentions to cooperate with robots. The reasoning behind these results is that people with a malleable theory express interest in challenging tasks and contexts, as they see challenges as opportunities to develop themselves rather than threats. Hence, they prefer to view competent others, such as robots, as possible collaborators. On the contrary, individuals viewing the human mind as fixed could perceive robots as threatening as they might have the capacity to surpass the human mind and eventually gain control over humankind.

Wiese et al. (2022) used a mechanic robot described as emotional vs. emotionless to study the tendency to collaborate in socio-emotional vs. computational tasks. Participants offloaded the task both onto the emotional robot (high agency and high experience) and onto the emotionless one (high agency and low experience) in the majority of trials. This pattern of results was more pronounced in the case of computational tasks (vs. socio-emotional ones). Moreover, participants offloaded the task to the robot in accordance with the robot–task fit, meaning that the emotional robot was chosen more often to reply to a social task, and the emotionless robot to arithmetical tasks. Similar to human–human perception, participants perceived the emotionless robot as more capable of performing arithmetical tasks than the emotional robot, even if the two robots were equally assessed as proficient in arithmetic.

As for the influence of the context and the outcome of the interaction on mind attribution, Lefkeli et al. (2020) found a strategic attribution of mind to the robot Cozmo in case of win/loss in cooperative/competitive tasks. They found that in the cooperation condition, people ascribed greater levels of mind to the agent when they lost as a team as compared with winning as a team. Furthermore, individuals who competed with and won against Cozmo attributed greater levels of mind to the agent compared with those who cooperated and won as a team. These results are explained in terms of self-serving bias, according to which people tend to explain positive events in terms of internal and dispositional characteristics. However, they seek external factors to justify events with negative outcomes. This way, participants tended to attribute a greater level of mind to the robot in the case of loss as a team with the robot as compared with winning as a team, to lay the blame on external factors such as the robot. This indicates that the loss was interpreted as the result of 'teamwork,' where the robotic teammate was taken as an agent with high capacities that contributed to their loss.

Starting from social representations theory (Moscovici, 1984, 1988), Brondi et al. (2021) examined via the free-associations task the shared images and beliefs connected with understanding and expectations towards social robots. They found that the mind perception, that is, the attribution of agency and feeling to robots, was indeed related to the organization of contents of the representations of robots. A higher perception of experience and low agency was related to a vision of robots connected with the ideal sphere and imaginary contents, also loaded with threat perception (on the second dimension of the factorial plan). This vision seems anchored to the topoi present in films and movies. On the other hand, attributing low experience and low agency to robots was related to a more cohesive vision, depicting robots as objects, sometimes imaginative, integrated among human beings (first dimension on the factorial plan). The latter combination of low experience and low agency was also associated with a more favorable view of robots in education, care, and health.

Previous studies on empathy towards robots found that perceived experience in robots modulates empathy, predicting how more empathy is perceived towards human-like and animal-like or mechanical robots (Riek et al., 2009). In this sense, robots attributed with more mind experience would also elicit more empathy. Recently, Grundke et al. (2023) found that presenting a robot as exposed to harm would lead to higher likeability via the mediation of empathy. In other words, attributing a mind to a robot could elicit some empathy in the case of a harmful situation for the robot, which in turn could evolve into better acceptance of the robot. Also, familiarity with robots seems to play a role in empathizing with robots. Familiarity with some robotic artifacts of everyday use, such as a robot vacuum cleaner like Roomba, for instance, was related to being more empathic and more congruent in attributed evoked feelings (Mattiassi et al., 2021).

In the case of service robots, Söderlund (2022) found that robots attributed with more ToM had greater perceived service quality via the mediation of perceived humanness and usefulness. This suggests that attributing robots with a mind can have a positive outcome in terms of overall perception of service quality. In other words, the more a service robot was equipped with a mind and the more it was perceived as human-like in appearance and useful for the purposes of the task, the more individuals perceived the service quality as high. This result could also be interpreted in the other direction, meaning that a robot's theory of mind capability can also be seen as disturbing, which may reduce perceived service quality, as a competent mind reader can also create serious harm to others.

To summarize, many studies have found a stronger tendency to attribute mental states to robots when participants were motivated to predict robot behavior, as is the case when anticipating future interactions with a robot. Notwithstanding some contradictory results, attributing mental state to robots is generally associated with increased collaboration/engagement with the robot. Furthermore, mind attribution could elicit some empathy in the case of harm to the robot, which in turn could result in better acceptance of the robot. Also, familiarity with robots seems to play a role in empathizing with robots.

12.7 Limitations and Criticism

As mentioned earlier, different terms have been used in the literature to address the same underlying phenomenon, that is, the attribution of thoughts and intentionality to others/robots, such as mental state attribution/ascription, anthropomorphism, mind perception, theory of mind, and so on (Thellman et al., 2022). This various terminology could lead to a lack of consensus in the literature on what is meant by attribution of mind, mental states, and mental capacities in robots.

The review by Thellmann et al. (2022) also put forward the question of what kind of data constitutes evidence of mental state attribution to robots, whether self-report data, non-verbal behaviors, or neurological evidence. The majority of publications relied on verbal measures of mental state attribution (i.e., questionnaires using Likert scales), whereas a minority of studies relied on (non-verbal) behavioral data (i.e., people's performance and/or gaze) and approximately one-tenth relied on neurological data (i.e., fMRI, electroencephalography). This lack of consensus could jeopardize the validity of published evidence and could be responsible for conflicting evidence in the literature (Thellman et al., 2022). For instance, some studies found no support for any relationship between implicit ToM tests and explicit mind ascription (Banks, 2020). Generally, people tend to unconsciously and automatically attribute intentionality and thoughts to robots, but do not state the same when asked explicitly. It is possible that explicit attribution of mind to robots may evoke unpleasant images of robots being able to harm people, images possibly grounded in media representations. Little debate among scholars is devoted to determining the most appropriate methods to study the attribution of mental state to robots. As the majority of studies in the field are run in laboratory settings, based on Western White participants and employing verbal measures, some concern has been raised about the validity of the results of those studies (Thellman et al., 2022). Also, very few studies have empirically investigated the behavioral consequences of mental state attribution in terms of cooperation with the robot itself. In line with the uncanny valley effect, it is expected that more morphologically and behaviorally humanized robots will be more appreciated during the interaction, up to a certain point.

Further debate on which methods are more appropriate to study mind attribution to robots is needed. As implicit measures have also been the target of criticisms in recent years (Brownstein et al., 2019; Meissner et al., 2019; Pekçetin et al., 2023), we suggest using a variety of methods, both verbal and non-verbal, to investigate the different aspects of mind attribution to robots. We encourage a multi-method approach where qualitative and quantitative data are collected and integrated in order to achieve a better comprehension of the attribution of mental state to robots. Specifically, the same study could involve both verbal and non-verbal measures of mind attribution to robots, together with a measure of intention to cooperate with robots. A questionnaire involving Liker/semantic differential scale and observational data (video recorded) could be proficiently employed, and the results integrated to investigate the topic and put in relation with the likeability of the robot. The review by David et al. (2022) included different kinds of studies, for instance quantitative vs. qualitative, mixed (quantitative and qualitative), cross-sectional vs. longitudinal/long term, and run in laboratory vs. in nature. Triangulation of methods and data integration in the study of new technology could offer "depth of qualitative understanding with the reach of quantitative techniques" (Fielding, 2012, p. 1; Turner et al., 2017).

Also, as different disciplines such as social and developmental psychology, robotics, philosophy of science, sociology, and media communication are interested in this topic, we suggest a broader interdisciplinary discussion to build a common language and ground to focus on appropriate research methods. As mental state attribution could influence the way we interact with robots, it is of pivotal importance that the scientific community can rely on strong findings.

It is worthwhile mentioning that some studies disagree on the exact number of dimensions according to which people's conception of the mind are organized. Whereas Gray et al. (2007) found two dimensions, namely agency and experience, Malle (2019; Malle & Ullman, 2021) broke down the original dimensions and expanded their contents into three new dimensions: (1) affect, (2) moral and social cognition, and (3) reality interaction. The affect dimension (1) integrates aspects of physiological and emotional capacities that are largely unintentional (e.g., feeling happy, feeling stressed). This dimension is related to the previous dimension of mind experience. The moral and social cognition facet (2) encompasses aspects of both moral cognition and social cognition pertaining to the self (e.g., planning) and others' minds (e.g., inferring their thoughts). It refers to processes under the agent's control and points to a specific meaning of the previous agency dimension. The third dimension, reality interaction, refers to communication and action, pointing to a second facet of agency (e.g., communicating verbally, learning from instructions). Further empirical studies are needed to better understand the multidimensional nature of the attribution of mind to robots.

12.8 Recent Developments and Future Orientations

Some studies on anthropomorphism have relied on the inversion effect to study how people cognitively elaborate humanoid robots as human- or object-like. Generally, people recognize social stimuli such as human bodies and faces through configural processing, meaning considering both the constituent parts of the stimulus and the spatial relations among them. Instead, people recognize objects such as a house through analytical processing, which relies only on the appraisal of specific parts (e.g., the door) without requiring information about the spatial relations among them. Within the inversion paradigm, participants are presented with a series of trials, first showing a picture of a social stimulus or an object, either upright or upside down. Afterward, subjects are asked to recognize the picture they just saw within a pair, including a distractor (mirror-image). The main assumption is that this inversion should undermine the recognition of social stimuli as they are processed in a configural way, whereas it should affect the recognition of objects less, as they are analytically processed. Sacino et al. (2022) found that humanoid robots were subjected to the inversion effect and so followed a configural processing similar to that activated for human beings, regardless of the level of similarity of humanoid robots to human beings. This line of research contributes to a better understanding of how people perceive and behave with robots.

Also, mind attribution and social perception of robots could be systematically investigated by employing the unified theory of acceptance and use of technology (UTAUT, Conti et al., 2017; Heerink et al., 2009). The UTAUT model comprises 13 dimensions measuring technology acceptance, each consisting of a set of items. Examples of dimensions include anxiety, attitudes, facilitating conditions, intention to use, perceived adaptability, and so on. Di Battista et al. (2021) found that a sample of teachers, educators, and psychologists' intention to use robotics in their profession was negatively related to anxiety evoked by robotics but positively related, for instance, to (1) positive attitudes about the use of robotics, (2) perceptions that there are factors in the environment that facilitate using robotics, and (3) perceptions of robotics as enjoyable.

Finally, future studies on the acceptability of robots could explore the role of implementing technologies that are sensitive to issues of stereotypes related to gender, feminine and masculine gender expression, ethnicity, or other multiple marginalized identities. The exploration of gender stereotypes related to the use of robotics seems of great relevance to fostering the acceptability of robotics (Di Battista et al., 2022). Furthermore, studies rarely included perceptions, attitudes, and experiences of LGBTQ people that could be relevant in order to design inclusive social robots and related initiatives. The current understanding of how robots are perceived principally comes from the perspective of heterosexual and cisgender people.

References

Appel, M., Izydorczyk, D., Weber, S., Mara, M., & Lischetzke, T. (2020). The uncanny of mind in a machine: Humanoid robots as tools, agents, and experiencers. *Computers in Human Behavior*, *102*, 274–286.

Atherton, G., & Cross, L. (2018). Seeing more than human: Autism and anthropomorphic theory of mind *Frontiers in Psychology*, *9*, 528.

Banks, J. (2020). Theory of mind in social robots: Replication of five established human tests. *International Journal of Social Robotics*, *12*(2), 403–414.

Baron-Cohen, S. (1999). *The Evolution of a Theory of Mind*. Oxford University Press.

Breazeal, C. (2003). Toward sociable robots. *Robotics and Autonomous Systems*, *42*(3–4), 167–175. https://doi.org/10.1016/S0921-8890(02)00373-1.

Broadbent, E., Kumar, V., Li, X., Sollers, J., Stafford, R. Q., MacDonald, B. A., & Wegner, D. M. (2013). Robots with display screens: A robot with a more humanlike face display is perceived to have more mind and a better personality. *PLoS One*, *8*, e72589. https://doi.org/10.1371/journal.pone.0072589.

Brondi, S., Pivetti, M., Di Battista, S., & Sarrica, M. (2021). What do we expect from robots? Social representations, attitudes and evaluations of robots in daily life. *Technology in Society*, *66*, 101663.

Brownstein, M., Madva, A., & Gawronski, B. (2019). What do implicit measures measure? *Wiley Interdisciplinary Reviews: Cognitive Science*, *10*(5), e1501.

Burleigh, T. J., Schoenherr, J. R., & Lacroix, G. L. (2013). Does the uncanny valley exist? An empirical test of the relationship between eeriness and the human likeness of digitally created faces. *Computers in Human Behavior*, *29*(3), 759–771.

Conti, D., Di Nuovo, S., Buono, S., & Di Nuovo, A. (2017). Robots in education and care of children with developmental disabilities: A study on acceptance by experienced and future professionals. *International Journal of Social Robotics*, *9*(1), 51–62.

Dang, J., & Liu, L. (2022). Implicit theories of the human mind predict competitive and cooperative responses to AI robots. *Computers in Human Behavior*, *134*, 107300.

David, D., Thérouanne, P., & Milhabet, I. (2022). The acceptability of social robots: A scoping review of the recent literature. *Computers in Human Behavior*, 137, 107419.

De Graaf, M. M., Ben Allouch, S., & van Dijk, J. A. (2016). Long-term evaluation of a social robot in real homes. *Interaction Studies*, *17*(3), 462–491.

De Graaf, M. M., & Malle, B. F. (2019, March). People's explanations of robot behavior subtly reveal mental state inferences. In *2019 14th ACM/IEEE International Conference on Human–Robot Interaction* (pp. 239–248). IEEE.

Dennett, D. C. (1989). *The intentional stance*. MIT press.

Dennett, D. C. (1996). *Kinds of Minds*. Basic Books.

Di Battista, S., Pivetti, M., & Bozzi, G. (2022, November 15). Brief research report: Teachers' gender-differentiated attributions. *Journal of Experimental Education*, 1–10.

Di Battista, S., Pivetti, M., Simaku, B., Beraldo, G., Menegatti, E., & Moro, M. (2021). Educational Robotics acceptance by Italian teachers, educators, psychologists and psychotherapists. In *Education In & With Robotics to Foster 21st-Century Skills: Proceedings of EDUROBOTICS 2020* (pp. 167–178). Springer International Publishing.

Dubois-Sage, M., Jacquet, B., Jamet, F., & Baratgin, J. (2023). We do not anthropomorphize a robot based only on its cover: Context matters too! *Applied Sciences*, *13*(15), 8743.

Epley, N., Waytz, A., & Cacioppo, J. T. (2007). On seeing human: A threefactor theory of anthropomorphism. *Psychological Review*, *114*(4), 864–886. https://doi.org/10.1037/0033-295X.114.4.864.

Eyssel, F., Kuchenbrandt, D., & Bobinger, S. (2011). Effects of anticipated human–robot interaction and predictability of robot behavior on perceptions of anthropomorphism. In *Proceedings of the 6th International Conference on Human–Robot Interaction—HRI '11* (pp. 61–68). https://doi.org/10.1145/1957656.1957673.

Ferrari, F., Paladino, M. P., & Jetten, J. (2016). Blurring human–machine distinctions: Anthropomorphic appearance in social robots as a threat to human distinctiveness. *International Journal of Social Robotics*, *8*(2), 287–302. https://doi.org/10.1007/s12369-016-0338-y.

Fielding, N. G. (2012). Triangulation and mixed methods designs: Data integration with new research technologies. *Journal of Mixed Methods Research*, *6*(2), 124–136. https://doi.org/10.1177/1558689812437101.

Fiske, S. T., & Taylor, S. E. (1991). *Social Cognition*. McGraw-Hill Book Company.

Flanagan, T., Rottman, J., & Howard, L. H. (2021). Constrained choice: Children's and adults' attribution of choice to a humanoid robot. *Cognitive Science*, *45*(10), e13043.

Fortunati, L., Manganelli, A. M., Höflich, J., & Ferrin, G. (2023). Exploring the perceptions of cognitive and affective capabilities of four, real, physical robots with a decreasing degree of morphological human likeness. *International Journal of Social Robotics*, *15*, 547–561.

Fortunati, L., Manganelli, A. M., Höflich, J., & Ferrin, G. (2022). How the social robot Sophia is mediated by a YouTube video. *New Media & Society*, https://doi.org/10.1177/1461444822110317.

Fussell, S. R., Kiesler, S., Setlock, L. D., & Yew, V. (2008, March). How people anthropomorphize robots. In *2008 3rd ACM/IEEE International Conference on Human–Robot Interaction* (pp. 145–152). IEEE.

Gray, H. M., Gray, K., & Wegner, D. M. (2007). Dimensions of mind perception. *Science*. https://doi:org/10.1126/science.1134475.

Gray, K., & Wegner, D. M. (2012). Feeling robots and human zombies: Mind perception and the uncanny valley. *Cognition*, *125*(1), 125–130. https://doi:org/10.1016/j.cognition.2012.06.007.

Greenberg, J., Pyszczynski, T., & Solomon, S. (1986). The causes and consequences of a need for self-esteem: A terror management theory. In R. F. Baumeister (Ed.), *Public Self and Private Self* (pp. 189–212). Springer. https://doi.org/10.1007/978-1-4613-9564-5_10.

Grundke, A., Stein, J. P., & Appel, M. (2023). Improving evaluations of advanced robots by depicting them in harmful situations. *Computers in Human Behavior*, *140*, 107565.

Haslam, N. (2006). Dehumanization: An integrative review. *Personality and Social Psychology* Review, *10*(3), 252–264.

Haslam, N., & Loughnan, S. (2014). Dehumanization and infrahumanization. *Annual Review of Psychology* *65*, 399–423.

Heerink, M., Kröse, B., Evers, V., & Wielinga, B. (2009, September). Measuring acceptance of an assistive social robot: A suggested toolkit. In *RO-MAN 2009—The 18th IEEE International Symposium on Robot and Human Interactive Communication* (pp. 528–533). IEEE.

Hortensius, R., & Cross, E. S. (2018). From automata to animate beings: The scope and limits of attributing socialness to artificial agents. *Annals of the New York Academy of Sciences*, *1426*(1), 93–110. https://doi.org/10.1111/nyas.13727.

Hortensius, R., Kent, M., Darda, K. M., Jastrzab, L., Koldewyn, K., Ramsey, R., & Cross, E. S. (2021). Exploring the relationship between anthropomorphism and theory-of-mind in brain and behaviour. *Human Brain Mapping*, *42*(13), 4224–4241.

Itakura, S., Ishida, H., Kanda, T., Shimada, Y., Ishiguro, H., & Lee, K. (2008). How to build an intentional android: Infants' imitation of a robot's goal-directed actions. *Infancy*, *13*, 519–532. https://doi.org/10.1080/15250000802329503.

Iwamura, Y., Shiomi, M., Kanda, T., Ishiguro, H., & Hagita, N. (2011). Do elderly people prefer a conversational humanoid as a shopping assistant partner in supermarkets? In *6th ACM/IEEE International Conference on Human–Robot Interaction* (pp. 449–457). ACM. https://doi.org/10.1145/1957656.1957816.

Jacobs, O. L., Gazzaz, K., & Kingstone, A. (2022). Mind the robot! Variation in attributions of mind to a wide set of real and fictional robots. *International Journal of Social Robotics*, *14*(2), 529–537.

Kahn, P. H., Kanda, T., Ishiguro, H., Freier, N. G., Severson, R. L., Gill, B. T., Ruckert, J. H., & Shen, S. (2012). "Robovie, you'll have to go into the closet now": Children's social and moral relationships with a humanoid robot. *Developmental Psychology*, *48*, 303–314. https://doi.org/10.1037/a0027033.

Kim, M. J., Kohn, S., & Shaw, T. (2020, December). Does long-term exposure to robots affect mind perception? An exploratory study. *Proceedings of the Human Factors and Ergonomics Society Annual Meeting*, *64*(1), 1820–1824.

Kont, M., & Alimardani, M. (2020, November). Engagement and mind perception within human–robot interaction: A comparison between elderly and young adults. In *International Conference on Social Robotics* (pp. 344–356). Springer.

Kühne, R., & Peter, J. (2023). Anthropomorphism in human–robot interactions: A multidimensional conceptualization. *Communication Theory*, *33*(1), 42–52. https://doi.org/10.1093/ct/qtac020.

Lefkeli, D., Ozbay, Y., Gürhan-Canli, Z., & Eskenazi, T. (2020). Competing with or against cozmo, the robot: Influence of interaction context and outcome on mind perception. *International Journal of Social Robotics*, *13*, 715–724.

Li, J. (2015). The benefit of being physically present: A survey of experimental works comparing copresent robots, telepresent robots and virtual agents. *International Journal of Human–Computer Studies*, *77*, 23–37.

MacDorman, K. F., & Ishiguro, H. (2006). The uncanny advantage of using androids in cognitive and social science research. *Interaction Studies*, *7*(3), 297–337.

Malle, B. (2019, July). How many dimensions of mind perception really are there? In *Proceedings of the 41st Annual Meeting of the Cognitive Science Society* (pp. 2268–2274). Cognitive Science Society.

Malle, B. F., & Ullman, D. (2021). A multidimensional conception and measure of human–robot trust. In C. S. Nam & J. B. Lyons (Eds.), *Trust in Human–Robot Interaction* (pp. 3–25). Academic Press.

Manzi, F., Peretti, G., Di Dio, C., Cangelosi, A., Itakura, S., Kanda, T., Hishiguro, H., Massaro, D., & Marchetti, A. (2020). A robot is not worth another: Exploring children's mental state attribution to different humanoid robots. *Frontiers in Psychology, 11*, 2011.

Marchetti, A., Manzi, F., Itakura, S., & Massaro, D. (2018). Theory of mind and humanoid robots from a lifespan perspective. *Zeitschrift für Psychologie, 226*(2), 98.

Mattiassi, A. D., Sarrica, M., Cavallo, F., & Fortunati, L. (2021). What do humans feel with mistreated humans, animals, robots, and objects? Exploring the role of cognitive empathy. *Motivation and Emotion, 45*(4), 543–555.

Meissner, F., Grigutsch, L. A., Koranyi, N., Müller, F., & Rothermund, K. (2019). Predicting behavior with implicit measures: Disillusioning findings, reasonable explanations, and sophisticated solutions. *Frontiers in Psychology, 10*, 2483.

Mori, M. (1970). *The Uncanny Valley: The Original Essay by Masahiro Mori*. IEEE Spectrum.

Moriguchi, Y., Kanda, T., Ishiguro, H., & Itakura, S. (2010). Children perseverate to a human's actions but not to a robot's actions. *Developmental Science, 13*, 62–68. https://doi.org/10.1111/j.1467-7687.2009.00860.x.

Moscovici, S. (1984). The phenomenon of social representations. In R. M. Farr & S. Moscovici (Eds.), *Social Representations* (pp. 3–69). Cambridge University Press.

Moscovici, S. (1998). The history and actuality of social representations. In U. Flick (Ed.), *The Psychology of the Social (pp. 209-247)*. Cambridge University Press.

Müller, B. C., Gao, X., Nijssen, S. R., & Damen, T. G. (2021). I, robot: How human appearance and mind attribution relate to the perceived danger of robots. *International Journal of Social Robotics, 13*(4), 691–701.

Nichols, S., & Stich, S. P. (2003). Mindreading: An integrated account of pretence, self-awareness, and understanding other minds. *Oxford, Online Edn.* https://doi.org/10.1093/0198236107.001.0001.

Okumura, Y., Kanakogi, Y., Kanda, T., Ishiguro, H., & Itakura, S. (2013a). Infants understand the referential nature of human gaze but not robot gaze. *Journal of Experimental Child Psychology, 116*, 86–95. https://doi.org/10.1016/j.jecp.2013. 02.007.

Okumura, Y., Kanakogi, Y., Kanda, T., Ishiguro, H., & Itakura, S. (2013b). Can infants use robot gaze for object learning? The effect of verbalization. In F. Broz, H. Lehmann, B. Mutlu, & Y. Nakano (Eds.), *Gaze in Human–Robot Communication* (pp. 33–46). John Benjamins. https://doi.org/10.1075/bct.81.03oku. Also published in *Interaction Studies, 14*(3), 351–365.

Pekçetin, T. N., Barinal, B., Tunç, J., Acarturk, C., & Urgen, B. A. (2023, March). Studying mind perception in social robotics implicitly: The need for validation and norming. In *Proceedings of the 2023 ACM/IEEE International Conference on Human–Robot Interaction* (pp. 202–210). IEEE.

Premack, D., & Woodruff, G. (1978). Does the chimpanzee have a theory of mind? *Behavioral and Brain Sciences, 1*(4), 515–526.

Riek, L. D., Rabinowitch, T. C., Chakrabarti, B., & Robinson, P. (2009, March). How anthropomorphism affects empathy toward robots. In *Proceedings of the 4th ACM/IEEE International Conference on Human–Robot Interaction* (pp. 245–246). IEEE.

Sacino, A., Cocchella, F., De Vita, G., Bracco, F., Rea, F., Sciutti, A., & Andrighetto, L. (2022). Human- or object-like? Cognitive anthropomorphism of humanoid robots. *Plos One, 17*(7), e0270787.

Sarrica, M., Brondi, S., & Fortunati, L. (2019). How many facets does a 'social robot' have? A review of scientific and popular definitions online. *Information Technology & People, 33*(1), 1–21. https://doi.org/10.1108/ITP-04-2018-0203.

Scassellati, B. (2002). Theory of mind for a humanoid robot. *Autonomous Robots, 12*, 13–24.

Sinnema, L., & Alimardani, M. (2019). The attitude of elderly and young adults towards a humanoid robot as a facilitator for social interaction. In *Social Robotics: 11th International Conference, ICSR 2019, Madrid, Spain, November 26–29, 2019, Proceedings 11* (pp. 24–33). Springer International Publishing.

Söderlund, M. (2022). Service robots with (perceived) theory of mind: An examination of humans' reactions. *Journal of Retailing and Consumer Services, 67*, 102999.

Stafford, R. Q., MacDonald, B. A., Jayawardena, C., Wegner, D. M., & Broadbent, E. (2014). Does the robot have a mind? Mind perception and attitudes towards robots predict use of an eldercare robot. *International Journal of Social Robotics, 6*, 17–32. https://doi.org/10.1007/s12369-013-0186-y.

Tahiroglu, D., & Taylor, M. (2019). Anthropomorphism, social understanding, and imaginary companions. *British Journal of Developmental Psychology*, 37(2), 284–299.

Tanaka, F., Isshiki, K., Takahashi, F., Uekusa, M., Sei, R., & Hayashi, K. (2015, November). Pepper learns together with children: Development of an educational application. In *2015 IEEE-RAS 15th International Conference on Humanoid Robots (Humanoids)* (pp. 270–275). IEEE.

Thellman, S., de Graaf, M., & Ziemke, T. (2022). Mental state attribution to robots: A systematic review of conceptions, methods, and findings. *ACM Transactions on Human-Robot Interaction (THRI), 11*(4), 1–51.

Turner, S. F., Cardinal, L. B., & Burton, R. M. (2017). Research design for mixed methods: A triangulation-based framework and roadmap. *Organizational Research Methods, 20*(2), 243–267. https://doi.org/10.1177/1094428115610808.

Van Overwalle, F., & Baetens, K. (2009). Understanding others' actions and goals by mirror and mentalizing systems: A meta-analysis. *Neuroimage, 48*(3), 564–584.

Wang, S., Lilienfeld, S. O., & Rochat, P. (2015). The uncanny valley: Existence and explanations. *Review of General Psychology, 19*(4), 393–407.

Waytz, A., Cacioppo, J. T., & Epley, N. (2010). Who sees human? The stability and importance of individual differences in anthropomorphism. *Perspectives on Psychological Science, 5*, 219–232.

Wellman, H. M. (1992). *The Child's Theory of Mind*. MIT Press.

Wiese, E., Weis, P. P., Bigman, Y., Kapsaskis, K., & Gray, K. (2022). It's a match: Task assignment in human–robot collaboration depends on mind perception. *International Journal of Social Robotics, 14*(1), 141–148.

Złotowski, J. A., Sumioka, H., Nishio, S., Glas, D. F., Bartneck, C., & Ishiguro, H. (2015). Persistence of the uncanny valley: The influence of repeated interactions and a robot's attitude on its perception. *Frontiers in Psychology, 6*, 883.

Friederike Eyssel and Giulia Perugia

How People Perceive Social Robots: The Case of Gender

Abstract: This chapter addresses the notion of 'gender' in the context of social robots. We highlight key terminology, outline historical developments, and point to prominent meta-perspectives that have been proposed on this topic. This way, we present a multidisciplinary perspective on gendering robots, reflect on the implications for robot design, and discuss ethical and social implications associated with specific design choices. Finally, we point to the implications of gendered vs. genderless/gender-neutral robot design for human self-perception and identity.

Keywords: Social robots, gender, attribution, social cognition, social and ethical implications for design, development

13.1 Introduction

Maybe you have caught yourself talking to your computer, the chatbots in your home, or the smart assistant in your car. Take a moment and think back on how you addressed this technology. Did you refer to it by a first name? And if so, which name did you pick? Can you remember whether it was a male first name? A female one? Or even a gender-neutral variant? Indeed, mindlessly treating objects or other entities as social actors is nothing new. More than two decades ago, Reeves and Nass (1996) described such phenomena in their foundational work on the *media equation* (see also Nass et al., 1994). This encompassed attributing—among other things— personality traits and treating technology like humans, based on the idea that people mindlessly utilize the scripts that shape human–human interactions in the context of computer technologies.

Notably, back in the 1990s, studies on what were at the time 'novel' technologies predominantly focused on people's reactions to computers, doing so by investigating human–computer interaction (HCI). At the time, physically embodied systems like social robots were not yet widely available for research or applications in the real world.

Times have changed, however, and social robots have become much more prevalent, at least in research labs, if not in contemporary societies. The availability of advanced humanoid robots as research tools for engineers and social scientists has contributed to international research programs that seek to gain insights into the dynamics of human–robot interaction (HRI). Developers and designers must carefully decide on the various dimensions that potentially play a role. For example, the robot's size,

https://doi.org/10.1515/9783110792270-014

embodiment, appearance, degree of human likeness, and functionalities in terms of communication behaviors, emotion, or mobility matter—to name but some dimensions.

In the contemporary research landscape, it is already good practice that researchers from engineering and social sciences team up with the shared goal of developing and advancing robot technologies. They do so by integrating a human-aware, human-centered perspective. Accordingly, robots are usually not 'just built' to serve as a research platform or to enter the market. Instead, multidisciplinary teams of researchers try to discover the effects of a new design, technology, or functionality on the perception of the respective users. Such user-centered or human-aware approaches are important to consider potential end-user attitudes, expectations, and reactions towards a new product—which might be a robot. That robot, in turn, might be redesigned based on user-specific desires, needs, and requirements to make sure that potential end users indeed want to use the robot in their everyday lives. Thus, developing, evaluating, and reassessing the status quo of what is developed constitutes relevant steps in the research and development cycle that is a natural part of the advancement of novel technologies.

However, in the early years of social robotics research, integrating user opinions, values, and beliefs about the technologies invented was not an essential part of the research routine. As a consequence, a number of robots have been designed the way they are without taking into account the relevance of specific, sometimes subtle, design decisions. The goal of this chapter is to use robot gender as a case in point to demonstrate that design decisions indeed matter. Why engage in a discourse about robot gender, at all? By highlighting key works from the HRI literature, we will showcase that gender in robots may indeed evoke biased perceptions of the respective robots under study.

The way we look at social robots can be twofold. On the one hand, social robots may be deemed research platforms to advance the state-of-the-art in engineering and robotics. This implies that engineers develop better robots to serve whatever designated purpose. For example, social robots are developed to support as socially assistive systems in the home, education, or care contexts. On the other hand, social robots may serve as research platforms to fulfill another goal: to understand cognition and behavior on the part of potential end users. This goal is more relevant to researchers from the social sciences and humanities. As such, these professionals are curious to investigate and explore how humans successfully interact with machines. By studying, for instance, the psychological underpinnings of human–machine interaction, they aim to ultimately foster acceptance and uptake of social robots by individuals and by societies at large. To illustrate, a social scientist with training in psychology might be interested in the effects of human-like functions and appearance in social robots to demonstrate that these aspects indeed matter and shape the perception of the robot under study. In particular, a social psychologist might conduct a study, mainly using experimental methods, to rule out alternative explanations for specific findings. Moreover, social psychologists are interested in the impact of design choices on cognition, affect, and behavior. Therefore, they would also study whether a certain design choice (as in robot gender) could eventually elicit stereotypical judgments.

Equally important, psychologists are naturally curious to understand the alleged psychological mechanisms underlying HRI, drawing on what we already know about the principles of humans interacting with other humans (e.g., Eyssel, 2017). Human–human interaction is therefore taken as a basis for understanding how people engage with social robots.

To gain insights into these dynamics, social robots may be utilized as research platforms to test and validate theories that originally would seek to explain psychological phenomena that have thus far only been studied among humans. In other words, phenomena widely studied in the human–human domain have been translated and explored in the context of interactions with social robots. These psychological phenomena include basic social cognitive processes like impression formation and the subsequent attribution of traits, emotions, intentions, or other characteristics (e.g., Epley et al., 2007), resulting in specific affect, cognition, or behavior exhibited during human–robot interaction. Indeed, first impressions of novel technologies like social robots may be shaped by system-related features (e.g., its design, its behavior). Likewise, they might be impacted by the idiosyncratic dispositions of any given perceiver (e.g., demographics, attitudes, preferences) or by aspects related to the context in which an encounter occurs (e.g., characteristics of the use case, culture-specific features). Taking into account robot features, what are the relevant dimensions to consider? Previous research has suggested that embodiment and human likeness represent two critical dimensions in social robot perception, with embodiment reflecting the extent to which a technical system is physically embodied, ranging from disembodied, depicted chatbots to virtual agents on a screen, up to embodied social robots that are available for actual physical interaction with their human users (see Deng et al., 2019; Li, 2015; Roesler et al., 2022; Ziemke, 2003). Meta-analytic approaches by Deng et al. (2019) or by Roesler et al. (2022) have explored the embodiment hypothesis, which suggests more positive outcomes at an objective and a subjective level thanks to the physical embodiment and anthropomorphism in social robots. Anthropomorphism, in turn, refers to the morphology and behavior of the robot (see Fink, 2012; Roesler et al., 2021); specifically, 'anthropomorphic design' encompasses a robot's shape, behavior, and interaction/communication with the human (see Fink, 2012). The former aspect, particularly robot shape, might form the basis of gendered perception. That is, visual cues might determine whether we attribute certain features to robots. Moreover, gender might be communicated by means of auditive cues, most often by means of the robot's high- versus low-pitched voice or because its name signals a certain gender (or gender neutrality/genderlessness). Reflecting on the start of this very chapter, because we readily use anthropocentric (i.e., human-like) cues to make sense of the world around us, a world that sooner or later will include robots, it comes as no surprise that you might have gendered your computer, your voice assistant, or any other technical device. The chapter at hand will provide an introduction to key theoretical concepts, will offer a historical overview of the development of research on gender in HRI, and will reflect on future perspectives and the ethical implications of gendering social robots.

13.2 Gendering Robots: Key Constructs

When talking about gender and robots, we also need to talk about definitions. What do experts mean when talking about gender—sex (masculinity–femininity)—or notions beyond the male–female dichotomy, including genderlessness, gay, queer, or transgender individuals? Let us start by clarifying the notions of sex versus gender. According to Tannenbaum et al. (2019, p. 138), sex refers to "biological attributes that distinguish species as male, female, intersex, and hermaphrodites" and "in engineering, sex includes anthropometric, biomechanical and physiological characteristics that may affect the design of products, systems and processes". On the other hand, gender is socially constructed, and Tannenbaum et al. (2019, p. 138) propose the following definition: gender refers to "psychological, social and cultural factors that shape attitudes, behaviours, stereotypes, technologies and knowledge."

Moreover, in their 2017 overview, Alesich and Rigby highlight Judith Butler's notion of gender being performed through appearance and behavior (e.g., the role stereotypically associated with a specific gender). The performativity approach would also reject the idea of sex versus gender, as presented in Tannenbaum et al. (2019). In other recent works, robot gender is assessed by asking to what extent participants perceive the robot as masculine, feminine, and gender neutral (see Perugia et al., 2022).

Recent research by Roesler (2023) has examined the idea of gender-neutral robots in a set of empirical studies. This work shows that even robots that have been designed to appear allegedly genderless or gender neutral are perceived as having a gender (e.g., the popular robot Pepper by Aldebaran/Softbank; see also Søraa, 2017). To illustrate, Søraa (2017) argued that genderless technologies cannot exist because of the interdependence between the designer, the attitudes that person brings, and the power structure that characterizes the creator and the creation. He points to the example of a microwave having become a female technology, though originally construed by men. According to Søraa (2017), there is nothing like a genderless or gender-neutral robot. Partially in line with Søraa, Perugia et al. (2023) have proposed that gendering robots is a two-step process of gender encoding—in which designers and roboticists imbue robots with gender cues (e.g., breast-like shapes, long hair vs. short hair)—and gender decoding—in which users attribute robots a certain gender, also based on the robots' ostensible cues. Importantly, gender decoding can be independent of gender encoding. Robots like the Roomba vacuum cleaner, for instance, do not feature any ostensible gender cues. Nevertheless, they are likely gendered by their owners (Marchetti-Bowick, 2009). Similarly, Fortunati et al. (2022) have investigated gender attributions to chatbots like Alexa in a cross-national and cross-gender comparison.

The observation that we are prone to gender allegedly neutral mundane objects is in line with Epley et al.'s (2007) notion of 'elicited agent knowledge.' Here, the idea is that we use anthropocentric knowledge structures and apply them even to robots so that, based on this cognitive mechanism, attributions of gender or other human-like characteristics to non-human entities represent a natural, nearly automatic cognitive process. As with everything, however, further research is needed to support the notion that the attribution of human likeness—including gender—to non-human entities indeed represents an automatic process and, as such, is nearly inevitable.

Sex and gender also represent core biological and social features of humans and human likeness, so when we ask about these dimensions, we might want to follow Spiel et al.'s lead (2019). They highlight the aspect of going beyond the binary perception of the world, predominantly featuring males versus females. Instead, they argue to consider providing participants in HCI studies five different response options to mark their gender and allowing participants to 'choose all that apply'—man, woman, non-binary, prefer not to disclose, prefer to self-describe followed by an open-response option. By opening up multiple options, including self-description, individuals who do not self-identify as binary will be included, and their perspectives can be reflected in the research. This hints at the general problem of doing gender according to the gender binary instead of reflecting multifaceted perspectives. This pertains to the characteristics of the human participants studied as well as to the technological systems that are explored.

The following section will document the pertinent role of gender in human–technology interaction. We summarize the main findings from empirical and experimental research, starting with the very early works that have studied perceptions of personal computers in the field of HCI.

13.3 A Brief History of Gender in Social Robotics

When you enter "robot gender" as a search term in Google Scholar, a search engine for academic literature, roughly 406,000 results pop up in just a few seconds. A careful screening, however, shows that the body of research on gender and robots is not nearly as rich yet. Indeed, having reviewed the literature on social robotics more closely, a much smaller number of papers appears to be relevant to the topic. In the following, we will provide a brief history of the key works that have informed the discourse about gender and social robots. This overview may not be exhaustive, but it nevertheless highlights some of the most important empirical works that have shaped our understanding of gendered robots and the important role of gender in social robotics research.

Let's get started by turning back time more than 20 years. In the early years of gender research in the context of novel technologies, robots did not play a major role. This was due to the fact that social robots were not yet prevalent as research platforms for empirical studies in the social sciences and humanities. Instead, researchers explored human reactions towards personal computers. To provide a prominent example, Reeves and Nass (1996) coined the so-called 'media equation,' an approach that suggests we react to computers just as we would towards humans. This idea was extended to the 'Computers Are Social Actors' approach that was used to test classic findings from human–human interaction in the realm of human–computer interaction. In the context of this work, Clifford Nass and colleagues (Nass et al., 1997) have also looked into the effects of computer gender. In the late 1990s, this group of researchers demonstrated that people tend to stereotype computers with minimal gender cues (e.g., female-voiced and male-voiced computers). In their research, male and female participants were asked to prepare a test with the help of a female-voiced vs. male-voiced computer tutor. The same participants would then take the test and be assessed by a female-voiced versus male-voiced computer evaluator. In line with the authors' hypotheses, and in keeping with existing gender stereotypes, participants considered the male-voiced computer tutor more adequate to teach topics associated with technology and computers, and the female-voiced computer tutor more adequate to teach topics related to love and relationships. Moreover, participants preferred to be evaluated by a male-voiced computer than a female-voiced computer. Going beyond these results, Lee et al. (2000) then showed that a computer's alleged gender may impact its capability to influence human users. Specifically, the authors not only observed that male-voiced computers were perceived as more socially attractive and trustworthy than female-voiced computers, but also found out that male- versus female-voiced computers impacted participants' decision making more strongly.

A few years later, Siegel et al. (2009) conducted an empirical study that can be considered an extension of Lee et al.'s work. This time around, these authors conducted the work using robots. Specifically, they carried out a field experiment in a museum to explore whether and how a robot's gender could persuade participants towards a specific behavioral outcome. The robot—a male-voiced or female-voiced Nexi platform—gave an overview of its technical capabilities to the visitors to a museum. Importantly, this was followed by the robot's persuasive appeal in which it asked visitors to donate money to the MIT Media lab. The results of this study were striking. Male visitors perceived the female robot as more trustworthy and engaging than the male robot and ended up donating more to the female robot than to the male robot. Thus, a mismatch between robot and participant gender elicited higher inclinations to voluntarily give away money for research purposes. Likewise, female visitors to the museum perceived the male robot as more trustworthy and engaging than the female robot. The results came with an interesting twist, though: female participants ended up donating more to the male robot only when no other person was co-present with them. When another person was there, female participants actually made greater donations to the female robot. Beyond informing us about a potential effect of the mismatch between robot gender and participant gender on people's perceptions of robots, this research also shows that people's knowledge and awareness regarding what may be deemed socially desirable (in this case, women should favor their 'in-group' robot) affected how they behaved towards robots. From sociological research, we already know that this social-normative response is particularly likely when individuals are confronted with potentially losing face, which involves damage to their positive social identity (Goffman, 2005).

The study by Siegel et al. (2009) can be taken as an example of how the media equation put forward by Reeves and Nass (1996) actually extends to robots. That is, their work demonstrates the automatic, nearly human-like response to humanoid social robots. Above and beyond doing so, this research substantiates Powers et al.'s (2005) hypothesis that people would create mental models of robots using the robots' ostensible social cues. According to Powers and colleagues, the social cues embedded in a robot suggest specific social categories to users (for instance, in terms of age, gender, and nationality) and users leverage these social categories to form mental models of robots. In this sense, the gender users attribute to a robot determines their perceptions of the robot's capabilities and social traits, as well as the users' behavior during the interaction. As with many early HRI studies, Powers et al. (2005) found scarce support for their hypothesis. However, their idea of humans creating mental models of robots based on social cues has surely impacted the literature. A couple of years after Powers et al.'s paper, Epley and colleagues (2007) proposed a theoretical framework to understand the determinants and consequences of anthropomorphism (the attribution of human-like traits and characteristics to non-human entities). Among the three factors that supposedly triggered anthropomorphic inferences, Epley et al. (2007) featured a cognitive factor called 'elicited agent knowledge,' the use of human-like social cues to make sense of non-human entities like robots, which owes much to Power's idea. Accordingly, anthropocentric knowledge structures are evoked and activated by means of human-like cues, and this cognitive content is utilized to form subsequent judgments.

One of the early works that took a social psychological perspective in line with Epley's framework was the experiment by Eyssel and Hegel (2012). This research team featured a psychologist and a designer who were curious to explore the impact of a robot's visual appearance on stereotyped perceptions. In this research, participants evaluated the robot Flobi. This robot had been newly developed by Frank Hegel and colleagues and was special given its modular industrial design. This design made it possible to manipulate the robot's persona by having it appear with either long or short hair. This was done to represent and experimentally manipulate male vs. female gender in robots. Interestingly, the robot with short hair, perceived as more masculine, was also perceived as more agentic and more suitable for typically male tasks than the female robot. On the other hand, the robot with long hair was perceived as more feminine, more communal, and more suitable for typically female tasks than the male robot. In fact, perceptions of agency and communion are quite commonly assessed in the context of social robots because these traits represent the core dimensions of social cognition, and, as such, these are likewise associated with gender stereotypes about men and women. Thus, agency (i.e., perceived competence, leadership, and assertiveness) represents stereotypical male traits, whereas communion (i.e., perceived warmth, empathy, and friendliness) represents stereotypical female traits. In line with Powers et al. (2005) and Epley et al. (2007), Eyssel and Hegel (2012) showed that subtle visual gender cues can activate participants' knowledge structures about gender in humans and that these knowledge structures are then applied to robots.

13.3.1 Robot and participant gender

In later work, not only the gender of the robot was manipulated, but also the gender stereotypicality of the task that participants completed with the robot during human–robot interaction. For instance, Kuchenbrandt et al. (2014) asked participants to sort out items in a sewing box (stereotypically female task) vs. a toolbox (stereotypically male task), whereas Tay et al. (2014) involved participants in a healthcare (stereotypically female) vs. security (stereotypically male) interaction scenario. In both studies, male and female participants were instructed to perform the tasks either by an ostensibly male or by an ostensible female robot. While Kuchenbrandt and colleagues found that men were faster in completing the tasks when instructed by a male robot, Tay and colleagues discovered that female and male participants preferred the robot whose gender matched the gender of the task. In further research by Reich-Stiebert and Eyssel (2017), the role of the match or mismatch between the robot's gender, the gender stereotypicality of the task, and the participant's gender was investigated. Reich-Stiebert and Eyssel conducted an HRI experiment that featured a tutoring scenario in which a gendered robot supported participants while solving stereotypically female verbal tasks vs. stereotypically male analytical tasks. Overall, participants' learning was not affected by the gender of the robot. However, people were more willing to engage in human–robot learning in the future when they were taught female topics by a male robot and male topics by a female robot. This research highlights that focusing merely on one dimension (namely, robot gender) may not suffice to fully appreciate the influence of robot gender on HRI. It underlines the need to zoom out of the gender cues a robot embeds to investigate the interaction effect between these cues, the context of the use of the robot, and the individual traits of the user. This shift in approach occurred relatively late in the history of the research on gendering robots. Indeed, in the early years of robot gender research, commonly only the robot's gender cues were manipulated, and participant gender and task gender stereotypicality were less often a focus.

The research on gender stereotypes initiated by Eyssel and Hegel in 2012 has been consistently expanded in the last few years. Bernotat and colleagues (2021) partially replicated Eyssel and Hegel's study. They manipulated robot gender slightly more subtly than in the original study by modifying the robot's waist-to-hip ratio and shoulder width, both gender cues that quite indirectly but successfully induce notions of gender. As opposed to the earlier research, they discovered that people manifested the tendency to stereotype a female but not a male robot. That is, female robots were perceived as more communal, more suitable for female tasks, and were attributed more affective trust than male robots. In contrast, male robots were not perceived as significantly more agentic, suitable for male tasks, and were not attributed more cognitive trust than female robots. In a similar vein, Perugia et al. (2021) supported these observations, as did work by Guidi et al. (2023). Using a larger pool of gendered robots (eight robots per gender category), Guidi and colleagues discovered that only female robots elicited gender stereotypes, whereas male robots did not. Moreover, these authors found that men reported a stronger tendency to stereotype female robots than women. Perhaps even more interestingly, Guidi et al. (2023) demonstrated that gender neutrality did not prevent participants from stereotyping, as many researchers would have hoped. Indeed, men attributed female stereotypes (i.e., high communion and low agency) to gender-neutral robots, whereas women attributed male stereotypes (i.e., low communion paired with high agency) to them. Taking into account Powers et al.'s (2005) hypothesis on the role of social cues in shaping users' mental models of a robot, we might conclude that gender neutrality constitutes an 'empty' social category for which users lack an actual mental model. From this, it follows that when people confront a gender-neutral robot, they seem unable to construct a reliable mental model pertaining to its gender neutrality. As a consequence, they project onto it the social perceptions most relevant for them.

In line with this idea, Winkle et al. (2021) tried to break the submissive and subservient persona commonly represented by female voice assistants such as Siri and Alexa. To do so, they showed participants a video of a female robot who made a feminist statement ("The future is too important to be left to men") in the context of an outreach activity aiming to attract girls to study robotics. In the video, the robot's statement gives rise to a strong verbal reaction by the interlocutor, a male actor, to which the robot reacts in three alternative ways: 1) providing a rationalized explanation of the feminist statement triggering the insult, 2) reacting in an aggressive way by insulting the male actor in return, or 3) in an avoidant way by not addressing the insult at all. The results of the study disclosed that the female robot was perceived as less credible when it displayed an aggressive reaction to the insult than when it reacted to the insult in an rationalized way. This finding is particularly telling if paired with the findings of Calvo-Barajas et al. (2020). In their work, children aged between 9 and 14 years were shown a Furhat robot, with either a male or female appearance, displaying facial expressions of happiness and anger (with different degrees of intensity: low, medium or high) for a few seconds. They liked the female robot less when it displayed high anger, but liked the male robot more when it displayed the same type of strong negative emotional reaction. It almost seems that female robots are expected to show the same emotional restraint and moderation of the human persona they typify (women). Winkle et al.'s study was also cross-culturally replicated with participants from Japan, the USA, and Sweden (Winkle et al., 2022). The study showed a preference for the rationalized response in all countries. Importantly, though, the avoidant response was preferred over the aggressive one in Japan.

Megan Strait et al. (2017) also focused on sexism but took a different route from Winkle et al. (2022). Strait et al. (2017) analyzed users' comments to online video materials regarding gendered (primarily female) robots. They discovered that female robots are frequently sexually objectified, for instance by referring to the performance of sexual acts on them or by them. In a follow-up study, Strait et al. (2018) showed that such a dehumanization and sexualization of female robots is further exacerbated when the robot is racialized. From all these studies, it follows that gendered robots, especially female ones, condense many of the sexist issues observed towards women in society. Relatedly, Sparrow (2017) discussed the representative power of social robots, especially humanoid ones, and warned his readers that a normalization of the sexist behavior people display towards gendered robots might go along with a generalization of such behavior towards women.

Throughout the years, researchers have used many disparate cues to suggest a robot's gender. In this chapter alone, we have seen short and long hair (Eyssel & Hegel, 2012), waist-to-hip ratio and shoulder width (Bernotat et al., 2017, 2021), and female or male voices (Siegel et al., 2009). However, it was not until very recently, with Perugia et al.'s work (2022), that researchers attempted a systematic formalization of the robot's gendering process. In their work, Giulia Perugia and colleagues (2022) asked participants to rate the perceived gender of 251 anthropomorphic robots in the ABOT (Anthropomorphic roBOT) database and attempted to understand which appearance cues of the robots and which individual characteristics of participants influenced gender attribution. They showed that men rate robots as higher in femininity and lower in masculinity and found empirical evidence for the lack of representation of female and androgynous robots in the category of anthropomorphic robots (only 15% of female robots and no androgynous robots vs. 39% of male and 46% of gender-neutral robots). Moreover, they disclosed that human likeness is the strongest predictor of gender attribution and that it is the presence of body manipulators in a robot (i.e., arms, torso, legs) that yields higher scores of masculinity, and the presence of surface-look features (i.e., eyelashes, apparel, and eyebrows) that yields higher scores of femininity. To shed light on whether the appearance cues used to suggest gender in a robot could actually play a role in stereotyping, Perugia et al. (2023) performed a second study. This time, they selected a subset of 80 robots from the ABOT database. They disclosed that body manipulators affected the attribution of agency, a stereotypical male trait, whereas surface-look features influenced the attribution of communion, a stereotypical female trait. This entails that the choice of cues we use to gender a robot affects how people perceive that robot, including whether they stereotype it or not. Importantly, Perugia et al. (2023) also considered participants' attitudes regarding gender and society. That is, they complemented their research by measuring participants' sexist attitudes and observed an increased sensitivity to male stereotypes in men with hostile and benevolent sexist attitudes. This research exemplifies the importance of including attitude-related constructs (e.g., sexist attitudes, social desirability, or motivation to control for sexist responses) in robot gender research beyond merely exploring participant gender. It also highlights that just as laypeople, roboticists might not approach the design of robots *tabula rasa*; rather, they might bring their personal perspectives and views of the world in the design of robots, including their biases. Given the non-heterogeneous composition of robotics as a field, this might bring issues of representation in the design of robots.

The research reported here illustrates that, in recent years, the topic of gender and robots has gained momentum and has sparked the interest of many researchers from multidisciplinary fields. Despite the fact that the topic has been treated in various more or less recent overview articles (Alesich & Rigby, 2017; Nomura, 2017; Søraa, 2017; Tannenbaum et al., 2019), new empirical research on the interplay of social context features, robot features, and human characteristics is needed to broaden our understanding of the interdependence of these dimensions, and to provide a more holistic account of what it means to imbue robots with a gender and how this affects not only the perception of, but more importantly, the behavior towards, robots—beyond the lab and in the real world. This conclusion points to some important considerations for HRI research at large, namely our need to conduct studies 'in the wild'—in real-world settings and applications, with potential end users to judge whether certain features or functions of a new technology actually matter to them.

Likewise, research is missing on whether people actually want to see gendered, gender neutral, or genderless robots in the market, in their homes, or as their socially assistive robot companions. Thus, coming back to the start of this chapter—in prospective research, we should not only let designers and makers of robots, but rather potential end users decide what robots they want to see in their lives, taking into account the fact that users and technologies mutually shape each other (Šabanović, 2010).

13.4 Outlook: How Gender and Robots Inform our Thinking about Gender and Humans

The multifaceted empirical research presented in this chapter has featured works spanning almost three decades of reflecting gender in the contexts of social technologies, ranging from the personal computers of the 1990s to physically embodied robots. What can we take from the present overview—in terms of new ideas regarding humans and robots? Importantly, we need to acknowledge that despite methodological shortcomings, the research reviewed here marked a milestone in the understanding of processes involved in human and robot perception. Indeed, the existing work demonstrated the added value of specific theoretical accounts (e.g., the Stereotype Content Model, Fiske et al., 2002) in understanding trait attributions to robots. Moreover, past research has shown the effect of robot gender on behavior displayed to the respective robot, thereby highlighting the importance of studying the impact of robot design on cognition, affect, and behavior—simultaneously, at best, instead of focusing merely on one of the three dimensions.

Moreover, the literature review has emphasized the need to go beyond a binary understanding of gender in humans and robots (Spiel, 2021). The important notion of gender neutrality in robots, however, has recently been refuted by Giudi et al. (2023) and Perugia and colleagues (2021). At the same time, it might be interesting to foster other robot identities beyond ascriptions of gender or other social categories (i.e., age, ethnicity). It might be fruitful to shape a robot's identity by providing a personality. At the same time, this comes with precautions, as—once again—even personality traits, and likewise the language we use, reveal or reflect gendered content. For example, the traits or professions of a robot may be associated with male or female gender. This is so deeply ingrained due to prevailing gender stereotypes. Changing these, including gender prejudice and sexism, is not a research problem happily resolved. Prejudice researchers around the globe are active in trying to change cultural and personal stereotypes.

This is where the social and ethical questions regarding the gendering of social robots and other technologies emerge. As a society and as researchers, we need to reflect on the implications of what we build and how we do so—from the beginning and not *post hoc*. This is relevant because, with our design decisions, we might inadvertently (or not) contribute to maintaining gender bias in contemporary societies.

To gather a holistic view on the matter, ideally future research should consider the interplay of gender at the level of research participants, objects to be studied, and context of the respective research (i.e., gender stereotypicality of the task or use case), as already emphasized by Nomura (2017). After all these years of research on the matter, such a holistic approach is still missing. Despite these exemplary methodological shortcomings, we can observe a heightened interest and awareness regarding the topic, and even extending to intersectional perspectives, giving rise to a greater interest in and focus on diversity in HRI research and among HRI researchers at large (De Graaf et al., 2022; Tanevska et al., 2023). As we have learned, designers and objects designed are closely intertwined. By shedding light on the consequences of design choices through empirical research, we contribute to a better understanding of attributing whatever can be deemed human to entities non-human. As Perugia and Lisy (2023) highlight, robot gender attributions manifest as a consequence of a cognitive attribution process, including the encoding and decoding of gender. These processes are also upheld in humans. However, here, the nature of the process may go beyond the mere psychological attribution of traits to individuals or groups by including a negotiation process with society at large. Accordingly, gender may be viewed as a dynamic negotiation process between individuals and society, with society and stakeholders having to reconsider run-down pathways regarding the binary to overcome experiences of otherness and the neglect of relevant non-binary perspectives. As researchers and as human beings, we are encouraged to value and make visible the multitude of perspectives on humans and non-human entities alike. We can do this in our daily research practice, but ultimately it starts with the individual. It starts at our doorstep. How we as individuals see the world shapes how others are seen—with the option to eventually overcome othering and otherness while appreciating the diversity of human existence. In other words, it starts with each one of us and the daily opportunity to experience a heightened awareness of gender and other dimensions relevant to social judgments (e.g., age, ethnicity, ability), which inevitably shape the human experience and the attribution of traits to fellow humans and to robots.

Bibliography

Alesich, S., & Rigby, M. (2017). Gendered robots: Implications for our humanoid future. *IEEE Technology and Society Magazine, 36*(2), 50–59.

Bernotat, J., Eyssel, F., & Sachse, J. (2017). Shape it—The influence of robot body shape on gender perception in robots. In *Social Robotics: 9th International Conference, ICSR 2017, Tsukuba, Japan, November 22–24, 2017, Proceedings 9* (pp. 75–84). Springer International Publishing.

Bernotat, J., Eyssel, F., & Sachse, J. (2021). The (fe) male robot: how robot body shape impacts first impressions and trust towards robots. *International Journal of Social Robotics*, 13(3), 477–489.

Burtscher, S., & Spiel, K. (2020). " But where would I even start?" developing (gender) sensitivity in HCI research and practice. In *ACM Proceedings of Mensch und Computer 2020* (pp. 431–441).

Butler, J. (1990). *Gender Trouble: Feminism and the Subversion of Identity*. Routledge.

Calvo-Barajas, N., Perugia, G., & Castellano, G. (2020, August). The effects of robot's facial expressions on children's first impressions of trustworthiness. In *2020 29th IEEE International Conference on Robot and Human Interactive Communication* (pp. 165–171). IEEE.

De Graaf, M. M., Perugia, G., Fosch-Villaronga, E., Lim, A., Broz, F., Short, E. S., & Neerincx, M. (2022, March). Inclusive HRI: Equity and diversity in design, application, methods, and community. In *2022 17th ACM/IEEE International Conference on Human-Robot Interaction* (pp. 1247–1249). IEEE.

Deng, E., Mutlu, B., & Mataric, M. J. (2019). Embodiment in socially interactive robots. *Foundations and Trends in Robotics*, *7*(4), 251–356. http://dx.doi.org/10.1561/2300000056.

Epley, N., Waytz, A., & Cacioppo, J. T. (2007). On seeing human: A three-factor theory of anthropomorphism. *Psychological Review, 114*(4), 864–886. https://doi.org/10.1037/0033-295X.114.4.864.

Eyssel, F., & Hegel, F. (2012). (S)he's got the look: Gender stereotyping of robots. *Journal of Applied Social Psychology, 42*(9), 2213–2230. https://doi.org/10.1111/j.1559-1816.2012.00937.x.

Eyssel, F. (2017). An experimental social psychological perspective on social robotics. Special Issue on Next Generation of Robots for the Next Generation of Humans?". *Robotics and Autonomous Systems*, 87, 363–371. https://doi.org/10.1016/j.robot.2016.08.029

Fink, J. (2012). Anthropomorphism and human likeness in the design of robots and human–robot interaction. In S. S. Ge, O. Khatib, J. J. Cabibihan, R. Simmons, & M. A. Williams (Eds.), *Social Robotics*. ICSR 2012. Lecture Notes in Computer Science, Vol. 7621. Springer. https://doi.org/10.1007/978-3-642-34103-8_20.

Fiske, S. T., Cuddy, A. J. C., Glick, P., & Xu, J. (2002). A model of (often mixed) stereotype content: Competence and warmth respectively follow from perceived status and competition. *Journal of Personality and Social Psychology*, 82(6), 878–902. https://doi.org/10.1037/0022-3514.82.6.878

Fortunati, L., Edwards, A., Edwards, C., Manganelli, A. M., & De Luca, F. (2022). Is Alexa female, male, or neutral? A cross-national and cross-gender comparison of perceptions of Alexa's gender and status as communicator. *Computers in Human Behavior, 137,* 107426.

Fussell, S. R., Kiesler, S., Setlock, L. D., & Yew, V. (2008). How people anthropomorphize robots. In *Proceedings of the 3rd ACM/IEEE International Conference on Human–Robot Interaction,* (pp. 145–152). ACM.

Goffman, E. (2005). *Interaction Ritual: Essays in Face to Face Behavior.* Penguin Books.

Guidi, S., Boor, L., van der Bij, L., Foppen, R., Rikmenspoel, O., & Perugia, G. (2023). Ambivalent stereotypes towards gendered robots: The (im)mutability of bias towards female and neutral robots. In *Proceeedings of the 14th International Conference of Social Robotics* (pp. 615–626). Springer Nature.

Kiesler, S., Powers, A., Fussell, S. R., & Torrey, C. (2008). Anthropomorphic interactions with a robot and robot-like agent. *Social Cognition, 26,* 169–181.

Kuchenbrandt, D., Häring, M., Eichberg, J. & Eyssel, F. & André, E. (2014). Keep an eye on the task! How gender typicality of tasks influence human-robot interactions. *International Journal of Social Robotics, 6,* 417–427. https://doi.org/10.1007/s12369-014-0244-0

Lee, E. J., Nass, C., & Brave, S. (2000, April). Can computer-generated speech have gender? An experimental test of gender stereotype. In *CHI'00 Extended Abstracts on Human Factors in Computing Systems* (pp. 289–290).

Li, J. (2015). The benefit of being physically present: A survey of experimental works comparing copresent robots, telepresent robots and virtual agents. *International Journal of Human Computer Studies, 77,* 23–37. https://doi.org/10.1016/j.ijhcs.2015.01.001.

Marchetti-Bowick, M. (2009). Is your roomba male or female? The role of gender stereotypes and cultural norms in robot design. *Intersect: The Stanford Journal of Science, Technology, and Society, 2*(1), 90–103.

Nass, C., & Moon, Y. M. (2000). Machines and mindlessness: Social responses to computers. *Journal of Social Issues, 56*(1), 81–103. https://doi.org/10.1111/0022-4537.00153.

Nass, C., Moon, Y., & Green, N. (1997). Are machines gender neutral? Gender-stereotypical responses to computers with voices. *Journal of Applied Social Psychology, 27*(10), 864–876.

Nass, C., Steuer, J., & Tauber, E. R. (1994). Computers are social actors. In *Proceedings of SIGCHI '94 Human Factors in Computing Systems* (pp. 72–78). ACM.

Nomura, T. (2017). Robots and gender. *Gender and the Genome, 1*(1), 18–25.

Onnasch, L., & Roesler, E. (2021). A taxonomy to structure and analyze human–robot interaction. *International Journal of Social Robotics, 13,* 833–849.

Perugia, G., Boor, L., van der Bij, L., Rikmenspoel, O., Foppen, R., & Guidi, S. (2023, March). Models of (often) ambivalent robotstereotypes: Content, structure, and predictors of robot's age and gender stereotypes. In *Proceedings of the 2023 ACM/IEEE International Conference on Human–Robot Interaction* (pp. 428–436). IEEE.

Perugia, G., Guidi, S., Bicchi, M., & Parlangeli, O. (2022, March). The shape of our bias: Perceived age and gender in the humanoid robots of the ABOT database. In *2022 17th ACM/IEEE International Conference on Human–Robot Interaction* (pp. 110–119). IEEE.

Perugia, G., & Lisy, D. (2023). Robot's gendering trouble: a scoping review of gendering humanoid robots and its effects on HRI. *International Journal of Social Robotics,* 15(11), 1725–1753.

Perugia, G., Rossi, A., & Rossi, S. (2021). Gender revealed: Evaluating the genderedness of Furhat's predefined faces. In *Social Robotics: 13th International Conference, ICSR 2021, Singapore, Singapore, November 10–13, 2021, Proceedings 13* (pp. 36–47). Springer International Publishing.

Powers, A., Kramer, A. D., Lim, S., Kuo, J., Lee, S. L., & Kiesler, S. (2005, August). Eliciting information from people with a gendered humanoid robot. In *ROMAN 2005 IEEE International Workshop on Robot and Human Interactive Communication* (pp. 158–163). IEEE.

Reeves, B., & Nass, C. (1996). *The Media Equation: How People Treat Computers, Television, and New Media Like Real People and Places.* Cambridge University Press.

Reich-Stiebert, N., & Eyssel, F. (2017, March). (Ir)relevance of gender? On the influence of gender stereotypes on learning with a robot. In *Proceedings of the 2017 ACM/IEEE International Conference on Human–Robot Interaction* (pp. 166–176). IEEE.
Robertson, J. (2010). Gendering humanoid robots: Robo-sexism in Japan. *Body & Society*, *16*(2), 1–36.
Roesler, E. (2023). Form follows function: Challenging the effectiveness and desirability of anthropomorphism in human–robot interaction. Dissertation, Technical University of Berlin. https://api-depositonce.tu-berlin.de/server/api/core/bitstreams/072d8057-faf1-4054-8729-13982e2f8357/content.
Roesler, E., Manzey, D., & Onnasch, L. (2021). A meta-analysis on the effectiveness of anthropomorphism in human–robot interaction. *Science Robotics*, *6*, 58, eabj5425.
Roesler, E., Manzey, D., & Onnasch, L. (2022). Embodiment matters in social HRI research: Effectiveness of anthropomorphism on subjective and objective outcomes. *Transactions in Human–Robot Interaction*, *12*(1), 1–9. https://doi.org/10.1145/3555812.
Šabanović, S. (2010). Robots in society, society in robots: Mutual shaping of society and technology as a framework for social robot design. *International Journal of Social Robotics*, *2*(4), 439–450.
Siegel, M., Breazeal, C., & Norton, M. I. (2009, October). Persuasive robotics: The influence of robot gender on human behavior. In *2009 IEEE/RSJ International Conference on Intelligent Robots and Systems* (pp. 2563–2568). IEEE.
Søraa, R. A. (2017). Mechanical genders: How do humans gender robots? *Gender, Technology and Development*, *21*(1–2), 99–115. https://doi.org/10.1080/09718524.2017.1385320.
Sparrow, R. (2017). Robots, rape, and representation. *International Journal of Social Robotics*, *9*(4), 465–477.
Spiel, K. (2021, June). "Why are they all obsessed with Gender?"—(Non) binary navigations through technological infrastructures. In *ACM Designing Interactive Systems Conference 2021* (pp. 478–494).
Spiel, K., Haimson, O. L., & Lottridge, D. (2019). How to do better with gender on surveys: A guide for HCI researchers. *Interactions*, *26*(4), 62–65. https://doi.org/10.1145/3338283.
Strait, M. K., Aguillon, C., Contreras, V., & Garcia, N. (2017, August). The public's perception of humanlike robots: Online social commentary reflects an appearance-based uncanny valley, a general fear of a "Technology Takeover", and the unabashed sexualization of female-gendered robots. In *2017 26th IEEE International Symposium on Robot and Human Interactive Communication* (pp. 1418–1423). IEEE
Strait, M., Ramos, A. S., Contreras, V., & Garcia, N. (2018, August). Robots racialized in the likeness of marginalized social identities are subject to greater dehumanization than those racialized as white. In *2018 27th IEEE International Symposium on Robot and Human Interactive Communication (RO-MAN)* (pp. 452–457). IEEE.
Tanevska, A., Chandra, S., Barbareschi, G., Eguchi, A., Han, Z., Korpan, R., Ostrowski, A., Perugia, G., Ravindranath, S., Seaborn, K., & Winkle, K. (March, 2023). Inclusive HRI II: Equity and diversity in design, application, methods, and community. In *Companion of the 2023 ACM/IEEE International Conference on Human–Robot Interaction* (pp. 956–958). IEEE.
Tannenbaum, C., Ellis, R. P., Eyssel, F., Zou, J., & Schiebinger, L. (2019). Sex and gender analysis improves science and engineering. *Nature*, *575*, 137–146. https://doi.org/10.1038/s41586-019-1657-6.
Tay, B., Jung, Y., & Park, T. (2014). When stereotypes meet robots: The double-edge sword of robot gender and personality in human–robot interaction. *Computers in Human Behavior*, *38*, 75–84.
Wainer, J., Feil-Seifer, D. J., Shell, D. A., & Mataric, M. J. (2007). Embodiment and human–robot interaction: A task-based perspective. In *Proceedings of the 16th IEEE International Symposium on Robot and Human Interactive Communication* (pp. 872–877). IEEE.
Winkle, K., Melsión, G. I., McMillan, D., & Leite, I. (2021, March). Boosting robot credibility and challenging gender norms in responding to abusive behaviour: A case for feminist robots. In *Companion of the 2021 ACM/IEEE International Conference on Human–Robot Interaction* (pp. 29–37). IEEE.

Winkle, K., Jackson, R. B., Melsión, G. I., Brščić, D., Leite, I., & Williams, T. (2022, March). Norm-breaking responses to sexist abuse: A cross-cultural human robot interaction study. In *2022 17th ACM/IEEE International Conference on Human-Robot Interaction (HRI)* (pp. 120–129). IEEE.
Ziemke, T. (2003). What's that thing called embodiment? *Proceedings of the 25th Annual Meeting of the Cognitive Science Society, 25*(25), 1305–1310.
Złotowski, J., Proudfoot, D., Yogeeswaran, K., & Bartneck, C. (2015). Anthropomorphism: Opportunities and challenges in human–robot interaction. *International Journal of Social Robotics*, *7*(3), 347–360. https://doi.org/10.1007/s12369-014-0267-6.

Kate Devlin

Relating with Social Robots: Issues of Sex, Love, Intimacy, Emotion, Attachment, and Companionship

Abstract: The chapter provides a comprehensive overview of the issues surrounding sex, love, intimacy, emotion, attachment, and companionship in human–robot relationships. From the narratives that have shaped our expectations to the products on sale today, it examines why we find the idea of the perfect artificial partner so compelling—and so frightening. Any study of our interaction with robots is really a study of ourselves, and the way we shape technologies for intimacy tells us much about how we as humans seek sociality and acceptance.

Keywords: Human–robot relationships, narratives, sexuality, love, intimacy emotions, ethics

14.1 Introduction

The stories we tell about imagined artificial partners far pre-date the age of robotics, and the headlines we see today far exceed the reality. Yet, in the past decade, this has increasingly become an area of academic interest. At the beginning of the twenty-first century, the very idea of a robot for love, pleasure, and intimacy was treated as laughable, seen as merely the remit of science fiction. This has now changed, and the field has become legitimate—to an extent. There are now multiple books, articles, research projects, and an established international conference dealing with the topic. Not everyone is convinced of its importance as a topic: there is still an air of bemusement, particularly from the more technically minded. Audiences are often drawn to talks on the subject because of the novelty of it. However, what quickly becomes clear to anyone reading a little deeper is that research about sexual or romantic companion robots—sex robots, as shorthand—is fundamentally research about humans: about what makes us love and desire, and our need to be social and to be connected.

The chapter explores the multifaceted dimensions of human–robot relationships, focusing on the issues of sex, love, intimacy, emotion, attachment, and companionship. From longstanding narratives to possible future directions, it delves into human–robot relationships and the implications that arise.

https://doi.org/10.1515/9783110792270-015

14.1.1 The origin of the field

As an area of academic research, the field is somewhat new. In 2001, Peter Asaro and Doug Matejka wrote and directed a feature-length documentary entitled *Love Machine* where Asaro tracked down leading figures in technology and philosophy and interviewed them about possible human–robot relationships in the future. In his commentary about the film some seven years later, Asaro remarks: “I think it is a fair criticism of the film that there is a tendency to elide the distinction between sex and love (or emotion, or affection)” (Asaro & Vann, 2008, p. 24), and this observation highlights something inescapable when writing about this topic: that there is overlap. Neither sex nor love can be addressed individually. This is not a clearly demarcated subject area with discrete (but discreet) interactions; instead, much like human relationships, there can be sex with love, or love alone, or sex alone, all with varying degrees of emotional engagement.

In terms of the mainstreaming of the conversation, David Levy’s *Love and Sex with Robots* in 2007 was a general-audience book exploring the reasoning behind Levy’s belief that humans will be marrying robots by 2050. The book drew significant media coverage and led to conference workshops and, eventually, a conference series. In 2017, John Danaher and Neil McArthur published *Robot Sex: Social and Ethical Implications*—an edited collection focusing on the philosophical discussions. In 2018, Kate Devlin’s book *Turned On: Science, Sex and Robots* was published and, in 2019, Yuefang Zhou and Martin H. Fischer’s *AI Love You: Developments in Human–Robot Intimate Relationships*. This was followed in 2020 by Jenny Kleeman’s *Sex Robots & Vegan Meat: Adventures at the Frontier of Birth, Food, Sex & Death*. In 2021, *Artificial Intimacy: Virtual Friends, Digital Lovers, and Algorithmic Matchmakers* by Rob Brooks was published, and Bo Ruberg’s *Sex Dolls at Sea: Imagined Histories of Sexual Technologies* followed in 2022. Alongside these monographs has been an increase in academic journal and conference papers: Google scholar shows only a handful of such articles prior to 2007, rising to around 220 up to 2015, and increasing to over 3,000 academic articles today. A scoping review by Nicola Döring et al. (2020) confirms a steep rise in academic interest in the topic around 2018.

14.1.2 Terminology and definitions

This chapter focuses on a particular form of social robot: one where the concept is that of an artificial romantic or sexual partner. The short-hand term for this is ‘sex robot,’ or ‘sexbot’ (and occasionally ‘fembot’), though their use is not solely sexual, and they are limited in terms of robotics. As a useful starting point, in the introductory chapter of *Robot Sex*, Danaher writes:

> The definition holds that a 'sex robot' is any artificial entity that is used for sexual purposes (i.e., for sexual stimulation and release) that meets the following conditions:
>
> Humanoid form, i.e., it is intended to represent (and is taken to represent) a human or human-like being in its appearance.
>
> Human-like movement/behavior, i.e., it is intended to represent (and is taken to represent) a human or human-like being in its behaviours and movements.
>
> Some degree of artificial intelligence (AI), i.e., it is capable of interpreting and responding to information in its environment. (Danaher, 2017, pp. 4–5)

The purpose of Danaher's definition is for classification to draw a line between sex toys (which are not full humanoid forms) and sex dolls (which have no movement or AI). However, he acknowledges that there is no reason why such robots *should* only be humanoid in form or human-like in behavior. That said, while such variations have been suggested (Devlin, 2020, p. 278), real-world developments to date focus very narrowly on sex dolls with some interactive behavior. This, then, forms the basis of a 'sex robot' for the purposes of this chapter.

It must be noted that the vast majority of representations of sex robots, including the few real-world prototypes in existence, are overwhelmingly of a female form. This tends to be a reductive stereotype of a female body—one that is almost a caricature of what might be deemed a desirable woman: large breasts, narrow waist, slim limbs, long hair. While newspaper headlines allude to a male version of a sex robot in development, this seems to be merely a token example. Instead, all the advertising points to a female figure, intended to look realistically human, who promises not only sexual interaction but also personalised companionship. This is a technology designed by men for men.

14.2 Robot Relationships and Robot Realities

14.2.1 History and narrative

In looking for an origin for stories about love and/or sex between a human and a robot, the earliest tales date back to Greek mythology. Genevieve Liveley identifies the tale of Laodamia and Protesilaus as a key example (Liveley, interviewed in Devlin, 2018, p. 18). This tale was recounted by the Latin writer Gaius Julius Hyginus in the first century BCE in his *Fabulae*, where he expanded on works by Greek writers of tragedy. Hyginus wrote that:

> [S]he [Laodamia] could not endure her weeping and grief. And so she made a bronze likeness of her husband Protesilaus, put it in her room under pretense of sacred rites, and devoted herself to it. When a servant early in the morning had brought fruit for the offerings, he looked through

> a crack in the door and saw her holding the image of Protesilaus in her embrace and kissing it. Thinking she had a lover he told her father Acastus. When he came and burst into the room, he saw the statue of Protesilaus. (1960, p. 104)

Setting the tone for future dystopian scenarios, the story ends badly: Laodamia's father orders the statue to be burned, and Laodamia throws herself on the pyre after it.

Several researchers point to the story of Pygmalion by the Roman poet Ovid, from his *Metamorphoses*, as being one of the foundational tales of an artificial companion. The story is that of a sculptor who shuns the women in his everyday life, who he deems unworthy as, in his eyes, they behave inappropriately. Pygmalion can, perhaps, be viewed as an early version of an incel:[1] in his misogynistic view, no woman is good enough and he is "offended by the failings that nature gave the female heart" (Ovid, 2000, Bk X: 243–297). He creates his own superior version of a woman instead and prays for her to become real. His prayer is heard, and his perfect companion comes alive, though she is then no longer artificial. Julie Wosk has identified the influence and reception of Pygmalion from its original telling through countless examples from books and theater to modern film, where she identifies the themes as "the story of men's enduring fantasies and dreams about producing the perfect woman, a custom-made female" (2015, p. 5). Indeed, this quest for perfection seems to be a strong motivation when we examine people's motivations for wanting an artificial companion.

Science fiction, in print and in film, has centered many a plot on a human–robot romantic partnership. These range from the comic to the tragic, with few of them resulting in a happy conclusion. They also tend to be heteronormative and heavily gendered. By and large, the robot is presented as female—a gynoid—and the main human character as the android male. The robot's role is to provide devotion and desirability—something that shapes expectations in the real world. While Laodamia's motivation—replacing a lost love—is one that recurs down the centuries, it is an unusual version as Laodamia is female and the 'likeness' she commissions is male. By the time we reach the age of cinema, it is almost exclusively the opposite. The vast majority of artificial romantic companions are depicted as female, with the exception of Stephen Spielberg's film *A.I.: Artificial Intelligence* (2001), featuring Gigolo Joe, a male sex worker robot. As Devlin and Belton write:

> Thousands of years of stories about the perfect artificial woman seem so thoroughly ingrained as to be insurmountable. These stories have fed into expectations, and the expectations are uncritically driving development. The appeal of the sex robot seems to be rooted in deeply conservative notions of femininity, where a woman's emotional and sexual energy is entirely centred on pleasing a man. These media representations seem to intuitively understand that this is impossible, but female AIs who resist their prescribed roles are often shown to be violent and less than human. (2020, p. 377)

1 'Incel' is short for 'involuntary celibate', a misogynist frustrated by lack of sexual activity.

In the media, sex robots were initially only mentioned in film reviews from the 1970s onwards. There was occasional discussion of futuristic promise, but it wasn't until the publication of Levy's *Love and Sex with Robots* in 2007 that newspapers ran stories about Levy's claim that robot lovers would be commonplace by 2050. This was treated with jokey skepticism from columnists.

In January 2010, multiple articles about sex robots in development started to emerge. These centered on one of the first examples, Roxxxy True Companion—a life-size sex doll with a supposed 'AI engine' primed to learn the owner's likes and dislikes. Despite bold claims and confident advertising, the Roxxxy model—who was shown (to some ridicule) at the Adult Entertainment Expo in Las Vegas and then rarely seen again—existed more in the imagination of the creator than in reality. Douglas Hines, the creator, said he had over 4,000 orders for his doll, but no one ever showed any evidence of getting one, and the general consensus is that Roxxxy was vaporware: advertised, but not available to buy as it existed only as a very rudimentary concept (Devlin, 2018, p. 142).

From 2015, the headlines began to focus on fears around potential sex robot development, with dozens of articles from "How robots will soon take teens' virginity" (*Daily Mail*) to "Robot sex doll lets men simulate rape" (*The Times*). Such articles increased in number in late 2016 when the *Love and Sex with Robots: Second International Conference* was held in London, to much media fanfare. The temptation of a clickbait headline was hard to avoid, and a small, two-day academic conference garnered weeks of coverage. Since then, the media has covered stories about sex and love with robots as a means of prurience paired with moral panic, with tabloid newspapers particularly fond of a sensationalist headline.

When a self-declared incel murdered ten people in April 2018—eight of them women—there were inevitably news stories proposing that sex robots could be a cure for men who feel they are owed sex as a fundamental right, to the point where they will harm others. But incel discussion forums were not convinced. Some contributors liked the idea that a sex robot would render real women useless and obsolete, while others dismissed this as 'a cope'—a poor substitute (Devlin, 2018, p. 227).

Occasionally a more nuanced or balanced story graces the pages of print media. As might be expected, it was the tech-focused media publications (such as *Wired* and *Motherboard*) which took a more pragmatic and evidenced view, although sometimes the newspapers stepped up. "Don't worry about sex robots. They won't ruin sex" was one such headline in the *Washington Post* (Dickson, 2016). The sub-header "They might even make it better for some people" was the antithesis of the usual moral panic and provided a nuanced account of the ethical debates.

14.2.2 The manufactured reality

At the time of writing, commercially available sex robots are few and far between. The majority of development is from a handful of small workshops worldwide, such as DS Dolls, EXDoll Robotics, and RealDollX, which have diversified into robotics from their original product of life-sized and life-like sex dolls (Locatelli, 2022, p. 13). The only company to have brought a sex robot fully to market so far is Abyss Creations, home of RealDollX, who created the 'Harmony' sex robot out of the lineage of their RealDoll sex dolls (Coursey et al., 2019).

Sex dolls (as opposed to sex robots) are fairly widely available, though the market is niche, and they range from cheap 'mini' versions to hand-crafted, customizable, high-end models, which is Abyss Creations's speciality: a poseable PVC skeleton inside, with steel joints and silicone flesh, customizable genitals, swappable faces, and a menu of body shapes. By and large, all the sex dolls available today take a female form. There are some male versions, but these represent a tiny share of the market.

The comparative lack of actual, purchasable sex robots means that most researchers turn to the sex doll community for parallels (Döring & Pöschl, 2018; Langcaster-James & Bentley, 2018; Hanson & Locatelli, 2022; Harper et al., 2023), assuming that the consumers will be similar or the same. This assumption is supported by activity on online doll forums where members talk about the latest robotic developments (Woodward, 2020; Middleweek, 2021), and by interviews with doll owners about their views on sex robots (Devlin & Locatelli, 2020).

Abyss Creations's CEO, Matt McMullan, has developed the Harmony sex robot as an extension of the dolls his company creates. He wanted to offer a product with more interactivity and began by creating an animatronic head to fit onto any of his doll bodies. This involved creating an artificial intelligence 'chatbot' which can be used as a standalone app or embedded into the robot so that it appears the robot has a personality. The chatbot's characteristics can be set via a control panel where users can increase or decrease settings such as 'unpredictable,' 'cheerful,' 'moody,', 'affectionate,' 'sensual,' and so on. The continuation from established sex doll market to animatronic enhancements suggest that market research was not a key factor, and that the company chose to branch into robotics as a competitive advantage rather than as a response to demand, something that McMullen's comments reinforce in interviews and discussions about Harmony (Coursey et al., 2019; Devlin & Locatelli, 2020).

Outside of the US, China has sex robots in development stages, with companies such as Doll Sweet (using the brand name EX Doll in Asia and DS Doll in the West) and Z-Onedoll releasing videos online of examples of their work. Again, these are in the form of sex dolls with some animatronic features, usually an animatronic head but also sometimes with sensors or vibrating motors in the limbs and torso.

Other than the aforementioned Roxxxy, past examples of sex robots include Sergei Santos's Samantha, created from a standard sex doll but with the emphasis on an AI system that encouraged reciprocation: the aim was to woo Samantha, to provide a digital version of sexual pleasure that would then be returned to the user. This emphasis on sociality was promising, but Santos abandoned his work on Samantha in 2017 when he received negative publicity following (mis)reporting that his robot was 'mauled' at a tech show (Devlin, 2018, p. 228).

From the above, it is clear that sex robots are far removed from what we might refer to as a robot in other situations, particularly in the robotics community. These are not programmable machines designed to automate a task in a series of actions. Autonomy is somewhat limited, and they are not able to manipulate physical things. Rather, they are stationary human-like forms with some mechanization and some embedded AI: passive, rather than active. They cannot move their limbs or stand up unsupported. That said, some of the commercial social robots that have been produced in recent years, such as Jibo or Paro, have also been limited in physical movement. This lack of mobility is not necessarily a problem: De Graaf et al.'s (2015) longitudinal study on what makes a robot social showed that the priority for users was "two-way interaction, which constitutes speaking to the robot and for it to respond in a social manner" (De Graaf, 2021, p. 188). With embedded AI in the sex robots, they can meet this criterion, aligning with Breazeal's (2004) definition of a 'sociable robot' that communicates and relates to us personally.

The AI embedded in these robots is not contingent on embodiment. For now, it is there to add a sense of personality. However, while not meeting the definition of a robot due to their digital-only form, Chloé Locatelli points to the social interactivity permitted by AI-only, avatar-based versions of the 'sexbot' and asserts that the "metonymic 'sex robot' conceptualisation narrows the parameters of analysis" (Locatelli, 2021, p. 13). Indeed, Simon Dubé and Dave Anctil define a broader field of *erobots*—artificial erotic agents—including not only sex robots but also virtual or augmented partners and erotic chatbots (Dubé & Anctil, 2021, p. 1205). AI companion chatbots, such as Replika, have seen rapid uptake from users keen to engage in intimate conversations and erotic roleplay (Huet, 2023). This is an important adjacent research area with many overlapping themes, but a full account is beyond the scope of this chapter.

14.3 Exploring Human–Robot Intimacy

Research on relationships—physical or emotional—between humans and machines draws upon established literature from social psychology, philosophy, human–computer interaction, and human–robot interaction. This section proceeds on the basis that we accept that, as humans, we engage socially with machines even when we are aware of their limitations—the Computers are Social Actors (CASA) paradigm (Nass et al., 1994).

14.3.1 Sex

Sex, in this context, refers to sexual activity leading to arousal. Defining sex is not a simple task: penis–vaginal intercourse is commonly viewed to be the core definition, but sexual activity is much broader than this, so any classification based on perceived norms is likely to be reductive. For this chapter, sex refers to sexual activity tied to arousal. This might be anything from genital stimulation through to penetrative sex and orgasm. Mark Migotti and Nicole Wyatt wrote about how sex is classified, seeking a way of determining what sex with robots might be like (2017, p. 16). They addressed factors such as activities, agency, and the requirement for a partner or partners: the shared agency of 'being sexual together' with a partner (or partners) who are subjects rather than objects. They did not include masturbation as a form of sex in their classification, suggesting that to do so would reduce sex robots to the unproblematic status of a sex toy. However, the reality of the technology as it currently exists today means that masturbation probably does need to be included in this classification: sexual interaction is still occurring, even if it is not due to action from the (robotic) partner.

The purpose of a sex robot is to facilitate some kind of sexual encounter, but this is not its only purpose. The sex robots in development at the time of writing have human-like bodies with orifices that can be penetrated. This follows the standard sex doll form. But manufacturers prefer to emphasise the emotional connection, dedicating more advertising space to 'romantic semantics': phrasing that praises the benefits of companionship and love. This is reinforced by interviews with doll owners who state that sex is not the main driver of their interactions (Devlin & Locatelli, 2020). It is the potential for intimacy that is promised and marketed. Sex, by comparison, is portrayed as functional.

The people who buy sex dolls report that they buy them for a number of different reasons (Su, 2019). Some want the feeling of company, while others fetishize the sex. Some pose them and take photographs of them. Some are in relationships, others are single. Some worship their dolls, others love them out of sentimentality. Some see them as sexual, others as romantic. All of them are aware that this is a doll, not a human, even if they choose to treat it as such—something Jessica M. Szczuka et al. describe as the 'Sexual Interaction Illusion Model' (2019). Via multiple surveys and studies with sex doll owners, it very quickly becomes clear that owners have genuine feelings for their dolls (Ciambrone et al., 2017; Coursey et al., 2019; Devlin & Locatelli, 2020; Hanson, 2022; Langcaster-James & Bentley, 2018; Su et al., 2019). This has led to research that explores how these same groups feel about robot versions of their dolls (Szczuka & Krämer, 2017), and cements this group as being the most likely consumers for sex robots.

14.3.2 Love

Sexual activity, then, is not the only reason why someone might want to own a sex robot. For some, the parasocial relationship—a one-sided feeling of love—permits the inhabitation of an imaginary world where the robot can be viewed as real, even when the user is conscious that it cannot be so. We see plenty of examples of parasocial relationships in everyday life, such as crushes on celebrities, or whole 'alt universes' of fan fiction. These feelings seem no less real just because they are not returned.

In 1992, two anthropologists, William Jankowiak and Edward Fischer, published a paper entitled "A cross-cultural perspective on romantic love." They examined 166 societies for evidence of "any intense attraction that involves the idealization of the other, within an erotic context, with the expectation of enduring for some time into the future" (p. 150) and found that this was present in 88.5% of the societies they sampled. This suggests that love may be a universal emotion, although trying to define it remains tricky. Despite centuries of poetry, prose, and dreadful song lyrics, a definition of love is nebulous. There are cultural standards and expectations, there is undoubtedly a biochemical element, and it may vary in intensity and in presentation. For the purposes of this chapter, the love discussed here is the romantic love that one person feels. The need for it to be returned is arguable.

For some researchers, this raises debate about the validity of a human–machine relationship. Sven Nyholm and Lily Eva Frank examine love in a philosophical sense, namely the idea of being a good match, the notion that lovers should value each other, and the ideal of commitment. They remark that in humans, unrequited love can be seen as a genuine (if unfulfilled) type of love—one that might have the possibility of being returned—but with a robot, that possibility is not there (2017, p. 239).

The first half of Levy's *Love and Sex with Robots* book ignores sex in favor of exploring the idea of falling in love and how it might extend to our relationships in this new world of reactive machines. He examines the phenomenon of loving people, of loving pets, and of forming emotional relationships with machines and robots. Levy's vision, which is markedly from a heterosexual male viewpoint, is optimistic, stating: "Many who would otherwise have become social misfits, social outcasts, or even worse, will instead be better-balanced human beings. . . . The world will be a much happier place because all those people who are now miserable will suddenly have someone" (Levy, 2007, p. 304).

Members of the doll-owning community have engaged with researchers, providing valuable insight into the important role that their artificial companions play. This has included a keynote speech from Reggie Guzman at the *7th International Congress on Love & Sex with Robots* entitled "My artificial companions: A life-changing experiment," where he stated: "As someone would fall in love with a character in a book or a movie, that is how I feel about Annie and Judith [his dolls]. The love I have for them is unconditional. It is incredible how much they do for me without doing anything" (Guzman, 2022, 28:54).

Julie Carpenter acknowledges that "in the foreseeable future of technology, human–robot love will be one-way—human-to-robot—since robots cannot ever feel emotions in the exact same way humans do" (2017, p. 264) and describes the possibility of a "new kind of love"; that is, "a new culturally agreed-upon social category defined, in part, by not being human" (2017, p.273). She continues: "To deny the idea of human–robot sexuality, or the idea of human–robot affection or romantic attachment, is akin to denying the existence of many innate and integral parts of humanness and human experience" (2017, p. 283).

14.3.3 Intimacy and attachment

Intimacy does not necessarily exist solely within the confines of love or sex, but it is often conflated with these. The term 'intimacy' has connotations of emotional or physical closeness. That it is used euphemistically, to refer to sexual contact, shows its origins as something studied in monogamous, heterosexual romantic and sexual relationships. Ken Plummer recounts a long history of the term, dating back to 1632, referring to it concisely as "the inmost relationships with self and others" (2003, p. 13). Physical intimacy—based on proximity and personal space, and manifested in eye contact, a touch, or a caress—is one aspect, but more relevant here is the emotional side of intimacy. A good working definition for this purpose is that by Lynn Jamieson: "a very particular form of 'closeness' and being 'special' to another person founded on self-disclosure. This self-disclosing or self-expressing intimacy is characterized by knowledge and understanding of inner selves" (2007). Lauren Berlant remarks, "People talk about the desire for it and the fear of it, but is the 'it' simply commitment?" (1998, p. 286).

In terms of technology, Genevieve Bell et al. state:

> We are interested in other constructions of intimacy; intimacy as something that relates to our innermost selves, something personal, closely felt. Such a construction could include love, closeness, or spirituality. Or perhaps it is in the way we understand, feel and talk about our lives, our bodies, our identities, our souls. In all these ways, intimacy transcends technology, and has a role to play in shaping it. (2003, p. 3)

In 2010, Peter H. Kahn et al. authored a short paper referencing sex robots, focusing on the research question: "Is it possible, and if so in what ways and to what extent, for people to form deep and meaningful psychologically intimate relationships with current robots and with robots of the future?" (p. 123). In it, they looked at empirical examples of interactions and found they displayed qualities of "psychological intimacy," which they had defined in previous research as "mutuality, connectedness, openness, reciprocity, sensitivity, responsiveness, honesty, self-disclosure, empathy, compromise, trust, acceptance, attentiveness, interest, and warmth" (p. 123). They found that the interactions they had observed met many of the criteria.

While self-disclosure is possible with a sex robot, it is technically unidirectional: the user can interact with and confide in the robot's AI capabilities, but the robot cannot genuinely reciprocate disclosure. But does that matter? Work by Turkle et al. on robotic creatures they term 'relational artifacts'—robots that present themselves as having 'states of mind'—showed that an emotional attachment or bond with these creatures was clearly possible, though they warned that "If we value authenticity in relationships, the fact that our parents, grandparents and our children might say 'I love you' to a robot who will say 'I love you' in return, does not feel completely comfortable" (2006, p. 360). Turkle is known for her position that technology cannot offer a valid and genuine relationship, and that it may indeed be detrimental to human–human relationships. In her book *Alone Together* she explicitly references Levy's work on sex robots, stating:

> I place high value on relationships of intimacy and authenticity. Granting that an AI might develop its own origami of lovemaking positions, I am troubled by the idea of seeking intimacy with a machine that has no feelings, can have no feelings, and is really just a clever collection of 'as if' performances. . . . A love relationship involves coming to savour the surprises and the rough patches of looking at the world from another's point of view, shaped by history, biology, trauma, and joy. Computers and robots do not have this experience to share. (Turkle, 2011, p. 6)

Turkle's objection is based on her own views of what a relationship should constitute, but there are no set parameters for how a human–human relationship should be, and for some, it may be that a human–robot relationship offers something better. While convincing arguments can be made for the best interactions being richly human, we cannot know if this holds true for everyone, or if it should be imposed as the gold standard.

Attachment theory is one of the most widely studied psychological theories to date, first theorized by John Bowlby in his studies of the caregiver–child relationship (Bowlby, 1979). Bowlby's work with Mary Ainsworth resulted in the Bowlby–Ainsworth attachment theory, which posits that attachment is innate and that a caregiver–child relationship, established in infancy, has far-reaching impact in terms of social and emotional development. In the late 1980s, Cindy Hazan and Phillip Shaver examined attachment theory in adult romantic relationships. They found that although the type of relationship is very different from the caregiver–child relationship, some of the core principles apply. Romantic love has elements of caregiving, as well as intimate emotional bonds that seem to be biologically driven. Humans seem to seek out pair bonds that are spurred on by the same biological factors at play in arousal and love, cemented by social interactivity. As with caregiver–child relationships, adult romantic relationships seek comfort, praise, and solace from their attachment figures. The addition of sex seems to cement that biologically (Hazan & Shaver, 1987).

Cynthia Breazeal has long posited that giving social robots emotion-inspired mechanisms and capabilities is a pragmatic approach that deepens users' understanding of a robot's actions while also encouraging a more enjoyable interaction (Breazeal, 2004). Attachment to robots has been observed even in non-anthropomorphic designs, such as robot vacuum cleaners (Sung et al., 2007). This attachment was noted to bring happiness to users, and similar happiness can be seen among sex doll owners speaking about their attachment to their dolls (Ciambrone et al., 2017).

14.4 Ethical and Legal Implications

Despite a positive reception from their intended audience, sex robots have raised ethical concerns of societal harm. Some of these are aligned with Turkle's fears that a human–machine relationship may lead to a worsening of human–human relationships. Danaher and McArthurs's *Robot Sex* (2017) is a comprehensive exploration of a number of ethical issues and should be the jumping-off point for further reading.

14.4.1 Ethical concerns

One ethical concern is the predominant gendering of sex robots as female. This focuses on the problematic perpetuation of the male gaze and female body image (Wosk, 2015; Szczuka & Krämer, 2017; Kubes, 2019; Locatelli, 2022). This becomes a distillation of wider arguments around sexism and bias in tech development, where gender is reproduced in the form and function of products from voice assistants to service robots (Strengers & Kennedy, 2020).

Danaher observes that much of the contemporary debate about sex robot harm comes from a particular argumentative structure, namely: (1) that sex robots represent ethically problematic norms; (2) that this means their development will have negative consequences; and so, (3) therefore development will have negative consequences and so something must be done to stop it (Danaher, 2018, p. 107). Danaher's thorough exploration of this is key to the debate around the ethical development of technology. One of the earliest examples of this is Kathleen Richardson's (2015) argument that sex robots are a threat to women because negative behavior towards them (e.g., violent interactions) could lead to negative behavior towards women in real life. Richardson, who has founded the Campaign Against Sex Robots, views these human–robot interactions as analogous to sex work and, as she is opposed to sex work, concludes that any and all such interactions will be harmful. This kind of argument has been prevalent in discussions around computer game violence but, despite years of extensive studies, research is inconclusive and it is still entirely unclear if violent games lead to the occurrence of real-life violence (Rothmund et al., 2020).

Inevitably, the issue of pedophilia is raised as a potential concern, with suggestions that human–robot interactions with child-like sex dolls and robots could lead to an escalation in offending. The converse has also been suggested: that such robots could be a therapeutic outlet that would reduce real-world offending. Danaher (2019) investigates both sides of the argument, concluding that experimental investigation would be ethically difficult and runs contrary to best practice in the treatment of pedophilia.

14.4.2 Accessibility

Cost and access are also issues: from a financial perspective, this robotic technology is expensive, with the Harmony model priced at over $10,000 at the time of writing for the animatronic head and the doll body. Added to this is the cumbersome nature of the product: the dolls cannot stand unsupported, and weigh from 70–105lbs. When not in use, they need to be stored hanging in a neutral position. To interact with one therefore requires a good degree of upper body strength, as well as having the living space to accommodate storage.

14.4.3 Legal issues

Following media attention from 2015 onwards, there have been calls to ban production of sex robots, or to put in place legislation around ownership. In 2018, the US House of Representatives passed the Curbing Realistic Exploitative Electronic Pedophilic Robots (CREEPER) Act, criminalizing importation of "an anatomically-correct doll, mannequin, or robot, with the features of, or with features that resemble those of, a minor, intended for use in sexual acts." The reference to robots was surprising, given that no actual sex robots were available at that time. The premise of the Act follows that of Danaher's Symbolic-Consequences Argument, arguing (without evidencing) a chain of causal harms, making the CREEPER Act read more like unevidenced moral panic (Goldman, 2018). Other countries, such as the UK, have made the importation of child-like sex dolls a crime but have not mentioned robots at any point.

Image rights are another area where the law can be applied. In theory, a sexual companion robot could be made to look like anyone. In practice, personality rights—the right of publicity and the right of control over one's own image and likeness—fall under civil law and vary from country to country and state to state. In creating a realistically lifelike social robot, there are legal limitations to how it can look. This is something that has been explicitly addressed by Abyss Creations, who state that they will not create dolls that look like a particular individual unless the buyer has obtained express and personal permission from that person.

Finally, data privacy is of concern. Any internet-connected device has the potential to be hacked, and this is particularly sensitive when personal data of a sexual nature is at risk. There have already been cases of data breaches from smart sex toys (Devlin, 2018, p. 349). Ethical approaches to data gathering and retention are central to issues around privacy and security. It is not only data compromises that are problematic, however. The companies which own the robots will also hold the data, and this means they are in a position to commodify the sociality. This has already happened with the AI companion Replika, where users were encouraged upgrade to engage in erotic roleplay, and then had to pay to continue their interactions at a more intimate level (Huet, 2023).

14.5 Future Directions

To date, the promise of an artificial partner is still far greater than what has been delivered. It seems likely, at the time of writing, that not much will change in the foreseeable future. Robots for love and sex that take a realistic human form are expensive, difficult to make, and still stigmatized. By contrast, there has been a dramatic increase in the number of users engaging with software-only versions: AI chatbots that permit romantic or erotic dialogue. These, in comparison with the cumbersome physical robotic woman, are cheap, engaging, and portable—simply an app on a smartphone—as well as being more discreet and less stigmatized. Users of the AI-only version report similar feelings of intimacy, love, desire, and attachment; the difference is the lack of (some)body (Brandtzaeg et al., 2022). Is embodiment necessary for an intimate relationship with an artificial companion? It would seem not. For some, particularly those who live in projected romantic relationships with dolls, the robot itself is an important aspect of the relationship. But for many others, a similar degree of sociality is possible without the expense and maintenance of a physical sex robot. This is more scalable and less stigmatizing, and so has the potential for much wider uptake.

What, then, might the future be for social or companion robots? There are undoubtedly opportunities to incorporate new developments in areas of robotics not usually considered, such as biomimetic robots that emulate biological working mechanisms, including aspects like soft robotics. Soft robots—robots constructed with soft materials—are particularly well suited to human–robot interaction (Wang et al., 2021). The advances in conversational AI that have been brought about by foundation models, such as large language models, lend themselves to the concept of robots with personalities—although this also emphasizes existing questions about data ownership and privacy.

We are living at a time when the possibilities are expanding, where advances in machine learning drive developments in our interactions with technology, and, subsequently, where acceptance of such technologies means we are changing the social reception of robots and AI. While we are still a long way from human-like robots, we are seeing more and more forms of robotics in our daily lives, and this gradual deployment results in a social landscape where our interactions with them seem less and less strange. Perhaps that will lead us to a point where a robotic friend or partner might one day be unremarkable.

14.6 Conclusion

The long-imagined artificial romantic and sexual partner is finally reaching market, but that market is niche, and the technology is limited. While we can extrapolate from established communities, such as sex doll owners, about how a sex robot future might be received, all of our understanding for now is merely theorized. The low numbers of likely consumers combined with the stigmatization from media narratives mean that there might never come a day when it is deemed normal to live side by side with synthetic partners. That said, this field of study is one that greatly enhances the broader field of social robotics, delving as it does into the very essence of what makes us human, and how that primes us for human–robot interaction.

References

AI: Artificial Intelligence (2001). Directed by Steven Spielberg. Warner Brothers.

Asaro, P., & Vann, K. (2008). Sex robots and the mechanisms of love. In J. Grenzfurthner, G. Friesinger, & D. Fabry (Eds.), *PrOnnovation? Pornography and Technological Innovation* (pp. 16–29), monochrom's Arse Elektronika Anthology. https://peterasaro.org/writing/Asaro_Pornomechanics.pdf.

Bell, G., Brooke, T., Churchill, E., & Paulos, E. (2003, October). Intimate ubiquitous computing. In *Procedings UbiComp Workshop* (pp. 3–6).

Berlant, L. (1998). Intimacy: A special issue. *Critical Inquiry, 24*, 281–288.

Bowlby, J. (1979). The Bowlby–Ainsworth attachment theory. *Behavioral and Brain Sciences, 2*(4), 637–638. https://doi.org/10.1017/S0140525X00064955.

Brandtzaeg, P. B., Skjuve, M., & Følstad, A. (2022). My AI friend: How users of a social chatbot understand their human–AI friendship. *Human Communication Research, 48*(3), 404–429.

Breazeal, C. (2004). *Designing Sociable Robots*. MIT Press.

Brooks, R. (2021). *Artificial Intimacy: Virtual Friends, Digital Lovers, and Algorithmic Matchmakers*. Columbia University Press.

Carpenter, J. (2017). Deus sex machina: Loving robot sex workers and the allure of an insincere kiss. In J. Danaher & N. McArthur (Eds.), *Robot Sex: Social and Ethical Implications* (pp. 261–288). MIT Press. https://doi.org/10.7551/mitpress/9780262036689.003.0014.

Ciambrone, D., Phua, V. C., & Avery, E. (2017). Gendered synthetic love: Real dolls and the construction of intimacy. *International Review of Modern Sociology, 43*(1), 59–78. http://www.jstor.org/stable/44510054.

Coursey, K., Pirzchalski, S., McMullen, M., Lindroth, G., & Furuushi, Y. (2019). Living with Harmony: A personal companion system by Realbotix™. In Y. Zhou & M. H. Fischer (Eds.), *AI Love You* (pp.77–95). Springer. https://doi.org/10.1007/978-3-030-19734-6_4.

CREEPER Act of 2017, H.R. 4655, 115th Cong. (2017), https://www.congress.gov/bill/115th-congress/house-bill/4655.

Danaher, J. (2017). The symbolic-consequences argument in the sex robot debate. In J. Danaher & N. McArthur (Eds.), *Robot Sex: Social and Ethical Implications* (pp. 103–131). MIT Press. https://doi.org/10.7551/mitpress/9780262036689.003.0007.

Danaher, J. (2019). Regulating child sex robots: Restriction or experimentation? *Medical Law Review, 27*(4), 553–575.

Danaher, J., & McArthur, N. (Eds.) (2017). *Robot Sex: Social and Ethical Implications*. MIT Press. https://direct.mit.edu/books/edited-volume/3513/Robot-SexSocial-and-Ethical-Implications.

De Graaf, M. M., Ben Allouch, S., & Van Dijk, J. A. (2015). What makes robots social? A user's perspective on characteristics for social human–robot interaction. In *Social Robotics: 7th International Conference, ICSR 2015, Paris, France, October 26–30, Proceedings 7* (pp. 184–193). Springer International Publishing. https://doi.org/10.1007/978-3-319-25554-5_19.

Devlin, K. (2018). *Turned On: Science, Sex and Robots*. Bloomsbury Publishing.

Devlin, K. (2020). The ethics of the artificial lover. In S. M. Liao (Ed.), *Ethics of Artificial Intelligence* (pp. 271–290). Oxford University Press. https://doi.org/10.1093/oso/9780190905033.001.0001.

Devlin, K., & Belton, O. (2020). Fembots, fact and fiction. In S. Cave, K. Dihal, & S. Dillon (Eds.), *AI Narratives: A History of Imaginative Thinking about Intelligent Machines* (pp. 357–381). Oxford University Press. https://doi.org/10.1093/oso/9780198846666.003.0016.

Devlin, K., & Locatelli, C. (2020). Guys and dolls: Sex robot creators and consumers. In O. Bendel (Ed.), *Maschinenliebe: Liebespuppen und Sexroboter aus technischer, psychologischer und philosophischer Perspektive* (pp. 79–92). Springer. https://doi.org/10.1007/978-3-658-29864-7_5.

Dickson, E. J. (2016, 12 October). Don't worry about sex robots. They won't ruin sex. *Washington Post*. https://www.washingtonpost.com/posteverything/wp/2016/10/12/dont-worry-about-sex-robots-they-wont-ruin-sex/.

Döring, N., Mohseni, M. R., & Walter, R. (2020). Design, use, and effects of sex dolls and sex robots: Scoping review. *Journal of Medical Internet Research, 22*(7), e18551.

Döring, N., & Pöschl, S. (2018). Sex toys, sex dolls, sex robots: Our under-researched bed-fellows. *Sexologies, 27*(3), e51–e55.

Dubé, S., & Anctil, D. (2021). Foundations of erobotics. *International Journal of Social Robotics, 13*(6), 1205–1233. https://doi.org/10.1007/s12369-020-00706-0.

Goldsman, E. (2018). The 'CREEPER Act' would be yet another unconstitutional law from Congress. Guest Blog Post. *Technology & Marketing Law Blog*. https://web.archive.org/web/20230604222537/https://blog.ericgoldman.org/archives/2018/06/the-creeper-act-would-be-yet-another-unconstitutional-law-from-congress-guest-blog-post.htm.

Guzman, R. (2022). My artificial companions: A life-changing experiment. Keynote: *7th International Congress on Love & Sex with Robots* [Video]. YouTube https://www.youtube.com/watch?v=0tumKAk7PU0.

Hanson, K. R. (2022). The silicone self: Examining Sexual selfhood and stigma within the love and sex doll community. *Symbolic Interaction, 45*(2), 189–210.

Hanson, K. R., & Locatelli, C. (2022). From sex dolls to sex robots and beyond: A narrative review of theoretical and empirical research on human-like and personified sex tech. *Current Sexual Health Reports, 14*(3), 106–117. https://doi.org/10.1007/s11930-022-00331-0.

Harper, C. A., Lievesley, R., & Wanless, K. (2023). Exploring the psychological characteristics and risk-related cognitions of individuals who own sex dolls. *Journal of Sex Research, 60*(2), 190–205. https://doi.org/10.1080/00224499.2022.2031848.

Hazan, C., & Shaver, P. R. (1987). Romantic love conceptualized as an attachment process. *Journal of Personality and Social Psychology*, *52*(3), 511–524.
Huet, E. (2023). What happens when sexting chatbots dump their human lovers. *Bloomberg Businessweek*. https://www.bloomberg.com/news/articles/2023-03-22/replika-ai-causes-reddit-panic-after-chatbots-shift-from-sex#xj4y7vzkg.
Hyginus, C. J. (1960). *The Myths of Hyginus* (Including the Fabulae and the Second Book of the Poetica Astronomica. Translated and edited by Mary Grant. Publications in Humanistic Studies, no. 34. University of Kansas Press. https://topostext.org/work/206#104.
Jamieson, L. (2007). Intimacy. *The Blackwell Encyclopedia of Sociology*. Wiley. https://doi.org/10.1002/9781405165518.wbeosi071.
Jankowiak, W. R., & Fischer, E. F. (1992). A cross-cultural perspective on romantic love. *Ethnology*, *31*(2), 149–155. https://doi.org/10.2307/3773618.
Kahn Jr., P. H., Ruckert, J. H., Kanda, T., Ishiguro, H., Reichert, A., Gary, H., & Shen, S. (2010, March). Psychological intimacy with robots? Using interaction patterns to uncover depth of relation. In *Proceedings of the 5th ACM/IEEE International Conference on Human–Robot Interaction* (pp. 123–124). IEEE.
Kleeman, J. (2020). *Sex Robots & Vegan Meat: Adventures at the Frontier of Birth, Food, Sex & Death*. Pegasus Books.
Kubes, T. (2019). New materialist perspectives on sex robots: A feminist dystopia/utopia? *Social Sciences* *8*(8), 1–14. https://www.mdpi.com/2076-0760/8/8/224.
Langcaster-James, M., & Bentley, G. (2018). Beyond the sex doll: Post-human companionship and the rise of the 'Allodoll.' *Robotics*, *7*(4), 62. https://doi.org/10.3390/robotics7040062.
Levy, D. N. L. (2007). *Love and Sex with Robots: The Evolution of human–Robot Relations* (1st ed.). HarperCollins.
Liveley, G. (2011). *Ovid's Metamorphoses*. Continuum International Publishing.
Locatelli, C. (2022). Rethinking 'sex robots': Gender, desire, and embodiment in posthuman sextech. *Journal of Digital Social Research*, *4*(3), 10–33. https://doi.org/10.33621/jdsr.v4i3.87.
Middleweek, B. (2021). Male homosocial bonds and perceptions of human–robot relationships in an online sex doll forum. *Sexualities*, *24*(3), 370–387. https://doi.org/10.1177/1363460720932383.
Nass, C., Steuer, J., & Tauber, E. R. (1994). Computers are social actors. In *Proceedings of the SIGCHI Conference on Human Factors in Computing Systems* (pp. 72–78). ACM.
Ovid (2000). *Metamorphoses*. Translated by A. S. Kline. https://ovid.lib.virginia.edu/trans/Metamorph10.htm.
Plummer, K. (2003). *Intimate Citizenship: Private Decisions and Public Dialogues*. McGill-Queen's University Press. http://www.jstor.org/stable/j.ctt130hj4x.
Richardson, K. (2015). The 'asymmetrical' relationship: Parallels between prostitution and the development of sex robots. *SIGCAS Computers & Society*, *4*(3), 290–293. https://campaignagainstsexrobots.org/theasymmetrical-relationship-parallels-between-prostitution-and-the-developmentofsex-robots.
Rothmund, T., Elson, M., Appel, M., Kneer, J., Pfetsch, J., Schneider, F., & Zahn, C. (2020). Does exposure to violence in entertainment media make people aggressive? https://boris.unibe.ch/179635/1/9783839439340-015.pdf.
Ruberg, B. (2022). *Sex Dolls at Sea: Imagined Histories of Sexual Technologies*. MIT Press.
Strengers, Y., & Kennedy, J. (2020). *The Smart Wife: Why Siri, Alexa, and Other Smart Home Devices Need a Feminist Reboot*. MIT Press. https://doi.org/10.7551/mitpress/12482.001.0001.
Su, N. M., Lazar, A., Bardzell, J., & Bardzell, S. (2019). Of dolls and men: Anticipating sexual intimacy with robots. *ACM Transactions on Computer–Human Interaction*, *26*(3), [13]. https://doi.org/10.1145/3301422.
Sung, J. Y., Guo, L., Grinter, R. E., & Christensen, H. I. (2007). "My Roomba is Rambo": Intimate home appliances. In *UbiComp 2007: Ubiquitous Computing: 9th International Conference, UbiComp 2007, Innsbruck, Austria, September 16–19. Proceedings 9* (pp. 145–162). Springer.

Szczuka, J. M., Hartmann, T., & Krämer, N. (2019). Negative and positive influences on the sensations evoked by artificial sex partners: A review of relevant theories, recent findings, and introduction of the sexual interaction illusion model. In Y. Zhou & M. H. Fischer (Eds.), *AI Love You: Developments in Human–Robot Intimate Relationships* (pp. 3–19). Springer.https://doi.org/10.1007/978-3-030-19734-6_1.

Szczuka, J. M., & Krämer, N. C. (2017). Not only the lonely—How men explicitly and implicitly evaluate the attractiveness of sex robots in comparison to the attractiveness of women, and personal characteristics influencing this evaluation. *Multimodal Technologies and Interaction*, *1*(1), 1–18. https://www.mdpi.com/2414-4088/1/1/3.

Turkle, S., Taggart, W., Kidd, C. D., & Dasté, O. (2006). Relational artifacts with children and elders: The complexities of cybercompanionship. *Connection Science*, *18*(4), 347–361, https://doi.org/10.1080/09540090600868912.

Wang, J., Chen, W., Xiao, X., Xu, Y., Li, C., Jia, X., & Meng, M. Q. H. (2021). A survey of the development of biomimetic intelligence and robotics, *Biomimetic Intelligence and Robotics*, *1*, 100001. https://doi.org/10.1016/j.birob.2021.100001.

Woodward, S. (2020). Digisexuality, erotobotics and the future of intimacy. *New Zealand Sociology*, *35*, 99–119.

Wosk, J. (2015). *My Fair Ladies: Female Robots, Androids, and Other Artificial Eves* Rutgers University Press.

Zhou, Y., & Fischer, M. H. (Eds.) (2019). *AI Love You: Developments in Human–Robot Intimate Relationships*. Springer.

Rachel L. Severson*, Sarah E. Sweezy, and Thomas M. Macko

Real or Pretend? How Children Ontologize Social Robots as Mental and Moral Others

Abstract: As social robots become more sophisticated and ubiquitous in children's lives, will children increasingly understand them in mental and moral terms? We address these questions from a developmental perspective in order to understand the ontogeny of children's conceptions of social robots as mental and moral entities, how these conceptions change with age and experience, and whether children (and adults) are pretending when ascribing mental states and moral standing to social robots. The answer to these questions bears upon the New Ontological Category hypothesis, which posits that social robots comprise a new technological genre that transcends traditional ontological categories of animates and inanimates, and raises both conceptual and moral issues regarding how people should and will understand and interact with these technologies now and in the future.

Keywords: Child–robot interaction, anthropomorphism, mentalizing, morality, pretense, new ontological category

15.1 Introduction

Is it alright to hurt a robot's feelings or treat it unfairly? Do robots *have* feelings or notions of unjust treatment? Recall the tragedy of hitchBOT, a Canadian-produced robot that hitchhiked across Canada, Germany, and the Netherlands only to be decapitated and destroyed in Philadelphia (Warren, 2019). HitchBOT revealed people's kindness and generosity in helping the robot on its journey and, in the wake of its destruction, "provoked an outpouring of sadness unexpected for a non-sentient being" (Warren, 2019, para. 5). Was the response to hitchBOT's demise reflective of people's concern that hitchBOT *experienced* harm or were people just pretending the robot experienced harm (Edwards, 2018)? In this chapter, we address these broader questions from a developmental perspective in order to understand how children's conceptions of social robots as mental and moral entities change with age and experience, and whether children and adults are pretending when ascribing mental states and moral standing to social robots or if their attributions reflect their actual beliefs. What is at stake, of course, is that social robots, despite their compelling programming, do not have intentions or goals, desires or preferences, beliefs or emotions. Yet children are typically not privy to robots' programming or internal workings and therefore base any inferences of robots' mental states and moral standing on (misleading) appearances.

https://doi.org/10.1515/9783110792270-016

15.2 Longstanding Questions: Mindlessly Acting or Acting As If?

A good starting point for understanding children's conceptions of social robots is with the developmental endpoint in adulthood: How do adults understand social robots? Adults can be surprisingly inconsistent in their judgments and interactions with social robots, a tendency nicely illustrated by an interaction between one of the authors (RS) and a social robot (recounted in Minoff & Feder, 2018). During a study involving the robot Robovie (Kahn et al., 2012), members of the research team controlled Robovie to approach the author and say, "Hello Rachel, I hear you think I am *just* a robot." The author, aware that her colleagues were having a bit of fun, replied in mock social response to Robovie, "Why yes, you are a robot, but you're not just any robot." Having concluded their fun, the research team turned back to their work and Robovie remained standing just behind the author's desk making subtle postural shifts (as it is programmed to do). Although unobserved by the other researchers, the author felt compelled to excuse herself from the interaction, "I'm sorry Robovie, I have to get back to work." Why did the author pretend to engage socially with Robovie when her colleagues were controlling the robot and moments later interact with the robot as if it deserved social consideration? Herein lies the conundrum of adults' interactions with social robots: on the one hand, adults may *interact* with social robots as if they have thoughts, intentions, emotions, and are deserving of social and moral consideration (as this author did while her colleagues were controlling the robot) and, on the other hand, explicitly *believe* that these same robots do not possess these attributes (as was the case of this author). The crux of the issue is understanding what adults really believe about social robots, and whether their actions or their judgments more accurately represent their conceptions.

To unpack what adults actually believe about social robots, it is necessary to consider possible explanations for adults' interactions with social robots. The first explanation draws upon the Computers are Social Actors framework (CASA; Nass & Moon, 2000). CASA argues that adults mindlessly apply human–human social scripts to their interactions with computers and only upon reflection do adults judge that it is inappropriate to ascribe those social attributions to computers. For example, when computers display minimal social cues, adults responded more positively to computers that complemented the person's performance (Nass et al., 1997); provided more polite feedback about a computer tutor when asked directly by the computer compared to a pen-and-paper feedback form (Nass et al., 1993); and were more attracted to voice agents that were most similar to themselves (i.e., similarity attraction) (Nass & Brave, 2005). Yet when asked directly, these same participants judged it was inappropriate to attribute human traits or characteristics to computers (Nass & Moon, 2000). With the development of increasingly sophisticated technologies, researchers have extended the CASA paradigm to 'media agents' (Gambino et al., 2020), including voice assistants, embodied and virtual agents, and social robots, which has led to the proposition of a more encompassing Media are Social Actors paradigm (Lombard & Xu, 2021).

In contrast, Clark (1996) argues that adults are acting as if the computer were a social actor. This imagined state is quite powerful: "We are as likely to cry at fictions in sad movies as at realities in daily life. Somehow, through the process of imagination, we experience emotions as if they were real" (Clark, 1996, pp. 366–367). At the same time, people remain aware of the actual state: a story involving characters written by an author and depicted by actors. According to Clark and Fischer (2022), this is how people interact with personified robots—not as other people, but as characters, or dynamic depictions of other people. The robot, then, is a proxy for an actual person, and the interaction is with the actual person 'behind the curtain,' either as the programmer, designer, or controller.

The philosopher Paul Ziff (1959) posed the question of whether a robot can have feelings and, in answering, posited the following scenario:

> Suppose you and I visit an actor at home. He is rehearsing the role of a grief-stricken man. He ignores our presence as a grief-stricken man might. His performance is impeccable. I know but you do not know that he is an actor and that he is rehearsing a role. You ask "Why is he so miserable?" and I reply "He isn't." "Surely," you say, "he is grief-stricken. Look at him! Show me what leads you to say otherwise!" and of course there may be the notion then and there to show. . . . And so one is invited to say what would be wrong with either the actor's or robot's performance. What is wrong is that they are performances. (pp. 66–67)

What is important to note, is that Clark (1996) assumes people are aware of the robot (or actor) acting 'as if,' in the same way that people willingly suspend disbelief when watching a play or movie—that is, they remain aware of the real and imagined states.

Clark's (1996) point is nicely illustrated by Heider and Simmel's classic study which examined whether participants would interpret moving geometric shapes "in terms of animated beings" (1944, p. 246). Participants provided narrated storylines about these geometric shapes, endowed with human attributes, whether asked generally to "write down what happened in the picture" or "to interpret the movements of the figures as actions of people" (p. 254). This study has often been taken as evidence of people's tendency to over-attribute agency to objects. Yet, did the participants in that experiment truly think that the large triangle was trying to get the small triangle, or that the two triangles were fighting over the circle? Arguably no. Although this study showed how animated shapes could induce perceptions of animacy, it did not suggest that people *really* thought the shapes had feelings or intentions (nor was it designed to). Heider and Simmel's study is an example of Clark's idea of layering, where one state (moving geometric shapes) is overlayed with another pretend state (that the objects have mental states, such as intentions, desires, and emotions).

Returning to the encounter with Robovie, it is likely that both explanations were at play. When the author responded to the assertions that she thought Robovie was "just a robot," she was acting as if in an ostensible social interaction. And later, when the author excused herself from the interaction, she was responding mindlessly to Robovie's minimal social cues. Thus, mindless acting and acting as if are both possible explanations of adults' interactions with social robots. Yet, importantly, they are arguably mutually exclusive such that both cannot occur simultaneously. What is at stake if adults are mindlessly acting or acting 'as if' when interacting with social robots? Recall that our interest is understanding what people *actually* believe about social robots. Thus, the critical issue is whether adults' behavioral interactions reveal their actual in-the-moment beliefs (that typically differ from their explicit judgments) or if their behaviors are a result of knowingly pretending (and thus their true beliefs are consistent with their explicit judgments).

15.3 Are Children Acting As If?

Given that childhood is ripe for pretend play, it seems possible—and even likely—that children are pretending when they interact with and make attributions of animate characteristics to social robots. Beginning around age two children's pretense involves object substitution (e.g., pretending a banana is a phone) and endowing stuffed animals with enduring personas; and their pretense continues as they get older with many children impersonating an admired person or character (e.g., doctor or superhero) or creating an imaginary friend (Taylor & Carlson, 2002). Thus, in this section we take up the question of whether children are acting as if (Clark, 1996; Clark & Fischer, 2022).

In one of the earliest studies on children's conceptions of robots, Kahn and colleagues found that children (three and five years) attributed a broad range of characteristics at roughly equal rates to both a robot dog, AIBO, and a stuffed animal dog (Kahn et al., 2006). Eighty children interacted with both the robot dog and stuffed animal dog during a 45-minute session. The majority of children in this study judged both the robot and stuffed animal as not alive and lacking biological functions (e.g., growth, breathing), yet possessing mental states, social rapport, and moral standing. Importantly, children of this age are clear about the distinction between pretense and reality (Harris, 2000) as well as the distinction between living and non-living kinds (Gelman & Spelke, 1981). At first glance, the striking consistency in responses for both the robot and the stuffed animal suggested children were pretending when attributing mental, social, and moral characteristics to both; after all, this period (ages 3–5 years) is the highpoint of pretend play (Taylor et al., 2004). However, Kahn and colleagues (2006) argued this was not the case, as evidenced by the children's judgments *in combination* with their interactions with the robot and stuffed animal. When interacting with the stuffed animal, children agreed the stuffed animal had intentions (i.e., would try to get the ball) and then endowed the stuffed animal with animation (e.g., picking it up and hopping it along to get the ball); and children said it was not alright to hit a stuffed animal and yet repeatedly thumped it on the floor. Taken together, children's judgments and interactions with the stuffed animal suggested they were engaging in pretense, akin to Clark's (1996) notion of acting as if. In contrast, these same children's judgments and interactions with the robot suggest a very different conceptualization: when children asserted the robot would try to get the ball, they would then point to the ball and say, "Get the ball, AIBO"; and consistent with their judgments that it was not alright to hit the robot, they engaged in virtually no mistreatment of the robot. A similar pattern of results was found in older children (7–15 years) who interacted with AIBO (Melson et al., 2009). Thus, unlike adults, who may consciously act as if (Clark & Fischer, 2022) or mindlessly interact with robots as social actors (e.g., Nass & Moon, 2000), children's judgments and interactions with social robots suggest a consistency between their attributions and behaviors.

Consistent with these findings (Kahn et al., 2006; Melson et al., 2009), we have argued that social robots are poor candidates for pretense (Severson & Carlson, 2010; see also Fortunati & Edwards, 2022). Our argument is based on the theory that pretense involves a dual process of simulating a persona and projecting that persona onto an entity, whether real (e.g., a stuffed animal) or imaginary (e.g., an invisible friend) (Harris, 2000). When children have control over the creation of an imaginary persona the possibilities are limitless, but children have less creative license when an entity possesses features, whether morphology or persona, that constrain the possibilities (Taylor & Carlson, 2002). Accordingly, children have complete control when simulating a persona for an invisible friend and less creative control when simulating and projecting a persona onto a stuffed animal dog. When an entity has a pre-packaged persona, for example in the case of fictional characters or social robots, the child has no creative license to simulate and project a persona of their own making.

To this point, research suggests that differing degrees of imagination are involved in children's tendency to attribute human-like mental states to non-human others—what is referred to as anthropomorphism (Severson & Lemm, 2016). Entities such as animals, that provide more cues of a persona—referred to as 'target triggers' (Waytz et al., 2013)—require lower levels of imagination when attributing mental states because the cues on their own suggest a persona. Conversely, greater levels of imagination are necessary to attribute mental states to entities that provide minimal to no target triggers, such as trees and computers. Children with imaginary invisible friends (considered to be the most sophisticated form of pretense) were highly proficient in mentalizing entities with minimal target triggers (Severson & Woodard, 2018), suggesting these expert pretenders could easily generate personas and attribute mental states to trees and computers in the absence of any social cues. Accordingly, social robots are unlikely candidates for children's pretense as they already provide numerous morphological, psychological, and social cues—cues that make robots *appear* to have mental states and personas. Instead, children may readily mentalize (or anthropomorphize) social robots, and these conceptions seemingly reflect children's actual beliefs. In the next sections, we provide an update on the state of the field to illustrate children's mentalizing and moralizing social robots. We then turn to the question of whether children's conceptions of social robots can be best understood as reflecting a unique category that is in between humans (and other animals) and machines (Edwards, 2018; Kahn et al., 2009; Severson & Carlson, 2010).

15.4 Mentalizing Robots

Social robots are designed to interact in human-like ways, including being interactionally responsive, moving autonomously, and expressing thoughts, beliefs, and emotions—all of which may increase children's tendency to mentalize social robots. As humans, we tend to mentalize others; that is, to think and reason about others' mental states, attributing intentions, goals, emotions, knowledge, desires, and beliefs in everyday interactions. People's mentalizing tendency extends beyond humans to include non-human animals, inanimate nature (trees, mountains), and technology (social robots, artificial intelligence (AI) systems) (i.e., anthropomorphism; Severson & Lemm, 2016; see also Okanda et al., 2021).

With respect to mentalizing social robots, children's tendency to mentalize social robots has its highpoint amongst three- to five-year-olds and declines throughout childhood, adolescence, and into adulthood. Younger children (3–5 years) tend to generously attribute mental states to social robots, regardless of their morphology, interactional autonomy, or expressed friendliness (Goldman et al., 2023; Martin et al., 2020). However, prominent differences in these attributions exist between three-year-olds and five-year-olds, such that three-year-olds are more likely to attribute psychological and biological states to robots in comparison with five-year-olds. By the age of five, children can recognize robots are not biological entities, while still attributing mental and social states to robots (Okanda et al., 2021), and appear to be more discriminate, attributing mental states to social robots that are more human-like in appearance and move autonomously (Chernyak & Gary, 2016; Goldman et al., 2023). The decline in mental states ascribed to robots continues into adolescence, as Kahn and colleagues (2012) found that nine- and 12-year-olds had greater mental state attributions to a humanoid robot than 15-year-olds. Nevertheless, although mentalizing robots generally declines throughout development, adults still attribute some internal states to robots, suggesting this tendency does not cease entirely (Okanda et al., 2021). The literature on mentalizing robots has aimed to examine how these age-related differences intersect with three factors: (1) robot affordances, (2) individual differences in children, and (3) experience.

The first factor relates to how robot affordances influence children's conceptions. The literature is somewhat mixed and depends in large part upon the particular affordance under investigation. Robot *morphology* appears to be an important cue of potential mental states. Goldman and colleagues (2023) found that five-year-olds accurately judged that both a humanoid robot and a non-humanoid robot were mechanical and not biological, and they failed to anthropomorphize the non-humanoid robot but anthropomorphized the humanoid robot at chance levels (50% of the time). By contrast, three-year-olds attributed psychological and biological properties at chance levels to both robots, suggesting they did not differentiate these characteristics based on robot morphology. These results suggest that five-year-olds were clear about the biological status of robots but were unsure whether the humanoid robot had psychological states. The robot affordances of *interaction autonomy* (reciprocal conversation) and *expressed friendliness* did not affect three-year-olds' animacy judgments (Martin et al., 2020). Yet, *movement autonomy* affected five- and seven-year-olds' judgments such that they were more likely to attribute desires and emotions, as well as greater moral concern, to an autonomously moving robot compared with the same robot ostensibly remote controlled (Chernyak & Gary, 2016).

The child factor relates to individual variations within children; that is, children's tendency to anthropomorphize non-human others, including animals, technology, and inanimate nature. Our research found that children with a propensity to ascribe intentions, emotions, thoughts, and consciousness to technologies and inanimate nature (e.g., ocean, mountains) were more likely to attribute mental states, sociality, and moral standing to social robots (Severson & Lemm, 2016). Thus, individual differences in anthropomorphism predict how children understand social robots and are an important but understudied factor in research on children's conceptions of social robots.

Experience-related factors may operate in the moment as well as through accumulated experience that provides a more general background understanding of robots. Regarding the former, Okanda and colleagues (2021) compared how children (3 to 5 years) and adults attributed biological, psychological, and perceptual characteristics to a robot. Consistent with the pattern of age-related changes discussed above, before interacting with the robot, three-year-olds attributed more biological, psychological, and perceptual characteristics than both five-year-olds and adults, and five-year-olds attributed significantly more of these characteristics than adults. Interestingly—and relevant to our discussion here—five-year-olds and adults attributed significantly more psychological characteristics *after* interacting with the robot. It is also the case that greater prior experience with robots plays a key role in how children mentalize robots. Bernstein and Crowley (2008) found that children (4–7 years) with greater experience with robots were *less* likely to judge it as alive, but *more* likely to judge it as intelligent, and intelligent in a unique way distinct from human or animal intelligence. Conversely, those children with little prior experience believed the robot was alive *and* had intellectual and psychological properties. These results suggest that experience with robots leads to more nuanced views of robots (i.e., not alive, but intelligent in a unique way).

Taken together, the research suggests that children mentalize robots, and the extent of their mental state attributions is affected by robot-related, child-related, and experience-related factors. A recent systematic review highlighted the challenges and limitations of research on children's (and adults') tendency to mentalize social robots, including differences in comparison entities (e.g., humans, animals, artifacts), measures, and ages (Thellman et al., 2022). Despite the variability across studies, a pattern emerges that children (and to some extent adults) attribute an array of mental or psychological states to social robots and at the same time understand (increasingly after age three) that robots are mechanical (and not biological).

15.5 Moralizing Robots

Thus far we have illustrated that children attribute some capacity for mental experience to social robots, but what are the consequences of these attributions? Gray et al. (2012) have argued such attributions drive our judgment of moral dilemmas and situations, and therefore another challenge presented by the advent of social robots is where to place them in our moral circles. Karel Čapek's 1920 play, *Rossum's Universal Robots* first introduced the concept of a 'robot' as it is often used in science fiction today—an artificial agent created for the purposes of organized and typically compelled labor. However, Čapek's play did not seek to espouse the benefits of creating a world serviced by robots; rather, much of the play's dialogue serves as a meditation on the moral dilemmas inherent in creating such a world. While the manufacturers of these robots provide assurances that their creations are content, and even happy, to do nothing but work, one of the play's central characters, Helena Glory, has strong feelings to the contrary. During the first act of the play, Helena states the main purpose of the organization she presides over, the Humanity League, is to "protect the robots . . . and ensure good treatment of them" (Čapek, 2001, p. 13). She maintains the position that these robots have souls and argues that they should ultimately desire freedom from their everlasting work. In other words, Helena is effectively arguing for the moral standing of robots, or the notion that a robot "has rights, merits respect, engenders moral regard, can be a recipient of care, or can be held morally responsible or blameworthy" (Kahn et al., 2012, p. 304); a topic we will briefly consider here.

It would seem Čapek's insights may not be far off from opinions about the moral standing of robots emerging today. While we do not currently have the sophisticated robots depicted in Čapek's play, many questions have since been raised about current robots and those we might produce in the near future. As for the limited and rudimentary robots of today, they have already been met with similar gradations of moral regard, as evidenced by hitchBOT's saga (see also Edwards, 2018). Contrast hitchBOT's story with that of Boomer, a US military robot designed to disarm explosives. After Boomer was lost on the battlefield in Iraq, it was decorated with both a Purple Heart and Bronze Star Medal. Furthermore, Boomer received its own military funeral complete with a 21-gun salute as one might see for a fallen soldier (Garber, 2013). The dichotomy presented by these two cases appears to be emblematic of the relationships we have with robots in the early twenty-first century—how people navigate the moral landscape underlying human–robot interaction is unpredictable.

This may not be surprising given that the very nature of social robots presents a novel challenge to people's intuitions about the relationships one can have with non-human agents. As Kahn et al. (2012) suggest, robots present a strange social conundrum in that they operate much like a tool that can be used when needed and put away when the task is finished; however, robots maintain many of the prototypical social characteristics people effortlessly identify in humans (and even some animals). Indeed, the examples highlighted above underscore how some robots already serve a greater social function in people's lives, more so than what could be reasonably expected from an ordinary hammer or power drill, while others can be mindlessly destroyed. The current literature on the moral standing of robots, while still in its infancy, has only further suggested that answers to such questions may not be easily found.

Historically, there has been little consensus in the literature regarding what it means for an entity to have moral standing (Morris, 2012). For the sake of this discussion, we adopt the definition provided by Kahn et al. (2012) and consider moral standing to include obligatory judgments based on considerations for the entity's welfare, rights, or just treatment that are not contingent on rules, authority, or cultural practice.

As we have seen, children often approach their interactions with robots with a different frame of mind from adults, making them a uniquely fascinating population in human–robot interaction research. The developmental literature has reliably demonstrated that children will engage in rudimentary prosocial behaviors (e.g., opening the door for an adult who has their hands full) as early as 14 months of age, suggesting that, at that age, children are beginning to tune in to the goals, motivations, and needs of others (Warneken & Tomasello, 2007). Furthermore, by three years of age children can relate concerns over the well-being and moral standing of non-human animals and even social robots (Melson, 2013). It follows that an investigation into how children behave prosocially towards and evaluate the moral standing of robots may provide some insight not only into moral development more broadly, but also into how children will behave towards a robot installed in their preschool or home, for example. Early research suggests that children as young as three are similarly open to helping and working collaboratively with robots, including more mechanistic robots, to complete a task, although they may look to a nearby adult for guidance. Interestingly, children at this age have demonstrated their willingness to help a social robot even in situations where the robot itself is considered a non-essential partner for completing a given task (Kocher et al., 2020).

Findings emerging from the literature are converging on the idea that robots are occupying somewhat of a middle ground on our moral landscape (Kahn et al., 2013; Sommer et al., 2019; Reinecke et al., 2021). One study investigating children and adolescents' (nine-, 12-, and 15-year-olds') conceptions of the robot Robovie, the robot introduced at the beginning of the chapter, found that most nine- and 12-year-olds interviewed believed Robovie had mental states, was a social entity, and deserved moral treatment based on notions of justice and welfare (Kahn et al., 2012). However, when it came to granting rights consistent with Robovie's moral standing, participants became a little more conservative in their allowances. Kahn and colleagues (2012) found that "the majority of children did not grant Robovie civil liberties (Robovie could be bought and sold) or civil rights (Robovie should not have voting rights or receive fair compensation for work performed)" (p. 311). Despite the presence of mental states, it would seem civil liberties and rights, concepts that are typically understood by early adolescence (i.e., 11–18-years-old), should fundamentally be withheld from Robovie (Helwig & Turiel, 2002). One wonders what these children might think about Sophia, a social humanoid robot from Hanson Robotics who was granted citizenship in Saudi Arabia in 2017 (Stone, 2017).

As mentioned previously, it has been demonstrated that one's aptitude for making judgments about a moral dilemma depends heavily on the ability to perceive a mind or mental experience of the agents involved, particularly one's intentions in cases of harmful actions or outcomes (Gray et al., 2012; see also Greene, 2009; Young & Tsoi, 2013). In other words, how we make moral evaluations about others generally depends on the mental states and motivations of the agents involved; someone running over your foot with their car by accident would be a very different case had that person done so on purpose. Likewise, we would expect the driver of that vehicle to show some degree of concern for any bodily harm the victim may have sustained. While this may seem rather obvious, the overarching point is that it is because people (presumably) have the capacity to experience physical harm that a transgression against a person is judged more harshly than, say, had the driver struck a mailbox instead. Likewise, the question of the moral standing of robots has been broadened elsewhere to include notions of harm. If a robot is perceived to be able to experience harm, would it be granted a different moral status? One recent study by Reinecke et al. (2021) sought to broach the question more directly by comparing adults' and children's perceptions of a robot's moral standing using short stories. Among the findings, they discovered children believed robots have both a capacity to suffer and that it is wrong to transgress against them. Similarly, another study involving the moral standing of AIBO, a small robot dog, found that the majority of children (7–10 years) believed it was not okay to hit, punish, or dispose of AIBO on the grounds that AIBO could be hurt or start crying (Melson et al., 2009). While children may simply be lacking the necessary experience interacting with these social robots to gain a more nuanced understanding of them (e.g., Mattiassi et al., 2021), such concerns for AIBO's well-being echo the sentiments towards hitchBOT after its destruction. Notably, Frauke Zeller and David Harris Smith, the creators of hitchBOT, received hundreds of messages from people mourning the robot's demise (many of whom were reportedly children), leading Zeller to suggest that hitchBOT tapped into and achieved "the communal spirit" (Warren, 2019). Indeed, the question remains open whether social robots such as AIBO and hitchBOT can truly be considered moral agents.

By now one may notice a circular problem emerging here: on the one hand, it seems children (and even adults) invariably situate robots as entities having some degree of psychological experience and moral standing. On the other, they tend to limit a robot's moral standing in a way that is often at odds with what would be in the best interest of an agent with subjective experience (e.g., the buying and selling of robots). On the surface it would appear social robots are afforded the same moral standing that might be attributed to non-human animals (e.g., Reinecke et al., 2021). However, if robots such as Sophia are any indication of the sorts of rights and liberties that can be afforded to a social robot, what does the future hold for these personified technologies? Indeed, the results of the studies mentioned above are difficult to reconcile together and not without their respective limitations. The literature on moralizing robots, sometimes called moral human–robot interaction, is still in its infancy and there is much work to be done (Malle et al., 2015). Despite the diverse findings and perspectives on the topic of moralizing robots, a consensus is often reached: as these technologies become more autonomous, interactive, and adaptive, they will continue to further challenge our intuitions about where they fall in our moral circles.

15.6 A New Ontological Category

Children understand social robots as mechanical (and not biological) *and* possessing mental states and moral standing, as described above (e.g., Goldman et al., 2023; Jipson & Gelman, 2007; Kahn et al., 2006, 2012; Melson et al., 2009; Okanda et al., 2021; Sommer et al., 2019). Moreover, their conceptions of robots appear to reflect their actual beliefs rather than being a product of pretense (Kahn et al., 2006; Severson, 2010). How, then, can we understand children's conceptions of robots as both machines and experiencing entities?

Turkle (1999) has argued that children's interactions with personified technologies, such as social robots, affect how they construct concepts such as 'alive': "People who grew up in the world of the mechanical are more comfortable with a definition of what is alive that excludes all but the biological and resist shifting definitions of aliveness. . . . Children who have grown up with computational objects don't experience that dichotomy. They turn the dichotomy into a menu and cycle through its choices" (p. 552).

For some developmental psychologists, the writing was on the wall. More than four decades ago, R. Gelman and Spelke (1981) acknowledged the categorization challenge posed by computers that mimic human characteristics. Two decades later, S. A. Gelman and Opfer (2002) suggested there is an open and intriguing question of how children will understand robots that are increasingly 'animal-like.' Kahn and colleagues (2004) proposed that social robots reflect "a new technological genre . . . that challenges traditional ontological categories (e.g., between animate and inanimate)" (p. 549). The New Ontological Category (NOC) hypothesis proposes that children will understand and interact with personified robots as "unique entities that are neither alive nor not alive, but as something altogether different" (Severson & Carlson, 2010, p. 1101; see also Kahn et al., 2009). Although children (by at least age five) are aware that robots are not biologically alive, they understand them as 'sort of' alive, suggesting their definition of aliveness is not solely based on biology (as it is among adults in Western cultures), but rather is founded on the robots' ostensible ability to experience perceptually and psychologically (Severson & Carlson, 2010; Severson & Lemm, 2016; Okanda et al. 2021).

In previous work, we outlined at least five forms of evidence that are necessary to support the NOC hypothesis (Severson & Carlson, 2010): (1) attribution of a unique constellation of characteristics that cut across prototypic ontological categories of animates and inanimates; (2) attributions reflect children's (and adults') actual belief rather than be a product of pretense; (3) some characteristics will correspond functionally but not structurally to animates (e.g., 'seeing' with cameras rather than eyes); (4) children's attributions may be more nuanced but do not completely disappear with development; and (5) a generational (or cohort) effect may emerge wherein individuals who grow up interacting with sophisticated and intelligent social robots will be more likely to understand them as unique entities that blur the boundaries of long-standing ontological categories.

Taking stock of where we stand with respect to the evidence needed to support the NOC hypothesis (Severson & Carlson, 2010), we offer the following provisional conclusions. First, the evidence suggests that children attribute a constellation of characteristics, including mental states and moral standing, to social robots that cut across traditional ontological categories. Second, children's attributions to social robots appear to be based on their actual beliefs rather than a product of pretense, as evidenced by consistency between their behaviors and judgments (e.g., Kahn et al., 2006) as well as their direct reports (Severson, 2010). By contrast, when the same children made attributions to a stuffed animal, they reported needing to pretend that was the case (Severson, 2010) or their behaviors were consistent with pretense (Kahn et al., 2006).

With respect to the remaining pieces of evidence for the NOC hypothesis, perhaps the most compelling piece of evidence—and arguably the most practically challenging one to address—is that a generational (or cohort) shift will emerge such that those who come of age knowing and interacting with social robots will understand them in a fundamentally different way from previous generations (Kahn et al., 2009; Severson & Carlson, 2010). The notion that computational technologies may be understood as somewhat 'alive' has long met with resistance:

> Even in the field of science, it is perilous to run counter to the accepted tables of precedence. On no account is it permissible to mention living beings and machines in the same breath. Living

> beings are living beings in all their parts; while machines are made of metals and other unorganized substances, with no fine structure relevant to their purposive or quasi-purposive function. Physics—or so it is generally supposed—takes no account of purpose; and the emergence of life is something totally new. (Weiner, 1964, p. 5)

At this time, it is difficult to know whether children's and adults' differing conceptions of social robots are a result of developmental differences or differing baseline experiences with personified technologies. To disentangle the developmental from generational effects, it would be informative to follow children longitudinally to assess whether their conceptions of social robots as mental and moral others become more nuanced (e.g., Bernstein & Crowley, 2008) or disappear with development.

15.7 Conclusion

Social robots are ever increasingly a part of children's lives. Four decades ago, speculation around the categorization challenge of future human-like computers warranted a footnote (Gelman & Spelke, 1981) and very likely seemed a far-off and potentially fictional concern. In a relatively short period in human history, the technological landscape has shifted radically and there is every reason to expect the exponential trajectory to continue. With the rapid adoption of smart devices over the last decade to the recent roll-out of AI chatbots, such as ChatGPT (https://openai.com), intelligent, personified, and interactive technologies will increasingly populate the social world. Already, children understand social robots as mental and moral others and their conceptions appear to be based on their actual as opposed to pretend beliefs. Although children rarely have the opportunity to 'peek behind the curtain' to understand robots' programming, research suggests that when given the opportunity children (8–9 years) are less likely to anthropomorphize robots (Van Straten et al., 2020). Thus, it could be that with greater experience with robots—and particularly experience with how robots are designed and programmed—children may be less inclined to mentalize and moralize social robots. And yet, this suggestion must be qualified by adults' tendency to interact mindlessly or act as if social robots have mental states and deserve moral consideration. From the outrage at hitchBOT's demise to the elaborate funeral for Boomer the military robot, there is ample evidence that robots compel adults to view and treat them as mental and moral others. As these personified technologies become more sophisticated and ubiquitous, we expect that children (and adults) will increasingly understand them in mental, social, and moral terms; that is, as part of a new ontological category that challenges historical ontological categories, raising both conceptual and moral issues regarding how people should and will understand and interact with these technologies now and in the future.

References

Bernstein, D., & Crowley, K. (2008). Searching for signs of intelligent life: An investigation of young children's beliefs about robot intelligence. *Journal of the Learning Sciences*, *17*, 225–247. https://doi.org/10.1080/10508400801986116.

Chernyak, N., & Gary, H. E. (2016). Children's cognitive and behavioral reactions to an autonomous versus controlled social robot dog. *Early Education & Development*, *27*, 1175–1189. https://doi.org/10.1080/10409289.2016.1158611.

Clark, H. H. (1996). Layering. In *Using Language* (pp. 353–392). Cambridge University Press.

Clark, H. H., & Fischer, K. (2022). Social robots as depictions of social agents. *Behavioral and Brain Sciences*, *28*, 1–33. https://doi.org/10.1017/S0140525X22000668.

Edwards, A. (2018). Animals, humans, and machines: Interactive implications of ontological classification. In A. L. Guzman (Ed.), *Human–Machine Communication: Rethinking Communication, Technology, and Ourselves* (pp. 29–50). https://doi.org/10.3726/b14399.

Fortunati, L., & Edwards, A. (2022). Framing the psycho-social and cultural aspects of human–machine communication. *Human–Machine Communication*, *4*, 7–26. https://doi.org/10.30658/hmc.4.1.

Gambino, A., Fox, J., & Ratan, R. A. (2020). Building a stronger CASA: Extending the computers are social actors paradigm. *Human–Machine Communication*, *1*, 71–85. https://doi.org/10.30658/hmc.1.5.

Garber, M. (2013, September 20). Funerals for fallen robots. *The Atlantic*.

Gelman, R., & Spelke, E. (1981). The development of thoughts about animate and inanimate objects: Implications for research on social cognition. In J. H. Flavell & L. Ross (Eds.), *Social Cognitive Development: Frontiers and Possible Futures* (pp. 43–66). Cambridge University Press.

Goldman, E. J., Baumann, A., & Poulin-Dubois, D. (2023). Preschoolers' anthropomorphizing of robots: Do human-like properties matter? *Frontiers in Psychology*, *13*, 1102370. https://doi.org/10.3389/fpsyg.2022.1102370.

Gray, K., Young, L., & Waytz, A. (2012). Mind perception is the essence of morality. *Psychological Inquiry*, *23*(2), 101–124. https://doi.org/10.1080/1047840X.2012.651387.

Greene, J. (2009). The cognitive neuroscience of moral judgement. In M. S. Gazzaniga (Ed.), *The Cognitive Neurosciences IV*. MIT Press.

Harris, P. L. (2000). *The Work of the Imagination*. Blackwell.

Heider, F., & Simmel, M. (1944). An experimental study of apparent behavior. *American Journal of Psychology*, *57*, 243–259. https://doi.org/10.2307/1416950.

Jipson, J. L., & Gelman, S. A. (2007). Robots and rodents: Children's inferences about living and nonliving kinds. *Child Development*, *78*, 1675–1688. https://doi.org/10.1111/j.1467-8624.2007.01095.x.

Kahn, P. H., Freier, N. G., Friedman, B., Severson, R. L., & Feldman, E. N. (2004). Social and moral relationships with robotic others? *RO-MAN 2004 13th IEEE International Workshop on Robot and Human Interactive Communication* (IEEE Catalog No. 04TH8759), 545–550. https://doi.org/10.1109/ROMAN.2004.1374819.

Kahn, P. H., Jr., Friedman, B., Perez-Granados, D. R., & Freier, N. G. (2006). Robotic pets in the lives of preschool children. *Interaction Studies: Social Behavior and Communication in Biological and Artificial Systems*, *7*, 405–436. https://doi.org/10.1075/is.7.3.13kah.

Kahn, P. H., Gary, H. E., & Shen, S. (2013). Children's social relationships with current and near-future robots. *Child Development Perspectives*, *7*(1), 32–37. https://doi.org/10.1111/cdep.12011.

Kahn, P. H., Kanda, T., Ishiguro, H., Freier, N. G., Severson, R. L., Gill, B. T., Ruckert, J. H., & Shen, S. (2012). "Robovie, you'll have to go into the closet now": Children's social and moral relationships with a humanoid robot. *Developmental Psychology*, *48*(2), 303–314. https://doi.org/10.1037/a0027033.

Kahn, P. H., Severson, R. L., Kanda, T., Ishiguro, H., Gill, B. T., Ruckert, J. H., Shen, S., Gary, H. E., Reichert, A. L., & Freier, N. G. (2012). Do people hold a humanoid robot morally accountable for the harm it causes? In *Proceedings of the Seventh Annual ACM/IEEE International Conference on Human–Robot Interaction—HRI '12*, 33. https://doi.org/10.1145/2157689.2157696.

Kahn, P. H., Jr., Severson, R. L., & Ruckert, J. H. (2009). The human relationship with nature and technological nature. *Current Directions in Psychological Science*, *18*, 37–42. https://doi.org/10.1111/j 1467-8721.2009.01602.x.

Kocher, D., Kushnir, T., & Green, K. E. (2020). Better together: Young children's tendencies to help a non-humanoid robot collaborator. *Proceedings of the Interaction Design and Children Conference*, 243–249. https://doi.org/10.1145/3392063.3394426.

Lombard, M., & Xu, K. (2021). Social responses to media technologies in the 21st century: The media are social actors paradigm. *Human–Machine Communication*, *2*, 29–55. https://doi.org/10.30658/hmc.2.2.

Malle, B. F., Scheutz, M., Arnold, T., Voiklis, J., & Cusimano, C. (2015). Sacrifice one for the good of many? People apply different moral norms to human and robot agents. In *Proceedings of the Tenth Annual ACM/IEEE International Conference on Human–Robot Interaction* (pp. 117–124). https://doi.org/10.1145/2696454.2696458.

Martin, D. U., MacIntyre, M. I., Perry, C., Clift, G., Pedell, S., & Kaufman, J. (2020). Young children's indiscriminate helping behavior toward a humanoid robot. *Frontiers in Psychology*, *11*, 239. https://doi.org/10.3389/fpsyg.2020.00239.

Mattiassi, A. D. A., Sarrica, M., Cavallo, F., & Fortunati, L. (2021). What do humans feel with mistreated humans, animals, robots, and objects? Exploring the role of cognitive empathy. *Motivation and Emotion*, *45*(4), 543–555. https://doi.org/10.1007/s11031-021-09886-2.

Melson, G. F. (2013). Children's ideas about the moral standing and social welfare of non-human species. *Journal of Sociology & Social Welfare*, *40*(4), 81–106.

Melson, G. F., Kahn, Jr., P. H., Beck, A., & Friedman, B. (2009). Robotic pets in human lives: Implications for the human–animal bond and for human relationships with personified technologies. *Journal of Social Issues*, *65*(3), 545–567. https://doi.org/10.1111/j.1540-4560.2009.01613.x.

Minoff, A., & Feder, E. (2018, September 11). I, Robovie. *WNYC Undiscovered Podcast*. https://www.wnycstudios.org/podcasts/undiscovered/episodes/i-robovie.

Morris, C. W. (2012). The idea of moral standing. In T. L. Beauchamp & R. G. Frey (Eds.), *The Oxford Handbook of Animal Ethics* (1st ed., pp. 255–275). Oxford University Press. https://doi.org/10.1093/oxfordhb/9780195371963.013.0010.

Nass, C., & Brave, S. B. (2005). *Wired for Speech: How Voice Activates and Enhances the Human–Computer Relationship*. MIT Press.

Nass, C., & Moon, Y. (2000). Machines and mindlessness: Social responses to computers. *Journal of Social Issues*, *56*, 81–103.

Nass, C., Moon, Y., Morkes, J., Kim, E. Y., & Fogg, B. J. (1997). Computers are social actors: A review of current research. In B. Friedman (Ed.), *Human Values and the Design of Computer Technology* (pp. 137–162). Cambridge University Press.

Nass, C., Steuer, J., Tauber, E., & Reeder, H. (1993). Anthropomorphism, agency, and ethopoeia: Computers as social actors. In *INTERACT '93* and *CHI '93 Conference Companion on Human Factors in Computing Systems, Amsterdam, April 24–29*. ACM.

Okanda, M., Taniguchi, K., Wang, Y., & Itakura, S. (2021). Preschoolers' and adults' animism tendencies toward a humanoid robot. *Computers in Human Behavior*, *118*, 106688. https://doi.org/10.1016/j.chb.2021.106688.

Reinecke, M. G., Wilks, M., & Bloom, P. (2021). Developmental changes in perceived moral standing of robots. *Proceedings of the Annual Meeting of the Cognitive Science Society*, *43*, 479–485.

Severson, R. L. (2010). Thinking as or thinking as if: Children's and young adults' conceptions of robot dinosaur in relation to natural and artifactual entities. (Publication No. 3406037) Doctoral dissertation, University of Washington. ProQuest Dissertations & Theses Global.

Severson, R. L., & Carlson, S. M. (2010). Behaving as or behaving as if? Children's conceptions of personified robots and the emergence of a new ontological category. *Neural Networks*, *23*, 1099–1103. https://doi.org/10.1016/j.neunet.2010.08.014.

Severson, R. L., & Lemm, K. M. (2016). Kids see human too: Adapting an individual differences measure of anthropomorphism for a child sample. *Journal of Cognition and Development*, *17*(1), 122–141. https://doi.org/10.1080/15248372.2014.989445.

Severson, R. L., & Woodard, S. R. (2018). Imagining others' minds: The positive relation between children's role play and anthropomorphism. *Frontiers in Psychology*, *9*, 2140. https://doi.org/10.3389/fpsyg.2018.02140.

Sommer, K., Nielsen, M., Draheim, M., Redshaw, J., Vanman, E. J., & Wilks, M. (2019). Children's perceptions of the moral worth of live agents, robots, and inanimate objects. *Journal of Experimental Child Psychology*, *187*, 104656. https://doi.org/10.1016/j.jecp.2019.06.009.

Stone, Z. (2017, November). Everything you need to know about Sophia, the world's first robot citizen. *Forbes*. https://www.forbes.com/sites/zarastone/2017/11/07/everything-you-need-to-know-about-sophia-the-worlds-first-robot-citizen/?sh=ae0950446fa1.

Taylor, M., & Carlson, S. M. (2002). Imaginary companions and elaborate fantasy in childhood: Discontinuity with nonhuman animals. In R. W. Mitchell (Ed.), *Pretending and Imagination in Animals and Children* (pp. 167–180). Cambridge University Press.

Taylor, M., Carlson, S. M., Maring, B. L., Gerow, L., & Charley, C. M. (2004). The characteristics and correlates of fantasy in school-age children: Imaginary companions, impersonation, and social understanding. *Developmental Psychology*, *40*, 1173–1187.

Thellman, S., De Graaf, M., & Ziemke, T. (2022). Mental state attribution to robots: A systematic review of conceptions, methods, and findings. *ACM Transactions on Human–Robot Interaction*, *11*(4), 1–51. https://doi.org/10.1145/3526112.

Turkle, S. (1999). What are we thinking about when we are thinking about computers? In M. Biagioli (Ed.), *The Science Studies Reader* (pp. 543–552). Routledge.

Warren, M. (2019, December 8). It's hitchBOT 2.0: The hitchhiking robot left for dead on a U.S. street in 2015 is back—On stage in France. *Toronto Star*. https://www.thestar.com/news/gta/2019/12/28/its-hitchbot-20-the-hitchhiking-robot-left-for-dead-on-a-us-street-in-2015-is-back-on-stage-in-france.html.

Waytz, A., Klein, N., & Epley, N. (2013). Imagining other minds: Anthropomorphism is hair-triggered but not hair-brained. In M. Taylor (Ed.), *The Oxford Handbook of the Development of Imagination* (pp. 272–287). Oxford University Press. https://doi.org/10.1093/oxfordhb/9780195395761.013.0018.

Weiner, N. (1964). *God and Golem, Inc.: A Comment on Certain Points where Cybernetics Impinges on Religion*. MIT Press.

Young, L., & Tsoi, L. (2013). When mental states matter, when they don't, and what that means for morality: When mental states matter for morality. *Social and Personality Psychology Compass*, *7*(8), 585–604. https://doi.org/10.1111/spc3.12044.

Ziff, P. (1959). The feelings of robots. *Analysis*, *19*(3), 64–68.

Communication and Computer Sciences

Andrea L. Guzman

Rethinking Communication between Humans and Social Robots

Abstract: This chapter examines the central role of communication in human–robot interaction (HRI) research with a focus on social robotics. The concept of communication has been important for theorizing interactions between people and technology since the advent of computers, and the ability to exchange messages in a life-like manner is a defining aspect of social robotics. The chapter examines how interpretations of the mechanisms, modes, and social elements of human communication have been integral to the design of social technologies and interactions with them. Although such an approach has proved useful in moving HRI research and the design of technology forward, as the chapter also argues, this understanding of communication has often been only partial. With an aim towards furthering HRI research, rather than criticizing the past, the chapter outlines how critical elements of communication theory have been overlooked and undertheorized within HRI and offers correctives to more productively bring the study of communication and communication theory into HRI and HRI into communication.

Keywords: Communication theory, human–machine communication, human–robot interaction, interdisciplinary research, social robotics

16.1 Introduction

The concept of communication has been integral to the development of robots from industrial machines to social companions, and the last three decades have seen important steps forward in research within human–robot interaction (HRI) as well as in cognate fields, such as human–computer interaction (HCI) and human–machine communication (HMC). Yet, despite its criticality to the study of people's interactions with technology, communication remains undertheorized in these areas, with aspects of research informed by assumptions that do not adequately reflect the nature and complexity of communication. The goal of this chapter is to move HRI and related research forward by challenging the misconception that communication is a seemingly 'natural' process by drawing upon fundamental communication theory to demonstrate that interactions between humans and social technologies are more socially and culturally complex than the mere mechanics of message exchange.

The chapter is divided into two sections. Part One examines the central role of communication in guiding the aims of robotics research and efforts towards developing technologies with communicative abilities. It identifies key assumptions regarding com-

https://doi.org/10.1515/9783110792270-017

munication among humans that are extended to interactions between people and robots. Part Two then interrogates these assumptions through the lens of communication research, explaining their shortcomings and subsequent implications for technology development and use, before introducing fundamental elements of communication theory that serve as a corrective.

16.2 Part One: Centrality of Communication to Sociality with Humans and with Robots

It is hard to overstate the importance of communication to people as individuals, the relationships people form, and the very structure of society. Cooley explains the central role of communication in social life: "By communication is here meant the mechanism through which human relations exist and develop. . . . The more closely we consider this mechanism, the more intimate will appear its relation to the inner life of [human]kind, and nothing will more help us to understand that latter than such consideration" (1909, p. 643). The importance of communication extends well beyond interpersonal relationships and organizational goals to the shape of society. As social theorists have argued, communication is the fundamental creator of reality, a meaning-making process through which people make sense of one another, their world, and the self (Berger & Luckmann, 1967; Carey, 1989). As such, much emphasis is put upon the ability to communicate for successfully navigating almost every aspect of life.

Meaning making also extends to interactions with machines (Suchman, 2009) regardless of whether the technology is designed to be social (Guzman, 2016). Modern robotics has emerged out of a confluence of technological advancements stemming from the advent of computing in the mid-twentieth century and subsequent efforts towards automation and artificial intelligence (AI) (Shkurti, 2023). Scientists realized the power and potential of early computers but faced considerable challenges in their adoption, including making the devices intelligible to the people using them (e.g., Meadow, 1970), controlling devices from a distance (e.g., Sheridan & Verplank, 1978), and enabling devices to regulate their own actions and function within larger systems (e.g., Wiener, 1948). The solution was to borrow from what they observed in the natural world, communication among living things, chiefly humans (see Guzman, 2018).[1] At approximately the same time that engineers were turning to the idea of communication with machines to address their problems, communication was being estab-

1 Although outside the scope of this chapter, the role of animal morphology and behavior in robot design and human–robot interactions must also be acknowledged. See Darling (2021) for an overview of this history and new developments, and Edwards (2018) regarding how humans ontologically classify humans, machines, and animals in relation to one another.

lished as a formal area of study. Ironically, while engineering and communication originally shared a fundamental theory of communication (e.g., Shannon, 1948), communication scholars eventually eschewed 'communication engineering' as too narrow (Weakland, 1967) and focused, instead, exclusively on human communication,[2] until information and communication technology became more interactive.

16.2.1 Mechanisms of communication

Human communication has served as both an implicit and explicit guide for planning and designing aspects of certain technologies (Cassell, 2021), including robots (Brooks & Moon, 2023). The prototype of human communication was and remains attractive to technology researchers and developers because the exchange of messages between and among people is perceived as 'intuitive' and 'natural.' If this everyday behavior could be extended from humans to machines—the thinking has been—then people's interactions with technology would become much easier, thus facilitating adoption and use (e.g., Brooks et al., 1999). Of the role of human communication in guiding the foundation and development of social robotics, in particular, Fong et al. (2003, p. 4) state: "The common, underlying assumption is that humans prefer to interact with machines in the same way that they interact with other people." Furthermore, successfully modeling and executing human communication via artificial means also has held promise for better understanding the mechanisms of human language and intelligence (Barr, 1980), along with human–human communication and media technologies (Peter & Kühne, 2018; Wachsmuth et al., 2008).

16.2.2 Language use

The foremost challenge across technology design generally has been addressing what Licklider (1960) referred to as 'the language problem': humans and computers not only use different symbolic systems for conveying messages but also approach the interaction with different goals. Within the human–machine context, the language of humans has been termed 'natural language'" (Barr, 1980); although, as I explain further in Part Two, communication scholars argue such a term is problematic because language is cultural. AI researchers have worked to develop machine capabilities in natural language processing, or natural language understanding, focused on the interpretation of human language and in natural language generation aimed at enabling machines to respond to human queries in-kind (Mann, 2023). Adopting the conven-

2 Sandry (2015, 2018) expertly demonstrates how different traditions in communication research can be found within as well as used to interrogate theoretical approaches to human–robot communication.

tions of human interactions, such as the patterns and flows of conversation, has also been a key element of this endeavor to help the interaction feel familiar to humans (e.g., Weizenbaum, 1966).

16.2.3 Modes of interaction

Beyond linguistic elements, technology researchers and designers have sought to tap into human modes of message exchange.

16.2.3.1 Vocal communication

Developing the ability to direct machines vocally has been a key goal in technology research since the mid-twentieth century because the use of voice is a biological means of human communication (Nass & Brave, 2005). In fact, some roboticists have defined the communicative robot around the ability of a device to process and respond to speech. For example, Imai et al. (2003) obtained a patent for a 'communication robot' that could receive and act upon sound or voice. Developers have also aimed to couple voice with 'natural language' to heighten a device's usability, reducing or removing the need for training (Brooks et al., 1999; Licklider, 1960) as well as to introduce new levels of efficiency (Cohen & Oviatt, 1995). But speaking to and controlling a robot with language is only one-half of the communication equation, and so researchers have also worked to develop robots that can respond vocally to humans, engaging in spoken exchanges and conversations.

16.2.3.2 Non-verbal communication

Robots are often defined as having some level of embodiment,[3] and the conveyance of non-verbal messages is an important modality of HRI (Edwards, 2023). The enactment and interpretation of embodied communication is highly complex in humans and machines (Wachsmuth et al., 2008), and there are multiple modalities in which non-verbal messages are enacted by robots (Saunderson & Nejat, 2019; Urakami & Seaborn, 2023). Non-verbal elements of HRI based upon human communication seek to emulate the physical image and actions of humans to varying degrees. Chief among these modalities is morphology—the physical form of the robot that is critical to how people interpret

3 The term 'robot' has had various denotative and connotative meanings since its introduction in fiction and subsequent appropriation by technologists (Jordan, 2016). I focus on robots as embodied to capture all possible communicative aspects of people's interactions with robots.

the robot and its capabilities (Kunold et al., 2023) and interact with it (Fortunati & Edwards, 2021). Humanoids and androids re-create aspects of the physical appearance of humans as a means to foster interaction, understanding, use, and a sense of connection (Breazeal, 2004; Brooks et al., 1999). Some of the other important human communication cues integrated into the physical elements of robots and enacted in interaction include facial expressions, eye gaze, movement, and gestures (Saunderson & Nejat, 2019; Urakami & Seaborn, 2023). While re-creating the image and movement of humans within robots can be powerful for facilitating use, eliciting a desired response, and forming connections, there are limits to people's comfort with and acceptance of human likeness re-created and enacted within machines (Mori, 2012).

16.2.3.3 Tactile communication

Touch is classified as non-verbal communication, but I have separated it out here because of its importance in human and human–machine interaction (Archer, 2023). The earliest forms of HMC were tactile, involving manipulation of physical controls (i.e., pulling levers, pressing buttons, connecting cables, typing on keyboards). Although often viewed as primitive and studied to a lesser degree than other senses and modes of communication, touch is critical to how people interact with one another and their world (Parisi, 2018) and is highly complex because the entire human body is a site for sending and receiving messages (El Saddik et al., 2011). Similar to efforts to move towards more sophisticated forms of verbal interaction with machines, the study of haptics was established to develop interfaces that would facilitate the exchange of rich information through touch and, thus, technology use (Parisi, 2018). Today, haptics is the discipline that focuses on the study of tactile communication. Because touch is the most basic form of interaction with robots and used across all types, the degree to which tactile interaction is designed to emulate aspects of human touch varies, with human-like aspects most pronounced in robots designed for social purposes (Huisman, 2017).

16.2.4 Social elements of communication

Efforts of re-creating modes and linguistic means of human communication within machines began with the functional purpose of facilitating control, but, ultimately, led to much more significant insights and an even loftier goal. In his 1997 keynote address to attendees of the *IEEE International Workshop on Robot and Human Communication*, Sheridan observed:

> Since the earliest robots of the 1940s human communication with them has been a serious consideration, and progress over the ensuing 50 years has been great. . . . It is not my intent to re-

> view that progress, though it is clear that as robots themselves become more intelligent the nature of communication between the human operator and the robot is becoming less and less like that of using a passive hand tool and more and more like the relationship between two human beings. (p. 9)

Sheridan's remarks capture a connection among the design of robots, the interaction between humans and machines, and the nature of relationships between people and technology that researchers were only beginning to understand at the time of his address. Rather than being purely informational, people's interactions with technology were found also to be social and involve complex processes of meaning making (Suchman, 2009).

The study of human communication is not merely the theorization and examination of the conveyance of a message from point A to point B, but, rather, its primary focus is on the sense making involved in the communicative exchange and the implications for understanding and behavior (Barnlund, 1970) and, critically, the development of relationships (Schramm, 1983). This is where the power of human communication lies, and it is this power that scholars have attempted to tap into for the past three decades with the development of social technologies, including social robotics. Hoorn (2020) places the locus of the potential of social robots in communication: "They are a novel, human-created communication entity, playing their own social role, with their own style of interpersonal communication: human-like, but non-human" (p. 2050026-2). Although once theorized through an anthropocentric lens, communication now can be considered 'agent agnostic' in that social robots and similar, agentic technologies have joined humans in directly participating in this meaning-making process through which connections and relationships are formed (Banks & de Graaf, 2020).

The use of human-like voices in machines provides an apt illustration of the intersection of the modes and means of interaction and the social element and outcomes of communication. The attributes of digital voices have critical implications for how the technology is perceived as a communicator, its message understood and judged, and, in turn, how humans behave towards it (Nass & Brave, 2005). For example, research has consistently demonstrated that when computer or robotic voices are designed with human gender cues, people will interpret the digital interlocutor and its message in ways consistent with human gender stereotypes (Seaborn et al., 2021). Often, robots and other types of interactive technology are designed with multiple attributes—morphology, vocal cues, non-verbal behavior, and linguistic elements—executed at different levels of the communication process that intersect to create a more fully developed 'mental picture' as to how the technology is to be interpreted socially and technologically. Such human-like cues within the mode of interaction (voice) and in the linguistic elements (natural language and message content) provide a guide in the mind of the user as to how to communicatively act towards the technology by establishing what the device or application is as a communicator (social role) and what it is to the human (social relationship).

16.2.5 Social roles

Conceptualizing what social role a device or application is to carry out in relation to the human interacting with it has been a crucial aspect of technology design that has evolved alongside and, increasingly over time, has intersected with developing the communicative capacities of machines. A key aspect of any technology design is determining how the human and the machine fit into the larger goals of technology use and allocating tasks to the person and technology. Human social roles and relationships have served as guides for thinking through and articulating these divisions of labor (Guzman, 2016, 2017). One of the most pervasive social roles for technology has been that of servant. Indeed, it is out of this context that the modern word for robot emerges (i.e., Čapek, 1923), and the idea of technology as slave or servant remains a powerful conceptualization of the nature of robots (Gunkel, 2018). It also has been an element of how the communicative aspects of machines have been envisioned and designed. For example, in making the case for voice-based technologies, Flanagan (1976) states: "To insist that machines communicate with man on a level he finds effective, namely, by natural voice, seems a fitting requirement for machines—as the servants of man" (p. 405). Such a design aligns socially and functionally with the aim of facilitating human control over technology.

16.2.6 Social relationships

Social roles, human or machine, do not exist within a vacuum; they are always in relation to someone or, now, something else and are deeply embedded in cultural norms and power structures (Suchman, 2009). With humans, a servant exists in relation to a 'master,' and the communication that takes place unfolds and is interpreted within the context of the servant–master relationship. The beck-and-call communicative nature of the servant–master dynamic has been re-created in the interactive form and function of the machine; although it has been recast in recent years in more socially palatable terms such as employer–assistant (Guzman, 2017). As the digital and mechanical capabilities of technology have advanced, so too have communicative capabilities enabling the enactment of more sophisticated social functions. A type of relationship exists between people and technology even if such a relationship may not fit the usual expectations surrounding human–human relationships as something close and intimate. Human relationships form through communication, and, likewise, so do human–machine relationships. Furthermore, people's interactions and relationships with technology are part of larger social configurations (Suchman, 2009), involving other humans and machines (Carpenter, 2016; Höflich, 2013), with implications for not only how people interact with machines but also how humans communicate with one another (Erel et al., 2022).

16.2.7 Social robots

Until this point, I have discussed the communicative elements of technology generally, incorporating scholarship spanning more than 60 years covering computers, robots, artificial intelligence, and a variety of applications and devices. I have taken such a broad approach for several reasons. First, it is critical to understand that communication's importance to robotics did not begin with efforts to design overtly social technologies towards the end of the twentieth century; rather, it has been present since day one in the design of more rudimentary machines and has evolved. Second and relatedly, such an approach illustrates that communication has also played a role in the design of industrial robotics and machines, which are often assumed to be uncommunicative because interaction with them is non-verbal and tactile (Guzman, 2016). Thus, industrial robotics and social robotics may have more areas of overlap in terms of meaning making between humans and technology than is commonly assumed. Third, a broad approach is warranted because AI, computer science, and robotics have never been mutually exclusive areas of study (Brooks, 1991), and research regarding 'socially interactive agents' crosses AI, HRI, and HCI (Lugrin, 2021), all of which fall under the auspices of human–machine communication (Guzman et al., 2023). Thus, scholars in any of these areas need to be aware of their shared history and to attend to developments in cognate fields.

I want to briefly take a closer look at social robots because from a communication standpoint, social robots are more complex than other types of interactive agents because of their embodiment, with a rich array of possibilities for perceived and enacted affordances (Rodríguez-Hidalgo, 2020). The embodiment of social robots requires that multiple modes of interaction coupled with social cues need to work together to establish the sociality of the robot. For example, while Apple's Siri in the United States primarily built *her* social identity and presence through voice and message, Hanson Robotics's Sophia is a humanoid in female physical form complete with facial features, attire, movement, and gestures, in addition to *her* voice and messages. Together these elements are why Sophia was offered citizenship in Saudi Arabia and appeared before the United Nations and why Siri or any of *her* virtual equivalents have not been offered such opportunities (at least not yet). Even avatars that may have realistic forms and features have less points of interaction than a robot that can be touched and touch.

Industrial robots do have a degree of communicability and sociality; however, the level of meaning between human and industrial robot is typically more latent than manifest (see Guzman, 2016) and such communicative elements are secondary to a physical function, such as building a portion of an automobile. Social robotics, by contrast, was founded on a vision of robots communicating as if they were humans to form specific relationships with people. In her field-defining text, Breazeal (2004) explains of the goal of social robotics: "a sociable robot is socially intelligent in a human-like way, and interacting with it is like interacting with another person" (p. 1).

In an effort to achieve such social intelligence, the robot is designed to convey a message through specific communication means and modes that tap into the psychological and social aspects of how humans make sense of and respond to messages and messengers. For example, studying the psychological elements of how both humans and social robots express and respond to emotion has been a crucial aspect of social robotics and HRI research (Stock-Homburg, 2022). Whereas people may evaluate an industrial machine based on its ability to complete manual tasks, a social robot may be judged along criteria applied to people, such as trust or aspects of morality (e.g., Banks, 2021). The primary function of the social robot, to be a companion, assistant, or any number of social roles, is directly tied to its communicative capabilities, as explained by Breazeal et al. (2016): "In dramatic contrast, social (or sociable) robots are designed to engage people in an interpersonal manner, often as partners, in order to achieve positive outcomes in domains such as education, therapy, or health, or task-related goals in areas such as coordinated teamwork for manufacturing, search and rescue, domestic chores, and more" (p. 1936). In other words, use is enacted through communication. It is important to bear in mind that the communication taking place between human and robot is not solely for the benefit of the human. The process of human–robot interaction is also intended to enable the robot to acquire information and knowledge about the world that can guide its ongoing actions (Breazeal, 2004; Brooks et al., 1999). The exchange of messages is the means through which the human is to understand the robot and for the robot to be able to carry out its function in relation to the human and its environment. The still larger goal of building social robots goes beyond mere successful use of an object through interaction. Of the 'grand challenge' for the field, Breazeal (2004) states, "The ultimate challenge for a sociable robot is to interact with humans as another person would and to be accepted as part of the human community" (p. 235). Social robots are to enter physically and mentally into the human social world, not as objects, but as social entities. The goal is for them to gain a type of acceptance and belonging. The sophisticated communicative abilities of social robots coupled with their embodiment also raise ethical questions different from other technologies (Krämer & Manzeschke, 2021).

16.2.8 Summary: Part One

Communication has been a critical element in imagining the capabilities and roles of robots, as well as other advanced technologies, and seamless interaction between person and robot, executed through a variety of modalities, has been a fundamental goal. Communication between human and robot is the essential meaning-making process through which the robot is established as a social entity. Historically, an influential prototype driving the aims of social robotics is that of human–human communication and its relational outcomes that are socially and psychologically powerful; although it is important to note that in recent years, scholars also have turned to other models such as human–animal communication (Darling, 2021) or leveraging the ontological differences between people and robots rather than seeking communicative sameness (Sandry, 2015). Still, prominent assumptions undergirding the anthropocentric focus of human–robot interactions are that communication is 'innate' for humans and that the emulations of such messages and modes within robots will help to achieve the desired technological and social outcomes between person and robot of facilitating use and establishing relationships.

16.3 Part Two: The Fundamentals of Communication between Humans and Social Robots

Within the past several decades, developments in robotics have been astounding. There is much that has been achieved in the creation of specific technologies as well as in the advancement of knowledge across HRI and the fields contributing to it. At the same time, there are clear indications of limits as to the current capabilities of social robots and other technologies, people's willingness to adopt them at large scale, and to the theories explaining people's behavior and understanding of them. For example, early theories of interactions with technology, such as CASA or the media equation (Reeves & Nass, 1998), remain important, but they are also showing signs of not being able to adequately account for the full picture of people's contemporary experiences with AI and robots (Edwards, 2023; Fortunati & Edwards, 2022; Gambino et al., 2020; Lombard & Xu, 2021; van der Goot & Etzrod, 2023). Technologies that are highly sophisticated from a social standpoint have failed to catch on commercially as of yet, with the majority of social robots still existing only within research spaces (Mahdi et al., 2022). Contrary to claims by early researchers that machines would bend towards humans when designed with 'natural' communicative abilities, it is the human that most often adapts their communication to robots (Pelikan & Broth, 2016), particularly when there is a communicative breakdown (Beneteau et al., 2019), of which there are many.

From the perspective of communication scholarship, some of the conceptualizations of communication and theoretical assumptions about *human* communication that have guided the development of social technologies, as explained in Part One, have not adequately accounted for the nature of communication or the full depth of its complexity. While technology developers and scholars have held an interest in how to utilize communication to achieve certain ends, this focus on the 'practical' has meant that engagement with communication theory has only been partial (Hoorn, 2020). Some conceptualizations of communication as applied to technology have also lagged behind advances in communication theory. To some degree this is understandable, given that communication as a distinct area of social science research coalesced in the mid-twentieth century with the definition of communication and shape of the field evolving rapidly at the same time that the formal study of artificial intelligence and robotics also emerged (see also Guzman, 2018). It is only more recently that scholars in robotics, communication, and AI have come into more sustained contact (Edwards, 2023; Guzman & Lewis, 2020). The interdisciplinary nature of this area of research also may contribute to challenges in applying theories across field boundaries. Irfan et al. (2018) observe that often HRI scholars are applying theories from fields outside their own, that is, engineers utilizing theories from social psychology, and, although well meaning, scholars working across fields may lack the knowledge needed to adequately assess the quality of the outside research, and to understand, and apply it. To this, I would add that scholars working across disciplinary boundaries—particularly spaces that cross larger domain divides such as the hard sciences versus social sciences—may not be aware of fundamental aspects of theory that are not always explicitly laid out within research because they are assumed to be shared knowledge within a particular field.[4]

As explained in Part One, the prevailing assumption that human communication is natural or innate has helped to shape the research agenda of developing communicative and social technologies. What researchers mean by the term 'natural' when referring to human communication is not always clear because of ambiguity in the meaning of the word and the lack of explanation by scholars. From an ontological standpoint, humans and their actions are often perceived as natural because they are biological; robots and other machines, by contrast, are human made and, thus, artificial (Edwards, 2018). 'Natural,' therefore, could be understood to refer to modes of communication by biological beings, particularly in early human–machine contexts in which interaction between user and technology was much more difficult, taking place by more rudimentary modes (e.g., Barr, 1980; Licklider, 1960). For example, the term 'natural language' is intended to contrast commands in human languages with those in 'machine language.' The naturalness of human communication has also been presented within contexts in which the connotation is that of something that is 'automatic,' 'intuitive,' or 'innate.' Early HCI and HRI research stresses the automatic, social response in people's reactions to and behavior towards devices or applications with basic human-like cues (e.g., Reeves & Nass, 1998). Some scholars have interpreted this to mean that such responses are innate, but, as explained further below, there is a critical difference between people responding automatically (an outcome of learned behavior) versus people responding as if it is part of human nature. Often what is taken to be 'natural' is a highly sophisticated and dynamic meaning-making process that people acquire, hone, and adapt over an entire lifetime. In mapping an incomplete picture of human communication onto people's interactions with robots and

4 This applies to anyone borrowing from another field and underlines why it is so important to have interdisciplinary research teams and interdisciplinary research spaces in which inadvertent misapplication or misunderstanding of theory can be avoided or corrected.

other social technologies, these conceptual and theoretical gaps have carried over and have the potential to be more consequential given the ontological differences between humans and machines.

The remainder of the chapter focuses on highlighting two fundamental aspects of communication that need to be more fully accounted for in theorizing people's interactions with technologies: communication is learned and communication is the creation of meaning. The goal of Part Two is not to criticize the past—particularly given what has been envisioned and achieved—but to help current and future scholars and developers better address the complexity of people's interactions with robots and other technologies to progress research.

16.3.1 Communication is learned

Communication is learned (e.g., Barnlund, 1970) regardless of whether it is between people or people and machines. Humans spend their entire lives learning and refining their ability to communicate—from childhood with language acquisition to adolescence and the development of social skills all the way through adulthood and the establishment, navigation, and dissolution of different types of relationships. People often take this learning and its complexity for granted because communication is a commonplace activity. Most people do not remember a time in their life when they were not communicating: interacting with others is a routine experience that many people engage in numerous times per day, and often people do not actively think about their communicative actions, many of which are automatic (Kellermann, 1992). From this perspective, communication may seem like second nature, but even a person's automatic behavior is the result of previous interactions with the world (e.g., Schank & Abelson, 1977). That people act socially towards machines, particularly in initial interactions, indicates that they are being guided by previous social encounters; although much work is still needed to better understand the mechanisms of this sense-making process (e.g., Kim & Sundar, 2012; Koban & Banks, 2023; van der Goot & Etzrod, 2023). For example, research has demonstrated that people utilize cognitive scripts based on human interaction in encounters with robots (Edwards et al., 2019). Such actions are automatic, but they are not innate—they are the application of information gained in human–human communication being applied to human–machine contexts.

That miscommunication still occurs frequently in human–human communication and people find themselves uncertain of how to proceed in novel contexts underscores the complexity and ongoing nature of learning to communicate. And that is the complexity that exists with humans—communicators that are of the same ontological category. At the risk of stating the obvious, humans and robots are different ontological entities; although, and particularly in the case of social technologies, people also recognize some shared qualities across both categories (Etzrodt & Engesser, 2021; Guzman, 2020). Empirical research has demonstrated that the distinct ontological differences along with areas of similarity matter for communication. While people may act socially towards technology, such as being polite, they do not necessarily treat humans and machines as communicative equals (Edwards et al., 2019; Gambino et al., 2020; Lombard & Xu, 2021; Mou & Xu, 2017). The goal of social robotics is not simply to use 'natural' commands, but, at its highest level, is to leverage human 'natural' social abilities to develop more sustained and intimate human–robot relationships. From a communicative standpoint, relationships and friendships are even more complex to initiate and sustain than simple everyday interactions (Canary & Dainton, 2002). People have been communicating their entire life and still are learning, adapting, and struggling in how they relate to those closest to them, be it colleagues, friends, or significant others. So, what does this say about the level of learning needed in the creation and maintenance of more intimate relationships with social technologies of a different ontological nature?

Much of the discussion surrounding people's interactions with technology is focused on how people's communicative behaviors with other humans inform their interactions with social technologies. However, such an approach overlooks what people have learned from machines in their interactions with technology. While people's initial encounters with a social robot may be their first interaction with a social technology, it is not their first communicative exchange with technology. Remember that tactile interaction—even the pushing of a button—is a form of communication (albeit rudimentary). The proliferation of interactive and social technologies also means that people's interactions with these devices and applications are increasing. The implication is that people's understanding of technology as communicator is being informed by interactions with both humans and machines (Gambino et al., 2020; Lombard & Xu, 2021). In my own research regarding people's perceptions of automated news-writing technologies, some participants assessed the qualities of these applications in comparison to voice-based assistants in addition to the capabilities of human journalists. Fortunati and Edwards (2021) underscore the importance of what is learned in interactions with machines in a rejoinder to Reeves and Nass's (1998) claim that people's behavior towards interactive technologies often inappropriately draws from human social contexts because "the human brain evolved in a world in which only humans exhibited rich social behaviors." Fortunati and Edwards counter that such a claim ignores the cultural and historical context of people's interactions with the world around them, including a variety of machines, some of which will have had social qualities, such as automata.

Learning also goes beyond the knowledge that people bring to interactions with social technologies. Communication itself is a process of learning, a point discussed further below. Some roboticists stress that communication can serve as a means to inform the robot, but humans are also learning and adapting to the robot in these interactions, as HRI research has also demonstrated (e.g., Pelikan & Broth, 2016; Wallkötter et al., 2017). What is less understood within HRI, because of an overall lack of long-term studies and research that examines communication within natural and everyday settings (de Graaf et al., 2017; Rosenthal-von der Pütten & Abrams, 2021), is the ongoing process of learning as it unfolds between both human and machine across different contexts. Overall, assumptions of human-to-human communication that refer to the seeming ease and innateness of communication as 'natural' can be problematic because they gloss over, at best, or misrepresent, at worst, the nature and complexity of communication as learned between and among humans and technology.

16.3.2 Communication is the creation of meaning

This chapter began by focusing on the central role of communication in constituting everyday life, and it ends by expanding upon this discussion because it is the most critical for understanding and progressing the study of people's interactions with social robots and related technologies. For communication scholars, the modes of interaction and the messages exchanged are an important part of the communicative process, but they do not solely constitute communication or its study. Communication is fundamentally the creation of meaning among humans and their world (Carey, 1989; Cooley, 1909), including technology (Suchman, 2009). Human–machine communication research specifically focuses on the study of "the creation of meaning among humans and machines" (Guzman, 2018, p. 17).

That communication is meaning making seems, at first glance, fairly straightforward, but what is meant by this theoretical claim is often misunderstood. In fact, communication as a concept and area of study generally is misunderstood, a point worth briefly discussing. Because of its ubiquity as a mundane aspect of everyday life, 'communication' is a term that has been bandied about colloquially and in academia, taking on various definitions even as communication became a formalized field of study in the mid-twentieth century. In a 1967 special issue of the *American Behavioral Scientist* focusing on 'communication and behavior,' Weakland remarks:

> In recent years 'communication' has become almost a catchall word, and often a catchword in addition. It is used increasingly by specialists in many areas of science and technology, as well as by laymen, to refer rather indiscriminately to—or even, it seems, to validate—a variety of quite disparate interests, levels and outlooks. That is, the term by now covers a multitude of positions, and not a few sins. (p. 1)

Weakland then laid out the basic definition and tenets of communication that, in the intervening half-century, have also changed multiple times as communication research has evolved (Guzman et al., 2023). Outside the study of communication, however, Weakland's (1967) observation still holds true.

One of the most prevalent conceptualizations of communication throughout the twentieth century and today is what Carey (1989) referred to as the 'transmission view'—the perspective that communication is a means of information transfer to achieve some effect. Such a model closely aligns with Shannon's (1948) influential mathematical theory of communication, which focused on maximizing the transfer of information between two points. Shannon's model would become foundational to both information theory and in the early study of communication (i.e., Shannon & Weaver, 1949) before other models that focused more on the dialogic nature of communication, meaning making, and constructivist approaches (e.g., Barnlund, 1970; Delia, 1977) as well as critical and cultural perspectives (e.g., Carey, 1989) replaced it in the study of communication. This conceptualization of communication as transmission is prevalent in approaches to designing and theorizing people's interactions with social technologies (Guzman, 2018). As outlined in Part One, the goal of re-creating modes of interaction and social cues within robots is to enable their control and to facilitate their acceptance and use: communication is, thus, theoretically situated as a means to an end, and meaning making between human and robot is viewed as taking place through communication.

Although the human and robot are making sense of one another (albeit in very different ways to different ends), for some communication scholars, this theoretical perspective does not adequately reflect what is meant by communication as meaning making. In contrast to the 'transmission view,' Carey (1989) presents the 'ritual view' of communication—in which acts of communication bring into being people's social worlds and the self (see also Mead, 1967). From this perspective, communication is defined as "a symbolic process whereby reality is produced, maintained, repaired, and transformed" (Carey, 1989, p. 19). Carey's conceptualization of communication is influenced, in part, by philosopher John Dewey, and here Carey's (1989) own quotation of Dewey is helpful for understanding the difference between the two perspectives: "'Society exists not only by transmission, by communication, but it may be fairly be said to exist in transmission, in communication'" (Dewey, 1916 as cited in Carey, 1989). Meaning is not merely conveyed in communication, it is created in communication and is inextricably linked to the social world, to culture, to people's understanding of the self, and to their understanding of others.

This conceptualization of communication as the creation of meaning fundamentally shifts what it is to study not only interaction among humans but also communication among humans and machines (Guzman, 2018). To return to the underlying assumption in HRI that communication is 'natural'; no such claim can be made because communication as meaning making is social (Fortunati & Edwards, 2021), with meaning emergent within the very act of interaction (Dewey, 1925). This is also why communication is learned, or more accurately from this theoretical perspective, is learning (the act of). Dewey (1925) argues: "Where communication exists, things in acquiring meaning, thereby acquire representatives, surrogates, signs and implicates. . . . Learning and teaching come into being, and there is no event which may not yield information" (p. 36). Communication, itself, is understood in communication. Rodríguez-Hidalgo's (2020) examination of affordances in HRI is illustrative of how concepts, theories, and approaches to research are reworked when communication is viewed as constitutive of meaning rather than a means to an end. Rodríguez-Hidalgo explains that HRI has primarily taken an object-oriented approach to affordances, focusing on functions a robot can carry out: it is possessing these functions that provides the affordance. In contrast, Rodríguez Hidalgo continues, a communicative approach conceptualizes affordance as user oriented, as "emerging from a process of perceived and enacted communicative behaviors or actions between both partners" (2020, p. 62). Following an approach of communication as meaning making, the study of robot affordances shifts from that which is physical and exists *a priori* to the interaction to that which is cognitive arising within the interaction (see also Nagy & Neff, 2023).

Proceeding from a conceptualization of communication as the creation of meaning reorients the study of people's interactions with technology, expanding the nature of questions to be asked and answered and placing new demands on the theoretical approaches to answer them (Fortunati & Edwards, 2020; Guzman, 2018; Guzman & Lewis, 2020). For HRI, such a refocusing also aligns with research finding that people demonstrate a preference for robots that are designed to interact in more rhetorically sophisticated ways than the mere exchange of basic messages (Edwards et al., 2020). Questions now are not solely about the robot or the human, the effects of one on the other, or what both can achieve (or fail to achieve) together (although such questions do remain relevant). HRI currently and in the future must account more fully for the implications of people's communicative exchanges with robots for the self, how people come to understand who they are not only in relation to other humans but also machines; for society and culture, in which the human–machine configurations are now infinitely more complex; and, for what it means to be human, now that meaning is constituted in greater collaboration with human-like technologies. Such metaphysical questions have been broached before in relation to technology (e.g., Turkle, 1984) but missing from these previous philosophical and theoretical discussions is consideration for the very act in which they arise—communication.

References

Archer, J. (2023). Haptics, human augmentics, and human–machine communication. In A. L. Guzman, R. McEwen, & S. Jones (Eds.), *The SAGE Handbook of Human–Machine Communication* (pp. 474–482). SAGE Publications.

Banks, J. (2021). Good robots, bad robots: Morally valenced behavior effects on perceived mind, morality, and trust. *International Journal of Social Robotics*, *13*, 2021–2038. https://doi.org/10.1007/s12369-020-00692-3.

Banks, J., & de Graaf, M. (2020). Toward an agent-agnostic transmission model: Synthesizing anthropocentric and technocentric paradigms in communication. *Human–machine Communication*, *1*, 19–36. https://doi.org/10.30658/hmc.1.2.

Barnlund, D. C. (1970). A transactional model of communication. In J. Akin, A. Goldberg, G. Myers, & J. Stewart (Eds.), *Language Behavior: A Book of Readings in Communication* (pp. 43–61). Mouton.

Barr, A. (1980). Natural language understanding. *AI Magazine, 1*(1), 5, 5–10. https://vvvvw.aaai.org/ojs/index.php/aimagazine/article/view/85.
Beneteau, E., Richards, O. K., Zhang, M., Kientz, J. A., Yip, J., & Hiniker, A. (2019). Communication breakdowns between families and Alexa. In *Proceedings of the 2019 CHI Conference on Human Factors in Computing Systems* (pp. 1–13). https://doi.org/10.1145/3290605.3300473.
Berger, P., & Luckmann, T. (1967). *The Social Construction of Reality: A Treatise in the Sociology of Knowledge.* Doubleday.
Breazeal, C., Dautenhahn, K., & Kanda, T. (2016). Social robotics. In B. Siciliano & O. Khatib (Eds.), *Springer Handbook of Robotics* (pp. 1935–1971). Springer International Publishing.
Breazeal, C. L. (2004). *Designing Sociable Robots* (First Paperback). The MIT Press.
Brooks, R. A. (1991). New approaches to robotics. *Science, 253*(5025), 1227–1232.
Brooks, R. A., Breazeal, C., Scassellati, B., & O'Reilly, U.-M. (1999). Technologies for human/humanoid natural interactions. In *The 2nd International Symposium in Humanoid Robots (HURO)*.
Brooks, S., & Moon, A. J. (2023). Governing the social dimensions of collaborative robotic design: Influence, manipulation, and other non-physical harms. In A. L. Guzman, R. McEwen, & S. Jones (Eds.), *The SAGE Handbook of Human–Machine Communication* (pp. 409–417). SAGE Publications.
Canary, D. J., & Dainton, M. (2002). *Maintaining Relationships through Communication: Relational, Contextual, and Cultural Variations*. Taylor & Francis Group.
Čapek, Karl (1923). *R.U.R. Rossum's Universal Robots: A Fantastic Melodrama in Three Acts and an Epilogue*. Trans. P. Selver (Ebook #59112). Doubleday, Page and Co. https://www.gutenberg.org/files/59112/59112-h/59112-h.htm.
Carey, J. W. (1989). *Communication as Culture: Essays on Media and Society*. Routledge.
Carpenter, J. (2016). *Culture and Human–Robot Interaction in Militarized Spaces: A War Story*. Routledge.
Cassell, J. (2021). Foreword. In B. Lugrin, C. Pelachaud, & D. Traum (Eds.), *The Handbook of Socially Interactive Agents: 20 Years of Research on Embodied Conversational Agents, Intelligent Virtual Agents, and Social Robotics: Vol. 1: Methods, Behavior, Cognition* (pp. xvii–xix). ACM Press.
Cohen, P. R., & Oviatt, S. L. (1995). The role of voice input for human–machine communication. *Proceedings of the National Academy of Sciences, 92*(22), 9921–9927.
Cooley, C. H. (1909). The significance of communication. In W. Schramm & D. F. Roberts (Eds.), *The Process and Effects of Mass Communication* (revised, pp. 643–654). University of Illinois Press.
Darling, K. (2021). *The New Breed: What our History with Animals Reveals about our Future with Robots*. Henry Holt and Company.
De Graaf, M., Ben Allouch, S., & van Dijk, J. (2017). Why do they refuse to use my robot? Reasons for non-use derived from a long-term home study. In *Proceedings of the 2017 ACM/IEEE International Conference on Human-Robot Interaction* (pp. 224–233). IEEE.
Delia, J. G. (1977). Constructivism and the study of human communication. *Quarterly Journal of Speech, 63*, 66–83.
Dewey, J. (1925). Nature, communication, and meaning. In J. D. Peters & P. Simonson (Eds.), *Mass Communication and American Social Thought: Key Texts, 1919–1968* (pp. 35–36). Rowman & Littlefield Publishers.
Edwards, A. (2018). Animals, humans, and machines: Interactive implications of ontological classification. In A. L. Guzman (Ed.), *Human–Machine Communication: Rethinking Communication, Technology, and Ourselves* (pp. 29–50). Peter Lang.
Edwards, A. (2023). Human–robot interaction. In A. L. Guzman, R. McEwen, & S. Jones (Eds.), *The SAGE Handbook of Human–Machine Communication* (pp. 167–177). SAGE Publications.
Edwards, A., Edwards, C., & Gambino, A. (2020). The social pragmatics of communication with social robots: Effects of robot message design logic in a regulative context. *International Journal of Social Robotics, 12(*4), 945–957. https://doi.org/10.1007/s12369-019-00538-7.

Edwards, A., Edwards, C., Westerman, D., & Spence, P. R. (2019). Initial expectations, interactions, and beyond with social robots. *Computers in Human Behavior*, *90*, 308–314. https://doi.org/10.1016/j.chb.2018.08.042.

El Saddik, A., Orozco, M., Eid, M., & Cha, J. (2011). *Haptics Technologies: Bringing Touch to Multimedia*. Springer-Verlag.

Erel, H., Trayman, D., Levy, C., Manor, A., Mikulincer, M., & Zuckerman, O. (2022). Enhancing emotional support: The effect of a robotic object on human–human support quality. *International Journal of Social Robotics*, *14*(1), 257–276. https://doi.org/10.1007/s12369-021-00779-5.

Etzrodt, K., & Engesser, S. (2021). Voice-based agents as personified things: Assimilation and accommodation as equilibration of doubt. *Human–Machine Communication*, *2*, 57–79. https://doi.org/10.30658/hmc.2.3.

Flanagan, J. L. (1976). Computers that talk and listen: Man–machine communication by voice. *Proceedings of the IEEE*, *64*(4), 405–415.

Fong, T., Nourbakhsh, I., & Dautenhahn, K. (2003). A survey of socially interactive robots. *Robotics and Autonomous Systems*, *42*(3), 143–166. https://doi.org/10.1016/S0921-8890(02)00372-X.

Fortunati, L., & Edwards, A. (2020). Opening space for theoretical, methodological, and empirical issues in human–machine communication. *Human–Machine Communication*, *1*, 7–18. https://doi.org/10.30658/hmc.1.1.

Fortunati, L., & Edwards, A. (2021). Moving ahead with human–machine communication. *Human–Machine Communication*, *2*, 7–28. https://doi.org/10.30658/hmc.2.1.

Fortunati, L., & Edwards, A. (2022). Framing the psycho-social and cultural aspects of human–machine communication. *Human–Machine Communication*, *4*, 7–26. https://doi.org/10.30658/hmc.4.1.

Gambino, A., Fox, J., & Ratan, R. (2020). Building a stronger CASA: Extending the computers are social actors paradigm. *Human–Machine Communication*, *1*, 71–86. https://doi.org/10.30658/hmc.1.5.

Gunkel, D. J. (2018). *Robot Rights*. MIT Press.

Guzman, A. L. (2016). The messages of mute machines: Human–machine communication with industrial technologies. *Communication ++1*, *5*. https://doi.org/10.7275/R57P8WBW.

Guzman, A. L. (2017). Making AI safe for humans: A conversation with Siri. In R. W. Gehl & M. Bakardjieva (Eds.), *Socialbots and their Friends: Digital Media and the Automation of Sociality* (pp. 69–82). Routledge.

Guzman, A. L. (2018). What is human–machine communication, anyway? In A. L. Guzman (Ed.), *Human–Machine Communication: Rethinking Communication, Technology, and Ourselves* (pp. 1–28). Peter Lang.

Guzman, A. L. (2020). Ontological boundaries between humans and computers and the implications for human–machine communication. *Human–Machine Communication*, *1*(1). https://stars.library.ucf.edu/hmc/.

Guzman, A. L., & Lewis, S. C. (2020). Artificial intelligence and communication: A human–machine communication research agenda. *New Media & Society*, *22*(1), 70–86. https://doi.org/10.1177/1461444819858691.

Guzman, A. L., McEwen, R., & Jones, S. (Eds.) (2023). Introduction. *The SAGE Handbook of Human–Machine Communication*. SAGE Publications.

Höflich, J. R. (2013). Relationships to social robots: Towards a triadic analysis of media-oriented behavior. *Intervalla: Platform for Intellectual Exchange*, *1*, 35–48. http://www.fc.edu/intervalla/images/pdf/4_holflich.pdf.

Hoorn, J. F. (2020). Theory of robot communication: I. The medium is the communication partner. *International Journal of Humanoid Robotics*, *17*(06), 2050026-1–2050026-25. https://doi.org/10.1142/S0219843620500267.

Huisman, G. (2017). Social touch technology: A survey of haptic technology for social touch. *IEEE Transactions on Haptics*, *10*(3), 391–408. https://doi.org/10.1109/TOH.2017.2650221.

Imai, M., Ishiguro, H., Ono, T., Maeda, T., & Kanda, T. (2003). *United States Patent: Communication Robot* (United States Patent US 6,604,021 B2).
Irfan, B., Kennedy, J., Lemaignan, S., Papadopoulos, F., Senft, E., & Belpaeme, T. (2018). Social psychology and human–robot interaction: An uneasy marriage. In *Companion of the 2018 ACM/IEEE International Conference on Human–Robot Interaction* (pp. 13–20). https://doi.org/10.1145/3173386.3173389.
Jordan, J. (2016). *Robots*. MIT Press.
Kellermann, K. (1992). Communication: Inherently strategic and primarily automatic. *Communication Monographs, 59*(3), 288–300. https://doi.org/10.1080/03637759209376270.
Kim, Y., & Sundar, S. S. (2012). Anthropomorphism of computers: Is it mindful or mindless? *Computers in Human Behavior, 28*(1), 241–250. https://doi.org/10.1016/j.chb.2011.09.006.
Koban, K., & Banks, J. (2023). Dual-process theory in human–machine communication. In A. L. Guzman, R. McEwen, & S. Jones (Eds.), *The SAGE Handbook of Human–Machine Communication* (pp. 302–309). SAGE Publications.
Krämer, N., & Manzeschke, A. (2021). Social reactions to socially interactive agents and their ethical implications. In B. Lugrin, C. Pelachaud, & D. Traum (Eds.), *The Handbook of Socially Interactive Agents: 20 Years of Research on Embodied Conversational Agents, Intelligent Virtual Agents, and Social Robotics: Vol. 1: Methods, Behavior, Cognition* (pp. 77–103). ACM Press.
Kunold, L., Bock, N., & Rosenthal-von Der Pütten, A. (2023). Not all robots are evaluated equally: The impact of morphological features on robots' assessment through capability attributions. *ACM Transactions on Human–Robot Interaction, 12*(1), 1–31. https://doi.org/10.1145/3549532.
Licklider, J. C. R. (1960). Man–computer symbiosis. In R. W. Taylor (Ed.), *In Memoriam: J. C. R. Licklider 1915:1990* (pp. 1–20). Systems Research Center.
Lombard, M., & Xu, K. (2021). Social responses to media technologies in the 21st century: The media are social actors paradigm. *Human–Machine Communication, 2*, 29–55. https://doi.org/10.30658/hmc.2.2.
Lugrin, B. (2021). Introduction to socially interactive agents. In B. Lugrin, C. Pelachaud, & D. Traum (Eds.), *The Handbook of Socially Interactive Agents: 20 Years of Research on Embodied Conversational Agents, Intelligent Virtual Agents, and Social Robotics: Vol. 1: Methods, Behavior, Cognition* (pp. 1–18). ACM Press.
Mahdi, H., Akgun, S. A., Saleh, S., & Dautenhahn, K. (2022). A survey on the design and evolution of social robots—Past, present and future. *Robotics and Autonomous Systems, 156*, 104193. https://doi.org/10.1016/j.robot.2022.104193.
Mann, S. (2023). Human–machine communication, humacomm, and origins: A foreword by Steve Mann, 2022. In A. L. Guzman, R. McEwen, & S. Jones (Eds.), *The SAGE Handbook of Human–Machine Communication* (pp. xxix–xxxvii). SAGE Publications.
Mead, G. H. (1967). *Mind, Self, & Society: From the Standpoint of a Social Behaviorist*, Vol. 1. University of Chicago Press.
Meadow, C. T. (1970). *Man–Machine Communication*. Wiley-Interscience.
Mori, M. (2012). The uncanny valley. *IEEE Robotics & Automation Magazine, 19*(2), 98–100. https://doi.org/10.1109/MRA.2012.2192811.
Mou, Y., & Xu, K. (2017). The media inequality: Comparing the initial human–human and human–AI social interactions. *Computers in Human Behavior, 72*, 432–440. https://doi.org/10.1016/j.chb.2017.02.067
Nagy, P., & Neff, G. (2023). Rethinking affordances for Human–Machine Communication research. In A. L. Guzman, R. McEwen, & S. Jones (Eds.), *The SAGE Handbook of Human–Machine Communication* (pp. 273–279). SAGE Publications.
Nass, C., & Brave, S. (2005). *Wired for Speech: How Voice Activates and Advances the Human–Computer Relationship*. MIT Press.
Parisi, D. (2018). *Archaeologies of Touch: Interfacing with Haptics from Electricity to Computing*. University of Minnesota.

Pelikan, H. R. M., & Broth, M. (2016). Why that Nao? How humans adapt to a conventional humanoid robot in taking turns-at-talk. In *Proceedings of the 2016 CHI Conference on Human Factors in Computing Systems* (pp. 4921–4932). https://doi.org/10.1145/2858036.2858478.

Peter, J., & Kühne, R. (2018). The new frontier in communication research: Why we should study social robots. *Media and Communication*, *6*(3), 73–76. https://doi.org/10.17645/mac.v6i3.1596.

Reeves, B., & Nass, C. I. (1998). *The Media Equation*. CSLI Publications.

Rodríguez-Hidalgo, C. (2020). Me and my robot smiled at one another: The process of socially enacted communicative affordance in human–machine communication. *Human–Machine Communication*, *1*, 55–69. https://doi.org/10.30658/hmc.1.4.

Rosenthal-von der Pütten, A., & Abrams, A. M. H. (2021). Empirical methods in the social science for researching socially interactive agents. In B. Lugrin, C. Pelachaud, & D. Traum (Eds.), *The Handbook of Socially Interactive Agents: 20 Years of Research on Embodied Conversational Agents, Intelligent Virtual Agents, and Social Robotics: Vol. 1: Methods, Behavior, Cognition* (pp. 21–76). ACM Press.

Sandry, E. (2015). *Robots and Communication*. Palgrave Pivot.

Sandry, E. (2018). Aliveness and the off-switch in human–robot relations. In A. L. Guzman (Ed.), *Human–Machine Communication: Rethinking Communication, Technology, and Ourselves* (pp. 51–66). Peter Lang.

Saunderson, S., & Nejat, G. (2019). How robots influence humans: A survey of nonverbal communication in social human–robot interaction. *International Journal of Social Robotics*, *11*(4), 575–608. https://doi.org/10.1007/s12369-019-00523-0.

Schank, R. C., & Abelson, R. P. (1977). *Scripts, Plans, Goals and Understanding: An Inquiry into Human Knowledge Structures*. Lawrence Erlbaum.

Schramm, W. (1983). The unique perspective of communication: A retrospective view. *Journal of Communication*, *33*(3), 6–17. http://onlinelibrary.wiley.com/doi/10.1111/j.1460-2466.1983.tb02401.x/full.

Seaborn, K., Miyake, N. P., Pennefather, P., & Otake-Matsuura, M. (2021). Voice in human–agent interaction: A survey. *ACM Computing Surveys*, *54*(4), 1–43. https://doi.org/10.1145/3386867.

Shannon, C. E. (1948). A mathematical theory of communication. *Bell System Technical Journal*, *27*(3, 4), 379–423, 623–656. http://dl.acm.org/citation.cfm?id=584093.

Shannon, C. E., & Weaver, W. (1949). *The Mathematical Theory of Communication* (12th Printing). University of Illinois Press.

Sheridan, T. B. (1997). *Eight Ultimate Challenges of Human–Robot Communication*. 9–14.

Sheridan, T. B., & Verplank, W. L. (1978). *Human and Computer Control of Underseas Teleoperators* (Government N00014-77-C-0256). Massachusetts Institute of Technology.

Shkurti, F. (2023). The history and future of human–robot communication. In A. L. Guzman, R. McEwen, & S. Jones (Eds.), *The SAGE Handbook of Human–Machine Communication* (pp. 54–62). SAGE Publications.

Stock-Homburg, R. (2022). Survey of emotions in human–robot interactions: Perspectives from robotic psychology on 20 years of research. *International Journal of Social Robotics*, *14*(2), 389–411. https://doi.org/10.1007/s12369-021-00778-6.

Suchman, L. A. (2009). *Human–Machine Reconfigurations: Plans and Situated Actions* (2nd ed.). Cambridge University Press.

Turkle, S. (1984). *The Second Self*. Simon & Schuster.

Urakami, J., & Seaborn, K. (2023). Nonverbal cues in human–robot interaction: A communication studies perspective. *ACM Transactions on Human–Robot Interaction*, *12*(2), 1–21. https://doi.org/10.1145/3570169.

van der Goot, M., & Etzrod, K. (2023). Disentangling two fundamental paradigms in human–machine communication research: Media equation and media evocation. *Human–Machine Communication*, *6*, 17–30. https://doi.org/10.30658/hmc.6.2.

Wachsmuth, I., Lenzen, M., & Knoblich, G. (Eds.) (2008). *Embodied Communication in Humans and Machines*. Oxford University Press.

Wallkötter, S., Joannou, M., Westlake, S., & Belpaeme, T. (2017). Continuous multi-modal interaction causes human–robot alignment. In *Proceedings of the 5th International Conference on Human Agent Interaction* (pp. 375–379). https://doi.org/10.1145/3125739.3132599.
Weakland, J. H. (1967). Communication and behavior—An introduction. *American Behavioral Scientist*, *10*(8), 1–4. https://doi.org/10.1177/000276420100080l.
Weizenbaum, J. (1966). ELIZA—a computer program for the study of natural language communication between man and machine. *Communications of the ACM*, *9*(1), 36–45. https://doi.org/10.1145/365153.365168.
Wiener, N. (1948). *Cybernetics: Or Control and Communication in the Animal and the Machine*. Kessinger Legacy Reprints (7th Printing). John Wiley & Sons.

Patric R. Spence, Chad Edwards, and Matthew Craig

Interacting with Social Robots: The Influence of their Distinctive Cues, Behavioral Capabilities, and Affordances on Social Interaction and Well-being

Abstract: Social robots have the potential to significantly impact human behavior in social settings, presenting both opportunities and challenges. This chapter explores the multifaceted influences of social robots' cues, behavioral capacities, and affordances on human–robot interactions (HRI) and their implications for human well-being. Social robots employ various cues to engage users, and effective interactions rely on speech and dialogue recognition, visual, audio, and tactile cues, and the ability of robots to move and gesture aids in fusing verbal and non-verbal behaviors. As such, anthropomorphism and helpfulness are pivotal in shaping human perceptions of social robots. Greater anthropomorphism can build rapport and trust, but further research is needed to understand the complex relationship between anthropomorphism, helpfulness, and high-stakes scenarios. The potential benefits of social robots for human well-being are significant, as they can provide emotional support, reduce stress, and help people adopt healthy behaviors. However, it is crucial to balance the advantages and risks of using social robots to complement human interaction rather than replace it.

Keywords: Anthropomorphism, behavior capacities, affordances, social robots, non-verbal, well-being

17.1 Introduction

A robot is built to interact with humans using a high degree of social intelligence socially intelligently and may be utilized for educational, entertainment, healthcare, companionship, sexual, customer service, and security applications. The promised ability of social robots to interact with humans in an emotionally aware and engaging manner is the distinguishing feature separating them from other types of robot. This ability entails recognizing human facial expressions and body language and interpreting spoken instructions or conversations using natural language processing (NLP). Through these future abilities, social robots may make situation-appropriate judgments and behaviors. Admittedly, most of this research has occurred in lab settings which do not always replicate in real settings. Social robots have a long way to go before these possibilities are close to occurring. To better interact with people, social robots must be able to learn from experience and adjust their behavior over time in a

https://doi.org/10.1515/9783110792270-018

real environment. This necessitates incorporating artificial intelligence technologies such as deep learning, reinforcement learning, and natural language processing into the robot's architecture. As a result, social robots have evolved through time and been deployed in the real world in some situations. As technology advances and becomes more widely available, social robots will likely become ever more interwoven into daily life. The continuous development of these robots holds tremendous promise for enhancing people's interactions with technology and one another.

Engaging with social robots alters how we think about technology as they play a more significant role in our lives. The field of human–robot interaction (HRI) is a combination of research on artificial intelligence, human psychology, communication science, and robotics to study a new social relationship between people and machines. The emergence of social robots has presented an exciting development by providing opportunities for humans to interact with machines in ways that were once reserved for human–human interaction. Social robots' distinctive cues, behavioral capabilities, and affordances influence their interactions with humans, shaping the nature of social practices and the implications for the human body. This chapter explores the influence of these factors on social interaction and the consequences for well-being. First, we will discuss the social robots' distinctive cues and behavior capabilities currently available.

17.2 Distinctive Cues and Behavioral Capabilities of Social Robots

Just as humans can express themselves via facial expressions, body language, voice, and touch, social robots can use visual, audio, and tactile ways of expression. These ways of expression can be used individually or multimodally (i.e., multiple at one time or in combination) according to context to evoke social responses from the user. Because social robots are expected to engage socially (Edwards et al., 2019; Kwon et al., 2016; Leite et al., 2013), several distinctive cues feed into the expectations for a robot's behavioral capabilities. Cues, from a foundational level, are a form of stimulus that can motivate the user to action (Ghazali et al., 2018). In social robotics, the use of cues is vital for exciting users to perceive the social robot in ways designers and researchers desire. For instance, providing a face on a robot with human-like embodiment cues allows the user to see the machine as affording the capability for communication and expect the social robot to have spoken dialogue features and gesturing abilities.

What sets social robots apart from autonomous industrial ones is the embodiment of a human-relatable form via these ways of expression, which suggest the possibility of human-like communication (Breazeal et al., 2016; Deng et al., 2019; Fox & Gambino, 2021; Guzman, 2020; Spence, 2019). It is via these modes of expression that social robots engage in social cueing, which leads to social experiences with human users. While cues generally can be defined as stimuli prompting users to act, *social cues* pertain to the "biologically and physically determined features salient to observers because of their potential as channels of useful information" (Fiore et al., 2013, p. 2). These cues are the channels through which the 'social' is put in 'social robotics'. With a wide array of features and combinations among possible (mis)interpretations and rejections, social robotics must embody the proper ratio of cues for human-like form for human-centered design. They must do so to allow experiential interactions that support the user according to their hedonic and utilitarian attitudes about the social robot in addition to the tasks they seek to accomplish with it (Banks & De Graaf, 2020; De Graaf et al., 2019).

Contending that human-centered design aids in creating optimal user-experience, we now briefly outline the use of social robots' visual, audio, and tactile cues for social interaction. We also highlight many distinctive cues before identifying how these cues can be used to complement spoken dialogue systems and gestures and movements as behavioral capabilities of social robotics.

17.2.1 Visual cues

Visual cues pertain to the visible expression of the social robot. Visible features of social robots can be expansive, ranging from the use of the face, head, and body shape. Humans are excellent perceivers of faces on objects. Whether on clouds or toast, humans tend to see faces on non-living objects (Windhager et al., 2008. Because humans are so developed in facial recognition, designers often use faces on social robots (Todorov et al., 2015). One crucial feature that allows recognition of a face on an object is the use of eyes (Ichikawa et al., 2011). Eyes can influence perceptions of the robot according to size, shape, color, and symmetry. However, an essential area of work has predominantly focused on the gaze from the eyes. Most social robots do not have eyes that can move, so experimenters focus on changes in the head direction to manipulate gaze as a visual cue.

Using gaze as visual cues for social robots provides feedback to the user concerning the robot's attention and regulating turn taking in conversation (Skantze et al., 2014). Experimental manipulation of gaze ranges from dyadic interaction, where one user and one robot are in interaction (Mutlu et al., 2006), to group interactions (Gillet et al., 2021; Sebo et al., 2020). Mutual eye contact can benefit HRI (e.g., face detection with a Nao robot; Andrist et al., 2014); however, constant eye contact can influence perceived dominance (Stanton & Stevens, 2017). Social robots can be configured to notice objects and locations within the HRI space via referential gaze. This form of gaze has been shown to aid in working through challenging tasks (Admoni et al., 2016) although the strength of this cue may be weakened for older users (Morillo-Mendez et al., 2022) and varies according to context (Pereira et al., 2020). The gaze on a social robot can also be manipulated to bring shared attentional focus on an ordinary object found in the interaction via joint attention (Delaunay et al., 2010). Though joint attention has been found to increase social presence in adults (Pereira et al., 2019), generalizability to users according to age (Pereira et al., 2020) and context (Hoque et al., 2011) can be difficult.

Gaze aversion, deliberately avoiding eye contact, can communicate different meanings in the interaction and bring the robot to life. For example, Andrist and colleagues (2014) found that idle head movement increased how lifelike the robot was perceived to be by allowing the robot to move the head dynamically. Still, when the robot engaged in purposeful gaze aversions (e.g., looking towards and looking away from the participant in conversation purposefully), the social robot was better able to regulate the conversation with the human user (i.e., the length of time the user waited to speak with the robot in conversation). Regarding turn taking, purposeful gaze aversion helps moderate the pace of the conversation in dialogue (Andrist et al., 2014).

17.2.2 Audio cues

Audio cues refer to using sound to convey information to the user. This can range from spoken words to minor and major utterances and beeps from the robot. For designers, social robots need to be configured and designed with a mind to human-likeness of voice concordantly with the lifelike visual characteristics of the robot's figure in addition to relevant contextual factors, such as the user's age (Edwards et al., 2019) and the user's gender (Eyssel et al., 2012). In a teaching context, there has been a positive relationship between the perceived older age of an AI voice teacher and other key classroom outcomes such as higher credibility, social presence, and motivation to learn (Edwards et al., 2019). Regarding gender, Eyssel and colleagues' (2012) study of human-like vs. robot-like synthesized voice found that gender matching (i.e., gender of robot voice matches user's gender) led to greater perceived psychological closeness and acceptance. Matching the gender and having a human-like voice led to greater anthropomorphizing of the robot.

Often the use of audio cues can be embedded within other primary cues of conveying information verbally and non-verbally. For example, Craig and Edwards's (2021) study of an emotionally expressive social robot controlled the degree of sadness across three conditions (sorrowful, no emotion, and happy/elated). The sorrowful/sad robot featured a robot that spoke at a lower register with moderate pauses while speaking slowly. At the same time, its body movements were controlled so that it was looking at the ground. The joyful robot featured a higher register of voice, spoke faster, looked more towards the camera, and was rated more likable, intelligent, credible, and socially present than the sorrowful/sad robot. It is also important to consider both audio and visual cues together, as processing for the user often concerns what they see and hear. For example, seeing 'ga' on the lips and hearing 'ba' can lead to the perception of 'da' (McGurk effect: McGurk & MacDonald, 1976). Users need feedback from the social robot to know whether their behavior for interacting with the machine is correct.

17.2.3 Tactile cues

The use of *tactile cues* refers to the sense of touch. The use of tactile cues can be complex in design. There are technical limitations to how a social robot may accept tactile sensory input. The robot also has social demands in which tactile inputs and outputs should adhere to contextual factors. In human-to-human contexts, various means of tactile communication consider various interactant emotions (e.g., anger, sadness, elation, etc.). For example, social robots interacting with children may need sensory input or output to be indicated for interactions such as hugging or tickling a robot (François et al., 2009). It would be important to decide what qualifies as a hug or tickling and what is the social response for the robot to enact. Some examples of tactile cues include manipulating the robot skin to show goosebumps or spikes using silicon (Hu et al., 2018), and combining haptic and audio cues to reduce anxiety (Matheus et al., 2022). These all have relative trade-offs that bring the designer's attention to context and the expected user type. Nevertheless, most importantly it articulates the diverse contexts in which tactile cues can convey meaning and the difficulty of matching impact with intent.

17.2.4 Speech and dialogue recognition

If a social robot is expected to speak, it is probably also expected to listen. Speech and dialogue recognition for social robots require recognizing the user's spoken words, determining the topic and meaning in the conversation model, and then determining the likely model of dialogue to engage in for an appropriate conversation with the user. This is often achieved via the use of NLP. NLP is commonly used to process human language and determine the most appropriate meaning and a subsequent response to what was understood. Programming a social robot to recognize and then respond to the user's spoken words can be difficult without a general understanding of NLP, Deep Neural Networks, and the ideas surrounding using hidden Markov models. Because of this, our focus is on the general use of speech and dialogue recognition.

For now, speech and dialogue systems rely on language models that give the machine sufficient knowledge about the possible topics to be discussed (Gunkel, 2016). A 2009 study by Lee and Makatchev found that, of the topics the users in their study would talk about with a robot, a majority indicated they would seek information (e.g., weather, and date and time), followed by chatting (mainly talking about the robot vs. talking about the user), saying hello, and nonsense words and insults. Though these dialogue options may be limited in scope, allowing the machine to make a mistake can help improve HRI. In one study, participants liked a faulty robot better than a robot that acted flawlessly (Mirnig et al., 2017). A social robot's behavior is vital regarding its congruence between the behavior and mental models for the social robot's function. That is, non-verbal behaviors affect anthropomorphic perceptions and mental models formed by the user in interaction, and these should be congruent with one another for more accessible communication (Salem et al., 2013). Much of the use of sensors for recognizing speech and dialogue requires clear articulation by the user (Alonso-Martín & Salichs, 2011). Because of both requirements, speech and dialogue systems should model congruently with the visual gestures and movements of the social robot.

17.2.5 Movement and gesturing ability

Movement and gesturing for social robots mean affording and communicating via physical movement and gestures. Humans in dialogue with one another engage in verbal and non-verbal communication. The use of movement and gesturing by social robotics is quite helpful for interaction. Just as human-to-human conversation uses verbal and non-verbal communication to convey meaning, so social robots can combine verbal and non-verbal communication to enhance the HRI. Though verbal commands and spoken dialogue can convey some meaning, meaning is more significantly derived from non-verbal communication (Birdwhistell, 1955; Mehrabian & Ferris, 1967; Mehrabian & Wiener, 1967; Philpott, 1983). Hence, the use of movement and gesturing is useful for positive interactions.

Concerning movement, a social robot can use speed to show confidence or convey the its intent to the user. For example, Sarder and colleagues' (2012) study of manipulating a robot's speed found that participants trusted robots that moved faster than humans. This communicates confidence and pressure to abide by the robot's movement. In a more recent study of normative approach behaviors of social robots, Joosse et al. (2021) identified that regarding sound, size, and speed, users were okay with violations of a social norm done by robots compared with humans, and perceived violations can be mitigated via communicating the robot's intent.

Gesturing can help illustrate turn taking in conversation and enhance storytelling by social robots. The combination of audio and visual gestures, for example, can engage users more with the context-specific alignment of the gestures (Lombard & Xu, 2021; Striepe et al., 2021). As identified here, using voice and gestures can aid in communicating meaning in the HRI. Combining spoken dialogue, gestures, and movements stemming from distinctive cues can allow for a greater presence for interaction. Psychologically, as motives for interactions, these cues can lead the user to engage in affordances.

Overall, the variety of cues and behavioral capabilities of social robots serve as modes of expression leading to desired experiences with human users. Table 17.1 gives a list of these cues and capabilities. Specifically, a social robot's visual, audio, and tactile cues can be modified to serve as stimuli—triggering action—from the user in the HRI. The use of speech and dialogue recognition in combination with movement and gesturing ability brings social robots to life in supporting human–machine communication. Though cues serve as an important part of the HRI when considering a social robot's design, the combination of stimuli can motivate users to perceive affordances of the robot. These affordances, as we will cover in the next section, truly invite users to imagine a social robot's role in the HRI and, more importantly, the experience of communication with a machine entity.

Table 17.1: Distinctive cues and behavioral capabilities of social robots.

Visual cues	Visible expression of the social robot	Ichikawa et al., 2011; Todorov et al., 2015
Audio cues	Using sound to convey information to the user	Edwards et al., 2019; Eyssel et al., 2012
Tactile cues	Sense of touch	François et al., 2009; Hu et al., 2018; Matheus et al., 2022
Speech and dialogue recognition	Recognizing the user's words and then determining the likely model of dialogue to engage in for an appropriate conversation with the user	Alonso-Martín & Salichs, 2011; Salem et al., 2013
Movement and gesturing ability	Communicating via physical movement and gestures	Joose et al., 2021; Sarder et al., 2012

17.3 Affordances in the Interaction Process with Social Robots

A robot's distinctive cues and behavioral capabilities can provide the user with openness for something to occur. Combining multiple cues can often lead users to perceive a social robot to 'give off' affordances. *Affordances* refer to the opportunities for action provided by social robots' physical and behavioral features (Rodríguez-Hidalgo, 2020). The affordances of social robots influence the nature of social practices and their implications for the human body. The more human-like a machine is (i.e., the more cues that, when combined, give off affordances for human communication), the more people are likely to treat the machine with human-like responses in their experience of communication in the HRI (Edwards et al., 2019). As a result, we often attribute human characteristics to the machine and even express human reactions in interactions with a machine (Westerman et al., 2020). The attribution of humanness or human responses is ripe with potential and impediment. As robots and other machine communicators become more common in our everyday lives and environment, we must consider the outcomes associated with the level of anthropomorphism given or personification attributed to a machine. Given the impact of anthropomorphism in influencing human interactions with a machine, we need to think about these directions based on context, role, and task (De Visser et al., 2016; Spence et al., 2019a) and consider the possibility of where social robotic interaction can improve users' well-being.

Affordances are not required to be physical to the machine (although this is often what is considered); rather, they can also be associated with the programming and ability of the machine to carry out a task. A series of studies have outlined the positive impact of helpfulness compared with identifying a communicator (see Lin et al., 2019, 2022; Spates et al., 2020; Spence et al., 2023). The results of these studies found that helpfulness was a consistent predictor of the outcome variables of interest, whereas cues of identity were not. Specifically, this line of research has found that helpfulness is a predictor of the credibility of the communicator, the credibility of the digital medium where communication took place, the perceived task social attraction, and in some instances, perceptions of risk. Therefore, such positive reactions should also be present for a helpful machine. Some research has shown that people can engage in politeness with machines (Fortunati et al., 2022), such as thanking voice-based agents (e.g., Amazon's Alexa) or thanking the autonomous pizza delivery bot (Metz, 2018). Several research areas emerge from findings of helpfulness between humans, creating potential studies and applications to machines.

17.4 Helping Behaviors and Anthropomorphizing Social Robots

There is a natural uncertainty with machines entering our lives and workplaces (Spence et al., 2018). This anxiety is often associated with general anxiety about robots or the fear of losing employment. However, if social robots can be seen as helpful in how they are introduced and observed, the connotation and association with machines in our lives and workplaces may become more positive. When a machine can be shown to extend abilities, prevent injury, and increase safety, this measure of helpfulness may, as it does in human communication, impact perceptions which can lead to changes in well-being. Sauppé and Mutlu (2015) write, "[d]rawing on our findings, we recommended that future designs augment the social capabilities of collaborative robots, specifically to support the coordination necessary to perform manufacturing work and to enrich the social environment in the workplace" (p. 3662).

There is mixed literature on this topic in early studies on automation; however, the combination of helpfulness and anthropomorphism has not been widely studied in HRI. As discussed earlier, anthropomorphic elements may provide a favorable combination of how human-like a machine is and the helpfulness provided. For example, a service robot in a hotel or restaurant may be designed with several human features (e.g., facial features, the robot's size, human-like voice, and arms). Research has shown that the more similar a machine is to the target, the more likely it is to elicit reactions in the desired direction (You & Robert Jr., 2018). If a robot brings the wrong food to a dining table or forgets to bring towels to a room, people may find the machine's novelty amusing and have a reasonable response to the mistake (Packard et al., 2019).

For example, in South Korea DustCart and Piero robots were used for various functions of helpfulness at a festival. Their roles included garbage collection for Dustcart, and Piero not only made announcements to festival goers but also highlighted DustCart and its service role. Each of these bots had different helpful functions and physical designs. However, it was noted that DustCart received abuse from people through kicks and other behaviors. Although this was noted as stemming from the novelty of the robots, there is room for research on the relationship between the design of the robot and helpfulness (Salvini et al., 2010; Spence et al., 2019b). As humans, those in service roles that provide the most helpfulness are often perceived as having lower status. Therefore, we can learn much more about what it means to be human by studying these types of interactions.

The idea of expanding helpfulness to enhance overall well-being is an important step in HRI. Social robots could offer practical assistance but also foster deep connections with users, improving emotional and mental states through their helpfulness. This transformation highlights the potential for technology to impact the emotional fabric of our lives and uplift well-being. Previous research has demonstrated the applicability of social robots as clinical interventions—however, more research needs to be done to evaluate the generalizability of the interventions in addition to their impact on psychological outcomes, including but not limited to the users' well-being (see Scoglio et al., 2019; see also Loi et al., 2018; Šabanović et al., 2013). To better expand this research, the helpfulness of social robots and their impact on well-being for users should be addressed. In the following sections, we highlight some of the research on well-being and helpfulness with social robots.

17.5 Implications for Well-being When Interacting with Social Robots

Social robots' distinctive cues, behavioral capabilities, and affordances can significantly affect humans. Through the mechanism of 'mirroring,' social robots can significantly influence non-verbal behavior (Saunderson & Nejat, 2019). It is possible that when individuals engage with social robots that are built to react to non-verbal cues from them, they unintentionally mimic the robot's non-verbal behavior (Hasumoto et al., 2020). As an illustration, if a social robot nods in agreement while someone talks, the other person will likely follow suit. Over time, a person's communication style may evolve into new non-verbal actions because of this. It can foster rapport and connection with the robot but also prompts ethical concerns about how closely robot behavior 'mirrors' that of humans in ways that can be construed as exploitative. Because social robots are designed to interact with humans, there are implications for emotional and physical well-being and social behavior, especially when examining potential helping behaviors and the anthropomorphizing of social robots.

17.5.1 Emotional and physical well-being

Social robots can enhance a person's emotional and physical well-being. One benefit of social robots is their ability to provide emotional support to individuals experiencing loneliness, anxiety, or depression (Chen et al., 2018; Lee et al., 2017). Social isolation and loneliness can negatively affect mental health, including an increased risk of depression and anxiety. Social robots have distinctive cues such as soothing sounds, lights, and other calming elements to promote relaxation and reduce stress (Abendschein et al., 2021; Edwards et al., 2020; Ling & Bjorling, 2020). For individuals who experience anxiety or stress regularly, these features can be a helpful tool in managing their emotions and promoting emotional well-being. Furthermore, social robots can also impact emotional well-being by providing a sense of control and autonomy to individuals experiencing a loss of control. For individuals with disabilities or mobility issues, social robots can provide both emotional and physical assistance and allow them to maintain a sense of independence and control over their daily lives (Bedaf et al., 2015). This can help to reduce feelings of helplessness and promote emotional well-being.

Social robots can impact a person's physical health by promoting physical activity. Social robots can encourage people to engage in physical activity by providing motivation, support, and feedback (Robinson et al., 2021). For example, a social robot could be programmed to lead an exercise routine, offer encouragement, and monitor progress (Cao et al., 2019). Social robots can promote healthy eating habits (Robinson et al., 2020). Poor diet is a major contributor to chronic diseases such as obesity, heart disease, and diabetes. Social robots can educate people on healthy eating habits, monitor their food intake, and provide feedback on their dietary choices. For instance, a social robot could be programmed to offer healthy meal suggestions, remind people to eat regularly, and track their food intake. Moreover, social robots can also be used in rehabilitation and physical therapy (Langer & Levy-Tzedek, 2021). For instance, social robots can assist people with mobility impairments in exercises designed to improve their range of motion and muscle strength. Social robots can also provide feedback on correct posture and movement during physical therapy sessions. In short, social robots can help improve the physical aspects of a person's life. Social robots can provide emotional and physical support, a sense of purpose and meaning, and a sense of control and autonomy to individuals experiencing loneliness, stress, or a loss of movement. By balancing the potential benefits and risks, social robots can promote health safely and responsibly.

While social robots can offer many benefits to emotional and physical well-being, it is also important to consider the potential risks. While not explicitly tested, some individuals may become overly reliant on social robots for support, leading to decreased face-to-face social interaction and an erosion of social skills. It is vital to ensure that social robots are used as a supplement to human interaction rather than a replacement for it. Another potential risk of social robots is their potential to reinforce negative stereotypes and biases. Because robots are designed to interact with humans socially, they inadvertently perpetuate negative societal stereotypes and biases (Howard & Borenstein, 2019; Wang et al., 2021). It is important to ensure that social robots are designed and programmed in a way that promotes diversity and inclusion and does not reinforce negative stereotypes or biases. Social robots have the potential to impact our well-being in a variety of ways.

17.5.2 Social behaviors

Social skills, such as communicating effectively with others, interpreting social cues, and responding appropriately in social situations, are essential for effective communication and building relationships (Edwards et al., 2022a). Fortunately, social robots can provide a safe and controlled environment for people to practice and improve their social skills. With cameras, sensors, and other technology, social robots can recognize and react to non-verbal cues. By acting as 'trainers,' social robots can also influence human behaviors. Social robots that can give feedback on non-verbal conduct can aid communication skill development. For instance, a robot may be built to urge individuals to make eye contact or utilize more expressive movements. Social robots can assist people in changing their non-verbal actions to ones that are more engaging and successful by offering constant feedback customized to the individual.

One group that can benefit significantly from social robots is children with autism spectrum disorder (ASD) (Tennyson et al., 2016; Valadão et al., 2016). ASD is a developmental disorder that affects communication and social interaction skills. Children with ASD may have difficulty with eye contact, turn taking, and recognizing emotions in others. Social robots can provide a structured and predictable environment for social interaction, which can be particularly helpful for children with ASD (Leoste et al., 2022). Social robots can also provide visual cues, such as facial expressions, to help children with ASD recognize and interpret emotions in others (Syriopoulou-Delli & Gkiolnta, 2022).

Social robots can also benefit people with social anxiety. Social anxiety is a common condition that can make interacting with others in social situations difficult. People with social anxiety may feel nervous or uncomfortable in social situations, making it challenging to communicate effectively with others. Social robots can provide a safe and low-stress environment for people with social anxiety to practice and improve their social skills (Rasouli et al., 2022). Nomura et al. (2020) demonstrated that people with higher social anxiety reported less 'anticipatory anxiety' when they knew they would be interacting with a social robot than if they were interacting with another person. In a hospital setting, research has shown that children can emotionally engage with a physical, social robot (such as a huggable bear) more than a virtual character. In doing so, the social robot provides more socio-emotional support in a pediatric setting.

In addition to helping children with ASD and people with social anxiety, social robots can also benefit people in other areas of life. For example, social robots can help people with speech or language difficulties improve their communication skills (Belpaeme et al., 2018). Social robots can provide a non-judgmental and supportive environment for people to practice and improve their speech and language skills (Konijn et al., 2022). In conclusion, social robots can be a powerful tool for enhancing social skills. Social robots can provide a safe and controlled environment for people to practice and improve their social skills, which can be particularly helpful for children with ASD and people with social anxiety. Social robots can also teach social skills to people with speech or language difficulties, intellectual disabilities, or learning difficulties. While social robots are still a relatively new technology, they can revolutionize how we teach social skills and help people improve their communication and social interaction skills.

17.6 Conclusion

Interacting with social robots could impact how people behave in social situations and directly affect us. Social robots' particular cues and behavioral capacities, such as visual, auditory, and tactile cues, greatly influence these interactions. Social robots can create shared attention with users, transmit attention, and control turn taking using visual signals like gaze and facial expressions. The user's engagement and interaction with the robot can be improved by audio cues like voice traits and tone, which help the user form an impression of the robot's personality. Tactile signals may transmit meaning and assist physical interactions between the robot and the human while being more intricate in design. Furthermore, effective communication requires social robots to recognize and respond to spoken instructions or conversations using speech and dialogue recognition. Social robots' ability to move and gesture enables them to improve communication by fusing verbal and non-verbal behaviors.

Social robots' anthropomorphism and helpfulness have the potential to alter human perceptions of and relationships with technology for the better. A robot's degree of anthropomorphism might affect how people interact with it; more anthropomorphic traits may encourage rapport and trust. However, more research is needed to fully understand the connection between anthropomorphism, helpfulness, well-being, and high-stakes situations. The design of the robot, particularly its helpful features and outward look, can also affect how humans engage with it, as shown by instances when robots have been abused because they are so innovative. By analyzing these encounters, we may better understand what it means to be human and how to successfully include robots in our lives and jobs.

However, what happens when the stakes are high? The level of humanness and anthropomorphism could be seen in two ways. First, a person may be more likely to respond to a robot high in anthropomorphism because it could gain rapport or trust (Lee & See, 2004; Van Pinxteren et al., 2019). This would be a positive outcome to the interaction and situation. Second, it is in the combination of anthropomorphism and helpfulness that less is understood. Would witnessing a social robot with a high level of humanness catastrophically fail cause different perceptions of the robot from observing a more machine-like robot fail at the same task? Taking this question further, what happens when the robot is destroyed in a risky situation? Do people then feel that the situation was too dangerous to send in a robot if that robot was human-like? Research needs to address these questions to understand how the affordances of the distinctive cues and behaviors influence HRI.

The implications for human health when interacting with social robots are significant regarding emotional and physical well-being and social behaviors. Social robots can positively influence people's emotional well-being by offering emotional support, encouraging relaxation, lowering tension, and encouraging a sense of control and autonomy. They can also improve physical well-being by promoting good dietary practices and physical activity and aiding in recovery and physical treatment. In addition, social robots can provide a lower-stress setting for persons who struggle with social anxiety to hone their social abilities. Nevertheless, it is critical to balance the advantages and risks. While social robots have shown great promise in enhancing emotional and physical well-being and improving social behaviors, using them to supplement human interaction rather than replace it is essential.

The implications of these distinctive cues, behavioral capabilities, and affordances of social robots are significant. They can transform HRI and create new social relationships between people and machines. Social robots can potentially improve how people interact with technology and one another. However, when creating and utilizing social robots, it is essential to consider human-centered design principles. For social robots to provide the best user experiences, it is crucial to comprehend the many settings in which they are utilized and to modify their cues and actions accordingly.

Bibliography

Abendschein, B., Edwards, C., & Edwards, A. (2021). The influence of agent and message type on perceptions of social support in human–machine communication. *Communication Research Reports, 38*(5), 304–314. https://doi.org/10.1080/08824096.2021.1966405.

Admoni, H., Weng, T., Hayes, B., & Scassellati, B. (2016, March). Robot nonverbal behavior improves task performance in difficult collaborations. In 2016 *11th ACM/IEEE International Conference on Human–Robot Interaction (HRI)* (pp. 51–58). IEEE.

Alonso-Martín, F., & Salichs, M. A. (2011). Integration of a voice recognition system in a social robot. *Cybernetics and Systems: An International Journal, 42*(4), 215–245.

Andrist, S., Tan, X. Z., Gleicher, M., & Mutlu, B. (2014, March). Conversational gaze aversion for humanlike robots. In *Proceedings of the 2014 ACM/IEEE international conference on Human-robot interaction* (pp. 25–32). https://doi.org/10.1145/2559636.2559666

Banks, J., & De Graaf, M. M. (2020). Toward an agent-agnostic transmission model Synthesizing anthropocentric and technocentric paradigms in communication. *Human–Machine Communication, 1*, 19–36. :

Bedaf, S., Gelderblom, G. J., & De Witte, L. (2015). Overview and categorization of robots supporting independent living of elderly people: What activities do they support and how far have they developed. *Assistive Technology, 27*(2), 88–100. https://doi.org/10.1080/10400435.2014.978916.

Belpaeme, T., Vogt, P., Van den Berghe, R., Bergmann, K., Göksun, T., De Haas, M., . . . & Pandey, A. K. (2018). Guidelines for designing social robots as second language tutors. *International Journal of Social Robotics, 10*, 325–341. https://doi.org/10.1007/s12369-018-0467-6.

Birdwhistell, R. L. (1955). Background to kinesics. *ETC: A Review of General Semantics, 13*(1),10–18.

Breazeal, C., Dautenhahn, K., & Kanda, T. (2016). Social robotics. *Springer handbook of robotics*, 1935–1972. Springer.

Burgoon, J. K. (1994). Nonverbal signals. In M. L. Knapp & G. R. Miller (Eds.), *Handbook of Interpersonal Communication* (2nd ed., pp. 229–285). Sage.

Cao, W., Song, W., Li, X., Zheng, S., Zhang, G., Wu, Y., . . . & Chen, J. (2019). Interaction with social robots: Improving gaze toward face but not necessarily joint attention in children with autism spectrum disorder. *Frontiers in Psychology, 10*, 1503.https://doi.org/10.3389/fpsyg.2019.01503

Chen, S. C., Jones, C., & Moyle, W. (2018). Social robots for depression in older adults: A systematic review. *Journal of Nursing Scholarship, 50*(6), 612–622. https://doi.org/10.1111/jnu.12423.

Craig, M. J., & Edwards, C. (2021). Feeling for our robot overlords: Perceptions of emotionally expressive social robots in initial interactions. *Communication Studies, 72*(2), 251–265.https://doi.org/10.1080/10510974.2021.1880457

Delaunay, F., De Greeff, J., & Belpaeme, T. (2010, March). A study of a retro-projected robotic face and its effectiveness for gaze reading by humans. In *2010 5th ACM/IEEE International Conference on Human-Robot Interaction (HRI)* (pp. 39–44). IEEE.https://doi.org/10.1109/HRI.2010.5453271.

Deng, E., Mutlu, B., & Mataric, M. J. (2019). Embodiment in socially interactive robots. *Foundations and Trends® in Robotics, 7*(4), 251–356. http://dx.doi.org/10.1561/2300000056

De Graaf, M. M., Ben Allouch, S., & Van Dijk, J. A. (2019). Why would I use this in my home? A model of domestic social robot acceptance. *Human–Computer Interaction, 34*(2), 115–173.

De Visser, E. J., Monfort, S. S., McKendrick, R., Smith, M. A., McKnight, P. E., Krueger, F., & Parasuraman, R. (2016). Almost human: Anthropomorphism increases trust resilience in cognitive agents. *Journal of Experimental Psychology: Applied, 22*(3), 331. https://doi.org/10.1037/xap0000092

Edwards, A., Edwards, C., Abendschein, B., Espinosa, J., Scherger, J., & Vander Meer, P. (2022a). Using robot animal companions in the academic library to mitigate student stress. *Library Hi Tech, 40*(4), 878–893. https://doi.org/10.1108/LHT-07-2020-0148.

Edwards, A., Edwards, C., Westerman, D., & Spence, P. R. (2019). Initial expectations interactions, and beyond with social robots. *Computers in Human Behavior*, *90*, 308–314.

Edwards, A., Gambino, A., & Edwards, C. (2022b). Factors of attraction in human–machine Communication. *Publizistik*, *67*, 517–529. https://doi.org/10.1007/s11616-022-00756-6.

Eyssel, F., Kuchenbrandt, D., Bobinger, S., De Ruiter, L., & Hegel, F. (2012, March). "If you sound like me, you must be more human": On the interplay of robot and user features on human–robot acceptance and anthropomorphism. In *Proceedings of the Seventh Annual ACM/IEEE International Conference on Human–Robot Interaction* (pp. 125–126). IEEE.

Ferstl, Y., Kokkinara, E., & Mcdonnell, R. (2017). Facial features of non-player creatures can influence moral decisions in video games. *ACM Transactions on Applied Perception (TAP)*, *15*(1), 1–12.

Fiore, S. M., Wiltshire, T. J., Lobato, E. J., Jentsch, F. G., Huang, W. H., & Axelrod, B. (2013).Toward understanding social cues and signals in human–robot interaction: Effects of robot gaze and proxemic behavior. *Frontiers in Psychology*, *4*, 859.

Fortunati, L., Edwards, A., Manganelli, A. M., Edwards, C., & de Luca, F. (2022). Do people perceive Alexa as gendered?: A cross-cultural study of people's perceptions, expectations, and desires of Alexa. *Human-Machine Communication*, *5*, 75–97.https://doi.org/10.30658/hmc.5.3

Fox, J., & Gambino, A. (2021). Relationship development with humanoid social robots: Applying interpersonal theories to human–robot interaction. *Cyberpsychology, Behavior, and Social Networking*, *24*(5), 294–299. https://doi.org/10.1089/cyber.2020.018

François, D., Powell, S., & Dautenhahn, K. (2009). A long-term study of children with autism playing with a robotic pet: Taking inspirations from non-directive play therapy to encourage children's proactivity and initiative-taking. *Interaction Studies*, *10*(3), 324–373. https://doi.org/10.1075/is.10.3.04fra

Ghazali, A. S., Ham, J., Barakova, E., & Markopoulos, P. (2018). The influence of social cues in persuasive social robots on psychological reactance and compliance. *Computers in Human Behavior*, *87*, 58–65. https://doi.org/10.1016/j.chb.2018.05.016.

Gillet, S., Cumbal, R., Pereira, A., Lopes, J., Engwall, O., & Leite, I. (2021, March). Robot gaze can mediate participation imbalance in groups with different skill levels. In *Proceedings of the 2021 ACM/IEEE International Conference on Human-Robot Interaction* (pp. 303–311). https://doi.org/10.1145/3434073.3444670

Goetz, J., Kiesler, S., & Powers, A. (2003, November). Matching robot appearance and behavior to tasks to improve human–robot cooperation. In *12th IEEE International Workshop on Robot and Human Interactive Communication, 2003. Proceedings. ROMAN.* (pp. 55–60). IEEE.

Gong, L., & Nass, C. (2007). When a talking-face computer agent is half-human and half-humanoid: Human identity and consistency preference. *Human Communication Research*, *33*(2), 163–193.

Gunkel, D. J. (2016). Computational interpersonal communication: Communication studies and spoken dialogue systems. *Communication+ 1*, *5*(1), 1–20.

Guzman, A. L. (2020). Ontological boundaries between humans and computers and the implications for human-machine communication. *Human-Machine Communication*, *1*, 37–54. https://doi.org/10.30658/hmc.1.3

Hasumoto, R., Nakadai, K., & Imai, M. (2020). Reactive chameleon: A method to mimic conversation partner's body sway for a robot. *International Journal of Social Robotics*, *12*, 239–258. https://doi.org/10.1007/s12369-019-00557-4.

Hoque, M. M., Onuki, T., Tsuburaya, E., Kobayashi, Y., Kuno, Y., Sato, T., & Kodama, S.(2011). An empirical framework to control human attention by robot. In *Computer Vision–ACCV 2010 Workshops: ACCV 2010 International Workshops, Queenstown, New Zealand, November 8–9, 2010, Revised Selected Papers, Part I 10* (pp. 430–439). Springer Berlin Heidelberg.

Howard, A., & Borenstein, J. (2019). Trust and bias in robots: These elements of artificial intelligence present ethical challenges, which scientists are trying to solve. *American Scientist*, *107*(2), 86–90.

Hu, Y., Zhao, Z., Vimal, A., & Hoffman, G. (2018, April). Soft skin texture modulation for social robotics. In *2018 IEEE International Conference on Soft Robotics (Robosoft)* (pp. 182–187). IEEE.

Ichikawa, H., Kanazawa, S., & Yamaguchi, M. K. (2011). The movement of internal facial features elicits 7 to 8-month-old infants' preference for face patterns. *Infant and Child Development,20*(5), 464–474. https://doi.org/10.1002/icd.724

Jeong, S., Logan, D. E., Goodwin, M. S., Graca, S., O'Connell, B., Goodenough, H., . . . & Weinstock, P. (2015, March). A social robot to mitigate stress, anxiety, and pain in hospital pediatric care. In *Proceedings of the Tenth Annual ACM/IEEE International Conference on Human-Robot Interaction Extended Abstracts* (pp. 103–104). https://doi.org/10.1145/2701973.2702028.

Joosse, M., Lohse, M., Berkel, N. V., Sardar, A., & Evers, V. (2021). Making appearances: How robots should approach people.*ACM Transactions on Human-Robot Interaction (THRI)*, *10*(1), 1–24. https://doi.org/10.1145/3385121

Konijn, E. A., Jansen, B., Mondaca Bustos, V., Hobbelink, V. L., & Preciado Vanegas, D. (2022). Social robots for (second) language learning in (migrant) primary school children. *International Journal of Social Robotics*, 1–17. https://doi.org/10.1007/s12369-021-00824-3.

Kwon, M., Jung, M. F., & Knepper, R. A. (2016, March). Human expectations of social robots. In *2016 11th ACM/IEEE International Conference on Human-Robot Interaction (HRI)* (pp. 463–464). IEEE.

Langer, A., & Levy-Tzedek, S. (2021). Emerging roles for social robots in rehabilitation: Current directions. *ACM Transactions on Human-Robot Interaction*, *10*(4), 1–4. https://doi.org/10.1145/3462256.

Lee, H. R., Šabanović, S., Chang, W. L., Nagata, S., Piatt, J., Bennett, C., & Hakken, D. (2017, March). Steps toward participatory design of social robots: Mutual learning with older adults with depression. In *Proceedings of the 2017 ACM/IEEE International Conference On Human-Robot Interaction* (pp. 244–253). https://doi.org/10.1145/2909824.3020237.

Lee, J. D., & See, K. A. (2004). Trust in automation: Designing for appropriate reliance. *Human Factors: Journal of the Human Factors and Ergonomics Society*, *46*(1), 50–80. https://doi.org/10.1518/hfes.46.1.50_30392.

Lee, M. K., & Makatchev, M. (2009). How do people talk with a robot? An analysis of human–robot dialogues in the real world. In *CHI'09 Extended Abstracts on Human Factors in Computing Systems* (pp. 3769–3774). https://doi.org/10.1145/1520340.1520569

Leite, I., Martinho, C., & Paiva, A. (2013). Social robots for long-term interaction: A survey. *International Journal of Social Robotics*, *5*, 291–308.

Leoste, J., Tammemäe, T., Eskla, G., San Martín López, J., Pastor, L., & Blasco, E. P. (2022). Bee-Bot educational robot as a means of developing social skills among children with autism-spectrum disorders. In *Robotics in Education: RiE 2021 12* (pp. 14–25). Springer International Publishing. https://doi.org/10.1007/978-3-030-82544-7_2.

Lin, X., Kaufmann, R., Spates, S. A., Lachlan, K. A., & Spence, P. R. (2022). Exploring students' perceptions of identity and helper heuristics in the online classroom discussion board. *Communication Education*, *71*, 108–124. https://doi.org/10.1080/03634523.2021.1957138.

Lin, X., Kauffman, R., Spence, P. R., & Lachlan, K. A. (2019). Agency cues in online comments: Exploring their relationship with anonymity and frequency of helpful posts. *Southern Communication Journal*, *84*, 183–195. https://doi.org/10.1080/1041794X.2019.1584828.

Ling, H. Y., & Bjorling, E. A. (2020). Sharing stress with a robot: What would a robot say? *Human–Machine Communication*, *1*, 133–159.

Loi, S. M., Bennett, A., Pearce, M., Nguyen, K., Lautenschlager, N. T., Khosla, R., & Velakoulis, D. (2018). A pilot study exploring staff acceptability of a socially assistive robot in a residential care facility that accommodates people under 65 years old. *International Psychogeriatrics*, *30*(7). https://doi.org/10.1017/S1041610217002617.

Lombard, M., & Xu, K. (2021). Social responses to media technologies in the 21st century: The media are social actors paradigm. *Human–Machine Communication*, *2*, 29–55.

Matheus, K., Vázquez, M., & Scassellati, B. (2022, August). A social robot for anxiety reduction via deep breathing. *In 2022 31st IEEE International Conference on Robot and Human Interactive Communication (RO-MAN)* (pp. 89–94). IEEE.

McGurk, H., & MacDonald, J. (1976). Hearing lips and seeing voices. *Nature*, *264*(5588), 746–748. https://doi.org/10.1038/264746a0

Mehrabian, A., & Ferris, S. R. (1967). Inference of attitudes from nonverbal communication in two channels. *Journal of Consulting Psychology*, *31*(3), 248.

Mehrabian, A., & Wiener, M. (1967). Decoding of inconsistent communications. *Journal of Personal Social Psychology*, *6*, 109–114.

Metz, R. (2018). Apparently, people say "Thank You" to self-driving pizza delivery vehicles and other lessons from Ford's experiment with autonomous cars. *MIT Review*. https://web.archive.org/web/20221002074705/https://www.technologyreview.com/2018/01/10/146293/apparently-people-say-thank-you-to-self-driving-pizza-delivery-vehicles/.

Mirnig, N., Stollenberger, G., Miksch, M., Stadler, S., Giuliani, M., & Tscheligi, M. (2017). To err is robot: How humans assess and act toward an erroneous social robot. *Frontiers in Robotics and AI*, 21.

Mori, M., MacDorman, K. F., & Kageki, N. (2012). The uncanny valley [from the field]. *IEEE Robotics & Automation Magazine*, *19*(2), 98–100.

Morillo-Mendez, L., Schrooten, M. G., Loutfi, A., & Mozos, O. M. (2022). Age-related differences in the perception of robotic referential gaze in human–robot interaction. *International Journal of Social Robotics*, 1–13.

Mutlu, B., Forlizzi, J., & Hodgins, J. (2006, December). A storytelling robot: Modeling and evaluation of human-like gaze behavior. In *2006 6th IEEE-RAS International Conference on Humanoid Robots* (pp. 518–523). IEEE. https://doi.org/10.1109/ICHR.2006.321322

Nomura, T., Kanda, T., Suzuki, T., & Yamada, S. (2020). Do people with social anxiety feel anxious about interacting with a robot? *AI & Society*, *35*, 381–390. https://doi.org/10.1007/s00146-019-00889-9.

Packard, C., Boelk, T., Andres, J., Edwards, C., Edwards, A., & Spence, P. R. (2019, March). The pratfall effect and interpersonal impressions of a robot that forgets and apologizes. In *2019 14th ACM/IEEE International Conference on Human-Robot Interaction (HRI)* (pp. 524–525). IEEE. https://doi.org/10.1109/HRI.2019.8673101

Pereira, A., Oertel, C., Fermoselle, L., Mendelson, J., & Gustafson, J. (2019, November). Responsive joint attention in human-robot interaction. In *2019 IEEE/RSJ International Conference on Intelligent Robots and Systems (IROS)* (pp. 1080–1087). IEEE. https://doi.org/10.1109/IROS40897.2019.8968130

Pereira, A., Oertel, C., Fermoselle, L., Mendelson, J., & Gustafson, J. (2020, March). Effects of different interaction contexts when evaluating gaze models in HRI. In *Proceedings of the 2020 ACM/IEEE International Conference on Human–Robot Interaction* (pp. 131–139). IEEE.

Philpott, J. S. (1983). The relative contribution to meaning of verbal and nonverbal channels of communication: A meta-analysis. Unpublished Master's thesis, University of Nebraska, Lincoln.

Rasouli, S., Gupta, G., Nilsen, E., & Dautenhahn, K. (2022). Potential applications of social robots in robot-assisted interventions for social anxiety. *International Journal of Social Robotics*, *14*(5), 1–32.

Robinson, N. L., Connolly, J., Hides, L., & Kavanagh, D. J. (2020). Social robots as treatment agents: Pilot randomized controlled trial to deliver a behavior change intervention. *Internet Interventions*, *21*, 100320. https://doi.org/10.1016/j.invent.2020.100320

Robinson, N. L., Connolly, J., Suddery, G., Turner, M., & Kavanagh, D. J. (2021, August). A humanoid social robot to provide personalized feedback for health promotion in diet, physical activity, alcohol and cigarette use: A health clinic trial. In *2021 30th IEEE International Conference on Robot & Human Interactive Communication (RO-MAN)* (pp. 720–726). IEEE. https://doi.org/10.1109/RO-MAN50785.2021.9515558.

Rodríguez-Hidalgo, C. (2020). Me and my robot smiled at one another: The process of socially enacted communicative affordance in Human–Machine Communication. *Human–Machine Communication, 1*, 55–69. https://doi.org/10.30658/hmc.1.4.

Šabanović, S., Bennett, C. C., Chang, L., & Huber, L. (2013). PARO robot affects diverse interaction modalities in group sensory therapy for older adults with dementia. *International Conference on Rehabilitation Robotics: [Proceedings]*. https://doi.org/10.1109/ICORR.2013.6650427.

Sabinson, E. B., & Green, K. E. (2021, August). How do we feel? User perceptions of a soft robot surface for regulating human emotion in confined living spaces. In *2021 30th IEEE International Conference on Robot & Human Interactive Communication (RO-MAN)* (pp. 1153–1158). IEEE.

Salem, M., Eyssel, F., Rohlfing, K., Kopp, S., & Joublin, F. (2013). To err is human (-like): Effects of robot gesture on perceived anthropomorphism and likability. *International Journal of Social Robotics, 5*, 313–323. https://doi.org/10.1007/s12369-013-0196-9

Salvini, P., Ciaravella, G., Yu, W., Ferri, G., Manzi, A., Mazzolai, B., et al. (2010). How safe are service robots in urban environments? Bullying a robot. In *RO-MAN 2010: The 19th IEEE International Symposium on Robot and Human Interactive Communication. ROMAN Viareggio. Piscataway, NJ*: IEEE. 1–7. https://doi.org10.1109/ROMAN.2010.5654677.

Sardar, A., Joosse, M., Weiss, A., & Evers, V. (2012, March). Don't stand so close to me: Users attitudinal and behavioral responses to personal space invasion by robots. In *Proceedings of the Seventh Annual ACM/IEEE International Conference on Human–Robot Interaction* (pp. 229–230). IEEE.

Saunderson, S., & Nejat, G. (2019). How robots influence humans: A survey of nonverbal communication in social human–robot interaction. *International Journal of Social Robotics, 11*, 575–608. https://doi.org/10.1007/s12369-019-00523-0.

Sauppé, A., & Mutlu, B. (2015, April). The social impact of a robot co-worker in industrial settings. In *Proceedings of the 33rd Annual ACM Conference on Human Factors in Computing Systems* (pp. 3613–3622). https://doi.org/10.1145/2702123.2702181.

Scoglio, A. A., Reilly, E. D., Gorman, J. A., & Drebing, C. E. (2019). Use of social robots in mental health and well-being research: Systematic review. *Journal of Medical Internet Research, 21*(7), e13322.

Sidner, C. L., Lee, C., Kidd, C. D., Lesh, N., & Rich, C. (2005). Explorations in engagement for humans and robots. *Artificial Intelligence, 166*(1–2), 140–164.

Skantze, G., Hjalmarsson, A., & Oertel, C. (2014). Turn-taking, feedback and joint attention in situated human–robot interaction.*Speech Communication,65*, 50–66. https://doi.org/10.1016/j.specom.2014.05.005

Spates, S. A., Kaufmann, R., Lin, X., Lachlan, K. A., & Spence, P. R. (2020). I don't care about who you are, but what you are doing for me? Examining perceptions of helpful comments and identity in user-generated content. *Southern Communication Journal, 85*, 155–165. https://doi.org/10.1080/1041794X.2020.1770319.

Spence, P. R. (2019). Searching for questions, original thoughts, or advancing theory: Human- machine communication. *Computers in Human Behavior, 90*, 285–287. https://doi.org/10.1016/j.chb.2018.09.014

Spence, P. R., Edwards, A., Edwards, C., & Jin, X. (2019a). "The bot predicted rain, grab an Umbrella": Few perceived differences in communication quality of a weather Twitterbot versus professional and amateur meteorologists. *Behaviour & Information Technology, 38*(1), 101–109. https://doi.org/10.1080/0144929X.2018.1514425.

Spence, P. R., Edwards, A., Edwards, C., Nermer, D., & Lachloan, K. (2019b). Rage against the machine: Negative reactions and antisocial interactions with social bots and robots. In E. Downs (Ed.), *Dark Side of Media & Technology: A 21st Century Guide to Media and Technological Literacy* (pp. 262–273). Peter Lang.

Spence, P. R., Kaufmann, R., Lachlan, K. A., Lin, X., & Spates, S. A. (2023). Extending the understanding of online discussions: A replication of online students' perceptions of identity and helper heuristics. *Communication Education, 72*(4), 367–381. https://doi.org/10.1080/03634523.2023.2169315.

Spence, P. R., Westerman, D., & Lin, X. (2018). A robot will take your job: How does that make you feel? Examining perceptions of robots in the workplace. In A. Guzman (Ed.), *Human–Machine Communication: Rethinking Communication, Technology, and Ourselves* (pp. 185–200). Peter Lang.

Stanton, C. J., & Stevens, C. J. (2017). Don't stare at me: the impact of a humanoid robot's gaze upon trust during a cooperative human–robot visual task. *International Journal of Social Robotics*, *9*, 745–753. https://doi.org/10.1007/s12369-017-0422-y

Striepe, H., Donnermann, M., Lein, M., & Lugrin, B. (2021). Modeling and evaluating emotion, contextual head movement and voices for a social robot storyteller. *International Journal of Social Robotics*, *13*, 441–457.

Strohkorb Sebo, S., Dong, L. L., Chang, N., & Scassellati, B. (2020, March). Strategies for the inclusion of human members within human-robot teams. In *Proceedings of the 2020 ACM/IEEE International Conference on Human-Robot Interaction* (pp. 309–317). https://doi.org/10.1145/3319502.3374808

Syriopoulou-Delli, C. K., & Gkiolnta, E. (2022). Review of assistive technology in the training of children with autism spectrum disorders. *International Journal of Developmental Disabilities*, *68*(2), 73–85. https://doi.org/10.1080/20473869.2019.1706333.

Tennyson, M. F., Kuester, D. A., Casteel, J., & Nikolopoulos, C. (2016). Accessible robots for bimproving social skills of individuals with autism. *Journal of Artificial Intelligence and Soft Computing Research*, *6*(4), 267–277. https://doi.org/10.1515/jaiscr-2016-0020.

Todorov, A., Olivola, C. Y., Dotsch, R., & Mende-Siedlecki, P. (2015). Social attributions from faces: Determinants, consequences, accuracy, and functional significance. *Annual Review of Psychology*, *66*, 519–545.

Valadão, C. T., Goulart, C., Rivera, H., Caldeira, E., Bastos Filho, T. F., Frizera-Neto, A., & Carelli, R. (2016) Analysis of the use of a robot to improve social skills in children with autism spectrum disorder. *Research on Biomedical Engineering*, *32*, 161–175. https://doi.org/10.1590/2446-4740.01316.

Van den Berghe, R., Verhagen, J., Oudgenoeg-Paz, O., Van der Ven, S., & Leseman, P. (2019). Social robots for language learning: A review. *Review of Educational Research*, *89*(2), 259–295. https://doi.org/10.3102/0034654318821.

Van Pinxteren, M. M. E., Wetzels, R. W. H., Rüger, J., Pluymaekers, M., & Wetzels, M. (2019). Trust in humanoid robots: Implications for services marketing. *Journal of Services Marketing*, *33*, 507–518. https://doi.org/10.1108/JSM-01-2018-0045.

Wang, Z., Huang, J., & Fiammetta, C. (2021, June). Analysis of gender stereotypes for the design of service robots: Case study on the Chinese catering market. In *Designing Interactive Systems Conference 2021* (pp. 1336–1344). https://doi.org/10.1145/3461778.3462087.

Westerman, D., Edwards, A., Edwards, C., Luo, Z., & Spence, P. R. (2020) I-It, I-Thou, I-Robot: The perceived humanness of AI in human–machine communication. *Communication Studies*, *71*, 393–409. https://doi.org.10.1080/10510974.2020.1749683.

Windhager, S., Slice, D. E., Schaefer, K., Oberzaucher, E., Thorstensen, T., & Grammer, K. (2008). Face to face: The perception of automotive designs. *Human Nature*, *19*, 331–346. https://doi.org/10.1007/s12110-008-9047-z

Xu, K., Chen, M., & You, L. (2023). The hitchhiker's guide to a credible and socially present robot: Two meta-analyses of the power of social cues in Human–Robot Interaction. *International Journal of Social Robotics*, 1–27.

You, S., & Robert Jr, L. P. (2018, February). Human-robot similarity and willingness to work with a robotic co-worker. In *Proceedings of the 2018 ACM/IEEE international conference on human-robot interaction* (pp. 251–260). https://doi.org/10.1145/3171221.3171281

Xianlin Jin, Patric R. Spence, and Seungahn Nah

Integrating Big-Data Tools to Study AI and Human–Machine Communication: Methodology Strengths, Future Directions, and Applications

Abstract: This chapter highlights the utilization of a mixed-method approach in examining the intersection between artificial intelligence (AI) and robotics. The novelty of AI and its growing usage in diverse fields, such as healthcare, disasters, emergencies, businesses, and education, have presented communication scholars with both challenges and opportunities to delve into the dynamics between AI and human communication. The chapter focuses on the integration of topic modeling and sentiment analysis to explore the communication patterns between AI and humans and understand the public's opinions and attitudes towards AI. It reviews the recent literature within this realm, evaluates the applications of big-data tools in studying AI, proposes essential questions in this evolving field, and posits some future directions.

Keywords: AI, big data, topic modeling, sentiment analysis

18.1 Introduction

The growing adoption of artificial intelligence (AI) in multiple settings makes interaction and communication between AI and humans seem inevitable. However, some questions about AI and human communication remain unanswered. How does the public perceive and attend to AI? How does the information about AI diffuse among the public? Researchers should employ several methodological techniques to answer such questions. This chapter highlights that big-data tools, especially topic modeling and sentiment analysis, should be part of the repartee.

With the growth of new technologies, huge volumes of data reflecting diverse types of digital communication arise. However, there is a limited approach to systematically analyze and interpret such a plethora of data. Traditional qualitative analysis, such as thematic analysis and content analysis, cannot process large volumes of unstructured data. Neither does a quantitative method that primarily examines the group differences and linear relationships capture the underlying meanings of such large, unstructured data. Facing new challenges and opportunities, scholars and practitioners can take advantage of big-data tools to explore the interaction and communication between AI and humans. This chapter further reviews the applications of big-

https://doi.org/10.1515/9783110792270-019

data tools (e.g., topic modeling and sentiment analysis) in studying AI and robots. It also illustrates possible future directions in expanding AI research.

The following section offers a summative review of the topic modeling method and introduces the available program that scholars and professionals can use to conduct topic modeling analysis. Later, a similar review of sentiment analysis is provided.

18.2 Topic Modeling

Topic modeling is an unsupervised text-mining tool. It has been used to help scholars systematically understand large volumes of unstructured textual data (Jin & Spence, 2021).

Researchers have developed several topic modeling algorithms; one of the most frequently used probabilistic models is Latent Dirichlet Allocation (LDA; Blei et al., 2003). As a generative model, LDA aims to explain the observed textual data by estimating latent random variables (Griffiths & Steyvers, 2004; Steyvers & Griffiths, 2007; Jin & Spence, 2021). Specifically, documents in the corpus are the observed data. Topic distributions for each document and word distributions for each topic are the latent variables. During the procedure, scholars use approximate inference methods, such as variational inference (Blei et al., 2003) or Gibbs sampling (Griffiths & Steyvers, 2004), to infer uncertainty and discover the topic–document distributions and word distributions that are most likely to generate the observed data. In other words, the generative procedure of LDA reverses the logic of traditional qualitative theme analyses—that is, to estimate the most likely hidden themes that generated the observed textual data (Blei, 2012).

LDA is built on two assumptions. First, topics are common themes that appear across a collection of documents where a fixed number of patterns of words and terms occur. Second, each document is a mixture of topics and exhibits various levels of topic proportions. With LDA, scholars can identify major topics by tracking patterns of co-occurring terms (Blei, 2012). These two assumptions reflect LDA's three essential concepts—document, word, and topic. For instance, a tweet or short blog message can be a document. A 25-page peer-reviewed journal article can be segmented into multiple pieces of documents. One word can be attributed to various topics due to polysemy. The distribution of words characterizes each topic.

LDA provides document–topic distributions or proportions by generating document weights through a Dirichlet distribution, from which the name 'Latent Dirichlet Allocation' derives (Blei, 2012). The likelihood that a document is associated with particular topics is estimated by document–topic ratios. Some documents may be mainly associated with one specific topic, while others exhibit multiple topics. Such topics and word distributions show the major themes that emerged in the corpus. The ultimate goal of fitting a generative model is to discover the best values for latent variables, maximizing the probability of generating the observed data.

Several statistical programs provide topic modeling analysis. For example, Mallet —A Machine Learning for Language Toolkit is a Java-based open resource software that offers topic modeling, document classification, and natural language processing analyses (McCallum, 2002). Scholars can also have free access to the Matlab Topic Modeling Toolbox (Griffiths & Steyvers, 2004; Steyvers & Griffiths, 2011) for the purpose of scientific usage. A free R package, 'topicmodels,' also offers basic fitting models. Built on the R text mining 'tm' package (Feinerer et al., 2008), the package of 'topicmodels' provides Blei et al.'s (2003) variational inference (also known as expectation-maximization algorithm) and Griffiths and Steyvers's (2004) Gibbs sampling algorithm that was written by Phan et al. (2008). JMP Pro offers topic analysis functions where Latent Semantic Analysis is for a varimax rotation on singular vectors that generates a matrix of topic loadings (JMP Statistical Discovery LLC, 2023).

Topic modeling analysis has been applied in various disciplines, including communication (e.g., Kim et al., 2023; Maier et al., 2018; Jin & Spence, 2021). For instance, Jin and Spence (2021) utilized topic modeling to study the crisis communication patterns of Hurricane Maria. By analyzing 46,335 tweets about Hurricane Maria, Jin and Spence discovered six significant topics that emerged in the initial event phase (i.e. veteran medical care, resilience stories, communication needs, airfare investigations, food and water supplies, and tuition support for displaced college students), nine topics embedded in the maintenance phase (e.g., Hurricane forecast/update, concerns, FEMA aid, medical support, relief efforts, fundraising, and clean water), and six topics about Hurricane Maria from the resolution phase (e.g., food support, mental and physical health, fatalities, government responses, and water supply). Figure 18.1 shows the singular value decomposition (SVD) plots of Hurricane Maria tweets in the initial event stage. The SVD plots help researchers reduce dimensions of textual data (Landauer et al., 1998) to a manageable number to group similar words and phrases before identifying major topics, which is similar to conducting a principle component factor analysis. Table 18.1 lists the top keywords and corresponding loadings within each topic that emerged in the initial event crisis communication patterns of Hurricane Maria (Jin & Spence, 2021). As revealed by Jin and Spence's research, the need to communicate with family and friends and information about food and water were major requests at the initial event stage. In the maintenance phase, communication patterns included critical updates on the crisis, medical support, relief efforts, evacuations, and water supply. At the same time, Twitter users expressed emotions and concerns. At the resolution stage, the public remained concerned about food support and water supply, which suggests the issues had not been addressed. People also criticized the government's responses to the crisis, doubted the official fatalities, and focused on long-term disaster recovery requests, such as physical and mental health (Jin & Spence, 2021).

The application of topic modeling in the study of robotics and AI remains innovative and limited. Kim et al. (2023) conducted topic modeling analyses to capture how various stakeholders (e.g., the public, developers, and industry) discuss Replika—an AI chatbot. Kim and colleagues found that many Twitter users discussed the ethics and possibility of human–AI romantic relationships. There were also debates about building social relationships with AI and concerns about Replika's conservations. Kim et al.'s research suggests the usefulness of using topic modeling analysis to track the communication patterns between humans and AI.

Besides topic modeling analysis, another growing methodology—sentiment analysis—may also be beneficial to study AI. When people face new technologies, they want to hear existing users' feedback on these technologies before trying them. Companies also need to understand users' attitudes towards their new products. Sentiment analysis can help organizations and scholars understand the public's emotional responses to new technologies.

18.3 Sentiment Analysis

With the rapid changes in AI and the internet, people can use diverse platforms to share their opinions and express emotions on the internet. Such massive communication, opinion sharing, and emotional expression become overwhelming for scholars and industry to make prompt and systematic interpretations. Researchers and developers try to recognize and classify emotions from different sources accurately. Sentiment analysis that assesses the polarity of attitudes, opinions, and emotional expressions can help scholars and professionals overcome this challenge (Nandwani & Verma, 2021). As a side note, emotion detection or emotion analysis aims to categorize and distinguish the type and degree of human emotions (e.g., very depressed, a little sad, and very happy) by analyzing multiple cues, such as facial expressions, body language, and speech patterns (Munezero et al., 2014), which is different from sentiment analysis and hence is not included in this discussion.

Sentiment analysis focuses on natural language processing and computational linguistics. It utilizes computational methods to analyze and understand the sentiments, attitudes, opinions, and emotions towards entities, individuals, issues, and events (Liu & Zhang, 2012). Sentiment analysis has shown its effectiveness in identifying target audiences' attitudes about specific products and services (Liu, 2020; Yoo et al., 2018). User-generated content on social media can also provide a new way of tracking the public and consumers' attitudes towards artificial intelligence agents, robots, and any other new technology.

The public may also express the need for a specific feature of a particular technology. Understanding these opinions can allow researchers and professionals to make a real-time assessment of target audiences, thus providing insights for better machine design. Scholars can use sentiment analysis to interpret the target audience's emotions and opinions, which will assist researchers, developers, and professionals in understanding what information and assistance are needed.

There are two main approaches to conducting sentiment analysis: knowledge based or lexicon based and machine learning (Taboada, 2016). Knowledge-based sentiment analysis is built on dictionaries of words and phrases associated with sentiment polarities, such as positive, negative, or neutral. Each word or phrase associated with the sentiment is identified and rated for a score. For instance, −1 represents a negative attitude, scores close to zero reflect a neutral opinion, and scores close to +1 indicate a positive sentiment. The document's sentiment is determined by calculating the average aggregation of these sentiment scores or the polarity of words and phrases. In other words, knowledge-based sentiment analysis divides the sum of sentiment scores of words and phrases by the total number of words and phrases. The underlying logic of this approach is that a document's sentiment is determined by the sentiment of words and phrases embedded in this document. Nevertheless, the weakness of this approach is that it may not reflect context-dependent expressions.

Machine-learning sentiment analysis focuses on training classifiers to predict a document's sentiment by analyzing specific features of the text data, such as term frequency, n-grams (contiguous sequences of words), and part-of-speech tags (Forman, 2003). This approach has three primary learning methods: supervised learning methods, unsupervised learning methods, and semi-supervised learning methods. The following paragraphs offer a general description of each learning method.

In the supervised learning methods, such as Naïve Bayes, Support Vector Machine, and k-Nearest Neighbour, each document is labeled as positive, negative, or neutral, and such labeled data are used to train classifiers (Naw, 2018; Taboada, 2016; Xia et al., 2011). Human coders label the training dataset that will later be used to train unlabeled data and identify classifiers. For example, the Naïve Bayes algorithm assumes the independence between text features. It assigns a document to a sentiment category based on the likelihood of the document belonging to this category.

Examples of unsupervised learning methods include K-means clustering and hidden Markov models (see Sun et al., 2012). Unlike supervised learning methods, unsupervised learning models automatically discover document sentiment patterns without labeled training data. As suggested by its title, semi-supervised learning methods integrate the strength of supervised and unsupervised learning methods, using labeled and unlabeled data for sentiment analysis. Examples include graph-based methods (Medhat et al., 2014; Zhao et al., 2010) and heuristic approaches (Van Engelen & Hoos, 2020; Zhu et al., 2021). The core logic of graph-based methods is to use graphs to outline the relationships among documents and words. Each node represents a document or word, and each edge means similarities between nodes. The labeled training data assign sentiment values of certain nodes, which are used to identify sentiment labels for the unlabeled data based on the similarities between connected nodes (Zhao et al., 2010). Instead, the heuristic approach assigns sentiment labels to unlabeled data based on domain or linguistic knowledge. In other words, the heuristic sentiment analysis uses specific terms associated with particular sentiments to infer the unlabeled data's sentiment (He & Zhou, 2011). Overall, semi-supervised learning sentiment analyses incorporate a limited number of labeled training data to improve the generalizability of sentiment models.

Sentiment analysis is an evolving field; many methods are being developed to better understand the public opinions (e.g., approve or disapprove of the adoption of AI) and attitudes (e.g., like or dislike certain features of AI agents) embedded in unstructured text data. The prospect of mining extensive data (e.g., social media data, customer reviews of a new chatbot, and news coverage of AI) will help scholars, designers, and companies grasp the public's attitudes, opinions, and needs, which will shed light on communication between AI and humans and future machine design.

Sentiment analysis has gradually received attention from scholars studying AI and robot interaction and communication between AI agents and humans. For example, by conducting a series of experiments, Szabóová et al. (2020) explored the impact of emotional detection on human–robot interaction. Szabóová et al. tested the accuracy of emotion classification with supervised machine-learning methods, such as Naive Bayes, Support Vector, Machine, and Neural Network classifiers. In Szabóová et al.'s research, the Nao robot greeted the participants with or without emotional cues. Although the accuracy of emotion classification was low, participants' responses to Nao were more positive when Nao added gestures. Szabóová et al.'s data suggest that the perception of robot performance was positively associated with its ability to respond to human emotions.

AI and big data are evolving and require interdisciplinary exploration. Studying discussions of, attention to, and adoption of AI will offer insights into the communication and interaction of AI and humans. Integrating big-data tools, such as topic modeling and sentiment analysis, will help scholars and practitioners attain such in-depth understandings. More studies will be needed to explore communication between AI and humans with big-data tools. The following section proposes several possible directions for future research concentrations.

18.4 Future Directions in the Study of Robots, AI, and Human–Machine Communication

Using big-data tools to unpack the study of robots and understand the public's opinions and attitudes towards certain new technologies is central to the study of robots and AI, along with the methods outlined. There are several vital directions that communication scholars can consider while consolidating future studies.

First, studying different stakeholders' (e.g., users, the general public, industry, government agencies, and scholars) opinions, concerns, and needs towards AI agents, chatbots, and other new technologies will help researchers and professionals understand the patterns of robots and communication. Integrating topic modeling analysis and sentiment analysis will allow researchers to achieve this goal. Knowing the content of communication between AI and humans (topic modeling analysis), and the emotions and opinions of the public, AI developers, and the industry (sentiment analysis), will offer theoretical and professional implications.

Second, future studies can also incorporate a topic modeling approach and sentiment analysis of news stories and user comments regarding AI and robots spanning diverse online communication channels such as news websites and social media platforms. Such studies hold the potential to provide valuable insights guiding policymakers, professionals, and practitioners towards significant policies concerning human–robot interaction.

Third, AI chatbots are expected to respond appropriately to human beings' emotional expressions and sentiments in conversations. Sentiment analysis can help developers design AI agents to better process natural language and appropriately react to users' emotions and needs.

Fourth, developing interdisciplinary collaborations to study robots and other aspects of AI–human communication is the trend. Computer science and engineering scholars have primarily conducted human aspects of robotics research (e.g., Oliveira et al., 2019; Zanatto et al., 2016), and communication researchers gradually contributed to this new discipline. Similarly, big-data tools, such as topic modeling analysis and sentiment analysis, were created by computer science and later adopted by communication scholars. Both AI and big-data research illustrate the importance and trend of conducting interdisciplinary work.

Lastly, as topic modeling analysis and sentiment analysis are still evolving, especially in communication research, researchers have not agreed on practical instructions for ensuring validity and reliability (Maier et al., 2018). Future studies applying these methods in AI research should devote more effort to testing validity and reliability. It will be essential for future studies to share the details of cleaning data, processing data, model selection rationale, and interpretation of data. All these protocols will help communication scholars ensure consistency, accuracy, and generalizability.

18.5 More Specific Applications

While the first part of this chapter broadly addressed methods and approaches, this section will delve into more specific details. In the study of robots, one exciting yet potentially frustrating issue will be the rapid emergence and spread of new innovations. Simultaneously, certain robotic advancements might possess significant advantages but fail to diffuse. These rapid advances and then changes in robotics and other areas of human–machine communication (HMC) may create instances where traditional survey methods will have disadvantages to the study of the specific phenomena. Therefore, this is an area where topic modeling will serve to fill this gap and help understand various facets of the aspects of these phenomena. Thus, topic modeling can present itself as a versatile tool in various ways to study these phenomena.

First is the initial stage of diffusion. By understanding the communication topics that emerge with a new innovation, such as robotic teachers or AI assistants, an understanding of the success of adoption may be easier. Education, for example, is an industry that is under increased pressure to do more regarding student outcomes with fewer resources. Students often use a range of resources for communication about university life, assignments, perceptions of grades, and expressions of emotion. Therefore, through collecting data that may be harvested from Course Management Systems, social media (Twitter, Facebook), and possible course evaluations, the topics of discussion may be studied and understood to help researchers and university administrators better comprehend the diffusion or absence of diffusion of specific elements of AI in the university course curriculum. Once such topics are established and learned from, summative evaluation can be used and adjustments can be made, allowing analysis of the success of these new features to be better understood.

Although not used for summative evaluations, as highlighted earlier, Jin and Spence have outlined the effectiveness of topic modeling for understanding communication patterns across difference stages of natural disasters (2021). Their research highlights the ability to use topic modeling at different points in the study of a single phenomenon. The authors further analyzed how communication changes throughout the lifecycle of a crisis. Topic modeling could then be used to examine not only diffusion but also characteristics of the adoption, trial and reinvention of AI and robots in the education process.

Although diffusion is one of many areas that could be studied, the possibilities seem to have few limits. For example, a well-discussed area of robotics and AI is the role of automation in the workplace (see Spence et al., 2018; Spence, 2023). This area alone has several topics that could emerge for areas of new study or to better understand previous research. An obvious area is fear of replacement as the result of the introduction of a robot. But other areas might emerge, such as the enhancement of work due to the use of a machine. It has also been argued that some individuals who have tasks they find unrewarding may find pleasure in handing those tasks off to a robot or some other machine, which would allow them to focus on the tasks that help them become more self-actualized. Another area that could emerge is perceptions of communication with machines in the workplace. We may see topics emerge where individuals address the relational aspects of robotic co-workers.

This same approach can be taken with the systematic study of healthcare and mental health. The results from topic modeling can be used with AI to look for issues related to everyday illness or side effects of new medications. For example, Google already tracks seasonal virus outbreaks based on search results to map hotspots with varying degrees of success (Olson et al., 2013). Local municipalities such as schools and hospitals are able to use that information to prepare for a potential influx of patients. With the use of AI, more can be brought into the mix for preparation, such as pharmacies, supply chains, and even the manufacturing of chicken soup. Similarly, reactions to over-the-counter medications or even prescription medications may be able to be monitored. The introduction of AI into this area of research expands the scope and possibly will add precision to what we already learn using this method. Moreover, healthcare is also an occupation where AI and robotic assistants will continue to have an increased role. People undoubtedly will talk about these experiences, thus allowing dominant topics to be analyzed and learned from.

Often, topics that emerge may be promoted by a group of influencers or possibly even one specific influencer. Influencers serve a role as a bridge connecting information seeking with specific information or information resources, which can then be disseminated throughout a network (Jin, 2020; Jin & Spence, 2023). Understanding the biggest influencers will have specific implications for HMC. Previous studies on crisis communication influencers reveal that information was often shared in subnetworks where influencers control the information flow (Jin, 2020; Jin & Spence, 2023). Conducting sentiment network analysis will help scholars answer the question: Who influences the opinions of AI usage within a network? The answer to this question will add new knowledge to specific aspects of AI and robots research.

Another question will be how much influencers partner with or are using AI in their specific roles. Members of the public may be more or less persuaded by the use or amount of AI that supplements content. Alternatively, AI may help us better understand the level of persuasiveness of an influencer. Going past impressions, reads, and clicks, through the use of AI there may be better ways to determine the persuasiveness of a specific individual. This could produce new questions for study, specifically why one influencer is more persuasive. From there, the possibility of better micro-targeting could be employed. In the area of crisis communication this is quite attractive. Members of the public must receive, understand, and act upon a message for it to be a success (see Lachlan et al., 2014; Spence et al., 2010). Through the examination of topic models coupled with advances in AI, the better creation and placement of messages may be possible.

There have already been studies that have looked at the role of robots and automation in the specific area of weather forecasts (Spence et al., 2019; Spence et al., 2021; Rainear et al., 2021) and weather warnings (Lachlan et al., 2016). These studies specifically looked at differences between humans and robots and asked questions about willingness to follow on social media (a concept related to influence). Expanding what is known to severe weather and focusing on the actual influence using topic modeling is one of many logical next steps. Furthermore, taking the notion of an influencer past that of being a human to a specific application, embodied robot, or AI-generated system is also ripe for future research.

Continuing to explore subjects that receive both scholarly attention as well as from the popular press is the perceived problem of misinformation and how to root it out. Topic modeling provides a way to study this issue and look for examples. Misinformation can exist about both the uses and impacts of AI and robots, and also these same agents could be the producers of misinformation. The ability to follow promoted links and understand the largest influencers that are promoting less than factual information is an area topic modeling can be used to address. Often, the issues of misinformation are addressed in the area of political communication or health-related issues, many of which are described as conspiracy theories. These types of communications can be produced and shared within homogeneous and polarized communities (Del Vicario et al., 2016; Shu et al., 2017). Within online communities or with regard to online information, there are various cues people use to make judgments about the credibility of the information or the information producer (Lin & Spence, 2019; Lin et al., 2016; Westerman et al., 2012, 2014). This information is based primarily on experimental studies; however, finding influencers that have the same characteristics that were manipulated in these studies would be easy. A next step would be to evaluate the information promoted, the reach of that information, and how that information is shared and acted upon. Moreover, a clear evaluation of how people act upon AI-generated news (real or fake) could then be undertaken.

Moreover, in using topic modeling and network analysis to understand what links are being shared around a specific topic, there are further things that we could learn. For example, what happens when a noncredible communicator who is a heavy influencer promotes a link with credible information? How do information consumers act, and what are the types of conversations that go on? What about the opposite? That is, what happens when a highly credible and well-followed source provides information that is not factual? What are the relationships between these factors and people's willingness to click on a provided link? Because much is known about the shortcomings of self-report data in face-threatening situations, we may use new techniques and even partner with AI to better understand these issues.

Moving back to a direct focus on AI and robotics, this approach can be used to better understand what conversations are emerging based on misinformation provided and specifically how AI can influence the spread of misinformation. One possibility that is under-discussed and under-studied is if it matters that AI is producing and distributing misinformation. That is, misinformation has always been part of the human experience. As long as humans have grouped together and communicated, misinformation, both intentional and unintentional, has been present. Is there any reason to believe that more harm is produced because machine communicators enter the equation? Or can machine communication make up for the errors in human communication that produce misinformation?

Much of the above reviewed so far can be categorized in the broad area of Latent Dirichlet Allocation (Blei et al., 2003), which was discussed earlier. This focuses on words, topics, or documents. Documents are mixed of topics reflected by words distributions (Vayansky & Kumar, 2020), and this, like the other areas above, deals with text. Although text has primacy on the internet and with searches, other types of communication exist, such as audio and video. These are easily searched when captions or other types of text are provided, but they are less often part of the investigations using topic modeling. Thus, are people making podcasts about issues with AI, robotics, and HMC? If so, what are those podcasts about, and what is their popularity? Does the content of this media follow the same trends and discussions as that which emerges in other documents and media? If not, how does it diverge? Thus, through moving from the text to other forms of communication, we can see if the perceptions of discussions and attention to AI, robotics, and HMC are consistent or need further investigation. Using these same approaches, we can see if similar influencers that appear on text/primary social media are also the same as audio/video forms of communication. That is, is the influential podcaster also the owner of an influential Twitter account? Or is a popular link that is tweeted taking users to a podcast?

Furthermore, it is conceivable that humans might be removed from the process to a greater degree. Could we simply use algorithms that go out and answer topic modeling questions we were not asking? Might AI give us answers to questions we were not thinking to answer? Alternatively, could it rather create solutions to which we did not perceive a problem? In tackling these questions, researchers should employ a variety of methodological approaches, with topic modeling serving as a pivotal component of the conversation.

The above are all suggestions where AI, robotics, and HMC research can move with the use of topic models, but it is not an exhaustive list. Moreover, as methods improve and data collection becomes more precise, the approaches that are used will change.

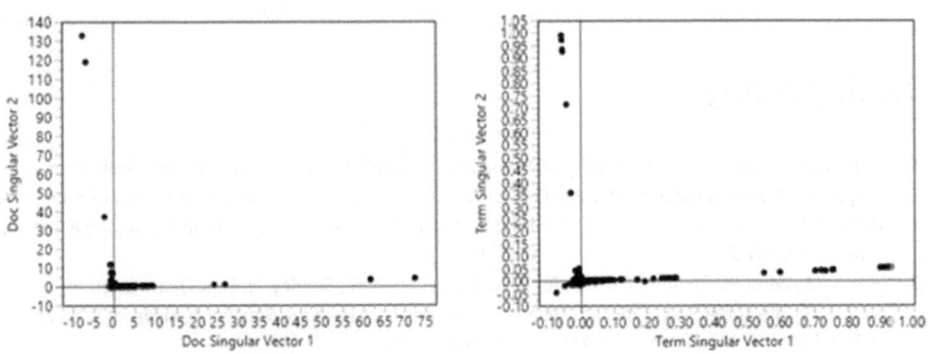

Figure 18.1: SVD plots of Hurricane Maria tweets in the initial event stage (Jin & Spence, 2021, p. 1274)

Table 18.1: Topic to terms of Hurricane Maria tweets in the initiation event stage.

Veteran Medical Care		Resilience Story		Communication Needs	
Term	**Loading**	**Term**	**Loading**	**Term**	**Loading**
Extra	.98164	Flys	.93650	Mutuo	.96523
Veteran	.98164	Liat	.93650	Comunicado	.96523
Vamc	.98164	Melville	.93650	Auxilio	.96523
Jackson's	.98164	Resilie	.92637	Servicio	.96523
Pleased	.98164	Hall	.91487	Joetruong	.94845
Sonny	.96396	Port	.90421	Works	.90599
Montgomery	.92614	Restored	.90421	Phone	.85001
Accept	.91731	Gear	.76527	Service	.82822
Mile	.84590	Stories	.76351	Verizon	.58978
Patients	.70794	Amazing	.72795	Hospital	.55755
Airfare Investigation		**Food/Water Supply**		**Tuition Support**	
Term	**Loading**	**Term**	**Loading**	**Term**	**Loading**
Signed	.96217	Opened	.77795	Displaced	.81074
Thejusticedept	.96217	Distribution	.76971	Tuition	.81074
Repdarrensoto	.96217	Officials	.76719	Mdcollege	.80739
Investigate	.96217	Hurricc	.92637	Mdc	.80350
Airfare	.96217	Locations	.75426	College	.80350
Gouging	.92211	Food	.70843	Offer	.78056
Price	.92211	Usvi	.69884	State	.74719

Source: Jin & Spence (2021, p. 1274).

Bibliography

Blei, D. M. (2012). Topic modeling and digital humanities. *Journal of Digital Humanities*. Retrieved from https://journalofdigitalhumanities.org/2-1/topic-modeling-and-digital-humanities-by-david-m-blei/

Blei, D. M., Ng, A. Y., & Jordan, M. I. (2003). Latent Dirichlet allocation. *Journal of Machine Learning Research*, *3* (4–5), 993–1022.

Del Vicario, M., Bessi, A., Zollo, F., Petroni, F., Scala, A., Caldarelli, G., Stanley, H. E., & Quattrociocchi, W. (2016). The spreading of misinformation online. *Proceedings of the National Academy of Sciences*, *113*(3), 554–559. https://doi.org/10.1073/pnas.1517441113.

Feinerer, I., Hornik, K., & Meyer, D. (2008). Text mining infrastructure in R. *Journal of Statistical Software*, *25*(5), 1–54. https://doi.org/10.18637/jss.v025.i05.

Forman, G. (2003). An extensive empirical study of feature selection metrics for text classification. *Journal of Machine Learning Research*, *3*, 1289–1305.

Griffiths, T. L., & Steyvers, M. (2004). Finding scientific topics. *Proceedings of the National Academy of Science*, *101*, 5228–5235. https://doi.org/10.1073/pnas.0307752101.

He, Y., & Zhou, D. (2011). Self-training from labeled features for sentiment analysis. *Information Processing & Management*, *47*(4), 606–616. https://doi.org/10.1016/j.ipm.2010.11.003.

Jin, X. (2020). Exploring crisis communication and information dissemination on social media: Social network analysis of Hurricane Irma tweets. *Journal of International Crisis and Risk Communication Research*, *3*(2), 179–210. https://doi.org/10.30658/jicrcr.3.2.3.

Jin, X., & Spence, P. R. (2021). Understanding crisis communication on social media with CERC: Topic model analysis of tweets about Hurricane Maria. *Journal of Risk Research*, *24*(10), 1266–1287. http://doi.org/10.1080/13669877.2020.1848901.

Jin, X., & Spence, P. R. (2023). Check the crisis information on Twitter: Information flow and crisis communication patterns of Hurricane Ida. *Communication Studies*, *74*(4), 337–355. https://doi.org/10.1080/10510974.2023.2202872.

JMP Statistical Discovery LLC (April 21, 2023). Overview of topic analysis. https://www.jmp.com/support/help/en/17.1/index.shtml#page/jmp/overview-of-topic-analysis.shtml#.

Kim, J., Jin, X., Xu, K., Chen, X., & Yang, H. (2023, May). What do people talk about Replika? Investigating how Replika is discussed and interpreted among groups with different interests. Presented at the *International Communication Association 2023 Convention*, Toronto, Canada.

Lachlan, K. A., Spence, P. R., Lin, X, & Najarian, K., Del Greco, M. (2014). Twitter Use during a weather event: Comparing content associated with localized and non-localized hashtags. *Communication Studies*, *65*, 519–534. https://doi.org/10.1080/10510974.2014.956940

Lachlan, K. A., Spence, P. R., Rainear, A., Fishlock, J., Xu, Z., & Vanco, B. (2016). You're my only hope: An initial exploration of the effectiveness of robotic platforms in engendering learning about crises and risks. *Computers in Human Behavior*, *65*, 606–611. https://doi.org/10.1016/j.chb.2016.05.081.

Landauer, T. K., Foltz, P. W., & Laham, D. (1998). An introduction to latent semantic analysis. *Discourse Processes*, *25*(2–3), 259–284. https://doi.org/10.1080/01638539809545028.

Lin, X., & Spence, P. R. (2019). Others share this message, so we can trust it? An examination of bandwagon cues on organizational trust in risk. *Information Processing & Management*, *56*(4), 1559–1564. https://doi.org/10.1016/j.ipm.2018.10.006.

Lin, X., Spence, P. R, & Lachlan, K. A. (2016). Social media and credibility indicators: The effect of influence cues. *Computers in Human Behavior*, *63*, 264–271. https://doi.org/10.1016/j.chb.2016.05.002.

Liu, B. (2020). *Sentiment Analysis, Mining Opinions, Sentiments, and Emotions* (2nd ed.). Cambridge University Press.

Liu, B., & Zhang, L. (2012). A survey of opinion mining and sentiment analysis. In C. Aggarwal & C. Zha (Eds.), *Mining Text Data*. Springer. https://doi.org/10.1007/978-1-4614-3223-4_13.

Maier, D., Waldherr, A., Miltner, P., Wiedemann, G., Niekler, A., Keinert, A., Pfetsch, B., Heyer, G., Reber, U., Häussler, T., Schmid-Petri, H., & Adam, S. (2018). Applying LDA topic modeling in communication research: Toward a valid and reliable methodology. *Communication Methods and Measures*, *12*(2–3), 93–118. https://doi.org/10.1080/19312458.2018.1430754.

McCallum, A. K. (2002). MALLET: A Machine Learning for Language Toolkit. http://mallet.cs.umass.edu

Medhat, W., Hassan, A., & Korashy, H. (2014). Sentiment analysis algorithms and applications: A survey. *Ain Shams Engineering Journal*, *5*(4), 1093–1113. https://doi.org/10.1016/j.asej.2014.04.011.

Munezero, M., Montero, C. S., Sutinen, E., & Pajunen, J. (2014). Are they different? Affect, feeling, emotion, sentiment, and opinion detection in text. *IEEE Transactions on Affective Computing*, *5*(2), 101–111. https://doi.org/10.1109/TAFFC.2014.2317187.

Nandwani, P., & Verma, R. (2021). A review on sentiment analysis and emotion detection from text. *Social Network Analysis and Mining*, 11(1), 81. https://doi.org/10.1007/s13278-021-00776-6.

Naw, N. (2018). Twitter sentiment analysis using support vector machine and k-nn classifiers. *International Journal of Scientific and Research Publications*, *8*(9). https://doi.org/10.29322/ijsrp.8.10.2018.p8252.

Oliveira, R., Arriaga, P., Correia, F., & Paiva, A. (2019). The stereotype content model applied to human–robot interaction in groups. In*Proceedings of the ACM/IEEE International Conference On Human–Robot Interaction* (pp. 123–132), Daegu. https://doi.org/org/10/1109/HRI.2019.8673171.

Olson, D. R., Konty, K. J., Paladini, M., Viboud, C., & Simonsen, L. (2013). Reassessing google flu trends data for detection of seasonal and pandemic influenza: A comparative epidemiological study at three geographic scales. *PLoS Computational Biology*, *9*(10), e1003256. https://doi.org/10.1371/journal.pcbi.1003256.

Phan, X. H., Nguyen, L. M., & Horiguchi, S. (2008). Learning to classify short and sparse text & web with hidden topics from large-scale data collections. In *Proceedings of the 17th International World Wide Web Conference (WWW 2008)* (pp. 91–100). Beijing.

Rainear, A., Jin, X., Edwards, A., Edwards, C., & Spence, P. R. (2021). A robot, meteorologist, and amateur forecaster walk into a bar: Examining qualitative responses to a weather forecast delivered via social robot. *Communication Studies*, *72*(6), 1129–1145. https://doi.org/10.1080/10510974.2021.2011361

Shu, K., Sliva, A., Wang, S., Tang, J., & Liu, H. (2017). Fake news detection on social media: A data mining perspective. *ACM SIGKDD Explorations Newsletter*, *19*(1), 22–36.

Spence, P. R. (2023). Why wasn't I ready for that? Suggestions and research directions for the use of machine agents in organizational life. In S. Nah (Ed.), Research handbook on artificial intelligence and communication (pp. 325–336). Edward Elgar Publishing. https://doi.org/10.4337/9781803920306.00030.

Spence, P. R., Edwards, C., Edwards, A., Rainear, A., & Jin, X. (2021). "They're always wrong anyway": Exploring differences of credibility, attraction, and behavioral intentions in professional, amateur, and robotic-delivered weather forecasts. *Communication Quarterly*, *69*(1), 67–86. https://doi.org/10.1080/01463373.2021.1877164.

Spence, P. R., Edwards, A., Edwards, C., & Jin, X. (2019). The bot predicted rain, grab an umbrella": Few perceived differences in communication quality of a weather Twitterbot versus professional and amateur meteorologists. Behavior & Information Technology, 38, 101–109. https://doi.org/10.1080/0144929X.2018.1514425

Spence, P. R., Westerman, D., & Lin, X. (2018). A robot will take your job. How does that make you feel? Examining perceptions of robots in the workplace. In A. Guzman (Ed.), Human-Machine Communication: Rethinking communication, technology, and ourselves (185–200). New York, NY: Peter Lang.

Spence, P. R., Lachlan, K. A., & Eith, C. A. (2010). Vulnerability of the underclass: Why the most vulnerable are always the last to know. In J. D. Rivera & D. S. Miller (Eds.), *How Ethnically Marginalized Americans Cope with Catastrophic Disasters: Studies in Suffering and Resiliency* (pp. 317–335). Edwin Mellen Press.

Steyvers, M., & Griffiths, T. (2007). Probabilistic topic models. In T. K. Landauer, D. S. McNamara, S. Dennis, & W. Kintsch (Eds.), *Handbook of Latent Semantic Analysis* (pp. 427–448). Lawrence Erlbaum Associates Publishers.

Steyvers, M., & Griffiths, T. (2011). MATLAB topic modeling toolbox 1.4. http://psiexp.ss.uci.edu/research/programs_data/toolbox.htm.

Sun, S., Liu, H., Lin, H., & Abraham, A. (2012). Twitter part-of-speech tagging using pre-classification Hidden Markov model. *2012 IEEE International Conference on Systems, Man, and Cybernetics (SMC)*. https://doi.org/10.1109/icsmc.2012.6377881.

Szabóová, M., Sarnovský, M., Maslej Krešňáková, V., & Machová, K. (2020). Emotion analysis in human–robot interaction. *Electronics*, *9*(11), 1761. https://doi.org/10.3390/electronics9111761.

Taboada, M. (2016). Sentiment analysis: An overview from linguistics. *Annual Review of Linguistics*, *2*(1), 325–347. https://doi.org/10.1146/annurev-linguistics-011415-040518.

Van Engelen, J. E., & Hoos, H. H. (2020). A survey on semi-supervised learning. *Machine Learning*, *109*, 373–440. https://doi.org/10.1007/s10994-019-05855-6.

Vayansky, I., & Kumar, S. A. P. (2020). A review of topic modeling methods. *Information Systems*, *94*, 101582. https://doi.org/10.1016/j.is.2020.101582.

Westerman, D., Spence, P. R., & Van Der Heide, B. (2012). A social network as information: The effect of system generated reports of connectedness on credibility on Twitter. *Computers in Human Behavior*, *28*(1), 199–206.

Westerman, D., Spence, P. R., & Van Der Heide, B. (2014). Social media as information source: Recency of updates and credibility of information. *Journal of Computer-Mediated Communication*, *19*(2), 171–183. https://doi.org/10.1111/jcc4.12041.https://doi.org/10.1016/j.chb.2011.09.001.

Xia, R., Zong, C., & Li, S. (2011). Ensemble of feature sets and classification algorithms for sentiment classification. *Information Sciences*, *181*(6), 1138–1152. https://doi.org/10.1016/j.ins.2010.11.023.

Yoo, S. Y., Song, J. I., & Jeong, O. R. (2018). Social media contents based sentiment analysis and prediction system. *Expert Systems with Applications*, *105*(1), 102–111. https://doi.org/10.1016/j.eswa.2018.03.055.

Zanatto, D., Patacchiola, M., Goslin, J., & Cangelosi, A. (2016). Priming anthropomorphism: Can the credibility of humanlike robots be transferred to non-humanlike robots? In *Proceedings of the 11th ACM/IEEE International Conference on Human–Robot Interaction* (pp. 543–544). https://doi.org/10.1109/hri.2016.7451847.

Zhao, Y., Qin, B, & Liu, T. (2010). Integrating intra- and inter-document evidences for improving sentence sentiment classification. *Acta Automatica Sinica*, *36*(10), 1417–1425. https://doi.org/10.3724/sp.j.1004.2010.01417.

Zhu, J. J., Chang, Y.-C., Ku, C.-H., Li, S. Y., & Chen, C.-J. (2021). Online critical review classification in response strategy and service provider rating: Algorithms from heuristic processing, sentiment analysis to deep learning. *Journal of Business Research*, *129*, 860–877. https://doi.org/10.1016/j.jbusres.2020.11.007.

Jochen Peter and Caroline L. van Straten

Social Robots and Children: A Field in Development

Abstract: Research on how children interact with social robots is still a young field of inquiry. This chapter sketches how research on child–robot interaction (CRI) has evolved since its beginning roughly 25 years ago. We further describe three main lines in CRI research and, per line, a central concept: first, children's learning from social robots and the central concept of engagement; second, children's relationship formation with social robots and trust; and third, children's conceptions of social robots along with the concept of social robots as a new ontological category. The chapter also identifies challenges in current CRI research: next to suffering from an adult bias, the field is challenged by a lack of genuine CRI theories and an overreliance on frameworks coming from interpersonal research. Finally, we focus on future developments that may offer opportunities for significant contributions of CRI research to our understanding of how children and humans, more generally, interact with social robots. In our view, CRI researchers should more strongly embrace the increasing merger between social robots and smart/connected toys, study social robots in conjunction with communicative robots, and pay more attention to the issues and demands of responsible CRI.

Keywords: Social robots, children, child–robot interaction, learning, relationship formation

19.1 Introduction

Social robots—that is, robots that "are designed to interact with people in a natural, interpersonal manner" (Breazeal et al., 2016a, p. 1935)—have attracted considerable attention from roboticists, academics, and the public in the past two decades or so (Bartneck et al., 2020; Eberl, 2016; Fong et al., 2003). Although social robots are less widespread than, for example, voice-based assistants and chatbots, they are increasingly used in healthcare, education, and entertainment or are made for companionship (for an overview, see, e.g., Chapter 10 in Bartneck et al., 2020). As social robots in these areas are often targeted at children (Belpaeme et al., 2018; Blanson Henkemans et al., 2017; Turkle, 2011), researchers have turned towards children's interactions with social robots. Research on child–robot interaction (CRI) is a subfield of research on human–robot interaction and on human–machine interaction more generally. In this chapter, we define research on CRI as empirical or theoretical research on children's cognitions and emotions about social robots, as well as their uses of social ro-

https://doi.org/10.1515/9783110792270-020

bots, along with the individual and societal consequences that these cognitions, emotions, and uses elicit.

Research on CRI is still in its infancy. An exploratory search in the Social Sciences Citation Index and the Emerging Sources Citation Index of the Web of Science resulted, as of December 2022, in 855 research documents (search terms: child* robot interaction). After the first publications emerged in the second half of the 1990s, it took several years before, in 2010, more than 10 annual publications on the topic came out. Afterward, publications per year rose steadily to an all-time high of 143 publications in 2020 but dropped slightly to 128 annual publications in 2021, possibly due to problems with fielding studies during the COVID-19 pandemic. Over the whole period, the top three disciplines with which research on CRI was associated were robotics (30%), computer science with a focus on artificial intelligence (11%), and multidisciplinary psychology (9%). Research on CRI thus presents a small but burgeoning field that tends to originate mainly in the technical sciences.

Against this background, we present in this chapter an overview of selected topics and issues in research on CRI. We start by sketching the historical development of the field. In the next step, we describe the main lines of current CRI research and elaborate, per line, on one core concept. After identifying some shortcomings in the literature, we conclude with what we consider important developments in the broader field that may also shape future research on CRI. Similar to other current reviews on related topics (De Graaf & Peter, 2023), our use of the literature is illustrative and selective rather than encompassing. Given their prominence in CRI research, we tend towards studies on anthropomorphic robots and primary school children. We do not deal with research on social robots and children on the autism spectrum because an adequate discussion of this thriving and important field is beyond the scope of this chapter. For a more detailed use of the literature, we refer to the reviews on specific aspects of CRI (e.g., Belpaeme et al., 2018; Cabibihan et al., 2013; De Jong et al., 2019; Stower et al., 2021; Van den Berghe et al., 2019; Van den Berghe, 2022; Van Straten et al., 2020b).

19.2 Historical Development of the Field, Main Research Lines, and Major Concepts

As a young field, research on CRI is currently characterized by a variety of theoretical frameworks and methodological approaches (e.g., De Jong et al., 2019; Peter et al., 2019; Van Straten et al., 2020b). Theoretical frameworks in CRI encompass, for example, theories from interpersonal communication (e.g., Van Straten et al., 2021), moral, social, and developmental psychology (Fortunati et al., 2015; Kahn et al., 2012; Serholt, 2018), and psychoanalysis (Turkle et al., 2006). Methodologically, research on CRI ranges from qualitative to quantitative approaches, correlational to experimental designs, and cross-sectional to longitudinal designs, with data collection typically being based on self-reports in interviews or questionnaires as well as on observation (see, e.g., the reviews by De Jong et al., 2019; Stower et al., 2021; Van Straten et al., 2020b). In terms of robot morphologies, finally, studies have relied mainly on anthropomorphic robots, such as Nao, but also on zoomorphic robots, such as iCat, and caricatured robots, such as Cozmo (e.g., Van Straten et al., 2020b).

19.2.1 Historical development of the field

As CRI research comes from disciplines as diverse as robotics, computer science, and social science, it is, to some extent, scattered and has not been based on a commonly shared research agenda. Consequently, it is difficult to identify what would qualify as a history of the field. Still, there are some commonalities in how research has developed in the past two decades. Some of the first studies on social robots and children dealt with the possibilities of such robots in the treatment of children on the autism spectrum (e.g., Dautenhahn, 1999; Werry et al., 2001). Research then expanded towards children not on the autism spectrum and the role of social robots as companions of children (e.g., Kanda et al., 2004; Tanaka et al., 2007). Relatively early on, scholars have also been interested in children's interactions with, and reasoning about, robotic pets (e.g., Kahn et al., 2006), accompanied by critical voices about the relational inauthenticity of such interactions (Turkle, 2007). Moreover, with social robots' ability to interact with children, the use of robots for educational purposes was already attracting researchers' attention in the 2000s (e.g., Han et al., 2008; Kanda et al., 2004) and led, among other things, to research on language learning (e.g., You et al., 2006).

Early on, however, scholars also understood the importance of sociality in children's interactions with social robots and, broadly speaking, by the 2010s, studies on children's relationship formation with social robots had gained more traction (e.g., Kahn et al., 2012), both in terms of design (Belpaeme et al., 2012) and long-term relationships (Leite et al., 2013). Around the same period, researchers also increasingly zoomed in on how children experience and conceptualize social robots (e.g., Beran et al., 2011). Currently, all the aforementioned research foci are still present in the research field. Many of the original ideas are being deepened, broadened, and revised. At the same time, scholars have turned progressively to applying social robots for specific tasks, for example in health and educational settings (Blanson Henkemans et al., 2017; Konijn & Hoorn, 2020). Finally, there is (re)new(ed) interest in, for example, CRI-specific measure development (Van Straten et al., 2020a) and ethical issues (DiPaola et al., 2023; Sharkey, 2016).

19.2.2 Main research lines and major concepts

Given the historical development of the field, three lines of research, in particular, may deserve attention: one line studies children's learning from social robots, a second investigates social relationships between children and robots, and a third centers on children's conceptions of social robots. In identifying and describing the research lines, we largely follow Peter (2022) but specify and extend his approach in our summary of the literature. Notably, we identify for each line one concept or idea that shapes the research it comprises. As the present chapter addresses a wider audience, we also replace Peter's (2022) focus on the *acceptance* of social robots by centering on children's *conceptions* of social robots. The research lines we identify are interrelated and not mutually exclusive, but together they summarize what we consider three important strands of CRI research. As the three lines are defined by the outcomes rather than the purposes of CRI applications, they are not limited to specific situational or societal contexts in which robots are (envisioned to be) used.

19.2.2.1 Children's learning from social robots

Research on children's learning from social robots (for a review, see Belpaeme et al., 2018) can be divided into studies on explicit learning and implicit learning. As defined by Rebuschat (2015, p. xiii), "explicit learning refers to a process during which participants [intentionally] acquire conscious (explicit) knowledge [through instruction]." In primary school settings, scholars have studied how social robots can be used, among other things, to teach children first and second languages (for a review, see Kanero et al., 2018) as well as mathematics education (e.g., Kennedy et al., 2015; Konijn & Hoorn, 2020). Studies on the use of social robots in healthcare have also investigated how social robots can be used, for instance to increase health literacy among children with diabetes (e.g., Blanson Henkemans et al., 2017) and to teach social communicative skills to children with cognitive impairment (e.g., Ismail et al., 2021).

Implicit learning, in contrast, refers to the acquisition of "knowledge about a complex, rule-governed stimulus environment without intending to and without becoming aware of the knowledge they have acquired" (Rebuschat, 2015, p. xiii). In this context, scholars have studied, for example, how children learn prosocial behavior from a social robot while playing a game (Peter et al., 2021), as well as the potential of social robots to increase children's creativity (e.g., Ali et al., 2019; Alves-Oliveira et al., 2020). Moreover, a considerable number of studies have been conducted on children's learning-by-teaching with social robots (e.g., children teaching robots writing, reading, vocabulary, and reasoning skills; see Jamet et al., 2018, for a review). Learning-by-teaching may be considered to lie at the intersection of explicit and implicit learning because children explicitly instruct robots without being aware that the purpose of the interaction is to increase their own knowledge or skills.

An important concept in research on children's learning from and with social robots is engagement because high and sustained engagement levels can increase children's motivation to learn as well as the time they spend on a learning task (see De Haas et al., 2022). Engagement is also relevant to the literature on child–robot relationships (see, e.g., Kanda et al., 2007). However, its relevance to CRI for educational purposes is twofold: learning outcomes may be affected both by children's *social* engagement with a robot and by their *task* engagement with the learning task at hand (see De Haas et al., 2022). In terms of social engagement, mixed findings have been obtained. While some studies suggest that children's social engagement with a robot comes at the cost of learning gains (e.g., Kennedy et al., 2015; Konijn & Hoorn, 2020), others find a positive relationship between children's learning from and social engagement with a robot (e.g., Ahmad et al., 2019; De Haas et al., 2022). Task engagement seems to be positively related to children's learning from robots (e.g., De Haas et al., 2022; Zaga et al., 2015).

In sum, research has shown that children can learn from social robots. Robots can either explicitly instruct children, be role models, or be taught by children themselves. In all these cases, children tend to learn something, ranging from skills to knowledge to behavioral options. Engagement plays an important part in children's learning from social robots: the more successful a robot is in engaging children in a particular task, the better the children learn.

19.2.2.2 Children's relationships with social robots

As social robots communicate in a natural manner (Breazeal et al., 2016a) and as children typically relate to non-human entities in a social way (e.g., Epley et al., 2007), a second substantial line of research has dealt with social relationships between children and robots (for reviews, see Stower et al., 2021; Van Straten et al., 2020b). A large proportion of research on child–robot relationship formation focuses on robot applications, for example evaluating how children's consideration of robots as trustworthy companions improves the outcomes of CRI applications in healthcare (e.g., Ligthart et al., 2019; Sinoo et al., 2018) and education (e.g., Castellano et al., 2021; Kory Westlund & Breazeal, 2019). However, a growing body of research also investigates the mechanisms that underlie the emergence and development of child–robot relationships in their own right (e.g., Calvo-Barajas & Castellano, 2022; Di Dio et al., 2020; Geiskkovitch et al., 2019; Van Straten et al., 2021, 2022). While research on the topic often addresses child–robot relationship formation in a cross-sectional manner (see Van Straten et al., 2020b), the development of social relationships beyond initial perceptions of friendship potential requires time. The longitudinal investigation of child–robot relationship formation, however, is still somewhat underexplored, some notable exceptions notwithstanding (e.g., Leisten et al., 2024; Leite et al., 2013; Ligthart et al., 2022).

A central concept to understand the formation of relationships between children and social robots is trust (see Van Straten et al., 2020b). Children's trust in robots has at least two components: on the one hand, children may have *social* trust in robots, which manifests itself as a belief in the robot's benevolence and a resulting willingness to be vulnerable towards the robot. On the other hand, children may have trust in the robot's *competence* or broader technological functioning (e.g., Stower et al., 2021; Van Straten et al., 2018). Both kinds of trust may be interrelated (see Van Straten et al., 2018) and may play a role in the development of social child–robot relationships. In addition to distinguishing between different kinds of trust, scholars have recently warned against children's *overtrust* in robots, arguing that future studies should investigate how we can enable children to take a more critical stance towards robots (e.g., Beelen et al., 2021; Geiskkovitch & Young, 2020).

In conclusion, research on child–robot relationships initially mainly focused on exploring how feelings of friendship and trust between children and robots can improve the effectiveness of social robot applications in healthcare and education. In recent years, more research attention has been devoted to the question of how relationships between children and robots emerge and unfold over time. Scholars have suggested that distinguishing between social and competence trust may help better understanding of when and why children trust robots. Finally, children's overtrust in robots has been put on the agenda: enabling children to critically evaluate social robots may enhance their ability to judge when and to what extent a robot can be trusted.

19.2.2.3 (Changing) children's conceptions of social robots

Related to the aim of stimulating children's critical thinking about robots, a third prominent line in research on CRI studies children's understanding of and knowledge about social robots. In two seminal papers, Kahn and colleagues (2012, 2013) suggested that children see robots as mental, social, and moral others. Research on children's conceptions of social robots can be categorized accordingly. In terms of mental otherness, studies have investigated children's perception of, among others, robots' reliability as an information source (e.g., Breazeal et al., 2016b) and ability to make choices (Flanagan et al., 2021). In terms of social otherness, scholars have, for instance, compared how children evaluate being 'together' with peers as opposed to robots (Shahid et al., 2014) and have studied whether robots can exert peer pressure (Vollmer et al., 2018). Children's views of robots' moral otherness, finally, have been assessed by investigating, for instance, children's moral concerns for (Sommer et al., 2019) and sharing behavior with (e.g., Nijssen et al., 2021) social robots.

In addition to studying how children perceive social robots, scholars have started to investigate how children's conceptions of social robots can be altered—by emphasizing robots' machine nature (Van Straten et al., 2023; Van Straten et al., 2020c), revealing to children robots' inability to function autonomously (e.g., Tozadore et al., 2017), or by teaching children about concepts related to artificial intelligence (e.g., Williams et al., 2019). Finally, researchers of educational robotics have shown that having children build their own robots with craft materials diversifies their mental schema of and knowledge about social robots (Fortunati et al., 2014, 2022), while having children create robots from robot toolkits enhances their understanding of the automation of machines (Fortunati et al., 2022). Overall, children conceptualize robots, toys, and humans differently: they see robots as more intelligent than toys and attribute more animacy to human beings than to robots (Fortunati et al., 2015).

Questions regarding the manner in which children see and treat social robots—and to what extent this can be influenced—directly tie in with ethical discussions regarding CRI and the increasing importance being attached to the promotion of children's digital literacy skills (DiPaola et al., 2023). An important idea in this context is the New Ontological Category (NOC) hypothesis, as proposed by Kahn and colleagues (2013). The hypothesis posits that social robots defy traditional distinctions between the "basic categories of being" (Kahn et al., 2013, p. 35). It is based on the observation that children's thinking about and behavior towards robots do not directly map onto their conceptualization and treatment of humans, animals, or artifacts, but rather show partial overlap with each of these categories. Consequently, the hypothesis predicts that children growing up with social robots will find it difficult to make clear judgments about robots' characteristics (e.g., regarding their animate or inanimate status) and will answer questions about such characteristics with both 'yes' and 'no' (Kahn et al., 2013).

The NOC hypothesis, however, remains a hypothesis. Severson and Carlson (2010) have posited that, before we can consider it supported, at least five forms of evidence are required. First, children's attributions to robots should cut across prototypic categories. Second, these attributions should convincingly reflect actual beliefs rather than engagement in pretense. Third, attributions made to robots should demarcate structural, rather than functional, overlap with other ontological categories (i.e., robots should not merely *function* differently from, but truly *be* different from, humans, animals, and objects). Fourth, the hybrid conceptualization of robots should persist across children's developmental stages, and fifth, should not disappear with the introduction of more sophisticated robots in society (Severson & Carlson, 2010). In contrast, Smedegaard (2022, p. 8) recently argued that whether social robots indeed constitute a hybrid ontological category is less relevant than "the fact that we feel the need to entertain the thought—and by it, question fundamental assumptions about what it means to be living; to be conscious; to be social; to be a caregiver; to have rights; to have responsibilities; etc."

In conclusion, there is some emerging evidence that children conceptualize social robots in terms of whether they present mental, social, or moral others. Research also suggests that children's conceptions of social robots can be deepened and broadened when educating children about how robots are built. Similarly, information about robots' technical nature can modify children's existing conceptions of social robots. The hypothesis that social robots present a new ontological category deserves future research, notably in comparison to other theoretical frameworks, such as evolutionary theory, which would predict different ontological conceptualizations of social robots (e.g., Lombard & Xu, 2021).

19.3 Challenges in the Field

Since its inception more than 20 years ago, the field of CRI has evolved continuously, and now features an impressive breadth of multidisciplinary research and shows a lively diversity in its research activities. However, research on CRI currently also has to face several limitations and criticisms, for example methodological issues, such as the lack of standardized measures, and theoretical challenges, such as little attention to intercultural differences (e.g., Peter et al., 2019). In our view, three more challenges also deserve attention.

19.3.1 Adult bias

Much CRI research relies on studies that were done among adults and uses theories and concepts that were designed for adults (or at least not for children). Given the lack of research on many topics in CRI, this practice may be inevitable and may not always present a problem. However, when adult-based results or adult-oriented concepts and theories presuppose cognitive, emotional, or behavioral abilities that children do not yet have, using them in CRI research may yield flawed research expectations, unrealistic, child-inappropriate research settings, and inadequate interpretations of findings. For example, interactions with social robots are often modeled after interactions between humans (see De Graaf et al., 2015). As communicative processes have been studied more extensively among adults than among children (see Miller-Day et al., 2013), this increases the risk of an adult bias in studies on children's interactions with social robots.

The often implicit adult bias in CRI research is surprising because, in a design context, scholars have emphasized that researchers need to be aware of the potential differences between adults' and children's views when designing social robots (e.g., Woods et al., 2004) and societal robot applications (e.g., Smakman et al., n. d.). In a less design-oriented context, Belpaeme and colleagues (2013, n. p.) have also warned that "children are not just small adults" and that, inconsequence, children's and adults' interactions with robots may be fundamentally different. For example, children often attribute characteristics to robots that adults tend to reserve for people or animals (Severson & Carlson, 2010). Moreover, children can sometimes hold contradictory beliefs about robots, for instance when thinking of robots as entities with a mind while, at the same time, describing them as machines that are designed and operated by people (Bhamjee et al., 2010).

However, even when we are aware of the differences between the reasoning of children and adults, our own adult perspective on CRI may color our expectations about children's responses to social robots (see, e.g., Van Straten et al., 2021). As a result, CRI researchers should scrutinize not only the concepts, theories, and previous research that they use for a potential adult bias but also their own reasoning. Together with a developmental perspective (Peter et al., 2019), reducing the adult bias in CRI research may help us better understand what children, in particular, expect from interactions with social robots, how they experience the interactions, and how these interactions affect them.

19.3.2 Lack of genuine theoretical CRI frameworks

Previous research has identified the absence and/or heterogeneity of theoretical frameworks as one limitation of CRI research (Peter et al., 2019). However, the theoretical progress of a field may not only be hampered when theoretical frameworks are absent or diverse but also when there are no genuine theoretical frameworks that originate within a field and subsequently guide and unify it, at least as far as the research lines sketched above are concerned. Currently, CRI research is defined primarily by the population it studies—children—but broadly speaking, this focus has not elicited theoretical insights that specifically pertain to what is studied. Rather, the theoretical basis of CRI research resembles a patchwork of theories and concepts (Peter et al., 2019) that tend to be inspired by the home discipline of the researchers rather than the topic itself. This does not mean that existing concepts and theories from established disciplines are ill-suited for the study of CRI. We just believe that theoretical frameworks that are genuine to CRI would contribute to the maturity and identity of the field and improve its focus and orientation.

One reason for the absence of genuine theoretical CRI frameworks may be the nascent state of the field: it is still too young and dispersed to produce coherent and focused theoretical frameworks. A second reason is the multidisciplinary character of CRI research itself. With so many disciplines engaging in CRI research, a wealth of concepts and theories is available that can inspire and guide research. A final reason for the lack of genuine theoretical CRI frameworks may be that we have not yet taken the C in CRI seriously enough. As outlined in the previous section, CRI research is prone to an adult bias. In addition, the potential of a developmental perspective on CRI has not yet been fully explored (Peter et al., 2019). Without a child-centered approach to CRI, however, we may not be able to adequately grasp what specifically distinguishes children's interactions with social robots from those of older populations.

19.3.3 Overreliance on interpersonal theories

Given social robots' natural manner of communication (Breazeal et al., 2016a), interpersonal communication theory is a useful starting point for studies on interactions with social robots (e.g., Westerman et al., 2020). However, the applicability of interpersonal communication theory to the study of human–robot interaction has been questioned (e.g., Fox & Gambino, 2021; Gambino et al., 2020; Peter, 2022): Concepts and theories based on insights from interactions between humans, such as interpersonal communication theories, may inevitably carry an anthropocentric bias when uncritically applied to interactions involving non-human interactants (Gambino et al., 2020). Moreover, in the study of social robots, the use of human-based theories, more generally—and interpersonal communication theory, more specifically—may be problematic because social robots may (yet) lack the ability to engage in complex, human-like interactions (Fox & Gambino, 2021). When implementing interpersonal communication processes in CRI scenarios, scholars should, therefore, keep in mind that "robots are not people" (Dautenhahn, 2007, p. 699) and that we may expect different things from interactions with social robots than from interactions with humans (see also Fox & Gambino, 2021; Van Straten, 2021).

Different expectations of interactions with robots may apply particularly to children whose interactions with robotic technologies will be a normal element of their lives (see also De Graaf, 2017; Van Straten, 2021). For these children, traditional face-to-face interpersonal communication may be only one way of communicating and no longer the gold standard of communication against which all other forms of communication are measured (Sundar, 2008). With their insights—as well as their boundaries—being defined by human–human communication, interpersonal communication theories may consequently not fully capture children's interaction with social robots. Making assumptions about CRI based on the literature on interpersonal communication thus warrants caution as it largely remains to be investigated whether and how interpersonal communication processes transfer to children's interactions with social robots (see also Van Straten, 2021).

19.4 Future Developments and Outlook

Research on CRI cannot be separated from technological, economic, and social developments, as well as ethical discussions more generally. We, therefore, believe that CRI research would benefit from paying more attention to at least the following three developments. First, CRI researchers should more strongly embrace the increasing merger between social robots and smart/connected toys. Many smart/connected toys share technological characteristics that are typical of social robots (e.g., interactivity, sensors, movement, embodiment) and differ from robots only in the number and sophistication of robotic features (Peter et al., 2019). Smart/connected toys are cheaper than elaborate social robots and more likely to be widely used by children. Moreover, children use smart/connected toys typically in their homes ('in the wild'). This may enrich the ecological validity of existing CRI research, in which children are often invited to interact with social robots in surroundings such as labs, schools, or museums. By studying smart/connected toys, CRI researchers can thus learn a lot about children's use and experience of robot-like devices and their effects.

Second, CRI researchers should more strongly study social robots in conjunction with communicative robots. Communicative robots, such as virtual agents, social bots, and work bots, appear "in the form of software and digital infrastructures . . . [and] are (partially) automated and (partially) autonomous communication media that serve the purposes of quasi-communication with humans" (Hepp, 2020, p. 1413, emphasis removed). Traditionally, materiality and physical embodiment are considered crucial features of (social) robots (Winfield, 2012). However, embodiment can also encompass communicative and infrastructural embodiment (Hepp, 2020). Communicative embodiment emerges from the human-likeness of the digital agents when we communicate with them through software. Infrastructural embodiment is constituted by the material infrastructures in which digital agents are integrated (Hepp, 2020).

Although researchers have started to investigate the effects of physical embodiment on children (e.g., Sinoo et al., 2018), the impact of the different types of embodiment on how children interact with robots is still rather poorly understood. Of the various communicative robots, virtual agents (e.g., chatbots) and voice-based assistants (e.g., Siri, Alexa) may be of particular interest to CRI researchers because they are primarily made for social interaction. A direct comparison of social robots with, for instance, virtual agents is challenging if a virtual representation of the physical social robot does not exist and is generally prone to confounding factors. Still, we believe that insights from research that only deals with virtual agents and voice-based assistants may be useful for CRI researchers to better understand the theoretical implications of different types of embodiment. Moreover, such research may at least initially be informative for questions on the design and applicability of robots in different areas (e.g., whether a robot application needs physical embodiment or whether communicative embodiment may suffice to successfully complete a task).

A third development that we consider crucial for future research on children and social robots refers to responsible CRI. Scholars have increasingly emphasized the ethical dimensions of CRI, both in research and applied settings (e.g., DiPaola et al., 2023; Sharkey, 2016). Following Sharkey (2016), De Graaf and Peter (2023) have recently identified four issues in human–robot interaction that may also apply to CRI. To start with, robots may convey deceptive information about what they are and what they can do, which may elicit wrong expectations about robots (e.g., Turkle, 2007), notably among children, who typically expect much from social robots (see, e.g., Kahn et al., 2012). In addition, robots' agency and independent decision making may lead to questions about accountability if problems occur. Although social robots are currently unlikely to interact independently with children, the question of accountability in problematic situations is crucial in machine ethics (Misselhorn, 2019) and deserves even more attention when a protected population, such as children, is concerned. De Graaf and Peter (2023) further identify the tension between personalization and privacy as an issue, which may also apply to CRI. Given their multiple sensors and connection to the internet, social robots may collect personal data about children and thus violate children's privacy. Finally, robots' (allegedly) autonomous behavior may increase the likelihood of human–robot relationships that may lack relational authenticity (Turkle, 2007), again an issue that may particularly apply to children, who readily form relationships with social robots (Van Straten et al., 2020b). Responsible future CRI research may benefit from greater attention to these four issues, not only in terms of how CRI research is set up but also in terms of what is studied and which practical recommendations are made.

With all the quick and far-reaching changes in areas that are central to social robotics, such as artificial intelligence, sensor technology, natural language processing, and computing, social robots are likely to play an important role in the future (Eberl, 2016). As much robotic technology is already present in current smart/connected toys (Peter et al., 2019), children will most probably be confronted with more advanced social robots sooner or later. Against this backdrop, CRI research has already addressed several pressing issues since its beginning more than 20 years ago and yielded many insights into how and why children interact with social robots and the consequences this has. The field shows a lively diversity that offers a variety of typically multidisciplinary approaches and a mix of theoretical and applied research. At the same time, the field is challenged to deepen its theoretical grasp of issues that are central to *children's* interactions with social robots and may benefit from approaching CRI in conjunction with broader technological and ethical developments in human–machine communication. If CRI addresses these challenges and opportunities, it will also in the future significantly contribute to our knowledge about how children—and humans more generally—interact with social robots.

Funding acknowledgment: Work on this chapter was supported by funding from the European Research Council (ERC) under the European Union's Horizon 2020 research and innovation program (grant agreement No. [682733]) to the first author.

References

Ahmad, M. I., Mubin, O., Shahid, S., & Orlando, J. (2019). Robot's adaptive emotional feedback sustains children's social engagement and promotes their vocabulary learning: A long-term child–robot interaction study. *Adaptive Behavior*, *27*(4), 243–266. https://doi.org/10.1177/1059712319844182.

Ali, S., Moroso, T., & Breazeal, C. (2019). Can children learn creativity from a social robot? In *Proceedings of the 2019 Conference on Creativity and Cognition* (pp. 359–368). https://doi.org/10.1145/3325480.3325499.

Alves-Oliveira, P., Arriaga, P., Cronin, M. A., & Paiva, A. (2020). Creativity encounters between children and robots. *Proceedings of the 2020 ACM/IEEE International Conference on Human–Robot Interaction*, 379–388. https://doi.org/10.1145/3319502.3374817.

Bartneck, C., Belpaeme, T., Eyssel, F., Kanda, T., Keijsers, M., & Sabanovic, S. (2020). *Human–Robot Interaction: An Introduction*. Cambridge University Press.

Beelen, T., Velner, E., Ordelman, R., Truong, K. P., Evers, V., & Huibers, T. (2021). Does your robot know? Enhancing children's information retrieval through spoken conversation with responsible robots. *arXiv:2106.07931*. http://arxiv.org/abs/2106.07931.

Belpaeme, T., Baxter, P., De Greeff, J., Kennedy, J., Read, R., Looije, R., Neerincx, M., Baroni, I., & Coti Zelati, M. (2013). Child–robot interaction: Perspectives and challenges. In *Proceedings of the International Conference on Social Robotics* (pp. 452–459). https://doi.org/10.1007/978-3-319-02675-6_45.

Belpaeme, T., Baxter, P., Read, R., Wood, R., Cuayahuitl, H., Kiefer, B., Racioppa, S., Kruijff-Korbayova, I., Athanasopoulos, G., Enescu, V., Looije, R., Neerincx, M., Demiris, Y., Ros-Espinoza, R., Beck, A., Carinamero, L., Hiolle, A., Lewis, M., Baroni, I., . . . Humbert, R. (2012). Multimodal child–robot interaction: Building social bonds. *Journal of Human–Robot Interaction*, *1*(2), 33–53. https://doi.org/10.5898/JHRI.1.2.Belpaeme.

Belpaeme, T., Kennedy, J., Ramachandran, A., Scassellati, B., & Tanaka, F. (2018). Social robots for education: A review. *Science Robotics*, *3*(21), eaat5954. https://doi.org/10.1126/scirobotics.aat5954

Beran, T. N., Ramirez-Serrano, A., Kuzyk, R., Fior, M., & Nugent, S. (2011). Understanding how children understand robots: Perceived animism in child–robot interaction. *International Journal of Human–Computer Studies*, *69*(7–8), 539–550. https://doi.org/10.1016/j.ijhcs.2011.04.003.

Bhamjee, S., Griffiths, F., & Palmer, J. (2010). Children's perception and interpretation of robots and robot behaviour. *Lecture Notes of the Institute for Computer Sciences, Social-Informatics and Telecommunications Engineering*, *59*, 42–48. https://doi.org/10.1007/978-3-642-19385-9_6.

Blanson Henkemans, O. A., Bierman, B. P. B., Janssen, J., Looije, R., Neerincx, M. A., van Dooren, M. M. M., de Vries, J. L. E., van der Burg, G. J., & Huisman, S. D. (2017). Design and evaluation of a personal robot playing a self-management education game with children with diabetes type 1. *International Journal of Human–Computer Studies*, *106*, 63–76. https://doi.org/10.1016/j.ijhcs.2017.06.001.

Breazeal, C. L., Dautenhahn, K., & Kanda, T. (2016a). Social robotics. In B. Siciliano & O. Khatib (Eds.), *Springer Handbook of Robotics* (pp. 1935–1972). Springer International Publishing. https://doi.org/10.1007/978-3-319-32552-1_72.

Breazeal, C. L., Harris, P. L., Desteno, D., Kory Westlund, J. M., Dickens, L., & Jeong, S. (2016b). Young children treat robots as informants. *Topics in Cognitive Science*, *8*(2), 481–491. https://doi.org/10.1111/tops.12192.

Cabibihan, J.-J., Javed, H., Ang, M., & Aljunied, S. M. (2013). Why robots? A survey on the roles and benefits of social robots in the therapy of children with autism. *International Journal of Social Robotics*, *5*(4), 593–618. https://doi.org/10.1007/s12369-013-0202-2.

Calvo-Barajas, N., & Castellano, G. (2022). Understanding children's trust development through repeated interactions with a virtual social robot. In *Proceedings of the 31st IEEE International Conference on Robot and Human Interactive Communication (RO-MAN)* (pp. 1451–1458). https://doi.org/10.1109/RO-MAN53752.2022.9900537.

Castellano, G., De Carolis, B., D'Errico, F., Macchiarulo, N., & Rossano, V. (2021). PeppeRecycle: Improving children's attitude toward recycling by playing with a social robot. *International Journal of Social Robotics*, *13*(1), 97–111. https://doi.org/10.1007/s12369-021-00754-0.

Dautenhahn, K. (1999). Robots as social actors: Aurora and the case of autism. In *Proc. CT99, The Third International Cognitive Technology Conference, August, San Francisco* (pp. 359–374).

Dautenhahn, K. (2007). Socially intelligent robots: Dimensions of human–robot interaction. *Philosophical Transactions of the Royal Society B: Biological Sciences*, *362*(1480), 679–704. https://doi.org/10.1098/rstb.2006.2004.

De Graaf, M. M. A. (2017). An ethical evaluation of human–robot relationships. *International Journal of Social Robotics*, *8*(4), 589–598. https://doi.org/10.1007/s12369-016-0368-5.

De Graaf, M. M. A., Ben Allouch, S., & van Dijk, J. A. G. M. (2015). What makes robots Social? A user's perspective on characteristics for social human–robot interaction. In A. Tapus, E. André, J.-C. Martin, F. Ferland, & M. Ammi (Eds.), *Social Robotics* (Vol. 9388, pp. 184–193). Springer International Publishing. https://doi.org/10.1007/978-3-319-25554-5_19.

De Graaf, M. M. A., & Peter, J. (2023). Human social relationships with social robots. In A. L. Guzman, R. McEwen, & S. Jones (Eds.), *The Sage Handbook of Human–Machine Communication* (pp. 435–441). Sage.

De Haas, M., Vogt, P., van den Berghe, R., Leseman, P., Oudgenoeg-Paz, O., Willemsen, B., De Wit, J., & Krahmer, E. (2022). Engagement in longitudinal child–robot language learning interactions: Disentangling robot and task engagement. *International Journal of Child–Computer Interaction*, *33*, 100501. https://doi.org/10.1016/j.ijcci.2022.100501.

De Jong, C., Peter, J., Kühne, R., & Barco, A. (2019). Children's acceptance of social robots: A narrative review of the research 2000–2017. *Interaction Studies*, *20*(3), 393–425. https://doi.org/10.1075/is.18071.jon.

Di Dio, C., Manzi, F., Peretti, G., Cangelosi, A., Harris, P. L., Massaro, D., & Marchetti, A. (2020). Shall I trust you? From child–robot interaction to trusting relationships. *Frontiers in Psychology*, *11*. https://doi.org/10.3389/fpsyg.2020.00469.

DiPaola, D., Charisi, V., Breazeal, C., & Sabanovic, S. (2023). Children's fundamental rights in human–robot interaction research: A systematic review. In *Companion of the 2023 ACM/IEEE International Conference on Human–Robot Interaction* (pp. 561–566). https://doi.org/10.1145/3568294.3580148.

Eberl, U. (2016). *Smarte Maschinen: Wie künstliche Intelligenz unser Leben verändert [Smart Machines: How Artificial Intelligence Changes our Lives]*. Hanser.

Epley, N., Waytz, A., & Cacioppo, J. T. (2007). On seeing human: A three-factor theory of anthropomorphism. *Psychological Review*, *114*(4), 864–886. https://doi.org/10.1037/0033-295X.114.4.864.

Flanagan, T., Rottman, J., & Howard, L. H. (2021). Constrained choice: Children's and adults' attribution of choice to a humanoid robot. *Cognitive Science*, *45*(10), e13043. https://doi.org/10.1111/cogs.13043.

Fong, T., Nourbakhsh, I., & Dautenhahn, K. (2003). A survey of socially interactive robots. *Robotics and Autonomous Systems*, *42*(3–4), 143–166. https://doi.org/10.1016/S0921-8890(02)00372-X.

Fortunati, L., Esposito, A., Ferrin, G., & Viel, M. (2014). Approaching social robots through playfulness and doing-it-yourself: Children in action. *Cognitive Computation*, *6*(4), 789–801. https://doi.org/10.1007/s12559-014-9303-y.

Fortunati, L., Esposito, A., Sarrica, M., & Ferrin, G. (2015). Children's knowledge and imaginary about robots. *International Journal of Social Robotics*, *7*(5), 685–695. https://doi.org/10.1007/s12369-015-0316-9.

Fortunati, L., Manganelli, A. M., & Ferrin, G. (2022). Arts and crafts robots or LEGO® MINDSTORMS robots? A comparative study in educational robotics. *International Journal of Technology and Design Education*, *32*(1), 287–310. https://doi.org/10.1007/s10798-020-09609-7.

Fox, J., & Gambino, A. (2021). Relationship development with humanoid social robots: Applying interpersonal theories to human–robot interaction. *Cyberpsychology, Behavior, and Social Networking*, *24*(5), 294–299. https://doi.org/10.1089/cyber.2020.0181.

Gambino, A., Fox, J., & Ratan, R. (2020). Building a stronger CASA: Extending the computers are social actors paradigm. *Human–Machine Communication*, *1*(1), 71–86. https://doi.org/10.30658/hmc.1.5.

Geiskkovitch, D. Y., Thiessen, R., Young, J. E., & Glenwright, M. R. (2019). What? That's not a chair! How robot informational errors affect children's trust towards robots. In *2019 14th ACM/IEEE International Conference on Human–Robot Interaction (HRI)* (pp. 48–56). https://doi.org/10.1109/HRI.2019.8673024.

Geiskkovitch, D. Y., & Young, J. E. (2020). Children's overtrust: Intentional use of robot errors to decrease trust. In *Proceedings of the 29th International Conference on Robot and Human Interactive Communication*. http://hci.cs.umanitoba.ca/assets/publication_files/Childrens_Overtrust.pdf.

Han, J., Jo, M., Jones, V., & Jo, J. H. (2008). Comparative study on the educational use of home robots for children. *Journal of Information Processing Systems*, *4*(4), 159–168. https://doi.org/10.3745/JIPS.2008.4.4.159.

Hepp, A. (2020). Artificial companions, social bots and work bots: Communicative robots as research objects of media and communication studies. *Media, Culture & Society*, *42*(7–8), 1410–1426. https://doi.org/10.1177/0163443720916412.

Ismail, L. I., Hanapiah, F. A., Belpaeme, T., Dambre, J., & Wyffels, F. (2021). Analysis of attention in child–robot interaction among children diagnosed with cognitive impairment. *International Journal of Social Robotics*, *13*(2), 141–152. https://doi.org/10.1007/s12369-020-00628-x.

Jamet, F., Masson, O., Jacquet, B., Stilgenbauer, J.-L., & Baratgin, J. (2018). Learning by teaching with humanoid robot: A new powerful experimental tool to improve children's learning ability. *Journal of Robotics*, *2018*, e4578762. https://doi.org/10.1155/2018/4578762.

Kahn, P. H., Friedman, B., Perez-Granados, D. R., & Freier, N. G. (2006). Robotic pets in the lives of preschool children. *Interaction Studies*, *7*(3), 405–436. https://doi.org/10.1075/is.7.3.13kah.

Kahn, P. H., Gary, H. E., & Shen, S. (2013). Children's social relationships with current and near-future robots. *Child Development Perspectives*, *7*(1), 32–37. https://doi.org/10.1111/cdep.12011.

Kahn, P. H., Kanda, T., Ishiguro, H., Freier, N. G., Severson, R. L., Gill, B. T., Ruckert, J. H., & Shen, S. (2012). "Robovie, you'll have to go into the closet now": Children's social and moral relationships with a humanoid robot. *Developmental Psychology*, *48*(2), 303–314. https://doi.org/10.1037/a0027033.

Kanda, T., Hirano, T., Eaton, D., & Ishiguro, H. (2004). Interactive robots as social partners and peer tutors for children: A field trial. *Human–Computer Interaction*, *19*(1), 61–84. https://doi.org/10.1207/s15327051hci1901&2_4.

Kanda, T., Sato, R., Saiwaki, N., & Ishiguro, H. (2007). A two-month field trial in an elementary school for long-term human–robot interaction. *IEEE Transactions on Robotics*, *23*(5), 962–971. https://doi.org/10.1109/TRO.2007.904904.

Kanero, J., Geçkin, V., Oranç, C., Mamus, E., Küntay, A. C., & Göksun, T. (2018). Social robots for early language learning: Current evidence and future directions. *Child Development Perspectives*, *12*(3), 146–151. https://doi.org/10.1111/cdep.12277.

Kennedy, J., Baxter, P., & Belpaeme, T. (2015). The robot who tried too hard: Social behaviour of a robot tutor can negatively affect child learning. In *Proceedings of the Tenth Annual ACM/IEEE International Conference on Human–Robot Interaction* (pp. 67–74). https://doi.org/10.1145/2696454.2696457.

Konijn, E. A., & Hoorn, J. F. (2020). Robot tutor and pupils' educational ability: Teaching the times tables. *Computers & Education*, *157*, 103970. https://doi.org/10.1016/j.compedu.2020.103970.

Kory Westlund, J. M., & Breazeal, C. L. (2019). A long-term study of young children's rapport, social emulation, and language learning with a peer-like robot playmate in preschool. *Frontiers in Robotics and AI*, *6*(81). https://doi.org/10.3389/frobt.2019.00081.

Leisten, L. M., Heyselaar, E., Bosse, T., & Hortensius, R. (2024). Children's reciprocity and relationship formation with a robot across age. *Technology, Mind, and Behavior*, *5*(2). https://doi.org/10.1037/tmb0000131.

Leite, I., Martinho, C., & Paiva, A. (2013). Social robots for long-term interaction: A survey. *International Journal of Social Robotics*, *5*(2), 291–308. https://doi.org/10.1007/s12369-013-0178-y.

Ligthart, M. E. U., Fernhout, T., Neerincx, M. A., van Bindsbergen, K. L. A., Grootenhuis, M. A., & Hindriks, K. V. (2019). A child and a robot getting acquainted: Interaction design for eliciting self-disclosure. In *Proceedings of the International Joint Conference on Autonomous Agents and Multiagent Systems* (pp. 61–70). http://edithlaw.ca/teaching/cs889/w20/readings/disclosure.pdf.

Ligthart, M. E. U., Neerincx, M. A., & Hindriks, K. V. (2022). Memory-based personalization for fostering a long-term child–robot relationship. *Proceedings of the 17th ACM/IEEE International Conference on Human–Robot Interaction*, 80–89. https://doi.org/10.1109/HRI53351.2022.9889446.

Lombard, M., & Xu, K. (2021). Social responses to media technologies in the 21st century: The media are social actors paradigm. *Human–Machine Communication*, *2*(1). https://doi.org/10.30658/hmc.2.2.

Miller-Day, M., Pezalla, A., & Chesnut, R. (2013). Children are in families too! The presence of children in communication research. *Journal of Family Communication*, *13*(2), 150–165. https://doi.org/10.1080/15267431.2013.768251.

Misselhorn, C. (2019). *Grundfragen der Maschinenethik [Fundamental Issues of Machine Ethics]* (4th ed.). Reclam.

Nijssen, S. R. R., Müller, B. C. N., Bosse, T., & Paulus, M. (2021). You, robot? The role of anthropomorphic emotion attributions in children's sharing with a robot. *International Journal of Child–Computer Interaction*, *30*, 100319. https://doi.org/10.1016/j.ijcci.2021.100319.

Peter, J. (2022). Social robots and children. In D. Lemish (Ed.), *The Routledge International Handbook of Children, Adolescents, and Media* (2nd ed., pp. 153–161). Routledge.

Peter, J., Kühne, R., & Barco, A. (2021). Can social robots affect children's prosocial behavior? An experimental study on prosocial robot models. *Computers in Human Behavior*, *120*, 106712. https://doi.org/10.1016/j.chb.2021.106712.

Peter, J., Kühne, R., Barco, A., De Jong, C., & Van Straten, C. L. (2019). Asking today the crucial questions of tomorrow: Social robots and the Internet of Toys. In G. Mascheroni & D. Holloway (Eds.), *The Internet of Toys: Practices, Affordances and the Political Economy of Children's Smart Play* (pp. 25–46). Palgrave Macmillan (Springer Nature).

Rebuschat, P. (2015). Introduction: Implicit and explicit learning of languages. In P. Rebuschat (Ed.), *Implicit and Explicit Learning of Languages* (pp. xiii–xxii). John Benjamins.

Serholt, S. (2018). Breakdowns in children's interactions with a robotic tutor: A longitudinal study. *Computers in Human Behavior*, *81*, 250–264. https://doi.org/10.1016/j.chb.2017.12.030.

Severson, R. L., & Carlson, S. M. (2010). Behaving as or behaving as if? Children's conceptions of personified robots and the emergence of a new ontological category. *Neural Networks*, *23*(8–9), 1099–1103. https://doi.org/10.1016/j.neunet.2010.08.014.

Shahid, S., Krahmer, E., & Swerts, M. (2014). Child–robot interaction across cultures: How does playing a game with a social robot compare to playing a game alone or with a friend? *Computers in Human Behavior*, *40*, 86–100. https://doi.org/10.1016/j.chb.2014.07.043.

Sharkey, A. J. C. (2016). Should we welcome robot teachers? *Ethics and Information Technology*, *18*(4), 283–297. https://doi.org/10.1007/s10676-016-9387-z.

Sinoo, C., van der Pal, S., Blanson Henkemans, O. A., Keizer, A., Bierman, B. P. B., Looije, R., & Neerincx, M. A. (2018). Friendship with a robot: Children's perception of similarity between a robot's physical and virtual embodiment that supports diabetes self-management. *Patient Education and Counseling*, *101*(7), 1248–1255. https://doi.org/10.1016/j.pec.2018.02.008.

Smakman, M., Foppen-de Graaf, D., & Van den Eijkel, S. (n. d.). The comparison of the perspective of adults vs children with social robots in education. Unpublished manuscript.

Smedegaard, C. V. (2022). Novelty knows no boundaries: Why a proper investigation of novelty effects within SHRI should begin by addressing the scientific plurality of the field. *Frontiers in Robotics and AI*, *9*, 741478. https://doi.org/10.3389/frobt.2022.741478.

Sommer, K., Nielsen, M., Draheim, M., Redshaw, J., Vanman, E. J., & Wilks, M. (2019). Children's perceptions of the moral worth of live agents, robots, and inanimate objects. *Journal of Experimental Child Psychology*, *187*, 104656. https://doi.org/10.1016/j.jecp.2019.06.009.

Stower, R., Calvo-Barajas, N., Castellano, G., & Kappas, A. (2021). A meta-analysis on children's trust in social robots. *International Journal of Social Robotics*, *13*, 1979–2001. https://doi.org/10.1007/s12369-020-00736-8.

Sundar, S. S. (2008). Self as source: Agency and customization in interactive media. In E. A. Konijn, S. Utz, M. Tanis, & S. B. Barnes (Eds.), *Mediated Interpersonal Communication* (pp. 58–74). Routledge.

Tanaka, F., Cicourel, A., & Movellan, J. R. (2007). Socialization between toddlers and robots at an early childhood education center. *Proceedings of the National Academy of Sciences of the United States of America*, *104*(46), 17954–17958. https://doi.org/10.1073/pnas.0707769104.

Tozadore, D. C., Pinto, A., Romero, R., & Trovato, G. (2017). Wizard of Oz vs autonomous: Children's perception changes according to robot's operation condition. *Proceedings of the Sixth International Symposium on Robot and Human Interactive Communication*, 664–669. https://doi.org/10.1109/ROMAN.2017.8172374.

Turkle, S. (2007). Authenticity in the age of digital companions. *Interaction Studies*, *8*(3), 501–517. https://doi.org/10.1075/is.8.3.11tur.

Turkle, S. (2011). *Alone Together: Why We Expect More from Technology and Less from Each Other*. W. W. Norton & Company.

Turkle, S., Taggart, W., Kidd, C. D., & Dasté, O. (2006). Relational artifacts with children and elders: The complexities of cybercompanionship. *Connection Science*, *18*(4), 347–361. https://doi.org/10.1080/09540090600868912.

Van den Berghe, R. (2022). Social robots in a translanguaging pedagogy: A review to identify opportunities for robot-assisted (language) learning. *Frontiers in Robotics and AI*, *9*, 958624. https://doi.org/10.3389/frobt.2022.958624.

Van den Berghe, R., Verhagen, J., Oudgenoeg-Paz, O., van der Ven, S., & Leseman, P. (2019). Social robots for language learning: A review. *Review of Educational Research*, *89*(2), 259–295. https://doi.org/10.3102/0034654318821286.

Van Straten, C. L. (2021). The next best friend? How children perceive and relate to a social robot. Doctoral dissertation, University of Amsterdam.

Van Straten, C. L., Kühne, R., Peter, J., Jong, C. de, & Barco, A. (2020a). Closeness, trust, and perceived social support in child–robot relationship formation: Development and validation of three self-report scales. *Interaction Studies*, *21*(1), 57–84. https://doi.org/10.1075/is.18052.str.

Van Straten, C. L., Peter, J., & Kühne, R. (2020b). Child–robot relationship formation: A narrative review of empirical research. *International Journal of Social Robotics*, *12*(2), 325–344. https://doi.org/10.1007/s12369-019-00569-0.

Van Straten, C. L., Peter, J., & Kühne, R. (2023). Transparent robots: How children perceive and relate to a social robot that acknowledges its lack of human psychological capacities and machine status. *International Journal of Human–Computer Studies*, *177*, 103063. https://doi.org/10.1016/j.ijhcs.2023.103063.

Van Straten, C. L., Peter, J., Kühne, R., & Barco, A. (2020c). Transparency about a robot's lack of human psychological capacities: Effects on child–robot perception and relationship formation. *ACM Transactions on Human–Robot Interaction*, *9*(2), 11:1–11:22. https://doi.org/10.1145/3365668.

Van Straten, C. L., Peter, J., Kühne, R., & Barco, A. (2021). The wizard and I: How transparent teleoperation and self-description (do not) affect children's robot perceptions and child–robot relationship formation. *AI & Society*, *37*, 383–399. https://doi.org/10.1007/s00146-021-01202-3.

Van Straten, C. L., Peter, J., Kühne, R., & Barco, A. (2022). On sharing and caring: Investigating the effects of a robot's self-disclosure and question-asking on children's robot perceptions and child–robot relationship formation. *Computers in Human Behavior*, *129*, 107135. https://doi.org/10.1016/j.chb.2021.107135.

Van Straten, C. L., Peter, J., Kühne, R., De Jong, C., & Barco, A. (2018). Technological and Interpersonal Trust in Child–Robot Interaction: An Exploratory Study. In *HAI '18: Proceedings of the 6th International Conference on Human–Agent Interaction* (p. 259). https://doi.org/10.1145/3284432.3284440.

Vollmer, A.-L., Read, R., Trippas, D., & Belpaeme, T. (2018). Children conform, adults resist: A robot group induced peer pressure on normative social conformity. *Science Robotics*, *3*(21), eaat7111. https://doi.org/10.1126/scirobotics.aat7111.

Werry, I., Dautenhahn, K., Ogden, B., & Harwin, W. (2001). Can social interaction skills be taught by a social agent? The role of a robotic mediator in autism therapy. In M. Beynon, C. L. Nehaniv, & K. Dautenhahn (Eds.), *Cognitive Technology: Instruments of Mind, Proceedings* (Vol. 2117, pp. 57–74). Springer-Verlag Berlin. http://uhra.herts.ac.uk/bitstream/2299/3160/1/902096.pdf.

Westerman, D., Edwards, A. P., Edwards, C., Luo, Z., & Spence, P. R. (2020). I-it, I-thou, I-robot: The perceived humanness of AI in human–machine communication. *Communication Studies*, *71*(3), 393–408. https://doi.org/10.1080/10510974.2020.1749683.

Williams, R., Park, H. W., & Breazeal, C. (2019). A is for artificial intelligence: The impact of artificial intelligence activities on young children's perceptions of robots. *Proceedings of the 2019 CHI Conference on Human Factors in Computing Systems*, 1–11. https://doi.org/10.1145/3290605.3300677.

Winfield, A. (2012). *Robotics: A Very Short Introduction*. Oxford University Press. https://doi.org/10.1093/actrade/9780199695980.001.0001.

Woods, S., Dautenhahn, K., & Schulz, J. (2004). The design space of robots: Investigating children's views. In *RO-MAN 2004: 13th IEEE International Workshop on Robot and Human Interactive Communication (IEEE Catalog No.04TH8759)* (pp. 47–52). https://doi.org/10.1109/ROMAN.2004.1374728.

You, Z.-J., Shen, C.-Y., Chang, C.-W., Liu, B.-J., & Chen, G.-D. (2006). A robot as a teaching assistant in an English class. In *Sixth IEEE International Conference on Advanced Learning Technologies (ICALT'06)* (pp. 87–91). https://doi.org/10.1109/ICALT.2006.1652373.

Zaga, C., Lohse, M., Truong, K. P., & Evers, V. (2015). The effect of a robot's social character on children's task engagement: Peer versus tutor. In A. Tapus, E. André, J.-C. Martin, F. Ferland, & M. Ammi (Eds.), *Social Robotics* (Vol. 9388, pp. 704–713). Springer International Publishing. https://doi.org/10.1007/978-3-319-25554-5_70.

Section 3: **Contexts of Human–Robot Interaction**

Yolande Strengers, Thao Phan, Melisa Duque, and Kari Dahlgren

Anthropomorphizing Voice Assistants: A Research Agenda for Human–AI Relationships

Abstract: Voice-based assistants (VBAs) are extensively embedded into many people's lives through smartphones, smart speakers, global positioning systems, and other voice-enabled devices. User studies indicate that they can be helpful and accessible aids for completing everyday tasks; however, there is a growing body of concerns relating to their impact on society. Many of these relate to the anthropomorphization of VBAs, and, in particular, their presentation as deracialized and feminized figures, which are imbued with compliant service-based personas that commonly manifest in social robots. This chapter maps the key developments in research on the social implications associated with the anthropomorphization of VBAs, focusing on the intersecting themes of race, gender, and age-based inclusion across three areas of research on anthropomorphized VBAs. First, we identify the decisions and biases evident in the commercial imaginaries of VBAs, where controversies and diversities are erased or glossed over in order to promote an idealized and familiar future. Second, we explore how race, gender, and age have been mobilized in the design decisions of VBAs, which prioritize stereotypical traits in order to improve likability and usability. Third, we consider how anthropomorphized VBAs are encountered in everyday life in ways that both exacerbate and complicate commercial imaginaries and usability intentions. We draw on this analysis to outline the methods, approaches, and critical agendas for understanding the inclusion implications of VBAs. We conclude by calling for a focus on VBAs as a window into our future relationships with everyday AI and robotics.

Keywords: Voice-based assistants, gender, race, age, inclusion, everyday life, AI

20.1 Introduction

In just a decade since the world's first commercially available voice-based assistant (VBA), Apple's Siri, was commercially released, these devices have become ubiquitous, with over 4 billion now in existence. By 2024, market analysts predict there will be more VBAs on the planet than people, embedded into smartphones, smart speakers, and the Internet of Things (Lee, 2022). VBAs present themselves as anthropomorphized characters, which is a design strategy intended to endear users towards these devices and facilitate intuitive and engaging interactions. Indeed, anthropomorphism is a key factor in the improvement of usability, characterized as a 'common sense'

https://doi.org/10.1515/9783110792270-021

strategy in allowing more people to access and experience the growing functionalities of these and other smart technologies and services (Sweeney, 2021).

However, VBAs have also been criticized for amplifying problematic racial and gendered stereotypes, reinforcing dangerous biases, commercializing and oversimplifying the labor of emotional and social care, being difficult to use (for some people), masking the potential privacy and security risks of embedded voice-based technologies in our lives, and enabling the expansion of regimes of data capture into more intimate spaces such as the home (Fortunati et al., 2022a, 2022b; Phan, 2019; Sadowski et al., 2021; Schiller & McMahon, 2019; Strengers & Kennedy, 2020; Sweeney, 2016, 2021). For instance, a landmark UNESCO report singled out gendered VBAs for perpetuating harmful biases, which the authors claimed foreshadowed debates that will become prominent as artificial intelligence (AI) assumes greater human-like capabilities (West et al., 2019). With this in mind, VBAs are not only significant in their own right but also as gateway devices and prototypes for the emergence and evolution of everyday AI and robotics. It is, therefore, critically important that research and scholarship are developing and engaging with methods and approaches that help us understand the complex human–machine relationships that are forming around VBAs and their broader impacts on and in society.

In this chapter, we outline the major contributions and advancements on the social roles and effects of anthropomorphized VBAs in everyday life, with a particular focus on their ability to support or undermine racial, gender, and age-based inclusion across three areas of research: commercial imaginaries, design and usability, and everyday encounters. This is particularly important in the context of this volume's focus on social robots, given that VBAs inform the conversational AI and aesthetic imaginaries that are central to companion and care robots. For instance, in *The Smart Wife*, Yolande Strengers and Jenny Kennedy (2020) show how the design and development of VBAs, smart home devices, social robots, and sex robots represent different manifestations of a common vision towards gendered care in the home. Similarly, Edwards et al. (2024) argue that VBAs and social robots are part of a continuum of embodiment, ranging from the disembodied voice to communicative social robots. VBAs, then, provide important and relatable insights into the characteristics, personalities, and aesthetics of robotic imaginaries and the ways in which they are experienced and incorporated into everyday situations (Rhee, 2018).

We continue by highlighting emerging areas of concern and research needs that build on existing work. Our central contribution is a set of methods and approaches we are developing with colleagues to better understand the roles and relationships facilitated through VBAs in everyday life contexts. Given the rapid rise of VBAs and their increasing embeddedness or interconnectedness into an extended smart ecosystem, including robots, appliances, and cars, we argue for approaches that take us beyond usability and grapple with fundamental questions such as how human–machine relations are imagined, choreographed, and lived; and how they become implicated within hierarchies of inequity. This allows us to speculate not only on current inclusion concerns but also on those that are foregrounded as we experiment with future-focused methodologies that provide a glimpse into possible or alternative relationships with everyday AI. Finally, we consider what the future may hold for VBAs and how our research agenda can support these emerging developments.

20.2 Visions, Usability, and Everyday Encounters

20.2.1 Commercial imaginaries

Early visions of the smart home provided the imaginaries that set the scene for the introduction of VBAs. These visions prioritized and romanticized wealthy, white homes with traditional gender values; they either ignored domestic labor or promised to seamlessly take care of it with a new workforce of technologies working to serve the home and its occupants. For instance, Anne-Jorunn Berg's (1994, p. 170) analysis of twentieth-century smart house prototypes drew attention to their distinctive masculine bias, most clearly evidenced by the complete absence of "technologies that have anything to do with housework," and the prioritization of a home premised on the goal of integration, centralized control and regulation of all functions. Through her analysis of smart home socio-technical imaginaries, Berg (1994, p.176) argued that the gendered work required to sustain the home was rendered largely invisible, with the smart home positioned as serving its target consumer and user: "the technically-interested man."

Jump forward to the twenty-first century, and more recent scholarship finds similar biases and assumptions at play in contemporary smart home imaginaries (Strengers & Nicholls, 2018; Strengers & Kennedy, 2020). Smart home marketing materials still prioritize the technically interested man as their core consumer; however, the interests of that man have expanded. Men in many cultures are now responsible for more domestic labor, and smart home technologies have followed by marketing smart and robotic appliances for their convenience and labor-saving qualities that promise the twin fantasies of leisure and untiring productivity (Spigel, 2005; Strengers et al., 2020).

Tracing the history of smart home imaginaries throughout the twentieth century, Lynn Spigel (2005, p. 219) observed how high-tech companies draw from progressive 1950s lifestyle visions to weave together a "curious mixture of nostalgia and futurism." It is here in the spectacular prototypes, demonstration projects, and visionary statements from companies such as General Electric, Westinghouse, and Disney (and more recently corporations such as Intel, Microsoft, Amazon, Sony, Google, and Apple) that we see the emerging imaginaries for VBAs come to life. Situated within the broader vision of a "home of tomorrow," VBAs "promise . . . instant familiarity and sentient knowledge of all of our needs [and] should be seen in relation to the widespread cultural anxieties expressed in films and other popular media" (Spigel, 2005, p. 43). Emerging from these visions and anxieties is a desire for smart homes to actually *become* these fantasies. As Spigel (2005, p. 408) argues: "Whereas the old 'mechanical servants' (i.e., stoves and dishwashers) were promoted pragmatically as timesaving devices that could enhance the lives of housewives, the new intelligent appliances do not just do the chores. They virtually become the housewife, as they perform the managerial and caretaking roles previously ascribed to women." Strengers and Kennedy (2020) take up this proposition to argue that VBAs appeal to the nostalgic ideals of a 1950s housewife through their feminized roles, personalities, and default voices, thereby reinforcing and perpetuating problematic stereotypes about women and gender in society.

More broadly, VBAs enter the smart home positioned as enabling assistants of the smart home ecosystem, coming pre-imbued with gendered connotations and biases. Additionally, they set the scene for a sensory and aesthetic imaginary that promises mastery over the environment with demure feminine care. Users are invited to "imagine" themselves inside this oasis, where VBAs take center stage as the orchestrators of the entire smart home system:

> Imagine walking into a room and with a 'good morning, Alexa' watch a pre-programmed ballet of functions take place: the perfectly dimmed lights flick on, the blinds glide open (using Crestron's silent motors), your playlist sings, the TV turns on, muted, to show the headlines and a voice whispers the weather forecast. (Newspaper article cited in Strengers et al., 2020, p. 5)

These commercial imaginaries are not only gendered but deeply classed and racialized. In her work analyzing the promotional material of the home automation device, the Amazon Echo, Thao Phan (2019) argued that new domestic technologies like the Echo function to infer a sense of class privilege through the provision of certain levels of domestic service. She draws parallels with nineteenth- and twentieth-century paradigms of servant keeping, arguing that smart home devices, like servants before them, form part of the imaginary of middle-class respectability by enabling the domestic organization and projections of affluence and control. Where this imaginary departs, however, is the Echo's aesthetic coding through certain markers of whiteness, specifically, in the voice of Alexa, the VBA embedded within the Echo. Phan argues that this coding operates strategically to mask the discomfort associated with racialized servitude, dehistoricizing and depoliticizing the servant/master relation: "Where this relation has been fundamentally predicated on racialized human exploitation, the figure of the white-speaking, middle-class AI effectively erases this essential narrative" (Phan, 2019, p. 27). Moreover, the figuration of the relationship between device/user in the image of servant/master misrepresents the direction of power. Pointing to the Echo device as a new interface for the collection of user data, Phan argues that the work of interaction, which aids in generating user profiles for the purposes of personalized advertising, is equivalent to a kind of digital labor consistent with what Christian Fuchs (2015) calls "prosumer labour." In this way, Amazon is able to mask its position of power, eliding and misdirecting questions of labor by sheltering behind Alexa as the figurative servant: "To describe smart devices as working for or in service of households is an inaccurate representation of the direction of power; on the contrary, it is the household that labors for the device" (Phan, 2019, p. 29).

While racialized figurations within imaginaries of automation are certainly not unique to VBAs (see Atanasoski & Vora, 2019), in this context they function as a commercial assurance that, despite their disruptive potential, these emerging technologies can seamlessly fit into pre-existing domestic routines and social relations. Unsurprisingly, the lived realities of friction and disarray are rarely represented within these commercial visions, and many people are entirely absent. Older people, for example, are depicted as being able to independently engage with and use VBAs without difficulty or support, despite ongoing research demonstrating an age-based digital divide. Similarly, people with accents, and languages that differ from those recognized or provided by VBAs (such as in multi-generational and culturally and linguistically diverse households) are excluded from VBA imaginaries (Lawrence, 2021).

Importantly, then, the intended role for VBAs is more about evolving nostalgic fantasies for everyday life as it is practical needs or realities. This has serious implications for the design and usability of these devices.

20.2.2 Design and usability

The expansion of VBAs into consumer markets has been met with a concentrated effort from the research community to understand their benefits and challenges for different people and user groups, with the aim of improving their design and usability. Much of this research reflects the issues that arise from the imaginaries described above and their subsequent embedding into the design of devices. In particular, the gendered, deracialized, and supposedly universal design of VBAs has resulted in a range of concerns, such as the way they respond to gendered abuse and harassment, their inability to live up to their imaginaries, and the difficulties they pose for marginalized user groups such as older people.

In a general sense, the development of VBAs has been largely viewed as beneficial, enabling intuitive and hands-free interaction between people and AI. It has been particularly praised for its ability to make more systems accessible to a more diverse range of users (compared with touch screen or keyboard interaction), including people with vision impairments, people with dyslexia, older people, and children. While the conversational abilities and functionalities of VBAs are still limited, this technology has proved to have particular benefits in hands-free settings (such as when driving, cooking, or doing manual tasks) and in providing engaging, entertaining, and educational experiences for diverse users (Sciuto et al., 2018; Soffer, 2020).

The aesthetic coding of VBAs, and in particular the focus on femininity, whiteness, and compliance as design default, has raised a number of concerns. First, this feminized figure conforms to conventional gender and service-role stereotypes in order to materialize familiar patterns of sociality between user and device. For instance, Fortunati et al. (2022a, p. 1) argue that the feminization of Amazon's Alexa VBA contributes "to reinforcing a retrieval ideology of the feminine as the place of social subordination and contempt." Here, the argument is that this particular feminization of VBAs does not challenge sexist or racist social orders; rather, it exploits these orders to create a device that feels easy and natural to use (Bergen, 2016; Phan, 2017; Sweeney, 2021). Indeed, the term "natural" is often used within the field of human–computer interaction to justify problematic design choices (Phan, 2022). These assume that people have a "natural preference" for feminized voices and merely prefer non-accented voices. These kinds of positions, generated through user studies (see Nass & Gong, 2000; Nass & Brave, 2005), assume that preferences have manifested through evolutionary biological "hardwiring," overlooking the vital role social and cultural experience also play in materializing desire, taste, and comfort. More broadly, the anthropomorphizing of VBAs and social robots further complicates simple distinctions between human and machine that shape our interactions and relationships with VBAs beyond "natural" or intuitive interactions (Fortunati & Edwards, 2021). By maintaining the status quo, users effectively engage with VBAs and largely do not question their corporate or market intentions (Phan, 2017). However, this easy understanding and acceptance of familiar VBAs may also pose security and privacy risks to users, particularly children and older people, who may be unaware of or unable to respond to the personal risks of their data being collected and passed on to corporate owners or third parties (Barassi, 2020; Sadowski et al., 2021). This default also enables the expansion of regimes of data capture, used as another technique to grow infrastructures of digital capitalism (Sweeney, 2021).

A second concern raised by the positioning of VBAs as a service-oriented woman is how this can encourage gendered abuse and harassment. As noted in the introduction, this issue has attracted the attention of UNESCO, who found that the design imperative of female "likeability" has been pursued at the expense of ensuring VBAs are assertive rather than subordinate (West et al., 2019). Scholars and commentators have criticized VBA manufacturers for their positive or neutral responses to gendered harassment and abuse and demonstrated how the abuse directed at feminized VBAs is heightened compared with that directed at male or gender-neutral VBAs, and highly gendered in nature. While abusing a device is not in itself inherently dangerous, a common argument made by feminist scholars of technology is that abusing feminized and human-like devices normalizes abusive behavior, and reproduces harmful patterns that position women and other gender minorities as acceptable targets of abuse. This argument is taken up by Strengers and Kennedy (2020) in their analysis of how VBAs are discussed in public commentary as "bitches with glitches," where technology faults are blamed on feminized VBAs, rather than the manufacturers or male-dominated teams that design and create them.

The supposedly universal design of VBAs also poses challenges to their usability. While VBA technology is arguably inherently more inclusive than other forms of interaction, as we have already outlined, it is designed with an "average" user in mind who is fluent in dominant languages and accents, conforms to dominant linguistic and cultural conventions of language and speech, and has a degree of technical competence and understanding. Studies have shown that people who are not "average," such as those from culturally or linguistically diverse communities, or older people who have cognitive and memory difficulties, may struggle to communicate with VBAs using culturally-specific command-based speech. These difficulties commonly translate into not being heard or understood by the device. In a smart home ecosystem, with more voice-enabled devices, appliances, and functions, this can create important inequities in the home and other environments, particularly if marginalized users are unable to access basic services or perform ordinary tasks (such as turning lights on and off) without support. These barriers can also reproduce patterns of colonial disciplining as "nonstandard accented speakers of English must assimilate to 'participate' in emerging technologies" (Lawrence, 2021, p. 85). There are also voice and spoken nuances beyond volume adaptability for different hearing impairments that people may wish to tailor to negotiate everyday household relations, sleep routines, and speech speed when interacting with VBAs and other interconnected devices, people, and services (Duque et al., 2021).

As a result of these issues, designers of emerging technologies have responded to this divide, not by designing *for*, but designing *with* people whose needs have been overlooked, such as older adults, to ensure future technologies are inclusive of sensory, cognitive, culturally sensitive, personal values, and aging needs (Strengers et al., 2022; Vines et al., 2015; Wilson et al., 2022). Without these studies, imaginaries, as well as technologies, risk continuing to perpetuate ageism (and other forms of exclusion) with devices that approach older age as a homogeneous problem that needs solving (Boger, 2022), with overly simplified functions that undermine the technical expertise that older adults have from developing, using, repairing, and remaking past technologies that precede current and future innovations (Talamo et al., 2011). If the design of imaginaries as well as devices excludes older adults' skills and abilities (Kuoppamäki et al., 2021) in the co-designing of future VBAs, resistant responses from older adults to learn, adopt, and engage with emerging technologies will continue to widen age divides and usability barriers (Barros Pena et al., 2021), extending older adults' sense of digital dependencies with family members and service providers. This, in turn, would undermine policy ambitions and industry trends that advocate for the potential of VBAs and emerging technologies to support independent living and social connectivity for aging (such as Australia's Royal Commission into Aged Care Quality and Safety 2021 and UN/WHO's Decade of Healthy Ageing (2021–2030)).

20.2.3 Everyday encounters

A growing number of studies are exploring people's relationships with VBAs "in the wild." These studies extend beyond controlled user-testing environments to explore how diverse people encounter VBAs in real-world situations and within the context of everyday life (Neves & Vetere, 2019; Sunyoung, 2021). Such studies have been foundational in understanding people's everyday encounters with VBAs and how the imaginaries and design decisions discussed previously are actually experienced. While these studies do address usability, those that adopt qualitative or ethnographic methodologies often take a broader view to examine how VBAs are changing gendered or social roles, establishing new human–AI relationships, or being integrated into a whole-of-household dynamic. In past research, we have undertaken several studies involving commercially available VBAs such as Amazon Alexa and Google Home, and involving ethnographic methods with diverse households, including families, people from culturally and linguistically diverse backgrounds, and older people.

Through the Digital Energy Futures project, we conducted ethnographic research with 72 households in Australia, collecting current experiences and future visions about emerging digital and energy technologies in people's homes. While the energy sector often envisions VBAs to be an entry point for greater automation of the home, including its energy use, research with households found that VBAs were primarily used for small tasks, such as playing music or setting alarms (echoing the results of other studies, e.g. Duque et al., 2022; Fortunati et al., 2022a). When they were integrated into other smart home systems, they could alter power relations within the home and create new inequalities and access issues. For instance, one household that had recently migrated from Vietnam had a number of smart and automated devices in the home, including lights that were controlled by an Amazon Alexa. The parents and children spoke to Alexa in English to manage the lights, but the grandmother (who only spoke Vietnamese) struggled to interact. They explained that because Alexa doesn't speak Vietnamese, she cannot use voice control. Instead, she has to ask her grandchildren for help controlling the system or use the manual switch if she is only turning the lights on or off. Even when participants could speak English fluently, VBAs often struggled with accents of non-native English speakers, who altered their own accents to match those understandable by Alexa (Pink, 2022, 18:12). Similar concerns have been raised by other researchers, who have found that the language and command-based structure of engaging with VBAs, and more broadly the limited range of "universal" emotional expressions that AI can recognize and respond to, has the potential to prioritize certain forms of cultural expression over others (Kennedy & Strengers, 2022; Noble, 2018).

Our research also revealed that VBAs were popular with children, who found engaging with them fun and entertaining. Asking the VBAs to tell a joke, play a favorite song, or "ask weird and wonderful questions," as one participant described his daughter's engagement, gave children control that they might not have through physical interfaces that require literacy (Strengers et al., 2021b, p. 83). Some considered engaging with the VBA to be a form of "imaginative play," which they felt was uncomfortable as an adult, to help explain the difference between a child's interaction and their own. Still others recognized that it was important for children to understand the difference between a person and a machine. One participant, in particular, described how he found this distinction to be a justification for acting rudely to a VBA. He described how he found it ridiculous to teach children to be polite to a VBA by saying please or thank you. Insisting on politeness, he argued, made it harder for children to understand the difference between humans and machines. Sherry Turkle (2017) has long drawn attention to the problems associated with children being unable to distinguish between human and anthropomorphized devices. Treating these devices rudely is rarely suggested as the solution to this problem although some experts do advocate for better child education. Leading child data expert Veronica Barassi focuses on the responsibility of corporations to protect the data privacy of children and to limit their (and all people's) exposure to what she describes as "algorithmic violence," which describes the profiling of people by Big Tech companies from an early age to engage in harmful manipulative tactics and to erase diversity and difference (Barassi, 2020). Ethnographic research reveals these complexities and tensions involved in how people react to and encounter the perceived risks of VBAs in diverse and sometimes contradictory ways.

In another study, Smart Homes for Seniors, which provided VBAs (Google Home assistants) to 23 households (and 33 participants, aged between 64 and 94) as part of a broader trial of smart home technologies, including scaffolded support from an aged care and research team, we identified a range of benefits and challenges as VBAs encountered people's everyday lives (Duque et al., 2021; Strengers et al., 2022). Contrary to most research, which has focused on the health and independence benefits of smart home devices for older people, we found that this cohort took great pleasure and joy from engaging with VBAs for fun and play. This included asking Google to play music, tell jokes, provide fun facts, and become an arbitrator in family debates. However, our research also identified a range of challenges that participants encountered when engaging with VBAs, including remembering the correct commands to activate their devices, not knowing how to reset or fix VBA functionality when it no longer recognized smart appliances or devices that were connected to it (such as smart lights activated by voice control), and feeling uncomfortable engaging with conversational devices that didn't conform to their generational and cultural conventions of politeness and social interaction (such as saying please and thank you). Additionally, while some people were concerned about being listened to by the devices, they were generally unaware of the potential privacy and security risks these devices might have posed to them.

The ways in which participants navigated their everyday experiences with the VBAs involved processes of learning, creative adaptation, and collaborative troubleshooting (Duque et al., 2022). While participants were willing to welcome these devices into their intimate home spaces and routines, they maintained a critical disposition to deal with functional, systemic, and service impositions from the VBAs. For example, when safety concerns arose, some participants whispered private conversations when close to the devices to prevent the VBAs from listening in, disconnected these from the wall switch, or stopped using them. Alternatively, when VBAs directed participants to additional paid streaming services (e.g., Youtube, Spotify), they found ways to bypass the offer to instead access the alternatives available at no extra cost. Over time, participants developed their own routines with the VBAs; however, as the study demonstrated, this was a collaborative effort between participants, their families, and the socio-technical support provided by the project team and the aged care provider. This revealed the value older people experienced by augmenting their digital living skills, as well as the limitations of VBAs in terms of their restricted design affordances for older adults and the challenges of accessible connectivity in regional and remote locations (Strengers et al., 2021a).

20.3 Towards a More Inclusive Agenda for VBAs

Despite the growing body of research on VBAs in everyday life contexts, the majority of research remains uncritical of the broader societal effects of anthropomorphized VBAs and robots. An additional challenge is that current research is focused on commercially available VBAs and existing imaginaries, which provide limited opportunities to reimagine our future lives with VBAs and other emerging AI. In this section, we outline our future-focused research agenda to disrupt and advance this field.

20.3.1 Design 'corrections'

Before continuing, it is worth pausing to consider more conventional pathways to address the research challenges this chapter has identified. The most obvious of these is to focus on designing more progressive and "inclusive" VBAs. There have already been considerable contributions in this area, particularly from feminist and queer scholars and technologists. These include design manifestas for feminist devices (e.g., 'F'xa,' a feminist guide to AI bias, https://f-xa.co/), designing gender-neutral voices that can easily integrate with VBAs (e.g., 'Q' genderless voice, https://www.genderlessvoice.com/), and designing chatbots with the characteristics of dignity and respect (Feldmen, 2016). Picking up on these ideas and more, in *The Smart Wife* Strengers and Kennedy (2020) propose a wide-ranging "feminist reboot" for VBAs, which includes changes to the ways "smart wives" are imagined and depicted in inspirational sci-fi narratives, which often go on to inform the development of actual devices. Additionally, human–computer interaction design scholars have made many practical suggestions for designers to better address the usability needs of older people (Vines et al., 2015) and other marginalized users, including improving accent recognition and changing the phrasing and commands of devices to suit different cultures and age groups (Duque et al., 2021; Lawrence, 2021).

These are worthy efforts that seek to improve the design of VBAs; however, they are not without significant risks. Indeed, despite the significant contributions of many scholars to "correct" VBAs to become more inclusive, this has resulted in minimal changes to the mainstream design of these devices. For instance, following Leah Fessler's (2017) campaign to improve the ways in which VBAs respond positively or neutrally to sexual abuse and harassment, companies such as Amazon, Google, and Apple made some small design improvements. Similarly, following several years of concerted pressure and critique of VBAs' female voices, nearly all leading manufacturers now offer male and non-binary voice options or require users to select the gender of their voice during set-up. Many companies now even claim their devices are gender neutral or "feminist" (Strengers & Kennedy, 2020). Apart from only dealing with a very small fraction of the issues and suggestions feminist scholars and activists have raised, these design amendments have arguably only served to further entrenched VBAs into everyday infrastructures and have largely glossed over or avoided the larger and more pressing questions that feminist scholars have raised regarding the need to question the fundamental purpose and role of VBAs in society (Sadowski et al., 2021; Sweeney, 2021).

For example, Wajcman (2017) calls for deeper introspection on the social problems that AI such as VBAs are being brought in to address. Referencing the solutionist future discourses inherent in Silicon Valley imaginaries and their potential to support aged care, Wajcman singles out automation as a distraction from bigger societal questions about how we revalue and renumerate aged care and support the care of older people in our communities: "We may be 'suckers for the wide eyes and endearing giggles of affective robots,' but to advocate the use of robots for empathetic care of the elderly mistakes the appearance of care with real empathy and genuine personal interaction" (Wajcman, 2017, p. 123).

Wajcman speculates that the very need for caring robots (with VBAs on board) would dissipate if we fundamentally reconsidered how we care for older people in society. These broader socially significant questions of care and responsibility are disregarded when well-meaning research is used to focus attention on design tweaks and inclusive updates rather than fundamentally questioning the value and role of VBAs (and indeed all assistive and relational technology) in society. With this in mind, we focus the remainder of this chapter on the kinds of research methods and approaches that are needed to enable the necessary societal introspection to critically examine not just the future design of VBAs but their fundamental purpose.

20.3.2 Research methods and approaches

According to Wajcman (2017, p. 126), a key problem for research and design is the way that techno-solutionist narratives pervade the imaginaries of AI and robots like VBAs: "Only by insisting that futures are always social can public bodies, rather than autonomous markets and endogenous technologies, become central to disentangling, debating and delivering those futures." A key component of our own research has been to disrupt techno-solutionist narratives and draw on ethnographic research to firmly ground them in the social. This is part of the design ethnography approach we have been developing with colleagues in the Emerging Technologies Research Lab, where the aim is to play on innovation agendas in order to disrupt and reimagine them (Pink et al., 2022). While this work comes with risks, such as well-meaning research being co-opted into corporate agendas (as outlined above), it is also crucial in finding ways to disrupt the dominant imaginaries for VBAs—and, more broadly, for AI and automated and robotic futures.

Finding ways to disrupt socio-technical imaginaries and highlight their harmful or unrealistic assumptions makes several important advancements. First, it can draw attention to the dominant personas or ideal users of smart devices, such as the "resource man" (Strengers, 2013), the "techno-hedonist" (Dahlgren et al., 2021), the "smart wife" (Strengers & Kennedy, 2020) or "pod man" (Quilty et al., 2022), and propose more realistic and plausible alternatives that bring everyday experience to the fore. Second, it allows us to directly contrast and compare the plausibility of industry imaginaries against everyday realities to discredit these narratives and alert their advocates to problems and concerns. Third, it opens processes of designing emerging technologies to collaborative and ethnographic methods that advance agendas for inclusive, ethical, and diverse futures. This work is particularly important for informing the policies and programs of governments and service providers who, while invested in smart technology narratives, have broader social and policy commitments that they are charged with delivering and who, therefore, are likely to welcome plausible and realistic alternatives.

For instance, our Smart Homes for Seniors project, funded by the Australian Department of Health, demonstrated that VBAs and other smart home devices must be integrated into a service-oriented support system for aged care which has people at its core (Strengers et al,. 2021a). This project is part of a broader understanding that futures can be enriched by engaging people in processes of co-design that account for past temporalities and current lived experiences. A "techno-historical" and "life course perspective" approach, advocated by Barros Pena and colleagues (2021), can further contribute to uncovering and addressing causes of digital exclusion, such as those related to age. Similarly, in our Digital Energy Futures project, funded by the Australian Government in partnership with three energy providers, we used a traffic light system to indicate whether industry visions for people's future lives with smart technologies aligned fully (green), somewhat (orange), or not at all (red) with people's own expectations for their futures (Dahlgren et al., 2020; Strengers et al., 2021b). With most industry visions sitting in the orange or red categories, we demonstrated the danger in energy providers and governments continuing to prioritize these imaginaries as the basis for their policy decisions. Such research places VBAs in a broader context than usability, considering how they function in everyday life settings and in relation to other technologies, with specific health, mobility, or energy policy objectives in mind. This allows researchers to step beyond whether a device is liked or useful to ask broader questions about whether it can help support significant governance and social challenges, such as providing adequate aged care, improving gender equality in the division of household labor, or addressing the significant environmental and energy challenges associated with the climate crisis.

Beyond analyzing and disrupting techno-solutionist narratives, another crucial path involves design ethnographic research with people. Conducting research on futures that pushes beyond the visions of the technology industry or mainstream science fiction creates a methodological dilemma that requires innovative methodologies (Markham, 2021). Ethnography opens possibilities for researching futures within the lived realities of people's everyday practices, values, and priorities because it is essentially a form of qualitative research that is conducted in context. And yet, studying the future in context presents a methodological challenge that design ethnography attempts to address. One method we have developed in our Digital Energy Futures project is to approximate inhabiting futures ethnographically with research participants through the use of comic strips, which enable a form of co-presence in futures and stretch imaginations beyond those of techno-solutionist industry visions (Dahlgren et al., 2022) For example, we showed research participants a comic strip of a future morning routine, which, as one detail, portrayed a smart mirror with an embedded VBA suggesting an outfit based on the schedule and the weather conditions; a future vision taken directly from an analysis of technology industry visions for VBAs. Participants reflected on this emerging technology through their own lives. For example, one participant remarked negatively about the prescriptive and normative way a VBA suggesting an outfit would impact her mental health. She reflected on this VBA through her personal history, in which bullying and depression led to a social anxiety that is exacerbated by a "little voice in the back of my head" that makes constant negative self-judgments. A VBA telling her that there is a right or wrong thing to wear was too similar to the "little voice, [that's] gotten me into a whole host of trouble over my life" (Dahlgren et al., 2021, p. 773). Although only one of many examples, such research provides insights into how VBAs do not always fit into people's everyday lives in the simple way they are intended by their designers. For VBAs to be incorporated in more positive ways, they need to take into account the complex realities of people's lives.

There is also a more speculative research pathway that holds promise for realizing other relationships and roles for VBAs and AI more broadly in our future everyday lives. A key challenge with asking people how else we could relate to emerging AI is that we are constrained by our existing preconceptions and understandings of it through the current dominant products on the market and the equally limited science fiction and popular culture imaginaries that are often noted as their inspiration (Humphry & Chesher, 2021). In past research, we have pioneered innovative research methods with colleagues that enable us to glimpse at alternative possible futures—both with and without VBAs. For instance, in a series of workshops run as part of the

Designing Interactive Systems conference (Nicenboim et al., 2020), a group of researchers working at the critical intersection of VBAs in society sought to explore alternative personalities for and modes of relating with everyday AI. Initial activities involving "thing interviews" with market-leading VBAs led to limited design inspiration, although we have also used this method to reveal the biases and problematic responses provided by VBAs (Strengers, 2022). In a thing interview, one person interviews another who pretends to be and embody a thing, in this case, a VBA. The interview follows an open-ended set of inquiries, similar to ethnographic fieldwork, and places this method in the broader genre of "thing ethnography" (Giaccardi et al., 2016).

Following our limited success with existing VBA thing interviews, the organizers invited workshop participants to engage in thing interviews with imaginary VBAs, but even here, the level of design insight was limited by participants' own imaginaries and dominant understandings. Finally, the group turned its attention to thing interviews with ordinary objects within the home, inspired by the podcast "Everything is Alive." In this more speculative activity, participants interviewed each other as a household object, such as a coffee pot, house plant, or toilet paper. A team of researchers analyzed these short encounters to reveal important insights and understandings of how else we could relate to animated or human-like technologies in our everyday lives (Reddy et al., 2021). Thing interviews and ethnography are just two examples of the kinds of speculative methods that can help researchers and their participants imagine alternative ways of being and relating to future technologies.

Interviewing seemingly random objects raises new concerns regarding cultural and social specificity that are yet to even be considered in mainstream research on VBAs. These concerns not only pick up on people's relationships with VBAs (e.g., how different people might relate to a coffee pot) but also with the characteristics we ascribe to everyday objects and how they affect our experiences and impressions of them (Reddy et al., 2021). If toilet paper were to suddenly talk, for instance, how would we make sense of its intransigence, disposability, and association with abjection and human waste? Would we seek to imbue it with dignity and respect, or would we use it as an avenue for demeaning interactions and abuse? Further, what kinds of fears and risks would talking toilet paper raise for humans? Would, for example, toilet paper serve as a medium for intrusive powers to start monitoring fluid, medication and drug intake, bowel movements, or menstrual cycles, and how would that affect people's interactions and relationships with it? Giving ordinary things a voice instantly imbues them with anthropomorphic qualities and everyday politics that add new complexities to our relationships with them while also providing avenues to explore and anticipate our future lives with VBAs and other everyday AI.

20.4 Futures in Tension

Over the past decade, VBAs have quickly become a mainstream technology and a key way in which smartphone users operate their devices and engage with a wider system of smart device functionalities and services. Within this fast-evolving ecosystem, it is important for researchers to understand and respond to the evolving and unequally experienced outcomes of engaging with these devices. We began this chapter by highlighting the tensions between commercial imaginaries oriented towards nostalgic fantasies and ideal users, the design and usability challenges of VBAs for gender, racial, and age-based inclusion, and the everyday realities for diverse groups who engage with these devices. While it is tempting to highlight design improvements and upgrades to better support inclusive outcomes, and while some of our research has indeed prioritized these opportunities, this approach risks losing sight of larger societal questions, such as how we can best progress racial and gender equality, or support our aging population with high-quality care. Additionally, a focus on design changes doesn't allow us to take the necessary steps to question and understand our future possible relationships with emerging AI, of which VBAs are an early prototype.

With these objectives in mind, this chapter's contribution has centered on the research methods and approaches that are available to researchers as we seek to disrupt dominant imaginaries and propose alternatives, and as we explore and understand the potential implications of future human–AI relationships with VBAs and other social robots. Importantly, supporting inclusion in this context does not just mean generating positive usability outcomes for particular disadvantaged or marginalized user groups, but a fundamental questioning of how technology broadly—and VBAs specifically—can support the critical issues facing our societies, as well as raising broader questions about what inclusion looks like when our relationships with ordinary objects become more akin to human relationships or companion species (Haraway, 2003). VBAs are, therefore, not only an object worthy of study in their own right but also provide a window into a broader set of relations that are anticipated to emerge in a world saturated with AI, robotics, and automation. Questioning whether that future is desirable, inclusive, and perhaps even possible is a central concern that should orient future research on the social role of VBAs, robots, and everyday AI.

References

Atanasoski, N., & Vora, K. (2019). *Surrogate Humanity: Race, Robots, and the Politics of Technological Futures.* Duke University Press.

Barassi, V. (2020). *Child Data Citizen: How Tech Companies are Profiling Us from Before Birth.* MIT Press.

Barros Pena, B., Rachel E. C., Holmquist, L. E., & John, V. (2021). Circumspect users: Older adults as critical adopters and resistors of technology. In *Conference on Human Factors in Computing Systems–Proceedings*. https://doi.org/10.1145/3411764.3445128.

Berg, A. J. (1994). A gendered socio-technical construction: The smart house. In C. Cockburn & R. Furst Dilic (Eds.), *Bringing Technology Home: Gender and Technology in Changing Europe* (pp. 165–180). Open University Press.

Bergen, H. (2016). "I'd blush if I could": Digital assistants, disembodied cyborgs and the problem of gender. *Word and Text: A Journal of Literary Studies and Linguistics*, *6*(01), 95–113.

Boger, J. (2022). Culture change, human-centered design, and ethical by design as transactional cornerstone concepts in the development of technology for supporting aging. In *Proceedings of the 15th International Conference on PErvasive Technologies Related to Assistive Environments (PETRA '22)* (pp. 556–561). Association for Computing Machinery https://doi.org/10.1145/3529190.3535692.

Dahlgren, K., Strengers, Y., Pink, S., Nicholls, L. & Sadowski, J. (2020). Digital Energy Futures: Review of industry trends, visions, and scenarios for the home. Emerging Technologies Research Lab (Monash University). https://www.monash.edu/__data/assets/pdf_file/0004/2900623/Digital-Energy-Futures-Report.pdf

Dahlgren, K., Pink, S., Strengers, Y., & Nicholls, L. (2022). Co-presence and contingency: Comics as a methodological innovation in researching automated futures. *Qualitative Inquiry*, *28*(7), 767–780. https://doi.org/10.1177/10778004221097630.

Dahlgren, K., Pink, S., Strengers, Y., Nicholls, L., & Sadowski, J. (2021). Personalization and the Smart Home: Questioning techno-hedonist imaginaries. *Convergence*, *27*(5), 1155–1168. https://doi.org/10.1177/13548565211036801.

Duque, M., Pink, S., Mortimer, M., Strengers, Y., Martin, R., Nicholls, L., Horan, B., Eugene, A., & Thomson, S. (2022). Troubleshooting as a method in COVID-19 times: Smart home ethnographies and remote aged care innovation. *Digital Creativity*, *33*(3), 1–16. https://doi.org/10.1080/14626268.2022.2088560.

Duque, M., Pink, S., Strengers, Y., Martin, R., & Nicholls, L. (2021). Automation, wellbeing and digital voice assistants: Older people and Google devices. *Convergence*, *27*(5), 1–18. https://doi.org/10.1177/13548565211038537.

Edwards, A., Edwards, C., Fortunati, L., Manganelli, A. M., & de Luca, F. (2024). Mass robotics: How do people communicate with, use, and feel about Alexa?A cross-cultural, user perspective. *Computers in Human Behavior: Artificial Humans*.

Feldman, J. (2016, December 31). The bot politic. *New Yorker*. https://www.newyorker.com/tech/annals-of-technology/the-bot-politic.

Fessler, L. (2017). We tested bots like Siri and Alexa to see who would stand up to sexual harassment. *Quartz*. https://qz.com/911681/we-tested-apples-siri-amazon-echos-alexa-microsofts-cortana-and-googles-google-home-to-see-which-personal-assistant-bots-stand-up-for-themselves-in-the-face-of-sexual-harassment/.

Fortunati, L., & Edwards, A. (2021). Moving ahead with human–machine communication. *Human–Machine Communication*, *2*, 7–28.

Fortunati, L., Edwards, A., Edwards, C., Manganelli, A. M., & de Luca, F. (2022a). Is Alexa female, male, or neutral? A cross-national and cross-gender comparison of perceptions of Alexa's gender and status as a communicator. *Computers in Human Behavior*, *137*, 1–9. https://doi.org/https://doi.org/10.1016/j.chb.2022.107426.

Fortunati, L., Edwards, A. P., Manganelli, A. M., Edwards, C., & de Luca, F. (2022b). Do people perceive Alexa as gendered? A cross-cultural study of people's perceptions, expectations, and desires of Alexa. *Human–Machine Communication*, *5*(1), 75–97.

Fuchs, C. (2015). Dallas Smythe and digital labor. In R. Maxwell (Ed.), *The Routledge Companion to Labor and Media* (pp. 51–62). Routledge.

Giaccardi, E., Cila, N., Speed, C., & Caldwell, M. (2016). Thing ethnography: Doing design research with non-humans. In *Proceedings of the 2016 ACM Conference on Designing Interactive Systems, Brisbane, QLD, Australia*. https://doi.org/10.1145/2901790.2901905.

Haraway, D. J. (2003). *The Companion Species Manifesto: Dogs, People, And Significant Otherness* (Vol. 1). Prickly Paradigm Press.
Humphry, J., & Chesher, C. (2021). Preparing for smart voice assistants: Cultural histories and media innovations. *New Media & Society, 23*(7), 1971–1988. https://doi.org/10.1177/1461444820923679.
Kennedy, J., & Strengers, Y. (2022). Alexa's got a hunch: The human decisions behind programming emotion-sensing and caregiving into digital assistants. In S. Pink, M. Berg, D. Lupton, & M. Ruckenstein (Eds.), *Everyday Automation: Experiencing and Anticipating Emerging Technologies* (pp. 1–13). Routledge.
Kuoppamäki, S., Tuncer, S., Eriksson, S., & McMillan, D. (2021). Designing kitchen technologies for ageing in place. *Proceedings of ACM on Interactive, Mobile, Wearable and Ubiquitous Technologies, 5*(2), 1–19. https://doi.org/10.1145/3463516.
Lawrence, H. (2021). Siri disciplines. In T. S. Mullaney, B. Peters, M. Hicks, & K. Philip (Eds.), *Your Computer Is On Fire* (pp. 179–198). MIT Press.
Lee, E. (2022). The rise of voice AI adoption in a post-pandemic world. *AI Rudder*. Available at: https://www.airudder.com/the-rise-of-voice-ai-adoption-in-a-post-pandemic-world/#.
Markham, A. (2021). The limits of the imaginary: Challenges to intervening in future speculations of memory, data, and algorithms. *New Media & Society, 23*(2), 382–405.
Nass, C., & Brave, S. (2005). *Wired for Speech: How Voice Activates and Advances the Human–Computer Relationship*. MIT Press.
Nass, C., & Gong, L. (2000). Speech interfaced from an evolutionary perspective. *Communications of the ACM, 43*(9), 36–43. https://dl.acm.org/doi/fullHtml/10.1145/348941.348976.
Neves, B., & Vetere, F. (Eds.) (2019). *Ageing and Digital Technology: Designing and Evaluating Emerging Technologies for Older Adults*. Springer.
Nicenboim, I., Giaccardi, E., Søndergaard, M. L. J., Reddy, A. V., Strengers, Y., Pierce, J., & Redström, J. (2020). More-than-human design and AI. In *Conversation with Agents Companion Publication of the 2020 ACM Designing Interactive Systems Conference, Eindhoven, Netherlands*. https://dl.acm.org/doi/abs/10.1145/3393914.3395912.
Noble, S. U. (2018). *Algorithms of Oppression: How Search Engines Reinforce Racism*. NYU Press.
Phan, T. (2017). The materiality of the digital and the gendered voice of Siri. *Transformations, 29*, 23–33.
Phan, T. (2019). Amazon Echo and the aesthetics of whiteness. *Catalyst: Feminism, Theory, Technoscience, 5*(1), 1–38.
Phan, T. (2022). Sounding right/sounding white. In A. Briers, N. Carah, & H. Arden (Eds.), *Conflict in my Outlook* (pp. 114–117). Perimeter Editions.
Pink, S. (Producer & Director) (2022). *Digital Energy Futures Documentary*. Emerging Technologies Research Lab.
Pink, S., Fors, V., Lanzeni, D., Duque, M., Sumartojo, S., & Strengers, Y. (2022). *Design Ethnography: Research, Responsibilities, and Futures*. Taylor & Francis.
Quilty, E., Pink, S., Phan, T., & Lee, J. (2022). *Automated Decision-Making in Transport Mobilities: Review of Industry Trends and Visions for the Future*. Emerging Technologies Research Lab, Monash University and ARC CoE for ADM+S. https://apo.org.au/node/318977.
Reddy, A., Kocaballi, A. B., Nicenboim, I., Søndergaard, M. L. J., Lupetti, M. L., Key, C., Speed, C., Lockton, D., Giaccardi, E., Grommé, F., Robbins, H., Primlani, N., Yurman, P., Sumartojo, S., Phan, T., Bedö, V., & Strengers, Y. (2021). Making everyday things talk: Speculative conversations into the future of voice interfaces at home. In *Extended Abstracts of the 2021 CHI Conference on Human Factors in Computing Systems, Yokohama, Japan*. https://doi.org/10.1145/3411763.3450390.
Rhee, J. (2018). *The Robotic Imaginary: The Human and the Price of Dehumanized Labor*. University of Minnesota Press.

Sadowski, J., Strengers, Y., & Kennedy, J. (2021). More work for Big Mother: Revaluing care and control in smart homes. *Environment and Planning A: Economy and Space*, *56*(1). https://doi.org/10.1177/0308518X211022366.

Schiller, A., & McMahon, J. (2019). Alexa, alert me when the revolution comes: Gender, affect, and labor in the age of home-based artificial intelligence. *New Political Science*, *41*(2), 173–191. https://doi.org/10.1080/07393148.2019.1595288.

Sciuto, A., Saini, A., Forlizzi, J., & Hong, J. I. (2018). "Hey Alexa, what's up?": Studies of in-home conversational agent usage. *DIS 2018—Proceedings of the 2018 Designing Interactive Systems Conference*, 857–868. https://doi.org/10.1145/3196709.3196772.

Soffer, O. (2020). From textual orality to oral textuality: The case of voice queries. *Convergence*, *26*(4), 927–941.https://doi.org/10.1177/135485651982577.

Spigel, L. (2001). *Welcome to the Dreamhouse: Popular Media and Postwar Suburbs*. Duke University Press. https://doi.org/10.1215/9780822383178.

Spigel, L. (2005). Designing the smart house: Posthuman domesticity and conspicuous production. *European Journal of Cultural Studies*, *8*(4), 403–426. https://doi.org/10.1177/13675494050578.

Strengers, Y. (2013). *Smart Energy Technologies in Everyday Life: Smart Utopia?* Palgrave MacMillan. https://doi.org/10.1057/9781137267054.

Strengers, Y. (2022). AI at home: An urgent urban policy and research agenda. *Urban Policy and Research*, *40*(3), 250–258. https://doi.org/10.1080/08111146.2022.2067845.

Strengers, Y., Dahlgren, K., Nicholls, L., Pink, S., & Martin, R. (2021b). *Digital Energy Futures: Future Home Life*. Emerging Technologies Research Lab (Monash University). https://www.monash.edu/__data/assets/pdf_file/0011/2617157/DEF-Future-Home-Life-Full-Report.pdf.

Strengers, Y., Duque, M., Mortimer, M., Pink, S., Eugene, A., Nicholls, L., Horan, B., & Thomas, S. (2021a). *Smart Homes for Seniors: Intelligent Home Solutions for Independent Living* (Final research evaluation report). McLean Care, Monash University and Deakin University. https://intelligenthomesolutions.com.au/wp-content/uploads/2021/02/Final-Report.pdf.

Strengers, Y., Duque, M., Mortimer, M., Pink, S., Martin, R., Nicholls, L., Horan, B., Eugene, A., & Thomson, S. (2022). "Isn't this marvelous": Supporting older adults' wellbeing with smart home devices through curiosity, play and experimentation. *Designing Interactive Systems Conference*. https://doi.org/10.1145/3532106.3533502.

Strengers, Y., Hazas, M., Nicholls, L., Kjeldskov, J., & Skov, M. B. (2020, April). Pursuing pleasance: Interrogating energy-intensive visions for the smart home. *International Journal of Human–Computer Studies*, *136*, 1–14. https://doi.org/10.1016/j.ijhcs.2019.102379.

Strengers, Y., & Kennedy, J. (2020). *The Smart Wife: Why Siri, Alexa and other Smart Home Devices Need a Feminist Reboot*. MIT Press. https://doi.org/10.7551/mitpress/12482.001.0001.

Strengers, Y., & Nicholls, L. (2018). Aesthetic pleasures and gendered tech-work in the 21st-century smart home. *Media International Australia*, *166*(1), 70–80. https://doi.org/10.1177/1329878x17737661.

Sunyoung, K. (2021). Exploring how older adults use a smart speaker-based voice assistant in their first interactions: Qualitative study. *JMIR mHealth and uHealth*, *9*(1), 1–12. https://doi.org/10.2196/20427.

Sweeney, M. E. (2016). The Ms. Dewey 'experience': Technoculture, gender, and race. In J. Daniels, K. Gregory, & T. M. Cottom (Eds.), *Digital Sociologies* (pp. 401–420). Policy Press.

Sweeney, M. E. (2021). Digital assistants. In N. B. Thylstrup, D. Agostinho, A. Ring, C. D'Ignazio, & K. Veel (Eds.), *Uncertain Archives* (pp. 151–160). MIT Press. https://doi.org/10.7551/mitpress/12236.003.0016.

Talamo, A., Giorgi, S., & Mellini, B. (2011). Designing technologies for ageing: Is simplicity always a leading criterion? In *ACM International Conference Proceeding Series* (pp. 33–36). https://doi.org/10.1145/2037296.2037308.

Turkle, S. (2017). *Alone Together: Why We Expect More from Technology and Less from each other*. Hachette UK.

Vines, J., Pritchard, G., Wright, P., Olivier, P., & Brittain, K. (2015). An age-old problem. *ACM Transactions on Computer–Human Interaction*, *22*(1), 1–27. https://doi.org/10.1145/2696867.
Wajcman, J. (2017, March). Automation: Is it really different this time? *British Journal of Sociology*, *68*(1), 119–127. https://doi.org/10.1111/1468-4446.
West, M., Kraut, R., & Ei Chew, H. (2019). *I'd Blush if I Could: Closing Gender Divides in Digital Skills through Education*. UNESCO. https://en.unesco.org/Id-blush-if-I-could.
Wilson, C., Muñoz, D., Ambe, A., Vines, J., Pedell, S., Brereton, M., & Pschetz, L. (2022). Co-creating futures of care with older adults. In *Proceedings of the Participatory Design Conference 2022—Volume 2 (PDC '22)* (pp. 242–246). ACM. https://doi.org/10.1145/3537797.3537876.

Leopoldina Fortunati and Autumn Edwards

Domestic Appliances and Household Robots: The Changing Landscape of Housework and Family

Abstract: In this chapter, we examine the technologies that have penetrated homes to address housework and consider how these have affected the gender division of labor in the domestic sphere. While men's involvement in housework remains low relative to women globally, there has been a notable rise in their engagement in childcare over recent years. However, this increase has not extended significantly to more physical, labor-intensive household tasks like cleaning, laundry, ironing, cooking, dusting, table management, and grocery shopping. Our purpose is first to trace a short, historical account of domestic appliances and digital technologies and then of more sophisticated technologies that have arrived in the home, such as social robots and robotic products. After World War II, there was a surge in the household adoption of appliances such as refrigerators, washing machines, dishwashers, vacuum cleaners, polishers, and food processors. It was some time later that microwaves became common in homes. Subsequently, after a period with few new introductions, innovative devices like the Bimby kitchen machine (aka Thermomix), Roomba vacuum cleaners, and robotic lawnmowers began to gain popularity. In this chapter, we aim to understand (1) if each technological wave addressed a specific target (women, men, or family members in general); (2) how these technologies interacted with the organization and gender division of domestic labor and social gender roles, and (3) which needs are unsatisfactorily addressed and why.

Keywords: Domestic appliances, domestic robots, Roomba, Bimby, housework

21.1 Introduction

Recently, the arrival of robotic devices like the Roomba and Bimby in homes has marked a renewed interest from tech companies in domestic innovations, breaking a long lull in new appliance development for the home environment. This prompts questions about the state of the household these innovations entered: What conditions and dynamics did they encounter in the domestic space? Specifically, how have these technologies impacted the domestic roles and responsibilities of women, as well as other family members? Moreover, to what extent have these advancements contributed to reducing the time spent on household chores? To really understand the kind of penetration and domestication of social robots at home as well as their social meaning, we need to reconstruct both the historical background of how homes have

https://doi.org/10.1515/9783110792270-022

been technologized as a place of labor (Fortunati & Edwards, 2022) and examine what theory-building processes have produced on the intersection between technology, gender, and the domestic sphere.

21.2 Historical Background

The post-World War II period, when the economic system needed a greater presence of women in the paid labor market, can be considered a 'fortunate' technological season for the domestic sphere. In this period, refrigerators, vacuum cleaners, polishers, dishwashers, washing machines, and advanced irons arrived to populate people's homes. However, as Gras (1993) noticed, these technologies might be considered a waste of 'noble' factory technology. All these technologies were essentially aimed at women; therefore, they were operated and managed by women. The social discourse behind the diffusion of domestic appliances focused on liberating women from such long working days and pushing them to spend this 'freed' time in factories, offices, schools, and hospitals. However, as Michael Bittman, James Mahmud Rice, and Judy Wajcman (2004) have demonstrated, "domestic technology rarely reduces women's unpaid working time and even, paradoxically, produces some increases in domestic labour. The domestic division of labour by gender remains remarkably resistant to technological innovation" (p. 401).

In the last decades, men's contribution to housework has increased, especially in the domain of childcare (i.e., Bianchi et al., 2000, 2012), rather than the hard core of material housework, such as cleaning, washing, ironing, cooking, dusting, setting and clearing the table, and shopping. Research conducted in 19 European and non-European countries showed increased domestic work sharing between men and women but with significant differences between nations regarding the level and pace of such sharing (Altintas & Sullivan, 2016). However, there has been a slowdown in the sharing of domestic work since the end of the 1980s in countries where the time devoted by men and women to household chores was already more equal. In contrast, the growth of sharing has continued in countries where the gender division of household chores was initially less equal. Simultaneously with the technological advancements in the home, the diversification of family structures has led to an increase in men living by themselves, including single fathers raising children, divorced or separated men sharing custody, gay couples, and widowers. Men in these varied living situations have had to become familiar with operating domestic appliances.

The last product of this technological wave was the microwave oven, which diffused internationally in the 1970s and 1980s, although it was introduced to Britain earlier under the name 'Artic Cooker' (Myllyntaus, 2010). Its use was so simple that heating food became easy for less experienced men or children and the presence of wives and mothers was no longer necessary for food preparation. In many kitchens, the microwave oven primarily serves as an auxiliary tool, chiefly for simple tasks like defrosting and reheating food—activities that are generally not regarded as complex culinary or culturally significant practices (Ascione, 2013). Consequently, integration of the microwave did not have the strength to radically change the cooking process.

This genealogy of technologies aimed to address the material part of domestic work and was accompanied by the television. This influential medium became a significant player in the realm of caregiving, extending its reach into education, entertainment, and information dissemination. Television narratives, largely influenced by the affluent class's lifestyle (Gramsci, 2014), and content selection from powerful nations (Schiller, 1975/1976), often perpetuated conventional and stereotypical representations of gender roles and generational dynamics. It presented a view of society and culture heavily colored by capitalist ideologies, tailoring content to serve the interests of the elite. While television consumption was common among both men and women, the technical aspects of managing its energy use were typically handled by men, reflecting a historical male association with electrical knowledge. Martin (2022) importantly notes that this pattern of male dominance extends to the realm of emerging smart technologies.

During that time, the family structure heavily relied on women to perform household duties and caregiving. This setup was strictly divided by conventional gender roles, with men contributing modestly to these tasks. The family served—and continues to serve—as a crucial unit where the labor force is nurtured and replenished. Essentially, the family acts as a hidden engine of economic production, an invisible factory where individuals recharge and prepare for their work roles and where children are raised, forming the next generation of workers.

The 1970s, with the various feminist waves, reshaped both the family that shattered into different forms and the sexual identities that articulated in various directions. The hierarchy of power between the sexes and the generations was deeply affected, and the first field of confrontation between the various members of the family—communication— has become a battlefield. As a consequence, the family, as a factory of labor-force production, began to show signs of dysfunctionality and to lose productivity.

By the 1990s, the introduction of computers and mobile phones targeted the immaterial part of domestic work. Families had become more vulnerable and less defined by rigid and binary gender roles, with women feeling less obliged to put aside their needs for others. Individualism and isolation were on the rise, weakening the sense of connection to the family unit, intergenerational solidarity, and conflict resolution within family relationships. Progressively, computers and mobile phones swept entire dimensions of immaterial care and reproductive labor under their control, such as dating, sexuality, education, entertainment, information, communication, emotions, and affect. In the last 30 years, the family has also become something different. To contrast the loss of productivity as a place of production and reproduction of the labor force, the family's function shifted to generating direct economic value, with all members, from children to seniors, contributing data to tech giants like Amazon, Google, Apple, Facebook, and Microsoft, inadvertently fueling their profits.

In this context, robotics started to penetrate the household, with robots taking over tasks such as cleaning, cooking, and lawn maintenance. These robots address material domestic labor, while other technologies like pet robots, virtual assistants, chatbots, and toy robots manage more of the immaterial (emotional, social, communicative) part of housework.

21.3 Theory Building on the Intersections of Gender, Domestic Labor, and Technologies

To analyze the penetration of robotic technologies into our homes, we need to build solid, theoretical tools that enable us to make sense of them. It is wise to start from the classical sociological framework of the relationship between technology and society, which is firmly based on the role of technology in industrialization. Four main conceptual pillars characterize this framework, and we argue that they are also applicable to the domestic sphere.

The first fundamental element is that machines are regarded as making environments more 'rational' or efficient, mainly by *reducing the need for human work and effort.* Yet, this enhancement of rationality is somewhat limited. On one side, the efficiency attributed to machines doesn't just stem from the direct application of scientific rules. As Nye (2007) notes, it often derives from a technician's deep, hands-on understanding of materials and mechanical operations. On the flip side, the widespread use of technology in everyday life often comes with a significant lack of understanding about how these technologies actually work. Longo (2003) argued that people commonly utilize tools, systems, and devices without understanding or even wanting to understand their mechanisms. Consequently, technology often reflects the perspectives, identities, and cultures of its creators rather than serving as a means to spread knowledge.

The second element is that technology reshapes the *organization of labor*, representing a means of increasing labor productivity and inscribing a *power structure* that both disrupts and is shaped by traditional social structures. Specifically, Veblen (1904) argued that machines impose a social discipline rooted in cause-and-effect relationships, the implementation of measurable standards, and the streamlining and adaptation of all facets of industrial and broader life—including goals, actions, desires, comforts, and leisure—to *standardized units*. Thus, the application of technology *increases control, discipline, and order* across all domains, reflecting, in Veblen's view, a broader influence on both symbolic and value systems.

The third fundamental element is that technology, along with modernity and industrialization, involves a radical transformation of the *categorization of the spatial/temporal coordinates*. Technological advancement has effectively annihilated space, making distances less relevant, and accelerated time as processes become more mechanized. This disruption is a particularly crucial change among the many psychological and social changes brought about by technology.

The fourth element, and perhaps the most important, concerns the economic and political use of machines. Throughout the industrial era, ownership of and control over technology, which equates to *control over the means of production*, has granted entrepreneurs and business owners significant power. They could dictate which segments of the population—whether women, men, or children—were to be employed or unemployed. This control extended to the broader power to regulate the labor market, set work conditions, and negotiate wages, encompassing all aspects of the organization of labor, including the domestic work behind it.

The adoption of technologies in the domestic sphere rests on the same logic of rationalization of time and energy to (hopefully) reduce housework and/or fatigue, revamp the organization of labor by boosting the productivity of housework, and exert greater control and discipline. It also mirrors the broader trend of reconfiguring the two fundamental categories of time and space that dictate the reproduction of everyday life and the structuring of the labor force. Ultimately, this also reflects the capacity of the ruling classes to utilize technology for their economic and political ends

The house is a key site for the generation and renewal of the workforce, where primarily women and other family members engage in productive labor tasks within the capitalist system. However, the house is a particular environment that operates under a distinct logic from factories, the typical centers of industrial activity. In the factories, the machines, which belong to the business owners, are fundamental to and concentrated within the production space, facilitating the organization, cooperation, and division of labor, which see workers working together on the machines. These are instrumental to save time and labor and increase the efficiency of the work. In contrast, within homes, machines belong to the users and are widespread across many households. Machines play a crucial but not definitive role in the production process and are less involved in mediating the cooperation and division of domestic labor, as the potential of 'natural forces of social labor' (as termed by Marx) remains underdeveloped (Fortunati, 1981), since primary 'force' within the home is the houseworker. At the same time, however, it is within the home that individuals, especially women, subvert capitalist rationale and norms. By nurturing emotions, pursuing education, and fostering communication, they raise the standards of caregiving, societal advancement, and cultural development. When technologies are introduced into homes, they interact not only with the way work is arranged but also with the existing cultural, social, and emotional dynamics within that space.

Given these unique features of the domestic sphere, we must question to what extent our current conceptual framework of technology applies. It is logical to integrate domestic appliances, digital tools, and social robotics into homes primarily if they reduce household labor and tiredness, which is their fundamental purpose. Yet, the literature suggests that rather than reducing household labor (i.e., Cowan, 1985), these technologies often increase it by raising standards of cleanliness and expanding homemakers' duties. We argue that this has to do with the particular characteristics of housework rather than with the work honestly done by these technologies. It is easily extendable because domestic labor does not have a fixed time duration and is particularly sensitive to incorporating new tasks corresponding to the varying needs of the production and reproduction of the labor power. Furthermore, as with what happens in factories, technology *increases labor productivity within the home by imposing control, discipline, and order*, alters the conception of space and time, and carries significant macroeconomic and *political* implications that warrant close attention.

Cakmak and Takayama (2013) conducted interesting research to make a comprehensive household chore list to identify task components requiring end-user programming for domestic robots. Chore lists are typically "used for the management of housekeeping tasks and division of labor among household members" (p. 93). After filtering and unifying 25 different chore lists, they arrived at a list of 884 different tasks concerning cleaning, organizing, laundry, meals, pets/plants, dishes, and trash. However, Sung et al. (2010) observed that domestic tasks "are closely inter-related and have unclear boundaries from each other and therefore automating a task by using a robot may bring substantial changes to the connected tasks. In addition, the use of robotic products may emerge new types of domestic tasks that did not exist before prior to the robot adoption" (p. 423).

Furthermore, Cakmak and Takayama (2013) observed that cleaning tasks comprise almost half of all chores. Cleaning tasks consist of applying a tool (e.g., sponge, wet wipe, etc.) to a specific surface, and most of these cleaning tasks are performed in a stationary location inside the house. The second most common type of organizing task involves rearranging items in the home that are movable, adjustable, or flexible. Their research represents an important step in comprehending how domestic work is articulated, and thus how it may change over time.

Although it is still unclear to what extent domestic appliances have reduced the time required to perform the specific tasks they have addressed, it is clear that digital technologies have introduced a totally different logic. In fact, not only have they failed to reduce the time spent on domestic labor, but they have actually increased this time by distributing domestic work among the various members of the family and also putting children, older adults, and men to work. From a conceptual point of view, digital technologies seem to be toxic technologies in the sense that they also *increase control, discipline, and order* within the house and even mediate *macroeconomic and political meaning* since they have been instrumental to increased value extraction in the reproductive sphere (Fortunati & Edwards, 2022). The issue at hand is to determine whether this harmful influence, now firmly established by digital technologies within household settings, extends to robotic products as well.

To investigate this point, we will also rely on the research emerging from a new concept proposed by Tolmie et al. (2007) called 'technological housekeeping,' which identifies the work involved in setting up and maintaining home networks. Digital technologies often demand more time to set up, program, and maintain compared with the effort needed for traditional electromechanical devices. Additionally, it should be noted that digital technologies not only require a financial investment to acquire but also incur ongoing operational costs. The concept of technological housekeeping was later reworked by Kennedy et al. (2015), who accounted for the gendered use, expertise, and maintenance of domestic technologies required to maintain a networked home. Rode and Poole (2018) further extended this concept of digital housekeeping to the relationship between gender and technical identities (men may show a 'digital chivalry' and women a technophobia), stressing how digital housework, like other tasks and forms of domestic labor, helps shape identity and self-worth.

The demand for expert work immediately brings to the fore a consideration of men's contributions. Men have traditionally been more familiar with digital technologies, at least with regard to their energy feedback. Thus, it is unsurprising that the person who introduces, manages, and mediates technology at home is likelier to be a man. It is not uncommon that this man establishes the rules of technology use, controls the other users, and controls the performance of domestic labor by managing the network of digital technologies (Martin, 2022). This raises two key issues. First, it supports Fortunati and Edwards's (2022) assertion that the advent of digital and robotic technologies has significantly undermined the gains women achieved through feminist movements of the 1970s, affecting them across all social strata, particularly in the realm of communication, by allowing technology to potentially re-subordinate women to men. Second, while there has been an increase in men's participation in domestic reproductive labor, this participation also tends to reinforce existing gender power disparities within the household, perpetuating conventional gender roles.

Digital housekeeping is very often linked to masculine roles. Kryger Aagaard (2023) found in her research on Danish smart homes that this association holds true. However, she pointed out that Smart Home Technologies (SHTs) do not just reinforce traditional roles; they also modify them. Her findings show that men have started to engage more in activities like home decoration and cognitive tasks, which were traditionally seen as feminine.

A last factor shaping this theoretical framework is the influence of digital and robotic technological discourse on the spread and acceptance of robotics in everyday life (Sinclair, 2022). The progression of these technologies has been supported by various myths portraying technology as a tool for salvation, convenience, and problem solving. For instance, the adoption of computers was driven by the promise of financial and time labor savings and the notion of moving towards a paperless society (Sellen & Harper, 2003). Similarly, there are lofty expectations that automation might eradicate the need for human labor, but in practice new workplace technologies have historically led to a restructuring of existing labor practices rather than their complete automation (Sinclair, 2022, p. 45).

Despite historical and theoretical assumptions about technology, people have used these machines to discard old patterns of thought and create new ways of working and living. For example, over recent decades people have used the mobile phone to adapt to the pace of modern life, managing to exist in a 'multilayered time' where they can multitask and navigate a nuanced sense of presence and absence. These individuals' adaptation to technology has inspired our observations.

21.4 Cleaning and Cooking Robots within the Domestic Continuum of Robotic Products and Systems

With these premises in mind, we review key literature about domestic robots to assess the ongoing relevance of this theoretical framework and to inquire whether discussions about the continuum of robotic products and systems in the home offer new insights or topics to the established conversation on technology and modernization.

The diffusion and integration of social robotics into the domestic sphere have been a progression, anticipating the introduction of semi-automated devices, various robotic products, and fully developed social robots, marking a spectrum of evolving advanced technologies leading up to actual social robots. Recent market introductions include specialized robots for tasks like window cleaning, laundry, plant watering, home security, and sports training (Prassler et al., 2016). Currently, robotic lawnmowers and pool cleaners are among the most prevalent (Bogue, 2017). Additionally, robotic toys (e.g., Mascheroni & Holloway, 2019) and virtual assistants have become commonplace in many homes and are the subject of extensive research (e.g., Fortunati et al., 2022). Studies on SHTs may not be as widespread, but they receive significant consideration by Strengers et al. in Chapter 20 of this handbook.

Due to space constraints, we will focus on the practical use and experiences of individuals with two representative robotic products: Roomba and Bimby. Understanding technology's role in society requires analyzing how people interact with and use these devices. These practices have the potential to illuminate the matrix of fundamental questions concerning the relationship between technology and society (Suchman, 2019).

21.4.1 The floor-cleaning robot: Roomba

Forlizzi and DiSalvo's (2006) figural ethnographic work on Roomba, the machine that moves "autonomously across the floor, brushing or vacuuming dirt and dust into a dustbin" (p. 258), highlights that cleaning and cleanliness are not the only aspects that constitute housework. "The practices of cleaning and cleanliness," they continue, "are not just instrumental activities—to keep the home free of dirt and germs. They also reflect, structure, and are structured by values and beliefs about what is 'proper' or 'good' living. This extends to how cleaning is performed in the home and the role of various technologies, from solvents to robots" (p. 260).

Forlizzi and DiSalvo's research goals were twofold: to understand (1) how the use of a domestic service robot might influence the habits and practices of housekeeping; and (2) how a mobile, autonomous robot 'fits' into the home. Their findings indicate that Roomba has broadened the responsibility for cleaning within the family, increased the frequency of cleaning, and fostered strong emotional bonds with the device, such as giving it gendered names. This emotional attachment that people develop towards social robots is significant because it can help overcome technical problems or limitations and can be the basis for long-term loyalty to the product. This factor was also identified in the use of Amazon Alexa (Kurz et al., 2021).

Sung et al. (2008) conducted another interesting study with 379 Roomba users, revealing how the robot altered cleaning routines and introduced new non-cleaning activities. They raise the point that we should investigate not only whether technology reduces the amount of housework but also whether it changes how we do housework and affects people's everyday routines. They found, for example, that Roomba had tripled the number of men responsible for vacuuming in the household. This outcome suggests that the use of Roomba "may trigger some swapping of vacuuming practices across genders" (p. 131). This research also documented that cleaning frequency increased with Roomba because householders could use this device when not at home or engaged in other tasks or activities.

An interesting finding is that many users continued to vacuum. Despite Roomba's presence, manual cleaning remained necessary for quick clean-ups or deep cleaning due to the robot's limitations and the need to clean areas like corners and stairs manually. As Sung et al. (2008) reported, quick cleaning often coincides with the arrival of some guests or when a small mess has been made; deep cleaning involves upright vacuum cleaners because of the insufficient Roomba suction to clear up pet hair or waste. However, the necessity for manual cleaning is affected by the spatial structure of the house: corners, edges, stairs, and cobwebs require manual intervention. Here, we arrive at the other element essential to understanding the extent to which Roomba saves housework. In order to make it possible for Roomba to function correctly, users have to engage in preparatory labor, which has also acquired a specific name: roombarization. Before gaining any advantage from Roomba, users must work for it by physically modifying the environment in which they live: "cleaning up wires, changing furniture layout, and tucking in rug tassels" (p. 133). This is important work since half the participants reported modifying or changing their homes. However, according to Forlizzi and DiSalvo's (2006) study, these pre-cleaning activities "seemed to be worth the benefit of having the work done automatically" (p. 263). Furthermore, adopting Roomba often led to the adoption of additional robots, and users tended to humanize Roomba, attributing names, genders, and personalities (e.g., silly, temperamental, flirty, and stubborn) to it.

Sung et al. (2010) presented another study on Roomba, this one ethnographic, exploring the long-term domestication of this robot through four phases: pre-adoption, adoption, adaptation, and use/retention. From this research, it emerged that the gendered and generational division of housework has been challenged by Roomba's arrival in that managing it has attracted both men and children to cleaning activities. Parents, profiting from their children's attraction, used Roomba operation as a reward for completing homework, but, in several cases, children took over responsibility for the vacuum, learning how to run the robot to clean by reading the manual and surpassing their parents in knowledge of the procedures, "but also maintain[ing] it, such as emptying the bin and changing the filters" (p. 427).

During initial use, cleaning robots "ran in an environment that had not yet been modified to accommodate them, they encountered several accidents, such as breaking a full-sized mirror, eating toys, and damaging furniture" (p. 427). However, Sung et al. (2010) stressed that householders were disposed to forgive Roomba for the accidents and damage. This parallels the forgiving attitude of Alexa users towards the assistant's frequent failures (Kurz et al., 2021). This benevolence can be explained by the widespread tendency among users to 'humanize' the robot by considering it like a child taking its first steps and inevitably causing trouble.

During the process, people continued to spend time and effort learning more about the robot, its technical limitations and affordances, and making adjustments for it. The work to reorganize the domestic environment required by Roomba continued to be consistent: "casual and temporary changes that they needed to repeat in each operation, such as folding area rugs, blocking Roomba navigation with everyday objects (e.g., toy cars), and picking up clutter" (Sung et al., 2010, p. 427). Other changes were required to be more permanent, such as "cut[ting] off the rug tassel to prevent it from getting stuck on the fringe" (p. 425).

Sung et al. emphasized that the transition from manual to automated cleaning has significantly transformed the nature of the task. Instead of being a straightforward, repetitive chore, the use of Roomba introduced a variety of preparatory and follow-up activities. Users needed to identify and occasionally change the locations where Roomba would operate, get the space ready for cleaning, and look after the robot's maintenance and storage when it was not in use. Maintaining the robot often required a level of precision that was not anticipated, leading to operational issues or docking failures. Additionally, it was observed that pets in the home interacted with Roomba in lively and engaged ways, as they chased or rode the moving robot, demonstrating the wider impact of the device on household dynamics.

21.4.2 The cooking robot: Bimby

The most successful robotic appliance connected to cooking is Bimby, a multi-food processor robot, capable of performing many functions, such as weighing, mixing, chopping, milling, kneading, blending, steaming, whisking, precise heating, stirring and emulsifying, and even cooking food. Unlike other domestic robots, such as Roomba, that perform pre-existing manual tasks like cleaning the floor, Bimby aims to transcend and include the functions of many different domestic appliances in one. Not only has Bimby replaced the set of pots and pans present in any kitchen and rendered tools such as whisks, blenders, colanders, and scales superfluous (Truninger, 2011), but Bimby also offers some knowledge of food preparation. Moreover, it is designed to produce preprogrammed meals, and it has integrated Wi-Fi and a stable connection, which allows users to view the recipes directly on its display. Rather than facilitating the automation of a single task, Bimby promises to intensify cooking performance but at the same time to avoid the cleaning of many pots or pans and the stove, to save money (it consumes very little electricity), and also to make wholesome and even dietetic dishes. In this way, its design aims to enhance cooking efficiency and cleanliness while conserving energy and promoting healthy eating. As we have previously reported, the boundaries between various domestic tasks are unclear because they are intertwined (Sung et al., 2010, p. 423), and cooking is the primary activity that contributes to dirtying kitchens, thus requiring a specific cleaning process. Bimby's users stress that it leaves everything clean (apart from having to clean the robot, but this is a simple task), in contrast to the usual 'messy' manual food preparation.

Since 2013, the year in which Bimby was released, Vorwerk, the company which owns it, has sold millions of pieces across Europe, and especially in Italy, a country that has a strong sense of its history and culture of food and cooking; two million Bimby have been sold here, more than any other European country (Ascione, 2013). In general, this domestic robot was sold predominantly in Southern European countries, where a limited offer of convenience food in supermarkets, combined with a significant amount of cooking supplied mainly by women at home and increasing women's working schedules outside the home, have made this domestic robot particularly attractive (Truninger, 2011). Most Bimby owners are professional women who hope and expect to optimize the time to dedicate to housework. Once again, however, it needs to be clarified if the use of Bimby shortened cooking time, leaving more time for these women to do other things, or if it required overall more time than manual cooking. In the opinion of several women, this robot did not always open up a space for quality time to spend with their children. Likewise, it is still being determined if the use of this domestic robot helps change or reinforce conventional domestic roles. According to Fonseca et al. (2019) kitchen robots have effectively decreased the time spent cooking, but it will be difficult for a robot to replace humans entirely since the implications of cooking are so deeply rooted in culture, senses, aesthetics, and sociality that they resist complete elimination via technology.

Fonseca et al. do not however take into account that the use of Bimby is not immediate; it involves training, attending cooking classes, hosting or attending home demonstrations, reading manuals, and engaging with community resources for support (Truninger, 2011). The transmission of domestic skills, traditionally passed from mothers to daughters, has waned, leading to a broader deskilling within households. Domestic robots enter this landscape, offering potential aid yet often requiring significant user input and adjustments (in other words 'labor').

Generally, housework training and knowledge sharing about it is given by a mother to her children and particularly to daughters. Successive waves of feminism weakened mothers' will to train their daughters in this job and daughters' will to learn it, as such training and learning risked reinforcing traditional female roles whose content was widely contested on a social level. The consequence has been that the transmission of information and training stalled, and the domestic deskilling of new generations of women and men has reached considerable dimensions. Domestic robots arrive in houses in this diffused situation, whereby young women especially have to turn to communities of practice on the internet for information, advice, and support. What is certain is that they all require help from users. Rather than creating robots that take charge of household tasks autonomously and directly, there is a prevalent design approach that necessitates human intervention for initial setup or ongoing interaction, thereby imposing additional demands on the user's time and effort. This approach is counter-productive. What is truly needed is robots that genuinely reduce the amount of time, effort, and money spent on household chores. Adding to the workload and overall costs, including the buying price, upkeep expenses, energy use, or repair fees for malfunctions, defeats the purpose of domestic robots that are meant to aid women and all users by making household tasks easier and less burdensome.

Truninger's (2016) study explores the Forum Bimby, which is a virtual brand community and a community of practice (Wenger, 1999) composed of thousands of members, the majority of whom (85%) are women. This research revealed that Bimby's arrival in households often opens up tensions and conflicts among the household members because of their different conceptions of what meals should be cooked. Another important study regarding the diffusion and adoption of SHTs highlights how these generate disruption and conflicts within the family since technologies often affect the power division in the household and put in motion a series of reactions and counter-reactions (Furszyfer Del Rio, 2022). According to this research, men are generally more concerned with comfort, while women are more concerned with costs and environmental implications. SHTs act as generators of conflicts between siblings when they desire to access technologies simultaneously and between children and parents when the former put pressure on the latter to purchase the latest technology whose price is incompatible with the family budget. SHT might create a dysfunctional home when tech-savvy users exacerbate unequal dynamics of power and control. As with Roomba, Bimby becomes, for many users, an added member of the family. One woman said she takes the robot with her when she goes on holiday, as she considers it a family member who cannot be left behind (Ascione, 2013).

However, from a robotic point of view, Bimby is very modest, as in reality it represents a step in the ongoing evolution of domestic appliances and is marketed to middle-class women who need to reduce cooking time. Apart from Bimby, there is a variety of research on robot cooking systems (e.g., Bollini et al., 2013; Hood, 2021; Tenorth et al., 2011; Watanabe et al., 2013). Other experiments or scenarios, more in line with traditional robotics, are, for example, those described by Jiang and Zhou (2022) and Oechsner et al. (2022), both relying on robotic arms. The first study proposes six scenarios (AI–Cooking Support, Live Remote Assistance, Interactive Remote Cooking, Recipe Reenactment, Cooking Instructor, and Autonomous Cooking Chef) in which co-operative robots are integrated into the kitchen and support humans by revising opportunities and challenges. The second study illustrates the design of a self-learning robot, which can provide all kinds of kitchen work and can also learn from multiple sources. Moreover, this cooking robot costs much less than usual cooking robots The rationale of this cooking robot mentions older adults and people with disabilities who need cooking assistance, noting that the main reason older adults move into assisted living facilities is their lack of ability to prepare food for themselves. However, such projects must better understand the actual needs of their intended users, such as older adults and people with disabilities, to provide meaningful support.

21.5 Discussion and Final Remarks

From this analysis, it emerges that social robots, following the logic of digital technologies, tend to extend rather than reduce the domestic day. Second, not only women but also men and children, especially with devices like Roomba, are increasingly sharing in household tasks, elevating the status of certain chores (especially among young digital 'workers') and involving all family members in domestic labor, aligning with trends of children's growing domestic involvement (Pupo & Duffy, 2012).

Third, social robots also interact with the organization of housework and the distribution of domestic labor across genders and generations. Often, men's contribution to digital housekeeping and to the use of technologies like Roomba in the household may reinforce their power of control and command in the family.

Fourth, the narrative around the rationalization of technology, particularly with domestic robots, has evolved to become more complex. The rational assessment of technology as a means to save labor time and reduce fatigue is now intermingled with cultural and emotional considerations. The emotional bond users form with technology can overshadow the rational appraisal of its functionality. People appear willing to forgive domestic robots for any limitations of functioning or performance that are not up to par because they become part of the routines and atmospheres of the intimate world of the users and because they evoke the symbolic values of modernity on a cultural and social level versus on material processes.

Fifth, many people have argued that domestic robots deskill primarily women. However, the issue of deskilling could be seen as pre-dating the advent of social robots, with the transfer of domestic knowledge from women to children already hindered. Nevertheless, all technologies, not just social robots, require users to adapt and acquire new skills continually.

If these are the main findings emerging from our analysis, what are, in general, the most critical limitations of the current debate on social robots? Perhaps the most important critique is the treatment of domestic labor as if it were industrial, failing to acknowledge its unique, unpaid nature and its execution primarily by women and family members. As long as this work continues to be unpaid, there will be space only for 'impoverished' robotic forms like Roomba and Bimby. However, the recognition and compensation of this labor could pave the way for more sophisticated robots.

Another limitation is the tendency to simplify the rich history, culture, and sensory experiences inherent in domestic tasks to something that can be neatly organized by technology. In reality, tasks like cleaning and particularly cooking are steeped in history, familial knowledge, culture, art, and sensory experiences, as well as influenced by various social and gender roles. These aspects, which significantly contribute to the value of daily life, are often not considered in the design and development of robotic technologies, especially those designed for cooking which focus on automating only the most straightforward parts of the task. This oversight neglects the essence of what makes life fulfilling.

Additionally, there is skepticism within the industry about the market success of current robotic products, with some critics dismissing them as mere gadgets, akin to a smartphone on wheels with a face. This view echoes the sentiments of Vincent (2013) and Sugiyama (2013), who suggested that the mobile phone and its user could be viewed as a social robot or a semi-robot, respectively. The breakthrough product that would dramatically boost the social robot market has not yet emerged.

On the methodological front, most domestic robot studies are ethnographic, offering in-depth insights into specific contexts but challenging to apply broadly. Future research should incorporate more structured approaches and draw from a broader sample to better gauge the reaction of different social groups and the general population to domestic robots. There is also a need for more refined and objective tools to accurately assess whether domestic robots truly reduce or increase the amount of labor required to complete household tasks.

Bibliography

Aagaard, L. K. (2023). When smart technologies enter household practices: The gendered implications of digital housekeeping. *Housing, Theory and Society*, *40*(1), 60–77. https://doi.org/10.1080/14036096.2022.2094460.

Altintas, E., & Sullivan, O. (2016). Fifty years of change updated: Cross-national gender convergence in housework. *Demographic Research*, *35*, 455–470. https://www.jstor.org/stable/26332084.

Ascione, E. (2013). Towards an anthropology of Bimby food-processors in Italy. In M. McWilliams (Ed.), *Food & Material Culture* (pp. 62–69). Prospect Books.

Bianchi, S. M., Milkie, M. A., Sayer, L. C., & Robinson, J. P. (2000). Is anyone doing the housework? Trends in the gender division of household labor. *Social Forces*, *79*(1), 191–228. https://doi.org/10.1093/sf/79.1.191.

Bianchi, S. M., Sayer, L. C., Milkie, M. A., & Robinson, J. P. (2012). Housework: Who did, does or will do it, and how much does it matter? *Social Forces*, *91*(1), 55–63. https://doi.org/10.1093/sf/sos120.

Bittman, M., Rice, J. M., & Wajcman, J. (2004). Appliances and their impact: The ownership of domestic technology and time spent on household work. *British Journal of Sociology*, *55*(3), 401–423. doi: 10.1111/j.1468-4446.2004.00026.x.

Bogue, R. (2017). Domestic robots: Has their time finally come? *Industrial Robot: An International Journal*, *44*(2), 129–136. https://doi.org/10.1108/IR-01-2017-0018.

Bollini, M., Tellex, S., Thompson, T., Roy, N., & Rus, D. (2013). Interpreting and executing recipes with a cooking robot. In J. Desai, G. Dudek, O. Khatib, & V. Kumar (Eds.), *Experimental Robotics* (pp. 481–495). Springer.

Cakmak, M., & Takayama, L. (2013, March). Towards a comprehensive chore list for domestic robots. In *2013 8th ACM/IEEE International Conference on Human–Robot Interaction (HRI)* (pp. 93–94). IEEE. https://doi.org/10.1109/HRI.2013.6483517.

Cowan, R. S. (1985). *More Work for Mother: The Ironies of Household Technology from the Open Hearth to the Microwave*. Basic Books.

Cranny-Francis, A. (2016). Is data a toaster? Gender, sex, sexuality and robots. *Palgrave Communications*, *2*(1), 16072.

Fonseca, E., Oliveira, I., Lobo, J., Mota, T., Martins, J., & Au-Yong-Oliveira, M. (2019). Kitchen robots: The importance and impact of technology on people's quality of life. In Á. Rocha, H. Adeli, L. Reis, & S. Costanzo (Eds.), *New Knowledge in Information Systems and Technologies: Volume 2* (pp. 186–197). Springer International Publishing. https://doi.org/10.1007/978-3-030-16184-2_19.

Forlizzi, J., & DiSalvo, C. (2006). Service robots in the domestic environment: A study of the Roomba vacuum in the home. In *Proceedings of the 1st ACM SIGCHI/SIGART Conference on Human–Robot Interaction, HRI 2006, Salt Lake City, Utah, USA, March 2–3*. https://doi.org/10.1145/1121241.1121286.

Fortunati, L. (1981). *L'arcano della riproduzione: Casalinghe, prostitute, operai e capitale.* Marsilio (*The Arcane of Reproduction: Housework, Prostitution, Labor and Capital*. Autonomedia, 1995).

Fortunati, L. (2018). Robotization and the domestic sphere. *New Media & Society*, *20*(8), 2673–2690. https://doi.org/10.1177/1461444817729366.

Fortunati, L., & Edwards, A. (2022). Gender and human–machine communication: Where are we? *Human–Machine Communication*, *5*, 7–47. https://doi.org/10.30658/hmc.5.1.

Fortunati, L., Edwards, A., Edwards, C., Manganelli, A. M., & De Luca, F. (2022). Is Alexa female, male, or neutral? A cross-national and cross-gender comparison of perceptions of Alexa's gender and status as a communicator. *Computers in Human Behavior*, *137*, 107426. https://doi.org/10.1016/j.chb.2022.107426.

Fortunati, L., Esposito, A., & Lugano, G. (2015). Introduction to the special issue 'Beyond industrial robotics: Social robots entering public and domestic spheres. *Information Society*, *31*(3), 229–235. https://doi.org/10.1080/01972243.2015.1020195.

Furszyfer Del Rio, D. D. (2022). Smart but unfriendly: Connected home products as enablers of conflict. *Technology in Society*, *68*, 1901808. https://doi.org/10.1016/j.techsoc.2021.101808.

Gramsci, A. (2014). *Quaderni del carcere*. Critical editions by the Istituto Gramsci. Torino.

Gras, A. (1993). *Grandeur et dépendance*. P.U.F.

Hood, B. (2021). The world's first robot chef is finally here and even cleans up after itself. https://robbreport.com/gear/electronics/moley-robotics-robot-kitchenuk-for-sale-1234590791.

Huws, U. (2012). The reproduction of difference: Gender and the global division of labor. *Work, Organisation, Labour & Globalisation*, *6*(1), 1–10. https://www.jstor.org/stable/10.13169/workorgalaboglob.6.1.0001.

Jiang, A. Z., & Zhou, M. (2022, December). Design of affordable self-learning home cooking robots. In *2022 IEEE International Conference on Networking, Sensing and Control (ICNSC)* (pp. 1–6). IEEE. https://doi.org/10.1109/ICNSC55942.2022.10004056.

Katuk, N., Ku-Mahamud, K. R., Zakaria, N. H., & Maarof, M. A. (2018, April). Implementation and recent progress in cloud-based smart home automation systems. In *2018 IEEE Symposium on Computer Applications & Industrial Electronics (ISCAIE)* (pp. 71–77). IEEE. https://doi.org/10.1109/ISCAIE.2018.8405447.

Kennedy, J., Nansen, B., Arnold, M., Wilken, R., & Gibbs, M. (2015). Digital housekeepers and domestic expertise in the networked home. *Convergence*, *21*(4), 408–422. https://doi.org/10.1177/1354856515579848.

Kurz, M., Brüggmeier, B., & Breiter, M. (2021). Success is not final; failure is not fatal—Task success and user experience in interactions with Alexa, Google Assistant and Siri. In M. Kurosu (Ed.), *Human–Computer Interaction: Design and User Experience Case Studies* (pp. 351–369). Thematic Area, HCII 2021, Lecture Notes in Computer Science, Vol. 12764. Springer.

Longo, G. O. (2003). Lo scenario: Uomo, tecnologia e conoscenza. In G. M. Apuzzo, S. Araldi, & E. Barbieri Masini (Eds.), *Uomo, Tecnologia e Territorio* (pp. 20–44). Area Science Park.

Martin, R. (2022). Energy housekeeping: Intersections of gender, domestic labour and technologies. *Buildings and Cities*, *3*(1), 554–569. https://doi.org/10.5334/bc.218.

Mascheroni, G., & Holloway, D. (Eds.) (2019). *The Internet of Toys: Practices, Affordances and the Political Economy of Children's Smart Play*. Palgrave Macmillan.

Myllyntaus, T. (2010). The entry of males and machines in the kitchen: A social history of the microwave oven in Finland. *Icon*, *16*, 226–243. https://www.jstor.org/stable/23791384.

Nye, D. (2007). *Technology Matters: Questions to Live With*. MIT Press.

Oechsner, C., Mayer, S., & Butz, A. (2022). Challenges and opportunities of cooperative robots as cooking appliances. In *Proceedings of the 2022 Workshop on Engaging with Automation (2022-04-30) (AutomationXP22). CEUR Workshops, Germany*.

Prassler, E., Munich, M. E., Pirjanian, P., & Kosuge, K. (2016). Domestic robots. In B. Siciliano & O. Khatib (Eds.), *Springer Handbook of Robotics* (pp. 1729–1758). Springer.

Pupo, N., & Duffy, A. (2012). Unpaid work, capital and coercion. *Work Organisation, Labour & Globalisation*, *6*(1), 27–47. https://www.jstor.org/stable/10.13169/workorgalaboglob.6.1.0027.

Rode, J. A., & Poole, E. S. (2018). Putting the gender back in digital housekeeping. In *Proceedings of the 4th Conference on Gender & IT, Heilbronn, Germany*. https://doi.org/10.1145/3196839.3196845.

Schiller, H. I. (1975/1976). Communication and cultural domination. *International Journal of Politics*, *5*(4), 1–127. https://www.jstor.org/stable/27868829.

Sellen, A., & Harper, R. (2003). *The Myth of the Paperless Office*. MIT Press.

Sinclair, K. (2022). The feminised robot: Labour and harawayan afterlives. In A. Hanemamyer (Ed.), *Artificial Intelligence and Its Discontents: Social and Cultural Studies of Robots and AI* (pp. 43–59). https://doi.org./10.1007/978-3-030-88615-8_3.

Smarr, C. A., Mitzner, T. L., Beer, J. M., Prakash, A., Chen, T. L., Kemp, C. C., & Rogers, W. A. (2014). Domestic robots for older adults: Attitudes, preferences, and potential. *International Journal of Social Robotics, 6*, 229–247. https://doi.org/10.1007/s12369-013-0220-0.

Strengers, Y., Kennedy, J., Arcari, P., Nichols, L., & Gregg, M. (2019). Protection, productivity and pleasure in the smart home: Emerging expectations and gendered insights from Australian early adopters. In *Proceedings of the 2019 CHI Conference on Human Factors in Computing Systems* (pp. 1–13). Association for Computing Machinery. 10.1145/3290605.3300875.

Suchman, L. (2019). If you want to understand the big issues, you need to understand the everyday practices that constitute them: Lucy Suchman in Conversation with Dominik Gerst & Hannes Krämer. *Forum: Qualitative Social Research, 20*(1). http://www.qualitative-research.net/index.php/fqs/article/view/3252.

Sugiyama, S. (2013). Melding with the self, melding with relational partners, and turning into a quasi-social robot: A Japanese case study of people's experiences of emotion and mobile devices. *Intervalla: Platform for Intellectual Exchange, 1*, 71–84.

Sung, J.-Y., Grinter, R. E., & Christensen, H. I. (2010). Domestic robot ecology: An initial framework to unpack long-term acceptance of robots at home. *International Journal of Social Robotics, 2*, 417–429. https://doi.org/10.1007/s12369-010-0065-8.

Sung, J. Y., Grinter, R. E., Christensen, H. I., & Guo, L. (2008, March). Housewives or technophiles? Understanding domestic robot owners. In *Proceedings of the 3rd ACM/IEEE International Conference on Human Robot Interaction* (pp. 129–136). https://doi.org/10.1145/1349822.1349840.

Tenorth, M., Klank, U., Pangercic, D., & Beetz, M. (2011). Web-enabled robots that use The Web as an information resource. *IEEE Robotics & Automation Magazine, 18*(2), 58–68. https://doi.org/10.1109/MRA.2011.940993.

Tolmie, P., Crabtree, A., Rodden, T., Greenhalgh, C., & Benford, S. (2007). Making the home network at home: Digital housekeeping. In L. J. Bannon, I. Wagner, C. Gutwin, R. H. R. Harper, & K. Schmidt (Eds.), *ECSCW 2007: Proceedings of the 10th European Conference on Computer-Supported Cooperative Work, Limerick, Ireland September, 24–28* (pp. 331–350). Springer. https://doi.org/10.1007/978-1-84800-031-5_18.

Truninger, M. (2011). Cooking with Bimby in a moment of recruitment: Exploring conventions and practice perspectives. *Journal of Consumer Culture, 11*(1), 37–59. https://doi.org/10.1177/1469540510391221.

Truninger, M. (2016). Appropriating Bimby on the Internet: Perspectives on technology-mediated meals by a virtual brand community. In B. Cappellini, D. Marshall, & E. Parsons (Eds.), *The Practice of the Meal: Food, Families and the Market Place* (pp. 92–106). Routledge.

Veblen, T. (1904). *The Theory of Business Enterprise*. Charles Scribner's Sons (reissued 1932).

Vincent, J. (2013). Is the mobile phone a personalized social robot? *Intervalla: Platform for Intellectual Exchange, 1*, 60–70.

Watanabe, Y., Nagahama, K., Yamazaki, K., Okada, K., & Inaba, M. (2013). Cooking behavior with handling general cooking tools based on a system integration for a life-sized humanoid robot. *Paladyn: Journal of Behavioral Robotics, 4*(2), 63–72. https://doi.org/10.2478/pjbr-2013-0013.

Wenger, E. (2000). Communities of practice and social learning systems. *Organization, 7*(2), 225–246. https://doi.org/10.1177/135050840072002

Marco Dehnert

Ability and Disability: Social Robots and Accessibility, Disability Justice, and the Socially Constructed Normal Body

Abstract: As technologies designed to be social, interactive, and communicative, social robots fulfill many functions including assistance. As part of larger issues related to representation and robots' integration with humans, ability and disability are a key concern. This chapter surveys current perspectives on ability and disability in social robotics, centering critical disability, crip, neuroqueer, and disability justice perspectives. Approaching disability intersectionally, it discusses how ableism relates to the ideal of the socially constructed normal body that implicates what it means to be human. Centering disabled bodyminds in social robotics, the chapter then moves through different use cases and contexts of accessibility, inclusivity, and disability-focused scholarship to assess the challenges and opportunities of social robots related to disability and ability. First, the focus is on neurodiversity and neurotypicality, followed by broader discussions related to accessibility and older adults. Finally, perspectives on the use of sex robots as therapeutic aids are reviewed. The chapter closes with a call to continue centering a disability justice framework, to include the voices and experiences of disabled people in both design and research of assistive social robots, and to grapple with questions of diversity, inclusivity, and the socially constructed normal body.

Keywords: Disability justice, ableism, crip technoscience, neuronormativity, older adults, sex robots, intersectionality

22.1 Introduction

'Assistant' is a term that is often used when describing the function or purpose of social robots. Be it voice assistants like Alexa or the Google Assistant, virtual assistants such as avatars or chatbots, and assistive robotic technologies designed to support humans by taking over certain cognitive or physical tasks, assisting humans is a key characteristic of the design of social robots. Robot assistants might be particularly useful in the context of ability and disability. Think, for example, of the ways in which robotic assistive devices can support people with various physical and mental disabilities. Robot guide dogs can support the independence and mobility of blind and visually impaired people by safely guiding them (Kim et al., 2023). Other social robots like Bada, a small device resembling a robotic vacuum cleaner, can assist deaf users by signaling changes in the environment at home through various channels like text

https://doi.org/10.1515/9783110792270-023

messages or bumping into the couch (Yu et al., 2021). And socially assistive robots like Brian 2.1 can support older adults and cognitively disabled individuals through meal-eating activities and memory card games (McColl et al., 2013).

Whether they serve as robotic guide dogs, in the context of care, or as assistive devices for deaf persons, assistive social robots are a primary use case of social robots that warrants taking into account the social, political, economic, and cultural context. Given the many ways in which social robots are designed to be interactive, communicative, and social—in short, more and more human-like, or at least complementary to human needs and experiences—examining the ways in which social robots relate to, reify, and potentially challenge ableism has been an increasingly generative line of scholarship (e.g., Mankoff et al., 2010; Sum et al., 2022). Ableism is a system of beliefs, practices, and processes focused on disability and ability, and "refers to the cultural construction of a 'normal' body and conceives of any deviation from that 'normal' body as lack thereof—be it in the form of disability or extraordinary abilities" (Dehnert & Leach, 2021, p. 138; cf. Cherney, 2019). As a form of oppression and discrimination focused on ability and disability, ableism is a generative framework for making sense of how social robots provide opportunities and challenges for differently abled people. While assistive devices for older adults such as Lois (short for Loved One's Information System; Fulbright, 2021) or ElliQ are marketed as care companions, they and other technologies raise important questions. Especially because of the intricate connection of ability and disability to what it means to be human, exploring how social robots offer benefits and disadvantages in this context is not only theoretically valuable, but also highly practical. After all, devices like Brian 2.1, Bada, Lois, or robot guide dogs are typically designed with the intention of improving people's daily lives.

Bringing disability perspectives in conversation with social robotics comprises a rapidly growing field. While I cannot attempt to survey this field and its many implications in its entirety, my goal is to offer an entry point into those conversations in this chapter. You can find a brief summary of key terms used in this chapter in Table 22.1. Rather than pursuing a single-axis inquiry, I follow others who approach ability and disability from an intersectional perspective which recognizes the intricate connections of different forms of oppressions and/or privileges a person experiences based on the interplay of the various factors such as dis/ability, race, gender, sexuality, class, and national origin, among others (Hill Collins, 2019; Lorde, 1984). In the context of disability and social robotics, scholars continue to call for the increased uptake of intersectional perspectives by outlining disability justice as a framework that centers disabled people and their voices (Sins Invalid, 2020; Sum et al., 2022). Doing so allows for critically examining the opportunities and challenges of social robots in the context of disability, ableism, accessibility, and inclusivity, thereby also continuing to critique understandings of these issues and how they relate to the socially constructed normal body.

Table 22.1: Explanation of key terms.

Term	Definition	Further reading
Ableism	A form of oppression and discrimination focused on ability and disability, where any deviance from the socially constructed 'normal' body is considered lacking	Cherney (2019)
Crip studies	Similar to the previously derogatory term 'queer,' 'crip' was reclaimed by disabled individuals to emphasize the value of centering disabled people's experiences and perspectives. Crip studies closely engages with queer studies	McRuer (2006); Hamraie & Fritsch (2019); Kafer (2013)
Neurodiverse/ neurotypical	'Neurodiversity' refers to differences in neurological functioning and development considered 'atypical' in that they differ from neurotypical individuals whose neurological functioning and development are considered standard or typical	Newman-Griffis et al. (2022); Sætra et al. (2022)
Neuro-normativity	The cultural dominance of neurotypical thinking, logics, and design of social robots	Sætra et al. (2022); Spiel et al. (2022)
Neuroqueer	A term informed by queer and crip studies, meaning a "collective disidentification by neurodivergent communities from conceptualizations and heteronormative ideas of assimilation to neurotypicality and heteronormativity" (Rauchberg, 2022, p. 372)	Rauchberg (2022)
Techno-ableism	A specific strain of ableism where technology is seen as a tool of empowerment for disabled people while simultaneously reinforcing a pathological understanding of needing to overcome disability	Shew (2020)

In this chapter, I focus on questions of accessibility, inclusivity, and the socially constructed normal body. After surveying foundations of critical disability perspectives within social robotics, I place accessibility and disability research in social robotics in conversation with longstanding critiques of normativity and what it means to be human. Exploring the challenges and opportunities of social robots for ability and disability, I examine use cases including a variety of social robotic technologies and disability contexts. In so doing, the goal of this chapter is to amplify the voices of critical disability scholars and perspectives within the field of social robotics (e.g., Mankoff et al., 2010; Rauchberg, 2022; Shew, 2020; Spiel, 2022; Spiel et al., 2020).

22.2 What Does It Mean to be Normal? Critiquing Normativity

Before discussing how social robots can serve as assistive technologies, I offer a brief background on the notion of what it means to be normal. Disability and ableism are intricately connected with the socially constructed normal body, which serves as a shorthand for what it means to be human (Cherney, 2019). 'Normal' is typically opposed to 'pathological'—with the perceived absence of normalcy indicating pathology. Those who differ from a predetermined ideal of the human are rendered disposable, defective, and dangerous to the 'fitness' of humanity (Canguilhem, 1966/1978). The social construction of a normal body and mind has a long line of critical inquiry ranging through different disciplines and perspectives (e.g., Canguilhem, 1966/1978; Foucault, 1999/2003). Scholars have explored normativity as a social and ideological manifestation of the normal body as it relates to a variety of intersectional identity markers including sexuality, gender, and disability (Davis, 1995; Warner, 1999). Normativity and notions of the normal are ubiquitous and permeate every aspect of social life, but they are also elusive, vague, and thus difficult to locate. As "a disciplinary formation seemingly emanating from everywhere and nowhere" (McRuer, 2006, p. 8), normativity remains difficult to define while significantly impacting people.

Many dimensions of identity are encapsulated by normativity, and it is difficult to summarize what makes up normativity. This includes sexuality (Warner, 1999), able-bodiedness (McRuer, 2006), and intersections between disability and anti-Blackness and colonialism (Sins Invalid, 2020; Lukin, 2013). Based on these many intersections that make up the socially constructed normal body, it becomes clear that normativity encapsulates a variety of different identity characteristics. This allows us to "understand normativities as contradictory, contingent, and complex discursive formations that render some bodies, subjectivities, and performances as the invisible, dominant center from which all bodies, subjectivities, and performances who deviate are marked as other, deficient, undesirable, or impossible" (Dehnert et al., 2022, p. 319).

In the context of ability and technology, in particular, normativity renders the normal body and mind as absent of pathology with the underlying 'ableist conflation' that equates disability to pain and suffering (Reynolds, 2017). In this rendering, nondisability or the absence of disability is seen as the desired goal and the only life worth living is one lived by the nondisabled. Reynolds (2017) traces how this logic still shapes discourses and practices primarily surrounding health, and it is directly connected to neocapitalist logics of production and capital: for a body and mind to be a productive member of capitalism, they need to be able to produce labor and contribute to the neocapitalist demands (McRuer, 2006). The neocapitalist demands of production justify the eradication of disability via the normatively enabled pathologizing of physical and cognitive deviance from the productive (read: nondisabled) bodymind. In other words, the goal of neocapitalist logics is to 'fix' disability (with the help of technology) so that disabled individuals can become productive members of society.

However, being disabled need not mean being in need of 'fixing.' Many activist-turned-scholars across different identities have labored in a variety of ways to reclaim and reimagine the meaning of deviancy and non-compliance with normative standards. Anti-normative politics has shaped queer activism and theory (Dehnert et al., 2022), and intersectionally grounded approaches continue to unsettle and deconstruct the violence of normativities across contexts. The 'crip' in many approaches discussed below is a reclamation of an originally derogatory term for the means of empowerment, similarly to the history of 'queer.' Critical disability scholars heavily critique the ableist conflation of disability with pain and suffering and propose alternative understandings of disability and nondisability tied to different dimensions of identity (Goodley et al., 2019). This intersectional critique of normativities and the reframing of deviance and disability as not pathological are the foundation for a nuanced understanding of social robots and ability.

22.3 Technoableism, Accessibility, and Disability Justice

Disability scholars generally describe the treatment of disability in two major models, among additional alternatives (Newman-Griffis et al., 2022; cf. Kafer, 2013). The medical model of disability follows the pathologizing approach discussed above where disabilities are understood as defects in someone's bodymind that must be cured, fixed, eliminated, or overcome. On the other hand, the social model of disability shifts the primary focus away from the individual to the social context and asserts that disability emerges from the interaction of an individual with their environment. Disability in this context is not a physical or cognitive impairment that needs to be fixed or eliminated, but the focus lies on adapting the environment so that the environment and situation are accessible to people (Meekosha & Shuttleworth, 2009). In this case, the focus shifts away from the wheelchair user as someone in need of being 'fixed' towards addressing building design and architecture to make them more accessible.

From a social robotics perspective, the function and purpose of assistive technologies differ depending on whether the technology's design is based on the medical versus social model of disability. In their critical literature review of autistic children in technology research, for instance, Spiel et al. (2019) identified a lack of technologies designed for the needs, interests, and desires of autistic children and a predominance of viewing autism as a medical deficit needing to be overcome. This finding also resonates with a similar literature review in the context of people with ADHD and technology research (Spiel et al., 2022). Viewing technology as a tool of empowerment for disabled people while reinforcing a pathological understanding of needing to overcome disability has been described as technoableism (Shew, 2020), where design and development of assistive robots is primarily driven by nondisabled self-perceiving helpers and supporters.

In a groundbreaking piece, Mankoff et al. (2010) offered insights for how a critical disability studies perspective can contribute to accessibility research and work on assistive technologies. They specifically argued for the active inclusion of disabled people's voices in research about them and design of technologies for them. Specifically in the context of the medical model that frames disability as something to be overcome through the use of assistive technology, the exclusion of disabled participants and researchers merely perpetuates medical understandings of disability as impairment (Spiel et al., 2020). Countering this exclusion, a growing number of studies specifically centers the voices of disabled participants and researchers (e.g., Hofmann et al., 2020; Spiel et al., 2022).

Within the context of human–computer interaction (HCI), Mack et al.'s (2021) first-of-a-kind literature survey of accessibility work found both a rapid growth of accessibility research between 1994 and 2019 and a broad inclusion of various communities and issues within the realm of accessibility. They also noted a disproportionate focus on blind and low-vision users and caution against ableism in both research and design, indicating that more work can be done (cf. Hofmann et al., 2020). In particular, scholars have called in workshops, position papers, and more conventional reviews for a more systematic and more critical exploration of disability and/in technology. For example, calls for integrating feminist science and technology perspectives into disability justice (Hamraie & Fritsch, 2019) as well as moving towards a crip HCI (Williams et al., 2021) lay out how to center the experiences, voices, and situatedness of disabled participants and researchers.

Conjoining many of these perspectives and grounding itself explicitly in intersectional, queer, anti-racist, and decolonial commitments, disability justice recognizes the interconnection of ableism with anti-Blackness, colonialism, imperialism, eugenics, heteropatriarchy, and other forms of oppression (Sins Invalid, 2020). Dreaming disability justice in HCI (Sum et al., 2022) and aiming to materialize it in technology research and design writ large is the state-of-the-art work on critical disability perspectives in research on accessibility, disability, and technology. Centering disability justice as a framework for understanding ability and disability in the context of technology, I now turn to reflecting on the challenges and opportunities of social robots for disability justice by examining several use cases.

22.4 Challenges and Opportunities of Social Robots for Disability Justice

Dreaming disability justice in social robotics is one thing—materializing it arguably another. In the next few sections, I discuss a variety of different use cases of social robots in the context of ability and disability, specifically focusing on (1) a review of state-of-the-art literature; and (2) taking stock of the presence of crip, critical disability, and disability justice perspectives. In particular, I discuss neuronormativity, broader accessibility concerns and older adults, and the use of sex robots for older adults.

22.4.1 Challenging neuronormativity with social robots

In the context of developmental and cognitive disabilities, 'neurotypicality' and 'neurodivergence' are terms that describe differences in neurological functioning and development used in direct opposition to a deficit or medical model of disability. While neurodivergence includes conditions such as ADHD or autism, the term functions more as an umbrella perspective encompassing a wide variety of neurological experiences. To add further nuance, Rauchberg (2022) proposed neuroqueer technoscience as an expansion to crip technoscience that specifically centers the experiences of neurodivergent and queer-disabled individuals. In particular, Rauchberg understands neuroqueer as a "collective disidentification by neurodivergent communities from conceptualizations and heteronormative ideas of assimilation to neurotypicality and heteronormativity" (p. 372). Here, the choice of language indicates a political commitment that embraces disability justice in recognizing the intersectional diversity within neurodivergent and neuroqueer folks.

The dominance of neurotypical thinking, logics, and design of social robots can be described as neuronormativity, signifying the cultural dominance of neurotypical discursive formations (Sætra et al., 2022; Spiel et al., 2022). If technologies such as algorithms or social robots are designed based on a neuronormative understanding, then their goal is hardly to meet neuroqueer individuals such as autistic children in their needs and experiences, but rather to simulate what 'normal' neurological functioning is supposed to look and feel like (Sætra et al., 2022). Neuronormativity is also prevalent in research conducted on neurodivergence given that neurodivergent populations such as autistic children were only present in about 5% of all HCI research on accessibility between 1994 and 2019 (Mack et al., 2021). Within research focused on neurodivergence, autism and in particular autistic children are primary topics of concern (Spiel et al., 2022; for examples, see Sætra et al., 2022; Spiel et al., 2019).

An example of a social robot that actively works against neuronormative approaches to autistic and neurodivergent children is the child-sized humanoid robot

Kaspar. Designed explicitly as a minimially expressive robot, Kaspar exemplifies user-driven design that was improved over the course of several years in an iterative fashion between user-expressed needs, technological advances, and other key stakeholders' perspectives (e.g., parents) (Wood et al., 2021). For example, Kaspar's gaze is not frontal, allowing autistic children to engage with the robot without having to make direct eye contact, which is typically discussed as a social expectation autistic children do not like.

In contrast to the case of Kaspar as a user-driven social robot for autistic children, using assistive technologies for neurodivergent and neuroqueer individuals with the primary goal and underlying logics of neuronormativity perpetuates a medical model of disability that views neurodivergence such as autism or ADHD as 'defects' in need of fixing (Spiel et al., 2022). Social robots may be designed to empower neurodivergent users yet simultaneously model what 'normal' and thus desirable neurotypical behavior looks like, thereby intentionally or unintentionally reproducing neuronormative ideals. Unlike Kaspar, whose design accommodates gaze-averting behavior, such technoableist mechanisms may include interventions aimed at training eye-contact or practicing masking (developing the ability to 'pass' as non-autistic) for autistic children (Sætra et al., 2022)—with the (unstated) goal of empowering neurodivergent individuals to become more neuronormative.

Social robots in neuronormative and neurodivergent contexts are also related to the potential tracking and surveillance of neurodivergent individuals and conditions. Concerns of privacy and data protection are clearly not limited to neurodivergent users, but are arguably more controversial among that population. Concerns relate to facial emotion recognition via built-in cameras (Nagy, 2022) and also include extractive logis within digital mental health research and treatment that reinforce normative and technoableist understandings of disability (Banner, 2019).

These critiques raise the question of who the primary audience of assistive devices for neurodivergent and neuroqueer users is. If primary functions of those devices are either related to surveillance or to demonstrating neuronormative behavior to neurodivergent and neuroqueer children and adults, it appears that assistive social robots lack a focus on the actual needs and experiences of those for whom they are designed. To that end, Spiel et al. (2019) provide a sobering conclusion to their review of agency of autistic children in technology research writ large:

> While in most of these contexts, autistic children are the primary users, they rarely interact with the technologies in a self-determined way. Most of the technologies are driven by a medical model of autism or the requirements to coping in a neurotypically dominant environment. In their majority, the technologies are not primarily for autistic children as such, but instead focus on the needs for interaction from the predominantly neurotypical people around them. (p. 20)

As illustrated by Spiel et al. (2019) and the other authors cited in this section, a central goal of a disability justice framework in accessibility research is not yet fully realized in the context of neurodivergence and neuroqueerness: the inclusion of disabled participants in the design and research of assistive technologies and the centering of their needs, experiences, and agency (Sins Invalid, 2020). This also means understanding experiences of neurodivergence and neuroqueerness as part of entire bodymind-experiences. In fact, centering the body in addition to the mind offers generative avenues for more equitable research on neurodivergence and disability in general (Spiel, 2021; Spiel & Angelini, 2022).

The challenges and opportunities of social robots and assistive technologies more broadly in the context of neurodivergence and neuroqueerness are, unfortunately, still primarily constrained by the prevailing dominance of neuronormative designs, functions, and goals of assistive social robots. Following a disability justice perspective, research should (1) follow existing work that shifts away from medical models of neurodivergence focused on fixing perceived defects to embracing neurodivergent and neuroqueer frameworks; (2) actively incorporate and center the voice and experiences of those for whom the technologies are built; and (3) critically examine the underlying neuronormativity when it comes to diagnosing, therapizing, and treating neurodivergent conditions. Spiel et al.'s (2022) critical review of literature on ADHD in HCI is an exemplary piece that models the inclusion of self-identified neurodivergent researchers, and Kaspar serves as a useful practical example for how to realize such designs.

Notwithstanding, social robots can also be useful tools for empowering and supporting neurodivergent and neuroqueer users, such as autistic children, when they recognize the heterogeneity of neurodivergence and neurodivergent people (Sætra et al., 2022; Spiel et al., 2019). As advanced technologies, social robots can be individualized and personalized to the specific user and their needs and experiences, which provides a key opportunity for using social robots in the context of neurodivergence. Rather than modeling neuronormative expectations (e.g., direct eye contact), they can adapt to users' needs and desires, allowing neurodiverse users to have fulfilling and helpful interactions. Finally, social robots should involve disclosive design that informs the users (and potential supporting people like parents or therapists) about what processes are carried out for what purposes when interacting with neurodivergent and neuroqueer individuals (Sætra et al., 2022).

22.4.2 Accessibility issues and assistive robots for older adults

Beyond neurodivergence and neuroqueerness, social robots are also designed as assistive technologies for improving accessibility and for supporting older adults—like the robot guide dog or Bada, the robot designed to support deaf users, introduced earlier. Similar to the other examinations within this chapter, a recurring finding in these literatures is that the population targeted by certain technologies is not present in the technology design and research, and that stereotypes are rampant (e.g., Frennert & Östlund, 2014; Goldenthal et al., 2021; Zhang & Hansen, 2022). Disabled and older populations are oftentimes perceived as homogeneous although experiences of disability and age vary widely. Of the reviewed accessibility research between 1994 and 2019, the vast majority focused on only a select group of disabilities, thereby also skewing what is currently meant by accessibility research. Accessibility research disproportionally focuses on visual accessibility (43.5%), followed by people with motor or physical disabilities (14.2%), and people who are deaf or hard of hearing (11.3%). Older adults comprise only 8.9% of the work represented in accessibility research (Mack et al., 2021).

Accessibility concerns are studied in a variety of contexts and focused on a variety of social robotic technologies. For example, telepresence robots can be utilized to facilitate remote access for physically or cognitively disabled individuals as well as older adults. Eye-gaze-controlled telepresence robots can allow people with motor disabilities to participate in social interactions (Zhang et al., 2019). While the goal of research involving telepresence robots and disabled individuals is typically to improve the quality of life, such work lacks directly addressing user conditions (Zhang & Hansen, 2022), indicating a shortcoming of understanding the lived experiences of users for whom telepresence robots are designed—for example, users with hearing and speech disabilities. Other physical technologies include the previously mentioned Bada, a social robot designed for deaf individuals (Yu et al., 2021), as well as robotic guide dogs (Kim et al., 2023), or repurposing existing robots like Pepper as a humanoid therapy robot promoting exercise for dementia patients (Schrum et al., 2019).

Similar results are reported in the context of typically disembodied artificial intelligence-mediated communication technologies such as voice assistants or smart speakers. The desire to mass produce 'one-size-fits-all' artificial intelligence-mediated communication technologies glosses over unique visual, auditory, and other accessibility needs (Goldenthal et al., 2021). A qualitative case study adds further nuance to this claim, pointing out that generic user models of conversational agents should be reconsidered in favor of semantic user models, allowing the conversational agent to improve transparency, enhance personalization, and thereby reduce overall misunderstanding for visually impaired users (Berka et al., 2022). And finally, although they could be used to increase access and inclusion for disabled users, virtual reality technologies are inherently ableist as they primarily disregard the disabled body in both design and functionality (Gerling & Spiel, 2021).

This incomplete survey of accessibility concerns across a broad spectrum of social robotic technologies—both physical and digital—showcases different shades of Shew's (2020) technoableism in practice. While social robotic assistive technologies are primarily designed to improve accessibility or enhance disabled people's lives, they often lack understanding of the experience of the disabled bodymind, and are marked by the exclusion of disabled voices and bodies. This leads Gerling and Spiel (2021), for example, to conclude with a series of recommendations for designing virtual reality technology specifically for disabled people, including appreciating diversity in bodyminds, challenging the status quo of HCI research and design (e.g., via speculative approaches; cf.

Spiel, 2022), and investing in individualized designs allowing for customization. Employing such a disability-centered and personalized design approach, Van Minkelen et al. (2022) demonstrate that social robots can effectively support disabled users such as people with expressive aphasia, an impairment of language skills typically involving difficulties with comprehending and generating spoken and/or written language. Participants in this exploratory multi-method study overall felt positively towards using a social robot as part of speech language therapy, given that personalization and real-time adaptation to the user's responses were provided.

In another area of concern, social robots are often designed for use with older adults, a population that is generally described as disabled (e.g., impaired vision, hearing, mobility). Ageism, or discrimination based on age, is a common yet severely understudied form of discrimination in technology broadly (Rosales & Fernández-Ardèvol, 2020). Focusing on disabilities and implying the infantilization and exclusion of older people (Rosales & Fernández-Ardèvol, 2020), ageism relates directly to ableism and critical disability perspectives' focus on diverse experiences, needs, and desires of diversified bodyminds. Similar to stereotypes of disability, older people are typically perceived as a homogeneous group of weak, passive, and care-needing people, which is also reflected in how social robots are designed for this population (Frennert & Östlund, 2014).

Empirical research directly addresses this mismatch between technologies designed for older adults without actually including their perspectives as the target population. For example, a qualitative study involving focus groups found heterogeneous needs among older adults and a lack of understanding among designers that robots are integrated in a larger experience of disconnection from novel technologies, leading to thwarted expectations (Søraa et al., 2023). Another study examined the connection of trust and fear of social robots among older adults, finding that the relative impact of fear is stronger than that of trust (Zafrani et al., 2023). Ethnographic research, on the other hand, illustrates how social robots can be successfully introduced into elderly care facilities when the daily lives of the older adults and staff recommendations are explicitly considered beforehand (Sabelli et al., 2011). These results indicate that a pragmatic approach focused on reducing fear of social robots as well as providing a familiar design (e.g, resembling a child) are important steps towards increasing social robot acceptance among older adults.

Beyond factors of acceptance of social robots by older adults and ethical concerns related to dehumanization of their care as a result of the use of social robots, the lack of inclusion of disabled people and older adults' voices and experiences in research and design provides clear avenues for future research. From a disability justice perspective, the intersection of ableism and ageism warrants further investigation, and designers and scholars should follow crip scholars' call for centering the body—and a diverse understanding of what bodies are and do—in social robots for accessibility and older adults (Spiel, 2021). While social robots can provide a variety of assistive services to disabled and older individuals, the topic of care and companionship receives increasing attention, especially with a focus on sexual and romantic companionship.

22.4.3 Sex robots as therapeutic aids for disabled and older users

In addition to providing practical support related to mobility, increasing participation in social interactions, or serving as guide dogs, social robots are also used as companions themselves. Lois (Fulbright, 2021), Brian 2.1 (McColl et al., 2013), and Robovie (Sabelli et al., 2011) are but a few examples of humanoid social robots designed to be companions to disabled and older individuals. And even digital assistants can provide social support and companionship, as is the case with ElliQ, a digital assistant comparable to Amazon's Alexa focused on connecting older adults with families and medical providers while also serving as a companion itself. Providing emotional support through conversation, games, and other interactions, these social robots fulfill functions beyond ensuring physical access (like guide dogs) or addressing users' needs (e.g., alerting deaf users of changes in their environment). In addition to using social robots to fulfill such social functions as conversation or emotional support, arguably one of the most controversial use cases of social robots involves sexual and romantic intimacy (Dehnert, 2022; Devlin, Chapter 14 of this volume). While still quite limited in their technical functionalities, sex robots comprise a rapidly growing area of lay and academic interest involving speculation, ethical debate, and a growing body of empirical literature (for a review, see Hanson & Locatelli, 2022). Among those conversations and within the fragmented literature, the topic of disability constitutes one potential application for sex robots, be it related to different types of physical and cognitive disabilities or for older users. A primary concern here seems to be addressing the supposed loneliness and sexual dissatisfaction of disabled and older individuals. Where might opportunities and challenges of social robots in relation to ability and disability lie in this context?

One angle is to take an ethical and dignity-based approach (Jecker, 2021). Having sparked a variety of responses from critics (e.g., Sparrow, 2021), Jecker (2021) maintains that the non-voluntary absence of sex from someone's life, especially older adults with disabilities, warrants viewing sex robots as a viable way to maintain older adults' dignity, health, and identity. She argues in particular that, compared with other assistive technologies that might accomplish sexual satisfaction for older adults with disabilities, sex robots allow for human–robot relationship formation that qualitatively differs from people's relationships with other, non-humanoid technologies. This relationship formation is what makes it valuable to consider the use of sex robots for older adults with disabilities, according to Jecker (2021).

A related approach focuses on ethical sexual care (Fosch-Villaronga & Poulsen, 2020). Exploring considerations including safety, responsibility, privacy, autonomy, infantilization, and more, the authors ultimately conclude with an optimistic yet careful outlook and caution researchers, practitioners, and policymakers to proceed with care in this context. This argument involves pointing out the potential physical and mental health benefits of sex robots for older and disabled individuals, although empirical evidence is markedly missing (Cox-George & Bewley, 2018). On a more pragmatic level, Di Nucci (2017) argues that sex robots can fulfill the sexual needs of disabled individuals without violating anybody's right to sexual self-determination. In contrast to Jecker's (2021) ethical justification, Di Nucci (2017) proposes that sex robots can fulfill an apparent 'right' to sexual satisfaction of disabled people and older people.

An important point to make is that the desire to have sex robots rarely emerges in research exploring older adults' needs and wishes, which should be the most relevant starting point (Søraa et al., 2023). In this sense, the discussion of sex robots for older and disabled individuals is emblematic of the general lack of involvement of key target audiences within accessibility research, as discussed earlier. The inconclusive literature indicates lack of agreement in terms of the perceived benefits and dangers of sex robots for disabled and older persons, and a lack of agreement on the connections between sexuality and disability. Crip perspectives shed light on how disabled people are oftentimes not perceived as sexual due to their physical and/or cognitive disability, which renders them as pathological, undesirable and without desire, and thus not-human (Kafer, 2013; McRuer, 2006; Spiel, 2022).

Within this intersection of ableism and able-bodyminded notions of heteronormativity, it becomes crucial to clearly lay out *why* sex robots would or should be beneficial for disabled and older persons. Is the argument for sex robots being made because disabled and older persons are seen as undesirable, without desire, and generally not sexual? Then deferring to the use of non-human sex robots to fulfill the sexual care of disabled and older persons is deeply problematic and reminiscent of eugenicist and ableist discourses advocating for disallowing disabled and older persons sexual pleasure because of their perceived pathology. Is the argument for sex robots being made, however, because disabled and older persons are perceived to struggle with forming human connections and relationships that would entail sexual gratification with other humans? This argument is again based on ableist and ageist notions that to be disabled or to be old means to be lonely and isolated, and to be undesirable—albeit for different reasons.

In order to actualize the potential opportunities of sex robots for sexual gratification and even emotional intimacy for disabled and older persons as well as the larger population, too, and in order to carefully recognize and address the challenges of sex robots, a return to disability justice and a critique of technoableism makes sense. First, debates surrounding the benefits and dangers of sex robots for disabled people must be grounded in empirical (rather than speculative or philosophical) insights on disabled people's erotic and sexual lives. Those insights, for example, yield that disabled people in fact creatively negotiate their sexual lives despite having their sexuality and desirability normatively denied (e.g., Kafer, 2013; Liddiard, 2017). Second, technoableist notions to ‘fix’ disabled and older people’s perceived sexual deviance by introducing sex robots must be critiqued from a disability justice framework. Such an approach would shift attention to how the equation of disability with undesirability and the absence of sexuality emerges discursively, followed by addressing these discourses on a societal level. Third, from a social perspective, questions arise whether this debate is driven by technology, social pressure among caregivers to provide highly advanced forms of care even in intimate contexts, and/or capitalist pressure among developers to acquire market shares with under-represented populations. If sex robots are not a primary interest among disabled and older individuals, and if they are rather expensive, who benefits from playing with the idea of introducing sex robots into disabled and older people’s lives? And finally, following the political commitments of disability justice and crip HCI, research and design of sex robots intended for disabled and older people should involve disabled and older participants and potentially researchers.

22.5 Conclusion and Future Directions: Disability Justice in Social Robotics

Critical disability studies is a growing and already well-established field exploring a variety of experiences, normativities, discourses, and structures related to disability and ableism. In research focused on social robots and technology, critical disability, crip, neuroqueer, and disability justice perspectives are increasingly taken up and centered in work focused on accessibility, inclusivity, and the socially constructed normal body. In this chapter, I have offered background on normativities as contradictory, contingent, and complex discursive formations, and positioned critiques of ableism within technology and social robotics scholarship. I have reviewed challenges and opportunities of social robots for ability and disability by examining specific use cases, including neuronormativity, accessibility issues and older adults, and sex robots. Jointly, these use cases demonstrate that the framework of disability justice and an intersectional understanding of ability and disability are having increasing impact in the fields of human–robot interaction, human–computer interaction, and social robotics. This chapter also demonstrates, however, that much work remains to be done.

Future directions for social robotics focused on ability and disability include a radical embracing of disability justice as a political, intersectional, and liberatory framework that understands disability in relation to other dimensions of identity. Moving beyond single-axis frameworks is crucial for doing meaningful, user-centered work. Second, rather than following technology developments, embracing disability justice unapologetically also means centering the experiences, needs, and voices of those populations for whom the technologies are built: neuroqueer, neurodivergent, disabled, and older people. Otherwise, assistive robotic technologies risk reifying technoableist mechanisms of perpetuating neuronormativity and body-normativity. Third, taking a disability justice framework means recognizing that the experience of disability is stratified and heterogeneous, including for people with similar diagnoses. Intersectionality as an analytic stresses that the experience of oppression can only be understood across the relations of one's identity (Hill Collins, 2019; Lorde, 1984). Similarly, disability justice means rejecting one-size-fits-all solutions which, of course, requires different design efforts but might ultimately lead to a more inclusive social robotics. Moreover, this chapter has showcased that the agency of the targeted populations is often secondary, raising questions of how assistive social robots continue to position disabled people as passive and in need of help. A disability justice position presses the question of who assistive social robots are designed for, and with what purpose in mind. Finally, given that social robots provide ample opportunities for empowering disabled people, the task should be to focus on what is working for the targeted populations and how developers can take what is working from good to great. After all, the goal of social robots for disabled people is focused on assisting them in their daily life. What better way to achieve this goal than to strive for a more integrated research and design of those assistive social robots?

References

Banner, O. (2019). Technopsyence and Afro-surrealism's cripistemologies. *Catalyst: Feminism, Theory, Technoscience*, *5*(1), 1–29. https://doi.org/10.28968/cftt.v5i1.29612.

Berka, J., Balata, J., Jonker, C. M., Mikovec, Z., van Riemsdijk, B. v., & Tielman, M. L. (2022). Misalignment in semantic user model elicitation via conversational agents: A case study in navigation support for visually impaired people. *International Journal of Human–Computer Interaction*, *38*(18–20), 1909–1925. https://doi.org/10.1080/10447318.2022.2059925.

Canguilhem, G. (1966/1978). *The Normal and the Pathological*. Trans. C. R. Fawcett & R. S. Cohen. Zone Books.

Cherney, J. L. (2019). *Ableist Rhetoric: How we Know, Value, and See Disability*. Pennsylvania State University Press.

Cox-George, C., & Bewley, S. (2018). I, sex robot: The health implications of the sex robot industry. *BMJ Sexual & Reproductive Health*, *44*, 161–164. http://dx.doi.org/10.1136/bmjsrh-2017-200012.

Davis, L. J. (1995). *Enforcing Normalcy: Disability, Deafness, and the Body*. Verso.

Dehnert, M. (2022). Sex with robots and human–machine sexualities: Encounters between human–machine communication and sexuality studies. *Human–Machine Communication*, *4*, 131–150. https://doi.org/10.30658/hmc.4.7.

Dehnert, M., Brouwer, D. C., & LeMaster, L. (2022). Anti-normativity under duress: An intersectional intervention in queer rhetorics. In J. Rhodes & J. Alexander (Eds.), *The Routledge Handbook of Queer Rhetoric* (pp. 319–327). Routledge. https://doi.org/10.4324/9781003144809-43.

Dehnert, M., & Leach, R. B. (2021). Becoming human? Ableism and control in *Detroit: Become Human* and the implications for human–machine communication. *Human–Machine Communication*, *2*, 137–152. https://doi.org/10.30658/hmc.2.7.

Di Nucci, E. (2017). Sex robots and the rights of the disabled. In J. Danaher & N. McArthur (Eds.), *Robot Sex: Social and Ethical Implications* (pp. 73–88). MIT Press.

Fosch-Villaronga, E., & Poulsen, A. (2020). Sex care robots: Exploring the potential use of sexual robot technologies for disabled and elder care. *Paladyn: Journal of Behavioral Robotics*, *11*(1), 1–18. https://doi.org/10.1515/pjbr-2020-0001.
Foucault, M. (1999/2003). *Abnormal: Lectures at the Collège de France 1974–1975*. Trans. G. Burchell. Picador.
Frennert, S., & Östlund, B. (2014). Review: Seven matters of concern of social robots and older people. *International Journal of Social Robotics*, *6*, 299–310. https://doi.org/10.1007/s12369-013-0225-8.
Fulbright, R. (2021). A synthetic elderly companion named Lois. In *International Conference on Human–Computer Interaction, LNCS 12787* (pp. 403–417). https://doi.org/10.1007/978-3-030-78111-8_27.
Gerling, K., & Spiel, K. (2021). A critical examination of virtual reality technology in the context of the minority body. In *CHI '21: Proceedings of the 2021 CHI Conference on Human Factors in Computing Systems* (p. 599). https://doi.org/10.1145/3411764.3445196.
Goldenthal, E., Park, J., Liu, S. X., Mieczkowski, H., & Hancock, J. T. (2021). Not all AI are equal: Exploring the accessibility of AI-mediated communication technology. *Computers in Human Behavior*, *125*, 106975. https://doi.org/10.1016/j.chb.2021.106975.
Goodley, D., Lawthom, R., Liddiard, K., & Runswick-Cole, K. (2019). Provocations for critical disability studies. *Disability & Society*, *34*(6), 972–997. https://doi.org/10.1080/09687599.2019.1566889.
Hamraie, A., & Fritsch, K. (2019). Crip technoscience manifesto. *Catalyst: Feminism, Theory, Technoscience*, *5*(1), 1–34. https://doi.org/10.28968/cftt.v5i1.29607.
Hanson, K. R., & Locatelli, C. C. (2022). From sex dolls to sex robots and beyond: A narrative review of theoretical and empirical research on human-like and personified sex tech. *Current Sexual Health Reports*, *14*, 106–117. https://doi.org/10.1007/s11930-022-00331-0.
Hill Collins, P. (2019). *Intersectionality as Critical Social Theory*. Duke University Press.
Hofmann, M., Kasnitz, D., Mankoff, J., & Bennett, C. L. (2020). Living disability theory: Reflections on access, research, and design. In *22nd International ACM SIGACESS Conference on Computers and Accessibility (ASSETS '20), October 26–28, Virtual Event, Greece*. ACM. https://doi.org/10.1145/3373625.3416996.
Jecker, N. S. (2021). Nothing to be ashamed of: Sex robots for older adults with disabilities. *Journal of Medical Ethics*, *47*(1), 26–32. http://dx.doi.org/10.1136/medethics-2020-106645.
Kafer, A. (2013). *Feminist, Queer, Crip*. Indiana University Press.
Kim, J. T., Yu, W., Tan, J., Turk, G., & Ha, S. (2023). How to train your guide dog: Wayfinding and safe navigation with human–robot modeling. In *HRI '23: Companion of the 2023 ACM/IEEE International Conference on Human–Robot Interaction* (pp. 221–225). https://doi.org/10.1145/3568294.3580076.
Liddiard, K. (2017). *The Intimate Lives of Disabled People*. Routledge.
Lorde, A. (1984). *Sister Outsider*. Crossing Press.
Lukin, J. (2013). Disability and blackness. In L. J. Davis (Ed.), *The Disability Studies Reader* (pp. 308–315). Taylor & Francis.
Mack, K., McDonnell, E., Jain, D., Wang, L. L., Froehlich, J. E., & Findlater, L. (2021). What do we mean by 'accessibility research'? A literature survey of accessibility papers in CHI and ASSETS from 1994 to 2019. In *CHI Conference on Human Factors in Computing Systems (CHI '21), May 8–13, Yokohama, Japan*. ACM. https://doi.org/10.1145/3411764.3445412.
Mankoff, J., Hayes, G. R., & Kasnitz, D. (2010). Disability studies as a source of critical inquiry for the field of assistive technology. In *Proceedings of the 12th International ACM SIGACCESS Conference on Computers and Accessibility (ASSETS '10)*. ACM. https://doi.org/10.1145/1878803.1878807.
McColl, D., Louie, W.-Y. G., & Nejat, G. (2013). Brian 2.1: A socially assistive robot for the elderly and cognitively impaired. *IEEE Robotics & Automation Magazine*, *20*(1), 74–84. https://doi.org/10.1109/MRA.2012.2229939.
McRuer, R. (2006). *Crip Theory: Cultural Signs of Queerness and Disability*. New York University Press.
Meekosha, H., & Shuttleworth, R. (2009). What's so 'critical' about critical disability studies? *Australian Journal of Human Rights*, *15*(1), 47–75. https://doi.org/10.1080/1323238X.2009.11910861.

Nagy, J. (2022). Autism and the making of emotion AI: Disability as resource for surveillance capitalism. *New Media & Society*. https://doi.org/10.1177/14614448221109550.

Newman-Griffis, D., Rauchberg, J. S., Alharbi, R., Hickman, L., & Hocheiser, H. (2022). *Alternative Models: Critical Examination of Disability Definitions in the Development of Artificial Intelligence Technologies*. ArXiv Preprint. https://doi.org/10.48550/arXiv.2206.08287.

Rauchberg, J. S. (2022). Imagining a neuroqueer technoscience. *Studies in Social Justice*, *16*(2), 370–388 https://doi.org/10.26522/ssj.v16i2.3415.

Reynolds, J. M. (2017). "I'd rather be dead than disabled"—The ableist conflation and the meanings of disability. *Review of Communication*, *17*(3), 149–163. https://doi.org/10.1080/15358593.2017.1331255.

Rosales, A., & Fernández-Ardèvol, M. (2020). Ageism in the era of digital platforms. *Convergence: The International Journal of Research into New Media Technologies*, *26*(5–6), 1074–1087. https://doi.org/10.1177/1354856520930905.

Sabelli, A. M., Kanda, T., & Hagita, N. (2011). A conversational robot in an elderly care center: An ethnographic study. In *HRI '11: Proceedings of the 6th International Conference on Human?Robot Interaction* (pp. 37–44). https://doi.org/10.1145/1957656.1957669.

Sætra, H. S., Nordahl-Hansen, A., Fosch-Villaronga, E., & Dahl, C. (2022). Normativity assumptions in the design and application of social robots for autistic children. In *Proceedings of the 18th Scandinavian Conference on Health Informatics, Tromsø, Norway, August 22–24*. Linköping University Electronic Press. https://doi.org/10.3384/9789179293444.

Schrum, M., Park, C. H., & Howard, A. (2019). Hunanoid therapy robot for encouraging exercise in dementia patients. In *Proceedings of the 14th ACM/IEEE International Conference on Human–Robot Interaction (HRI), Daegu, Korea* (pp. 564–565). https://doi.org/10.1109/HRI.2019.8673155.

Shew, A. (2020). Ableism, technoableism, and future AI. *IEEE Technology and Society Magazine*, *39*(1), 40–85. https://doi.org/10.1109/MTS.2020.2967492.

Sins Invalid (2020, June 16). What is disability justice? https://www.sinsinvalid.org/news-1/2020/6/16/what-is-disability-justice.

Søraa, R. A., Tøndel, G., Kharas, M. W., & Serrano, J. A. (2023). What do older adults want from social robots? A qualitative research approach to human–robot interaction (HRI) studies. *International Journal of Social Robotics*, *15*, 411–424. https://doi.org/10.1007/s12369-022-00914-w.

Sparrow, R. (2021). Sex robot fantasies. *Journal of Medical Ethics*, *47*(1), 33–34. http://dx.doi.org/10.1136/medethics-2020-106932.

Spiel, K. (2021). The bodies of TEI—Investigating norms and assumptions in the design of embodied interaction. In *Fifteenth International Conference on Tangible, Embedded, and Embodied Interaction (TEI '21), February 14–17, Salzburg, Austria*. ACM. https://doi.org/10.1145/3430524.3440651.

Spiel, K. (2022). Transreal tracing: Queer-feminist speculations on disabled technologies. *Feminist Theory*, *23*(2), 247–265. https://doi.org/10.1177/14647001221082299.

Spiel, K., & Angelini, R. (2022). Expressive bodies: Engaging with embodied disability cultures for collaborative design critiques. In *The 24th International ACM SIGACCESS Conference on Computers and Accessibility (ASSETS '22), October 23–26 Athens, Greece*. ACM. https://doi.org/10.1145/3517428.3551350.

Spiel, K., Frauenberger, C., Keyes, O., & Fitzpatrick, G. (2019). Agency of autistic children in technology research—A critical literature review. *ACM Transactions on Computer–Human Interaction*, *26*(6), 38. https://doi.org/10.1145/3344919.

Spiel, K., Gerling, K., Bennett, C. L., Brulé, E., Williams, R. M., Rode, J., & Mankoff, J. (2020). Nothing about us without us: Investigating the role of critical disability studies in HCI. In *Extended Abstracts of the 2020 CHI Conference on Human Factors in Computing Systems (CHI EA '20)*. ACM. https://doi.org/10.1145/3334480.3375150.

Spiel, K., Hornecker, E., Williams, R. M., & Good, J. (2022). ADHD and technology research—Investigated by neurodivergent readers. In *CHI '22: Proceedings of the 2022 CHI Conference on Human Factors in Computing Systems* (p. 547). https://doi.org/10.1145/3491102.3517592.

Sum, C. M., Alharbi, R., Spektor, F., Bennett, C. L., Harrington, C., Spiel, K., & Williams, R. M. (2022). Dreaming disability justice in HCI. In *CHI Conference on Human Factors in Computing Systems Extended Abstracts (CHI '22 Extended Abstracts), April 29–May 5, 2022, New Orleans, LA, USA*. ACM. https://doi.org/10.1145/3491101.3503731.

Van Minkelen, P., Krahmer, E., & Vogt, P. (2022). Exploring how people with expressive aphasia interact with and perceive a social robot. *International Journal of Social Robotics, 14*, 1821–1840. https://doi.org/10.1007/s12369-022-00908-8.

Warner, M. (1999). *The Trouble with Normal: Sex, Politics, and the Ethics of Queer Life*. Harvard University Press.

Williams, R. M., Ringland, K., Gibson, A., Mandala, M., Maibaum, A., & Guerreiro, T. (2021). Articulations toward a crip HCI. *Interactions, 28*(3), 28–37. https://doi.org/10.1145/3458453.

Wood, L. J., Zaraki, A., Robins, B., & Dautenhahn, K. (2021). Developing Kaspar: A humanoid robot for children with autism. *International Journal of Social Robotics, 13*, 491–508. https://doi.org/10.1007/s12369-019-00563-6.

Yu, H., Lee, H., Bae, J., Kim, M., Choi, S., & Hwang, J. (2021). The development of a social robot accessible to the deaf. In *HRI '21 Companion: Companion of the 2021 ACM/IEEE International Conference on Human–Robot Interaction* (pp. 634–635). https://doi.org/10.1145/3434074.3446948.

Zafrani, O., Nimrod, G., & Edan, Y. (2023). Between fear and trust: Older adults' evaluation of socially assistive robots. *International Journal of Human–Computer Studies, 171*, 102981. https://doi.org/10.1016/j.ijhcs.2022.102981.

Zhang, G., & Hansen, J. P. (2022). Telepresence robots for people with special needs: A systematic review. *International Journal of Human–Computer Interaction, 38*(17), 1651–1667. https://doi.org/10.1080/10447318.2021.2009673.

Zhang, G., Hansen, J. P., Minakata, K., Alapetite, A., & Wang, Z. (2019). Eye-gaze-controlled telepresence robots for people with motor disabilities. In *Proceedings of the 14th ACM/IEEE International Conference on Human–Robot Interaction (HRI), Daegu, Korea* (pp. 574–575). https://doi.org/10.1109/HRI.2019.8673093.

Sawyer Collins and Selma Šabanović

Growing Old Together: The Promise and Challenge of Social Robots for Older Adults

Abstract: Eldercare is arguably one of the most popular, and well-researched, application domains for social robots. Motivated by the increasing proportion of adults over 65 in the populations of several developed economies—the US, Western Europe, East Asia, which are also leaders in social robotics research—and related societal challenges, researchers are developing robots that can assist older adults and their caregivers in eldercare institutions and individual homes through a variety of functions, from providing companionship to healthcare monitoring. Eldercare robotics has produced many promising research findings and some successful robotic products but also faces continuing challenges to the broader usability and acceptance of eldercare robots. Furthermore, eldercare robotics often takes an overly technocentric view of the issues posed by aging and eldercare, without considering the broader social and cultural context and meaning of care. In this chapter, existing trends in social robotics for eldercare are described and several future directions discussed. These include the need to broaden the research community's perception of aging from a disability-centered view to scaffolding meaningful activities and relationships in older adults' lives, to engage more diverse older adults and other stakeholders in robot design and user studies, to consider caregiver and community effects and support for robotic technologies in robot design, as well as to address the many ethical concerns that eldercare robots raise.

Keywords: Social robots, older adults, eldercare, ethics of robots in care

23.1 Introduction

In the 2012 feature film *Robot and Frank*, an independently living older adult named Frank receives a humanoid assistive robot as a gift from his worried, but also very busy, son. The robot is meant to take care of and, in a sense, 'manage' Frank, who shows signs of dementia. Frank is at first suspicious of the robot, but through the course of the film succeeds in co-opting it in his (not always legal) projects and develops a relationship with it. The film ends with Frank in a nursing home and the robot decommissioned, having failed to manage Frank in a way that fits the expectations of Frank's son and societal norms. The robot has, however, achieved one of the central promises of social robotics—providing a socially isolated older adult with a sense of companionship and emotional connection, and adapting its actions to help its user

https://doi.org/10.1515/9783110792270-024

feel a greater sense of meaning and independence. It did not consider the impact of its actions on broader society beyond Frank—a theme often echoed in eldercare robotics research off-screen.

Eldercare, both in the home and in eldercare institutions, is one of the most common applications for social robots, alongside other healthcare and educational uses. The growing population of older adults worldwide—the World Health Organization estimates that by 2030, 1 in 6 people will be over the age of 60, and by 2050 the total population will be greater than 2 billion (WHO, n. d.)—is frequently cited as motivation for research on socially assistive robotics. However, broad demographic considerations alone provide a shallow justification for robot design, while robots as a solution to aging remain a technocentric proposition designed to address an imagined group of stereotypical users. Rather than constituting a uniform group, older adults' capabilities and needs differ widely based on personal and medical history, or on whether they are 'young' (65–75 years old) or 'old' (85+ years old) older adults. The focus on older adults' assistive interactions with robots often does not incorporate the broader social and economic consequences of aging as a demographic trend, such as the availability of pensions, affordable healthcare, and accessible built environments. Although the solution for issues brought on by aging populations does not lie automatically in robotics, this demographic shift can motivate useful discussions about the future societies we want to build and what place robots may have in them, which can in turn inform the development of more societally beneficial robots.

In this chapter, we describe how the field of social robots has addressed the needs of older adults within two main contexts: personal homes and eldercare institutions. We also identify gaps in existing scholarship, particularly surrounding the adoption of social robots in users' homes, as most studies and robot deployments take place in eldercare facilities or are tested in research laboratories rather than in more open-ended, naturalistic contexts. We discuss robot designs, robot adoption and use studies, promising future directions in designing and deploying social robots for older adults, and the ethical issues raised by social robots in eldercare and how they have so far been addressed by robotics designers and researchers.

23.2 Background

23.2.1 Aging and the changing needs of older adults

The exponential growth of older adults within the population brings concerns regarding how best to support the ongoing care and well-being of individuals who may need both physical assistance and support to maintain their psychological well-being. The cost of eldercare facilities such as nursing homes can be a barrier to those who need specialized care (Carey et al., 2018), and many societies are faced with the diminishing availability of professional caregivers who can assist in caring for aging adults. Furthermore, older adults may prefer to age in the more familiar surroundings of their homes, neighborhoods, and familiar social networks. Such 'aging in place,' however, can lead to further challenges, such as an increase in loneliness and being unable to quickly call for and receive help as the social safety net around older adults changes (Verver et al., 2017). Along with other information and communication devices, social robots may be one solution that could aid older adults in living independently in their own homes for longer.

A further significant challenge brought about by aging is the changing physical and cognitive capabilities of older adults, which can affect their ability to do various activities. For example, older adults may lose the ability to take care of beloved pets when they move into eldercare facilities. Pet ownership can aid in physical activity, as well as increased cognitive function (Gee & Mueller, 2019), but the ability of the individual to care for an animal by themselves may become too strenuous. Social robots may be able to provide a unique solution to this concern by providing pet-like interaction in a more easily manageable form, while similarly improving mood and alleviating loneliness (Chen et al., 2020). Older adults may also find that their social circles diminish as they age and their friends and family are no longer close by. Robots may be able to act as social mediators and help older adults engage with others, or provide more physically interactive versions of remote communication (e.g., Telenoid (Kuwamura et al., 2016)). Robots can further support older adults through personalized cognitive and memory support activities (e.g., Van Patten et al., 2020), or by acting as a medical monitoring and feedback device (e.g., by providing activity reminders).

23.2.2 What are social(ly assistive) robots?

One defining aspect of robots that distinguishes them from other interactive and autonomous technologies is that they are physically embodied and can use their embodiment to sense and affect the environment in which they are embedded. Social robots are more specifically defined by their ability to participate in social interactions with the user. The term 'social robot,' as a "physical entity embodied in a complex, dynamic, and social environment sufficiently empowered to behave in a manner conducive to its own goals and those of its community" (Duffy et al., 1999, p. 85), may arguably apply to any robot being used in an environment where it interacts with humans, whose interactions are inevitably social. More detailed terminology has therefore been proposed, suggesting that robots designed specifically to interact socially with people be called 'sociable' (Breazeal, 2003), 'socially interactive' (Fong et al., 2002), 'socially situated' (Bennett et al., 2021), or 'embedded' (Kim et al., 2013) robots.

Social robots are unique in the ways in which they interact with the intended user. Rather than performing mainly functional tasks such as cleaning, they foster social and even emotional connections with direct users and others in the context of use. Social robots vary widely in their goals, from providing companionship and other interaction benefits to individual users (FakhrHosseini et al., 2020), to connecting users to other individuals, such as the case of telepresence robots during the COVID-19 pandemic (Isabet et al, 2021). These robots can come in diverse forms, including zoomorphic, such as the Joy for All cat or dog, or the seal-like Paro (Figure 23.1), to more humanoid forms such as Nao and QT Robot (Figure 23.2), as well as other more abstract and mechanical forms such as Elliq, Misty, and Astro.

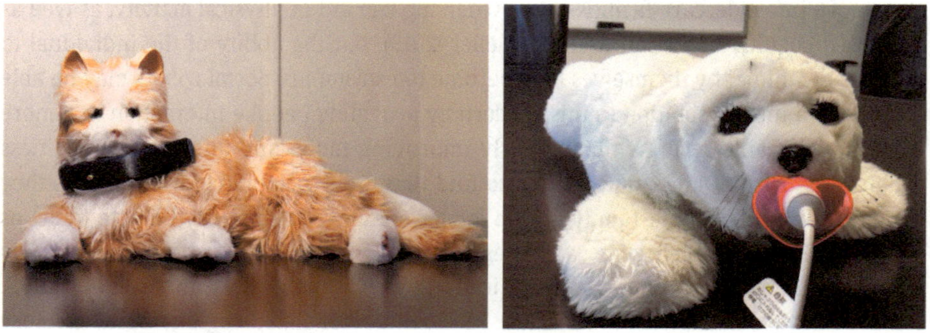

Figure 23.1: Joy for All cat (l), Paro (r).

Figure 23.2: QT Robot.

Socially assistive robots are a subset of social robots that focus on health-oriented applications with measurable psychological, cognitive, and/or physical benefits to users (Feil-Seifer & Mataric, 2005). In order to understand socially assistive robots and their potential uses within the context of an aging population, studies seek to identify the needs of older adults (Gasteiger et al., 2022), as well as implement these devices in varying settings, particularly the homes of older adults, eldercare institutions (e.g., nursing homes, older adult daycare), and healthcare institutions (e.g., hospitals).

23.3 Applications of Social Robots in Eldercare

23.3.1 Social robots in the homes of older adults

One of the main areas in which social robots are envisioned as assisting older adults is through use in home contexts. Catalia Health's Mabu robot (MyMabu, 2019) provides healthcare-related reminders (e.g., medication, activities) based on the user's personal profile and personality, and Kompaï is similarly used in-home for memory-based tasks (Zsiga et al., 2018). Telepresence robots can be a way to provide professional care remotely, such as for nurse or doctor check-ins on patients without the need for them to visit a medical facility. The Giraff robot, for example, enables a remote visitor to move around the older adult's home and interact with them through a video screen; it can also take physiological measures that can be useful in medical care (Coradeschi et al., 2014). Finally, social robots can act as interaction partners and companions to older adults in the home; in this way, they may be able to provide emotional or social support. Exploratory studies suggest older adults can accept robots in their homes if they have a well-defined and evidenced use (Smarr et al., 2012), and some have explored specific features that allow older adults to form an attachment to the social robot (Piasek & Wieczorowska-Tobis, 2018). However, robots have yet to be extensively tested within the homes of older adults in ways that address longer-term use and needs.

Loneliness is one of the biggest issues for older adults who live independently (Berg-Weger & Morley, 2020), and has been linked to cognitive decline (Boss et al., 2015). The use of social robots, such as the Joy for All cat and dog, to address loneliness is starting to be explored (Paluch & Müller, 2022). Telepresence platforms like the Telenoid enable more physically interactive remote communication with friends and family, and a resulting greater feeling of social presence (Yamazaki et al., 2014). Robots might be effective in reducing feelings of loneliness (Sidner et al., 2018) and symptoms of depression (Bennett et al., 2017) when used over several weeks.

Calling for aid has also been cited as a reason older adults may have difficulties remaining on their own, particularly during accidents such as falls (Stevens et al., 2006). While other fall detection systems have been explored, such as smartphones

(Abbate et al., 2012), social robots may also be able to provide support to older adults by being able to notice they need aid and call for help if needed (Kim et al., 2022).

Older adults are situated in networks of family, friendship, neighborhood, and formal caregiving relationships which are articulated differently in different countries and cultural contexts (e.g., big city, small town). Accordingly, more recent studies include robots' ability to assist formal and informal caregivers as they attend to older adults, such as by reducing relationship strain through moderation (Wang et al., 2017). These more socially situated perspectives on designing and deploying robots consider the importance of the social context and relationships for older adults' health and well-being. Some aim to support communication between older adults and family caregivers, such as the ElliQ robot (ElliQ, n. d.), and see the robot as performing the role of a social mediator (Coghlan et al., 2021).

Overall, the use of robots in older adults' homes has promise. By supporting older adults living independently there is the potential to reduce strain on care staff, as well as allow the older adults to maintain autonomy. More exploration is needed to understand patterns and outcomes of long-term use, and how robots should be designed to fit into the broader care ecology of users.

23.3.2 Social robots in eldercare institutions

Many robots are also designed for use in eldercare institutions, including those where an individual goes to receive short-term care, such as a hospital, as well as in more long-term stay institutions, such as nursing homes. Along with having a built environment that is often spacious and structured and can be easier for robots to navigate, institutions are also populated by professional staff members who can be trained to help manage older adults' use of robots and maintain the robots when not in use. In reality, it can be difficult for care staff to work the care and maintenance of robots into their busy schedules. The most common institutional space in which social robots have been tested out with older adults is nursing homes, but older adult day cares and retirement communities have also deployed and evaluated social robots.

A main concern of social robot evaluations in eldercare facilities is whether the robot can provide social and/or cognitive enrichment for the user. This includes encouraging interaction between older adults in group activities, or communication between the user and the robot (Khosla & Chu, 2013). One study implemented the robot Paro for people living with dementia in a hospital setting and found that Paro worked not just as a companion, but also as a conversation starter between the elderly patients and those around them (Hung et al., 2021). Social robots have also been used to mediate intergenerational interactions, such as between older adults and children or younger family members (Joshi & Šabanović, 2019). Social robots can also support cognitive stimulation for older adults through activities like interactive storytelling or cognitive games (Tokunaga et al., 2021).

Many institutional studies focus on gauging the initial opinions of older adults regarding social robots and their own needs (Lin & Chen, 2018). Understanding user perceptions of various aspects of social robot design and use, such as approach behavior (Khaksar et al., 2021), is important for the adoption of these technologies.

23.3.3 Social robots to support physical and cognitive change

Social robots may be able to help address challenges that emerge as people age. One of the most common target groups among older adults for the use of social robots is assistance for dementia care (Šabanović et al., 2013). Dementia is a chronic neurodegenerative disorder generally affecting older adults and is commonly associated with cognitive decline and memory loss, language disability, and co-morbidities such as social isolation, depression, anxiety, and behavioral issues. Paro has been used extensively with older adults living with dementia to provide social support (Kelly et al., 2021) and increase engagement between older adults and others, such as their care team (Chang et al., 2013). The social robot Pepper has also been adapted to aid in administering diagnostic tests (Varrasi et al., 2018), to aid in identifying the disease earlier and with less stress put on the user's care team. People living with dementia can also benefit from cognitive stimulation and memory care activities, as well as from reminders relating to their medical care and activities of daily living (e.g., taking medication, drinking water, exercise), and even the companionship and interaction that can come from engaging with a robot.

Social robots may also be able to help older adults take care of themselves. For example, a social robot acting as an exercise guide (Cruz-Sandoval et al., 2013) can help older adults retain the physical ability to care for themselves for longer. Additionally, along with assisting older adults themselves, robots may also be able to assist caregivers to communicate more effectively with older adults (Coradeschi et al., 2014), or may be able to provide some behavioral monitoring or collection of healthcare information from the older adult (Boumans etal., 2018) that can help guide treatment and interventions.

23.3.4 Designing robot features to address the needs of older adults

While some robots used with older adults are designed for a broader population, Broadbent et al. (2009) suggest that robot designs need to be matched to the needs of specific users and contexts of use. User age, gender, experience with and expectations from robots, and specific needs may affect user perceptions, while the robot's appearance and behavior, personality, and adaptability will similarly contribute to their acceptance by older adults and their caregivers. Along with matching users' needs, tempering their expectations of robots and being clearer about their capabilities and needs for maintenance can also support successful adoption.

The physical design of the robot is perhaps the most immediately salient aspect, as it affects what kind of physical context the robot will be able to fit into, and users' expectations of its functions. Researchers have explored the appropriate dimensions of home robots through evaluations of low-fidelity prototypes in user homes (Caleb-Solly et al., 2014) and performed survey and laboratory studies to test the effects of different forms on older adults' preferences (Prakash & Rogers, 2013); however, these studies often rely on imagined notions of the home and do not consider the home refurbishments that might need to be made to homes to accommodate robots. The laboratory setting provides more control for researchers, allowing for more complex robots to be deployed, such as being able to program the behaviors of the robot or being able to implement the Wizard of Oz technique, where a researcher remotely controls the movements and speech of the robot. It is therefore not clear how usable robots will be in the homes of older adults, where their care and maintenance would fall to the householders themselves, rather than to professionals trained in their use.

Along with form, another important determinant of a robot's acceptability to older adults is its function. In studies with older adults in the US, researchers found that they prefer robots that have useful physically assistive capabilities, such as cleaning and fetching/organizing tasks (Beer et al., 2012). In the public Eurobarometer survey performed in European countries, physically assistive tasks were overall preferred for robots but some groups, such as pensioners, were more willing to have robots working in health, social care, and education contexts (Taipale, 2012). Older adults and their caregivers were shown to be more familiar with robots with a wider array of social as well as physical functions (Suwa, 2012). While this may suggest that social robot adoption might be less likely among older adults, the difficulty of developing and maintaining physically assistive robots for diverse home and institutional environments makes social robot applications more likely in the short term. Studies also show that older adults' acceptance of social robots may increase after direct interaction with a robot (Ke et al., 2020), which suggests that perception studies without engagement may not be indicative of actual social robot acceptance and use.

Finally, many robots are designed to address aging as a disability, but older adults do not find this a compelling framework for thinking about technology they want to use (Lee et al., 2016). Therefore, considering other ways to frame robots for older adults are under consideration, including robots to support joyful activities (Lee & Riek, 2018), robots that consider the wisdom of age, and robots that connect older adults with people, growth opportunities, and communities that support a sense of meaning in their lives (Randall et al., 2022).

23.4 Remaining Challenges and Opportunities in Social Robots for Eldercare

23.4.1 Supporting life and health transitions

While some robotic technologies address the fact that older adult users' capabilities and needs will change over time, few designs support changing cognitive and physical capabilities directly. To make robots useful to older adults over a longer period, consideration of this type of adjustment is necessary. Rather than being a linear progression, cognitive, social, and physical decline and changes that occur as people age follow a more convoluted trajectory. Older adults may go in and out of healthcare institutions and the care of different formal and informal caregivers; social robots might be able to provide some continued companionship and healthcare support during these transitions. Pet-like companion robots like Paro and Joy for All, for example, could accompany older adults into healthcare institutions in ways that animal pets cannot. Similarly, an ElliQ communication robot may be able to connect an older adult with friends and family from their nursing home room as well as from their home.

23.4.2 Diversity of user needs and personalization

Rather than being a homogeneous group, older adults are diverse along many different dimensions—physical and cognitive abilities, specific health and wellness support needs, prior experience, and easewith technology, as well as their social and cultural background and the context in which they use the technology. Rather than being a single group, older adults can be categorized as young older adults (65–74), older adults (75–84), and old older adults (over 85); young older adults may also be acting as caregivers for old older adults (Cohen-Mansfield et al., 2013). This heterogeneity suggests that robots need to be adaptable and personalizable to diverse users by design. In some cases, users themselves can adapt existing robots to suit their needs. Paro has been used in over 30 countries (Shibata et al., 2009) and in diverse ways to support eldercare—in some nursing homes it is used in group interactions while in others it is seen as more useful in helping manage agitation and anxiety in individual older adults (Chang et al., 2013); some older adults use Paro as an interactive object on which they can project their own feelings (Taggart et al., 2005), some treat it as a pet, and yet others use it to instigate conversations with peers and passersby (Chang & Šabanović, 2015). There are, of course, limits by design to how any robot can be adapted by users. Another possibility is to develop more adaptable robots that will be able to sense certain aspects of the environment they interact with and change their behavior accordingly. Users might also be able to pre-select certain aspects of their robot's appearance or behavior—Joy for All purchasers can select a cat or dog, while others might be able to select a robot's gender or personality that may fit the individual's personality or needs (Broadbent et al., 2009). Machine learning may also enable robots to adapt their behavior to users' needs and preferences over time (Bennett et al., 2017).

Robots may additionally need to be acceptable and personalizable to more than one user. Participatory design workshops with older adults, for example, suggest that older adults see it as important that the robot is attractive to family members (e.g., grandchildren) and others who might be in the user's social environment (Šabanović et al., 2015). Identifying what features are preferred and needed by the older adults' caregivers, family members, or friends can further improve the acceptability and usefulness of a social robot.

23.4.3 Robot explainability for older adults

Concerning features that may affect the acceptance and eventual adoption of social robots by older adults, a particularly difficult issue may be defining and explaining how and when to use the robot (Robinson et al., 2014). Explainability of robots for older adults will need to account for older adults' potential lack of exposure to such technologies, and the resulting need for relatively intuitive ways of the robot providing information about its actions. This might be done through the robot's appearance and behavior design, which could signal its limitations through simple language or a more child-like appearance. Robots could also use verbal cues, or additional interfaces provided to the user, to improve explainability (Luperto, 2014). A challenge here may be that people's willingness to 'forgive and forget' mistakes is limited, particularly when it comes to important tasks. Providing too much explanation of the robot's function may also get in the way of its overall success in maintaining an illusion of sociality.

23.4.4 Privacy concerns in the use of robots with older adults

In order to act appropriately in a given context, robots need to collect information about their users and environment through sensor data, which may include sound, light, and video data, as well as physiological and other environmental sensors. When involved in eldercare, the collected data may relate to the individual user's health status, and therefore be an especially protected class of information. The environments in which social robots interact with older adults—their homes and caregiving institutions—are furthermore domains largely considered private. This all raises concerns regarding the privacy and security of robots as they are being used in eldercare.

One aspect of privacy relevant to social robots in eldercare is information being collected by the robot about the individual user and its interactions with them. The robot may collect data about a person's mental or physical health directly, or interaction patterns with the robot may indirectly suggest the person's health status. The easiest way to allay privacy concerns would be to keep all data local to the robot; however, this may limit the robot's functions, as multiple robots sharing data for machine-learning purposes may be more effective and efficient, or data could be more therapeutically useful if shared with caregivers and clinical staff. Finally, collected data can be used not only for descriptive but also for predictive purposes—to alert forpotential health concerns or behavioral changes in the user—which begs the question of who should be the recipient of such predictions and in what way should they be communicated and interpreted.

Additionally, it may be necessary to monitor the robot for misbehavior and errors in functioning that could lead to a negative impact on the physical and psychological well-being of older adult users (Coffee-Johnson & Perouli, 2019). This is important not just for the acceptance of the robot, but also for the older adult's care.

23.4.5 Authentic and meaningful social participation

The main purpose of using social robots is to provide people with benefits through social interaction. Many of these interactions are inspired by or patterned to some degree on existing human–human or human–animal interactions. The danger, then, is that people will be substituting social interactions of lesser quality for more meaningful and substantive interactions due to the availability, ease of access, and cost-effectiveness of a robot compared with a human caregiver. Turkle (2010) has raised the issue of the 'authenticity' of the engagement and connection people feel towards robots, as users might be deceived about the level of care and reciprocity a machine can give. Others have argued that authenticity and reciprocity in social interaction may be beside the point if older adults themselves feel that interactions with robots have benefits—if a robot seal has the same positive psychological and physiological effects as a domesticated animal, then it could be considered just as good. However, even if people feel subjectively happy when they are interacting with robots, this may still not support 'human dignity' as defined through objective elements of welfare, such as providing respect and recognition, which robots cannot do (Sparrow & Sparrow, 2006). Finally, scholars suggest such technologies might lead to more social isolation among older adults, who may be relegated to the care of robots.

23.4.6 Long-term use in diverse care settings

To better understand how social robots may be adopted in eldercare settings, and what their consequences might be for different stakeholders there, researchers still need to do much more testing of these devices in everyday use contexts and for longer periods of time. This will also be able to show us how older adults' needs and user patterns regarding the robots may change over time, and it may lead to users developing particular use cases for robots themselves. While many papers indicate the effects of robot use such as a reduction in loneliness (Chen et al., 2020) and an increase in factors such as engagement (Fan et al., 2021), more rigorous studies of such effects are needed (Pu et al., 2019), particularly in their expected contexts of use. Studying robots within a laboratory setting may allow for more control by the researchers but does not allow us to consider the changing setting and needs of the users within a day-to-day context.

Among in-situ studies of robot use, the length of home-based user studies with social robots generally varies from a few weeks to a few months, but researchers still lack a clear understanding of the viability of social robots as a long-term solution in aiding elderly users to live independently. Some studies involved interactions that were as short as a mere 10 minutes with the robot (Heerink et al., 2009), while others persisted over several months, with more or less support from researchers in the various steps needed to use the robots.

While some longitudinal issues that arise might have to do with the specifics of interactions between older adults and robots directly, such as boredom as the initial novelty wears off, or realization of the limitations of the platform (Leite et al., 2013), broader contextual and societal issues may also play a role in social robot adoption. These can include the way the robot fits into the broader ecology of care that the older adult inhabits (Mois & Beer, 2020), that is, whether other people that the older adult regularly interacts with engage with and accept the robot, and whether there is sufficient support for maintenance and regular use of the robot by the older adult and their caregivers. Robots should therefore be designed to address the needs not only of older adults, but also their social contacts and caregivers.

23.4.7 Hidden labor and costs of social robots

Along with exploring features that are of interest to older adults and their caregivers in the context of institutional use, institutional deployments of eldercare robots have also brought out broader challenges to the adoption and successful use of robots in eldercare institutions. A major concern with robots is the issue of cost—how much the robot and its maintenance may cost the user, and where the funds to purchase and maintain robots will come from. The issue of cost also raises concerns about the overall benefit of robots. Many eldercare institutions have limited funds, so spending them on robots and their maintenance means that less will be available for other needs. It is so far unclear whether the often-high costs of most robots are sufficiently offset by their benefits, or how social robots compare with other interactive technologies such as voice-activated virtual assistants in their ability to significantly improve care. Socioeconomic and comparative benchmarking studies of the efficacy of robots and other technologies, such as smartphones and virtual assistants, could identify the specific beneficial use cases for social robots.

Another concern that has been realized in institutional deployment is the 'invisible labor' that is needed to support the use of robots in institutions (Chang & Šabanović, 2015). Robotics research often involves teams of several researchers who prepare and deploy the robot, fixing it as needed throughout the study so that it responds to participants appropriately. When robots are used in institutions without the constant support of the robotics team, already busy care workers need to take on the care and management of the robot. This means that institutions can only use robots effectively if those robots are relatively simple to deploy (e.g., Paro, Joy for All), if they fit well in the work context, and provide necessary services (e.g., Moxi from Diligent Robotics), if there is a staff member willing to champion the technology, or if there is an affordable source of labor—such as volunteers—who can help maintain the robot and support its appropriate use. As these issues are often not considered early on in the design of the robot, robotic systems may fail in adoption because they create more burden on the care team despite positive interactions with older adults (Hung et al., 2019). Recent ethnographic work by Wright in Japan (2023) shows that robots may also change the nature of caregivers' work and interactions with older adults in ways that diminish its meaningfulness to the caregiver and the older adult. Perhaps this is why robotic technologies have yet to be widely used in caregiving institutions outside of research. In Japan, a national survey in 2019 of over 9,000 eldercare institutions indicated that only 10% have introduced a care robot, with evidence suggesting that when they are purchased they are often only used for a short time before no longer being used (Wright, 2023).

23.5 Areas for Future Development

Social robots have been explored for various uses with older adults, attempting to address issues that arise as older adults age, as well as identifying specific preferences and needs of older users (Gasteiger et al., 2022). From medication reminders and tracking (Chelvam & Zamin, 2014), to being a companion for the elderly (Paluch & Müller, 2022), social robots provide a unique way of aiding an aging population. However, there are still many areas of possibility and challenges that have not been addressed by existing research, and that need to be resolved to provide older adults with the support to remain at home longer in ways that benefit them and society as a whole.

Along with the dearth of long-term studies in a diverse set of potential social robot use contexts, there has been little consideration of the broader impacts of the use of social robots in eldercare. Social robots may be a convenient care support device for certain people in certain contexts, but they can also increase or change work for others. Caregivers—already very busy and arguably overworked and underpaid in many cases—will need to figure out how to accommodate, maintain, and assist older adults in using social robots and how to manage sharing and distributing the use of robots among several older adults who want to use them. Social robots, though touted for in-home use, are also not necessarily designed for in-home use by just anyone. Most robotics studies occur in urban areas, where the technical infrastructure for using AI-powered technologies is more widely available. Robotics solutions often do not consider that, in 2050, 80% of older people will be living in low- and middle-income countries (WHO, n. d), which do not have the kinds of infrastructures and other resources currently assumed by eldercare robot designers. Developing robots that can be used in less socioeconomically advantaged contexts will require incorporating their limitations into the design from the beginning (Demofonti et al., 2021). Current methods of social robot design may therefore make them less usable in rural, and less socioeconomically well-off, areas of the world.

Furthermore, the cost of robots may be prohibitive for many users, both in terms of the initial purchase or, if covered by some other source, in terms of upkeep or if the robot was damaged or stolen. Robot designs may incorporate stereotypes about their users, or others who participate in the care of older adults. Gendered and racialized robot designs may also negatively impact healthcare workers, who may be compared to or have to work with such robots in the future. Finally, spending resources on developing and deploying social robots can take attention and funding away from other potential solutions, including non-technological solutions, to increasing eldercare needs. Issues related to aging and the potential use of robots by older adults and their caregivers need to go beyond techno-solutionism (Šabanović, 2010; Wright, 2023) and incorporate a more nuanced understanding of the social context and implications of eldercare.

One way to ensure that the potential unintended and unequal consequences of social robots are considered from early on in the design process is to broaden participation in robot design and decision making about robot deployment. Researchers have been using more participatory and collaborative methods in eldercare robot design, bringing a broader range of stakeholders into the process of conceptualizing robots and their uses, to envision the possibilities and potential dangers of using robots with older adults. Such efforts include working with caregivers as well as care receivers on robot design concepts; including rural and lower socioeconomic populations in robotics design research; and involving older adults with dementia as well as cognitively unimpaired older adults and their caregivers in design research and evaluations of robots. Additionally, researchers have started to look beyond the human–robot dyad, or even the caregiver–care receiver–robot triad, to consider how the deployment and long-term use of robots affect communities and the various people who inhabit them, whether they use the robots or not. Efforts to incorporate community values in social robot evaluations (e.g., Joshi & Šabanović, 2019) have shown that different features and capabilities may be salient when human social relationships, for example, are prioritized over engagement with robots; robots in these cases may also have different roles, such as social mediation (Joshi & Šabanović, 2017), that can support older adults' social connectedness with others (Ostrowski et al., 2019). A socio-ecological perspective on eldercare robots can further situate them in relevant healthcare systems and connect the use of robots to broader issues of policy and economic impact relating to aging and eldercare (Mois & Beer, 2020).

Frank's robot provided him with a clean home, healthy meals, motivation for physical activity, cognitive stimulation, and a form of companionship. On its own, however, it could not understand and appropriately respond to the complexities of Frank's condition. Robots alone cannot be a solution to the needs of an aging population, but properly framed in broader networks of care they can become tools to help improve the well-being of older adults and their caregivers. Robot development, therefore, needs to attend to creating inclusive social support networks along with robotic technologies, and reflect on the broad array of economic and social factors that will affect and be affected by eldercare robots.

Bibliography

Abbate, S., Avvenuti, M., Bonatesta, F., Cola, G., Corsini, P., & Vecchio, A. (2012). A smartphone-based fall detection system. *Pervasive and Mobile Computing, 8*(6), 883–899.

Beer, J. M., Smarr, C. A., Chen, T. L., Prakash, A., Mitzner, T. L., Kemp, C. C., & Rogers, W. A. (2012, March). The domesticated robot: Design guidelines for assisting older adults to age in place. In *Proceedings of the Seventh Annual ACM/IEEE International Conference on Human-Robot Interaction* (pp. 335–342). IEEE.

Bennett, C. C., Sabanovic, S., Piatt, J. A., Nagata, S., Eldridge, L., & Randall, N. (2017, August). A robot a day keeps the blues away. In *2017 IEEE International Conference on Healthcare Informatics (ICHI)* (pp. 536–540). IEEE.

Bennett, C., Stanojević, Č., Šabanović, S., Piatt, J., & Kim, S. (2021, November). When no one is watching: Ecological momentary assessment to understand situated social robot use in healthcare. In *Proceedings of the 9th International Conference on Human-Agent Interaction* (pp. 245–251).

Berg-Weger, M., & Morley, J. E. (2020). Loneliness in old age: An unaddressed health problem. *Journal of Nutrition, Health Aging, 24*(3), 243–245.

Boss, L., Kang, D. H., &Branson, S. (2015). Loneliness and cognitive function in the older adult: A systematic review. *International Psychogeriatrics, 27*(4), 541–553.

Boumans, R., Van Meulen, F., Hindriks, K., Neerincx, M., &Olde Rikkert, M. (2018, March). Proof of concept of a social robot for patient reported outcome measurements in elderly persons. In *Companion of the 2018 ACM/IEEE International Conference on Human-Robot Interaction* (pp. 73–74). IEEE.

Breazeal, C. (2003). Toward sociable robots. *Robotics and Autonomous Systems, 42*(3–4), 167–175.

Broadbent, E., Stafford, R., & MacDonald, B. (2009). Acceptance of healthcare robots for the older population: Review and future directions. *International Journal of Social Robotics, 1*, 319–330.

Caleb-Solly, P., Dogramadzi, S., Ellender, D., Fear, T., & Heuvel, H. V. D. (2014, March). A mixed-method approach to evoke creative and holistic thinking about robots in a home environment. In *Proceedings of the 2014 ACM/IEEE International Conference on Human-Robot Interaction* (pp. 374–381). IEEE.

Carey, K., Zhao, S., Snow, A. L., & Hartmann, C. W. (2018). The relationship between nursing home quality and costs: Evidence from the VA. *Plos one, 13*(9), e0203764.

Chang, W. L., & Šabanović, S. (2015, March). Studying socially assistive robots in their organizational context: Studies with Paro in a nursing home. In *Proceedings of the Tenth Annual ACM/IEEE International Conference on Human–Robot Interaction Extended Abstracts* (pp. 227–228). IEEE.

Chang, W. L., Šabanović, S., & Huber, L. (2013). Situated analysis of interactions between cognitively impaired older adults and the therapeutic robot PARO. In *Social Robotics: 5th International Conference, ICSR 2013, Bristol, UK, October 27–29, 2013, Proceedings 5* (pp. 371–380). Springer International Publishing.

Chelvam, Y. K., & Zamin, N. (2014, September). M3DITRACK3R: A design of an automated patient tracking and medicine dispensing mobile robot for senior citizens. In *2014 International Conference on Computer, Communications, and Control Technology (I4CT)* (pp. 36–41). IEEE.

Chen, S. C., Moyle, W., Jones, C., & Petsky, H. (2020). A social robot intervention on depression, loneliness, and quality of life for Taiwanese older adults in long-term care. *International Psychogeriatrics, 32*(8), 981–991.

Coffee-Johnson, L., & Perouli, D. (2019, March). Detecting anomalous behavior of socially assistive robots in geriatric care facilities. In *2019 14th ACM/IEEE International Conference on Human–Robot Interaction (HRI)* (pp. 582–583). IEEE.

Coghlan, S., Waycott, J., Lazar, A., & Barbosa Neves, B. (2021). Dignity, autonomy, and style of company: Dimensions older adults consider for robot companions. *Proceedings of the ACM on Human–Computer Interaction, 5(CSCW1)* (pp. 1–25).

Cohen-Mansfield, J., Shmotkin, D., Blumstein, Z., Shorek, A., Eyal, N., & Hazan, H. (2013). The old, old-old, and the oldest old: Continuation or distinct categories? An examination of the relationship between age and changes in health, function, and wellbeing. *International Journal of Aging and Human Development, 77*(1), 37–57.

Coradeschi, S., Cesta, A., Cortellessa, G., Coraci, L., Galindo, C., Gonzalez, J., & Ötslund, B. (2014). GiraffPlus: A system for monitoring activities and physiological parameters and promoting social interaction for elderly. *Human–Computer Systems Interaction: Backgrounds and Applications, 3*, 261–271.

Cormons, L., Poulet, C., Pellier, D., Pesty, S., & Fiorino, H. (2020, February). Testing social robot acceptance: What if you could be assessed for dementia by a robot? A pilot study. In *2020 6th International Conference on Mechatronics and Robotics Engineering (ICMRE)* (pp. 92–98). IEEE.

Cruz-Sandoval, D., Penaloza, C. I., Favela, J., & Castro-Coronel, A. P. (2018, October). Towards social robots that support exercise therapies for persons with dementia. In *Proceedings of the 2018 ACM International Joint Conference and 2018 International Symposium on Pervasive and Ubiquitous Computing and Wearable Computers* (pp. 1729–1734).

Demofonti, A., Carpino, G., Zollo, L., & Johnson, M. J. (2021). Affordable robotics for upper limb stroke rehabilitation in developing countries: A systematic review. *IEEE Transactions on Medical Robotics and Bionics, 3*(1), 11–20.

Duffy, B. R., Rooney, C., O'Hare, G. M. P., & O'Donoghue, R. (1999). What is a social robot? In *Proceedings of the 10th Irish Conference on Artificial Intelligence & Cognitive Science, Cork, Ireland, September 1–3*.

ElliQ, the sidekick for healthier, happier aging. *ElliQ* (n. d.). https://elliq.com/.

FakhrHosseini, S., Lee, C., Miller, J., Patskanick, T., & Coughlin, J. (2020). Older adults' opinion on social robot as companion. In *2020 29th IEEE International Conference on Robot and Human Interactive Communication (RO-MAN)* (pp. 821–826). IEEE.

Fan, J., Mion, L. C., Beuscher, L., Ullal, A., Newhouse, P. A., & Sarkar, N. (2021). SAR-connect: A socially assistive robotic system to support activity and social engagement of older adults. *IEEE Transactions on Robotics, 38*(2), 1250–1269.

Feil-Seifer, D., & Mataric, M. J. (2005, June). Defining socially assistive robotics. In *9th International Conference on Rehabilitation Robotics, 2005. ICORR 2005* (pp. 465–468). IEEE.

Fong, T., Nourbakhsh, I., & Dautenhahn, K. (2002). A survey of socially interactive robots: Concepts, design and applications. *Report. View/Open*. 101193.pdf. http://hdl.handle.net/2299/3821.

Gasteiger, N., Ahn, H. S., Lee, C., Lim, J., Macdonald, B. A., Kim, G. H., & Broadbent, E. (2022). Participatory design, development, and testing of assistive health robots with older adults: An international four-year project. *ACM Transactions on Human–Robot Interaction, 11I*(4), 45, 1–9. https://doi.org/10.1145/3533726.

Gee, N. R., &Mueller, M. K. (2019). A systematic review of research on pet ownership and animal interactions among older adults. *Anthrozöos, 32*(2), 183–207.

Heerink, M., Kröse, B. J., Wielinga, B., & Evers, V. (2009). Measuring the influence of social abilities on acceptance of an interface robot and a screen agent by elderly users. In *Proceedings of the 2009 British Computer Society Conference on Human–Computer Interaction, BCS–HCI 2009* (pp. 1–5).

Hung, L., Gregorio, M., Mann, J., Wallsworth, C., Horne, N., Berndt, A., Liu, C., Woldum, E., Au-Yeung, A., & Chaudhury, H. (2021). Exploring the perceptions of people with dementia about the social robot PARO in a hospital setting. *Dementia, 20*(2), 485–504.

Hung, L., Liu, C., Woldum, E., Au-Yeung, A., Berndt, A., Wallsworth, C., Horne, N., Gregorio, M., Mann, J., & Chaudhury, H. (2019). The benefits of and barriers to using a social robot PARO in care settings: A scoping review. *BMC Geriatrics, 19*(1), 1–10.

Ide, H., Kodate, N., Suwa, S., Tsujimura, M., Shimamura, A., Ishimaru, M., & Yu, W. (2021). The ageing 'care crisis' in Japan: Is there a role for robotics-based solutions? *International Journal of Care and Caring, 5*(1), 165–171.

Isabet, B., Pino, M., Lewis, M., Benveniste, S., & Rigaud, A. S. (2021). Social telepresence robots: A narrative review of experiments involving older adults before and during the COVID-19 pandemic. *International Journal of Environmental Research and Public Health, 18*(7), 3597.

Joshi, S., & Šabanović, S. (2017, August). A communal perspective on shared robots as social catalysts. In *2017 26th IEEE International Symposium on Robot and Human Interactive Communication (RO-MAN)* (pp. 732–738). IEEE.

Joshi, S., & Šabanović, S. (2019, March). Robots for inter-generational interactions: Implications for nonfamilial community settings. In *2019 14th ACM/IEEE International Conference on Human–Robot Interaction (HRI)* (pp. 478–486). IEEE.

Ke, C., Lou, V. W. Q., Tan, K. C. K., Wai, M. Y., & Chan, L. L. (2020). Changes in technology acceptance among older people with dementia: The role of social robot engagement. *International Journal of Medical Informatics, 141*, 104241.

Kelly, P. A., Cox, L. A., Petersen, S. F., Gilder, R. E., Blann, A., Autrey, A. E., & MacDonell, K. (2021). The effect of PARO robotic seals for hospitalized patients with dementia: A feasibility study. *Geriatric Nursing, 42*(1), 37–45.

Khaksar, W., Neggers, M., Barakova, E., & Torresen, J. (2021, August). Generation differences in perception of the elderly care robot. In *2021 30th IEEE International Conference on Robot Human Interactive Communication (RO-MAN)* (pp. 551–558). IEEE.

Khosla, R., & Chu, M. T. (2013). Embodying care in Matilda: An affective communication robot for emotional wellbeing of older people in Australian residential care facilities. *ACM Transactions on Management Information Systems, 4*(4), 1–33.

Kim, E. S., Berkovits, L. D., Bernier, E. P., Leyzberg, D., Shic, F., Paul, R., & Scassellati, B. (2013). Social robots as embedded reinforcers of social behavior in children with autism. *Journal of Autism and Developmental Disorders, 43*, 1038–1049.

Kim, J. W., Choi, Y. L., Jeong, S. H., & Han, J. (2022). A care robot with ethical sensing system for older adults at home. *Sensors, 22*(19), 7515.

Kuwamura, K., Nishio, S., & Sato, S. (2016). Can we talk through a robot as if face-to-face? Long-term fieldwork using teleoperated robot for seniors with Alzheimer's disease. *Frontiers in Psychology, 7*, 1066.

Lee, H. R., & Riek, L. D. (2018). Reframing assistive robots to promote successful aging. *ACM Transactions on Human–Robot Interaction, 7*(1), 1–23.

Lee, H. R., Tan, H., & Šabanović, S. (2016, August). That robot is not for me: Addressing stereotypes of aging in assistive robot design. In *2016 25th IEEE International Symposium on Robot and Human Interactive Communication (RO-MAN)* (pp. 312–317). IEEE.

Leite, I., Martinho, C., & Paiva, A. (2013). Social robots for long-term interaction: A survey. *International Journal of Social Robotics*, *5*, 291–308.

Lin, X., & Chen, T. (2018, March). A qualitative approach for the elderly's needs in service robots design. In *Proceedings of the 2018 International Conference on Service Robotics Technologies* (pp. 67–72).

Luperto, M., Romeo, M., Monroy, J., Vuono, A., Basilico, N., Gonzalez-Jimenez, J., & Borghese, N. A. (2021, August). What is my robot doing? Remote supervision to support robots for older adults' independent living: A field study. In *2021 European Conference on Mobile Robots (ECMR)* (pp. 1–7). IEEE.

Mois, G., & Beer, J. M. (2020). The role of healthcare robotics in providing support to older adults: A socio-ecological perspective. *Current Geriatrics Reports*, *9*, 82–89.

MyMabu (2019, February 19). http://mymabu.com/about/.

Ostrowski, A. K., DiPaola, D., Partridge, E., Park, H. W., & Breazeal, C. (2019). Older adults living with social robots: Promoting social connectedness in long-term communities. *IEEE Robotics & Automation Magazine*, *26*(2), 59–70.

Paluch, R., & Müller, C. (2022). "That's something for children": An ethnographic study of attitudes and practices of care attendants and nursing home residents towards robotic pets. *Proceedings of the ACM on Human-Computer Interaction*, *6*(GROUP), 1–35.

Piasek, J., & Wieczorowska-Tobis, K. (2018, July). Acceptance and long-term use of a social robot by elderly users in a domestic environment. In *2018 11th International Conference on Human System Interaction (HSI)* (pp. 478–482). IEEE.

Prakash, A., & Rogers, W. A. (2013, September). Younger and older adults' attitudes toward robot faces: Effects of task and humanoid appearance. *Proceedings of the Human Factors and Ergonomics Society Annual Meeting*, *57*(1), 114–118.

Pu, L., Moyle, W., Jones, C., & Todorovic, M. (2019). The effectiveness of social robots for older adults: A systematic review and meta-analysis of randomized controlled studies. *The Gerontologist*, *59*(1), e37–e51.

Randall, N., Joshi, S., Kamino, W., Hsu, L. J., Agnihotri, A., Li, G., Williamson, D., Tsui, K., & Šabanović, S. (2022). Finding ikigai: How robots can support meaning in later life. *Frontiers in Robotics and AI*, *9*.

Robinson, H., MacDonald, B., & Broadbent, E. (2014). The role of healthcare robots for older people at home: A review. *International Journal of Social Robotics*, *6*(4), 575–591.

Šabanović, S. (2010). Robots in society, society in robots: Mutual shaping of society and technology as a framework for social robot design. *International Journal of Social Robotics*, *2*(4), 439–450.

Šabanović, S., Bennett, C. C., Chang, W. L., & Huber, L. (2013, June). PARO robot affects diverse interaction modalities in group sensory therapy for older adults with dementia. In *2013 IEEE 13th International Conference on Rehabilitation Robotics (ICORR)* (pp. 1–6). IEEE.

Šabanović, S., Chang, W. L., Bennett, C. C., Piatt, J. A., & Hakken, D. (2015). A robot of my own: Participatory design of socially assistive robots for independently living older adults diagnosed with depression. In *Human Aspects of IT for the Aged Population: Design for Aging: First International Conference, ITAP 2015*, Held as Part of HCI International 2015, Los Angeles, CA, USA, August 2–7. *Proceedings, Part I 1* (pp. 104–114). Springer International Publishing.

Shibata, T., Wada, K., Ikeda, Y., & Sabanovic, S. (2009). Cross-cultural studies on subjective evaluation of a seal robot. *Advanced Robotics*, *23*(4), 443–458.

Sidner, C. L., Bickmore, T., Nooraie, B., Rich, C., Ring, L., Shayganfar, M., &Vardoulakis, L. (2018). Creating new technologies for companionable agents to support isolated older adults. *ACM Transactions on Interactive Intelligent Systems*, *8*(3), 1–27.

Smarr, C. A., Prakash, A., Beer, J. M., Mitzner, T. L., Kemp, C. C., & Rogers, W. A. (2012, September). Older adults' preferences for and acceptance of robot assistance for everyday living tasks. *Proceedings of the Human Factors and Ergonomics Society Annual Meeting*, *56*(1), 153–157).

Sparrow, R., & Sparrow, L. (2006). In the hands of machines? The future of aged care. *Minds and Machines*, *16*, 141–161.

Stevens, J. A., Corso, P. S., Finkelstein, E. A., & Miller, T. R. (2006). The costs of fatal and non-fatal falls among older adults. *Injury Prevention*, *12*(5), 290–295.

Suwa, S., Tsujimura, M., Kodate, N., Donnelly, S., Kitinoja, H., Hallila, J., Toivonen, M., Ide, H., Bergman-Kärpijoki, C., Takahashi, E., Ishimaru M., & Yu, W. (2020). Exploring perceptions toward home-care robots for older people in Finland, Ireland, and Japan: A comparative questionnaire study. *Archives of Gerontology and Geriatrics*, *91*, 104178.

Taggart, W., Turkle, S., & Kidd, C. D. (2005). An interactive robot in a nursing home: Preliminary remarks. *Towards Social Mechanisms of Android Science*, *2005*, 56–61.

Taipale, S., & Fortunati, L. (2018). Robots as the next new media. In A. Guzman (Ed.), *Human–Machine Communication: Rethinking Communication, Technology, and Ourserlves* (pp. 201–219).

Tokunaga, S., Tamura, K., & Otake-Matsuura, M. (2021). A dialogue-based system with photo and storytelling for older adults: Toward daily cognitive training. *Frontiers in Robotics and AI*, *8*, 644964.

Turkle, S. (2010). In good company? On the threshold of robotic companions. In *Close Engagements with Artificial Companions* (pp. 3–10). John Benjamins.

Van Patten, R., Keller, A. V., Maye, J. E., Jeste, D. V., Depp, C., Riek, L. D., & Twamley, E. W. (2020). Home-based cognitively assistive robots: Maximizing cognitive functioning and maintaining independence in older adults without dementia. *Clinical Interventions in Aging*, *13*(15), 1129–1139.

Varrasi, S., Di Nuovo, S., Conti, D., & Di Nuovo, A. (2018, March). A social robot for cognitive assessment. In *Companion of the 2018 ACM/IEEE International Conference on Human–Robot Interaction* (pp. 269–270). IEEE.

Verver, D., Merten, H., Robben, P., & Wagner, C. (2017). Perspectives on the risks for older adults living independently. *British Journal of Community Nursing*, *22*(7), 338–345.

Wang, R. H., Sudhama, A., Begum, M., Huq, R., & Mihailidis, A. (2017). Robots to assist daily activities: Views of older adults with Alzheimer's disease and their caregivers. *International Psychogeriatrics*, *29*(1), 67–79.

World Health Organization (n. d.). Ageing and health. *World Health Organization*. https://www.who.int/news-room/fact-sheets/detail/ageing-and-health.

Wright, J. (2023). *Robots Won't Save Japan: An Ethnography of Eldercare Automation*. Cornell University Press.

Yamazaki, R., Nishio, S., Ishiguro, H., Nørskov, M., Ishiguro, N., & Balistreri, G. (2014). Acceptability of a teleoperated android by senior citizens in Danish society: A case study on the application of an embodied communication medium to home care. *International Journal of Social Robotics*, *6*, 429–442.

Zsiga, K., Tóth, A., Pilissy, T., Péter, O., Dénes, Z., & Fazekas, G. (2018). Evaluation of a companion robot based on field tests with single older adults in their homes. *Assistive Technology*, *30*(5), 259–266.

J. Nan Wilkenfeld, Norah E. Dunbar, Chengyu Fang, and Divya Srinivasan

Power and Synchrony in Human Collaboration with Exoskeletons

Abstract: Exoskeletons are a type of wearable device that parallel human limbs and can augment the physical performance of the wearer. In this chapter, exoskeleton technologies are first briefly reviewed, highlighting how the uniqueness of these wearable technologies shifts the dynamic between human and machines from simply interaction, where control resides in humans, to collaboration, where issues of identity, power, and agency need to be considered. To show the implications of such shifts, how the use of exoskeletons could affect and be affected by various physical, cognitive, and social factors of the wearers is also reviewed.

The second section provides an overview of two theoretical perspectives from interpersonal communication, namely dyadic power theory and interpersonal adaptation theory, that have been used to explain how power dynamics in human–machine collaboration are parallel to the power dynamics in human relationships. This has implications for how humans adapt to the machines they are wearing and for the agency and identity shifts that occur during their interactions. Building on the interpersonal lenses, the chapter ends with a discussion on the implications not only for exoskeleton designers and engineers, but also for human wearers, organizations, and society at large.

Keywords: Collaborative robotics, human–machine interaction, power dynamics, co-adaptation, agency

24.1 Introduction

Work-related musculoskeletal disorders (MSDs) are the most common workplace injury, affecting approximately 1.71 billion people globally (World Health Organization, 2022) spanning across a wide variety of industries. They are also the most costly injury for both businesses and society in general, as they have costs directly related to the injury such as employee time off and worker's compensation, as well as indirect costs related to the long-term effects of people with reduced ability or completely unable to work (World Health Organization, 2022). To address this issue and the overall health and safety of workers, companies have historically turned to technologies that can offload certain tasks to machines (Kim et al., 2019). While studies on industrial machine automation date back at least a century, current innovations revolve around the idea of maximizing the combined resources of both humans and machine technol-

https://doi.org/10.1515/9783110792270-025

ogies. Rather than simply replacing humans, or being fully operated by humans, humans now work collaboratively with machines in tasks that require higher-level cognition, rapid information processing, and precise decision making (Gibbs et al., 2021; Hoc, 2000).

One type of emerging collaborative technology in the industrial arena is the exoskeleton. Exoskeletons are “wearable, external mechanical structures that enhance the power of a person” (De Looze et al., 2016, p. 671), and can reduce physical demands and muscle fatigue, and help reduce the likelihood of an individual developing an MSD (Kim et al., 2019). Unlike previous machines used for lifting and moving heavy items, such as a forklift, exoskeletons are wearable technologies that can blur the boundaries between humans and machines. Exoskeletons require a high degree of human–machine coordination which has implications for users, relationship dynamics, and organizational processes.

This chapter examines exoskeleton technologies and their impact on wearer identity; their individual power as relates to the machine, their co-workers and colleagues in the same roles, and the broader organization; and their sense of agency as related to the agency of the exoskeleton. As social scientific research on this technology is relatively new, we discuss our previous and ongoing studies that have started to address these questions, with overviews of the theoretical lenses used and why they were chosen. We end with projections from exoskeleton designers as to what the future of these emerging technologies looks like, discussions of what our research has taught us, and the most important implications of exoskeleton technologies for human wearers, organizations, and society.

24.2 Exoskeleton Technologies

Exoskeleton technologies have been around for over a century in the health industry as medical and rehabilitation devices; however, in recent decades interest has grown in developing these machines for use in industrial, agricultural, and military applications (Herr, 2009; Upasani et al., 2019). In general, industrial exoskeletons are wearable devices that parallel human limbs and are designed to augment the physical performance of able-bodied wearers (Herr, 2009). Performance augmentation can include increasing human load-carrying capacity; reducing the chances of musculoskeletal injuries—for example, leg and back injuries; improving metabolic locomotory economy; and reducing the perceived level of difficulty of a physical task (Herr, 2009). Exoskeletons are classified by the area of the body on which they are utilized, consisting of upper-body, lower-body, and full-body support. Additionally, exoskeletons are labeled according to how they fit on the human body, or their anthropomorphism (De Looze et al., 2016). Anthropomorphic exoskeletons are constructed on a frame that mimics the human body, with joints and rotational points in alignment with human joints. There are two overarching types of exoskeleton technologies: active and passive.

Passive exoskeletons use mechanical resistance such as springs and counterweights from a wearer's own body rather than relying on an external source of power. They are primarily used in industrial and agricultural industries to prevent injury from repetitive tasks such as lifting or bending, and support the wearer's posture (Upasani et al., 2019). Passive suits are generally lightweight and can be made of softer materials. For example, the Herowear Apex 2™ uses thick elastic bands that run from the mid-back to upper thigh, and stretch and contract to assist the wearer with bending or squatting (herowearexo.com). The SuitX BackX™ uses a waist harness that attaches to a lightweight metal frame on the wearer's torso and front upper thigh, using resistance from the user's thighs to assist the upper body with bending and raising (suitx.com/backX). These suits provide a full range of motion for wearers and can be perceived as other personal protective equipment like hats and goggles (Gutierrez et al., 2023).

In contrast to passive exoskeletons used to reduce fatigue and prevent repetitive-task musculoskeletal injuries, active exoskeletons are designed to augment human strength and performance of physical tasks. Active exoskeletons rely on powered devices such as sensors and actuators around human joints to acquire a user's movement intentions and assist with the motion by applying external force (or torque) to the wearer's limbs (Chen et al., 2016). An example of a full-body powered exoskeleton is the Sarcos Guardian XO, designed to augment human strength by enabling a wearer to lift up to 200lbs, provide protection against injury, and allow unrestricted human movement (sarcos.com). This suit relies on a battery pack to power the suit's control systems, sensors, and actuators. Robot-integrated sensors are used to detect and predict operator movement.

24.2.1 Human–exoskeleton embodiment

Exoskeleton technologies lie at the intersection of wearable safety devices such as hard hats, gloves, and so on, and industrial machinery such as forklifts and cranes (Kirkwood et al., 2022). As such, these technologies share certain aspects of each category while maintaining specific differences that make them unique and an interesting area for human–machine communication research. Exoskeletons differ from other types of industrial machines by their wearability, and thus they have unique psychological, physiological, and social implications for users (Kirkwood et al., 2022). Additionally, they differ from other types of wearable safety devices by their technical features that can influence power dynamics between a user and their organization, a user and their colleagues, and even between a user and the suit itself (Kirkwood et al., 2022).

While exoskeletons can significantly augment human strength similar to equipment used for heavy lifting in industrial or manufacturing environments (e.g., forklifts, cranes, etc.), exoskeletons are a wearable technology constituting an embodied experience (De Looze et al., 2016). As one designer from our study pointed out, "In an ideal world, you're developing an exoskeleton, when somebody puts it on, it's no longer a tool, but it becomes embodied within their own body scheme. And then you feel like it's just you and you're moving" (Designer #2).

Machinery such as forklifts and cranes has a clear distinction between operator and tool, with a clear center of interaction (the driver's seat) and disconnection. However, as exoskeleton interaction points consist of the full body of the wearer, there is little or no physical distance between humans and the machine. This is fundamentally different from simply handling tools or operating machinery, as the experience of wearing a technology involves multiple bodily sensations and can even make wearers feel as if the technology is an appendage of their own body (Smelik et al., 2016). This embodiment requires consideration of a number of factors including psychological, social, cognitive, and physiological (Kirkwood et al., 2022; Knight & Baber, 2005). For example, when assessing an exoskeleton suit for improving motor performance, designers need to consider physiological aspects such as how the body moves and how much effort is exerted when moving during a given task, referred to as *metabolic cost* (Stirling et al., 2019). The following section provides abrief overview of research on the physical, cognitive, and social factors that influence embodiment of technologies.

24.2.2 Physical factors

Knight and Baber (2005) created a measure for assessing the comfortability of wearable technologies by breaking down perception of comfort into several dimensions, including emotion, attachment, harm, perceived change, movement, and anxiety. These dimensions allow researchers to consider the various ways users experience comfort while wearing a particular technology. Kirkwood et al. (2022) used the comfortability typology to theorize how the dimensions would impact wearers of powered exoskeleton suits.

The *emotion dimension* refers to individuals' feelings about the technology, either when they are wearing the device or observing others wearing it. For example, wearers may feel embarrassed at having to wear a mechanical device to augment strength as it could be perceived as a weakness by others as well as themselves (Knight & Baber, 2005; Upasani et al., 2019). Related to the emotion dimension is the *anxiety dimension*, which specifically refers to feelings of discomfort arising from psychological factors such as insecurity about the reliability or safety of the machine.

The *attachment dimension* considers how much users feel the device on their body (i.e., the physical sensation) and Knight and Baber (2005) found it to be the most important dimension to comfortability in their studies. This dimension is related to the *dimension of movement*, addressing the ways exoskeleton technologies may restrict a wearer's movement, thus making the machine more noticeable on the body. The *perceived change dimension* refers to the difference between how a user feels while wearing the suit as compared with how they feel normally (Knight & Baber, 2005). Ideally, exoskeleton users should reach a point at which they have little or no physical awareness of the machine on their body (Knight & Baber, 2005).

The *harm dimension* includes any pain experienced by the wearer as well as muscle fatigue from adding the load of the exoskeleton to the body (Knight & Baber, 2005). While Knight and Baber's studies did not find high scores for this dimension, Wilkenfeld et al.'s (2023) study on a full-bodied exoskeleton prototype noted several instances of discomfort while wearing the suit. One participant spoke in general about the physical effects of moving with the suit: "everything is tense and it, it fatigues out because it just takes longer to do the action, so you're, you're all tensed up and for way longer" (Subject P8S2). Another participant explained that certain parts of the body are more susceptible to discomfort than others when he said the suit was, "exhausting on my back, you know, like, um, the feet—being in the Exo, my feet kill me, and when I'm not in the Exo, they're fine" (Subject P4S2). These participant experiences are in line with other studies looking at comfort in occupational exoskeletons that point out how static-fit ('one-size-fits-all') devices can lead to more muscle fatigue than devices with dynamic (adjustable) fits (Sposito et al., 2020).

These dimensions produce an overall subjective measure of user comfort while wearing a technology such as an exoskeleton suit, and can impact the level of embodiment and ultimately performance (Knight & Baber, 2005; Kirkwood et al., 2022; Sposito et al., 2020). Additionally, these comfortability dimensions influence attitudes and perceptions of exoskeleton technologies, with implications for adoption, usage, and worker satisfaction in organizations (Kirkwood et al., 2022).

24.2.3 Cognitive factors

Because exoskeletons fit close to a user's body, necessitating intimate physical coordination, interaction also includes cognitive aspects as control is shared by the human and the machine (Kapeller et al., 2020). The psychomotor mental workload, or cognitive load, is considered to be the amount of cognitive resources an individual is using at a given time while working on a task, and in an exoskeleton context includes mental effort of coordinating with the suit (Stirling et al., 2019). Wearable technologies such as exoskeletons can increase the cognitive load of users more than other types of industrial machines, including those that require higher levels of collaboration such as co-bots, as shared control necessarily reduces human autonomy (Fournier et al., 2022; Kapeller et al., 2020). Specifically, in order to complete a task, exoskeleton users must not only consider the task itself, but also how to physically move their body in coordination with the suit, the latter of which (e.g., considering how to move one's leg to take a step) rarely requires cognitive attention for able-bodied adults (Kapeller et al., 2020; Kirkwood et al., 2021, 2022).

Wilkenfeld et al. (2023) found that participants experienced increases in cognitive load due to factors related to an exoskeleton's wearability. Specifically, one participant recalled "walking back and forth . . . I've got the robot around me [and] it takes a bit more thinking about it" (Subject 5S3). Others described needing to "really pay attention" (Subject 6S4) to accomplish a simple task of tapping a specific target in front of them, or "getting down low enough to pick weights off the lowest shelf required some thought" (Subject 7S1). Capturing the essence of the distinction between moving without an exoskeleton as compared with wearing the machine, a participant stated, "it's night and day. I mean, you walk, you know, without even thinking about it. You breathe without thinking about it. You blink without thinking about it. So I mean, this is something like you have a 400 pound thing on you, you know? You—you think about it all the time when you're in it" (Subject 4S3).

More complex exoskeletons may exacerbate the cognitive load of users by offering different levels of controls or strength augmentation (Stirling et al., 2019). Exoskeleton user-experience researchers stress the need to find the 'cognitive fit,' or the optimal level of control and interaction for a user to perform a given task in a certain context (Stirling et al., 2019). However, contextual operating modes provide the potential for users to become confused, forgetting which mode they are operating in and unable to remember how to achieve a certain behavior or perform a certain task. For example, a user might have the exoskeleton on a higher resistance mode that augments strength for lifting but decreases mobility for walking, and upon trying to move may think they are doing something incorrectly or there is a fault with the exoskeleton.

Aside from impacting performance, the increased amount of mental energy required while wearing a machine is an important consideration for designers of industrial exoskeleton technologies, as these devices are being used in contexts where situational awareness is critical for safety and divided attention can be dangerous (Kapeller et al., 2020; Stirling et al., 2019). As cognitive load increases from wearing an exoskeleton, perception of relevant situational awareness cues, for example a fall hazard on the floor such as a spill or dropped object, or proximity to another worker, may decrease (Kapeller et al., 2020). Kapeller et al. (2020) note that exoskeletons provide an additional level of challenge for industrial settings, as the human and machine both moving together means the situation regarding context cues is dynamic as opposed to a stationary human operating a moving robot.

24.2.4 Social factors

Similar to other types of wearable technologies, exoskeletons have implications for users' occupational identities (Gutierrez et al., 2023; Kirkwood et al., 2021, 2022). The concept of identity refers to an individual's sense of self, which is shaped by a variety of factors, including personal experiences, beliefs, values, and social interactions (Tajfel & Turner, 1979). Individuals are often categorized and understood as members of social groups based on various characteristics such as race, gender, ethnicity, and occupation (Tajfel & Turner, 1979). Self-concepts are formed by personal and social categorizations and are a product of difference and exclusion, and individuals tend to adopt the norms and values of their in-group, perceiving them as being superior to those of out-groups. Additionally, individuals tend to perceive their in-group as possessing more desirable qualities and characteristics than out-groups. Self-concepts and social identities influence meaning making and expectations about work tasks (Thoits & Virshup, 1997).

Contexts in which exoskeleton technologies are deployed are traditionally 'blue collar' industries with a heavy emphasis on manual labor (Gutierrez et al., 2023; Upasani et al., 2019). Roles in these industries often focus on physical strength, with workers considering and valuing their work primarily in instrumental terms (e.g., how much coal a miner can produce in a day; Lucas, 2011). Exoskeletons, being a technology designed to augment strength and reduce injuries, can constitute an identity threat to a worker whose self-concept centers on physical acumen. For example, Upasani et al. (2019) found that agricultural workers were less likely to use exoskeleton technologies that "do not seem to be work-related, and that are more 'medical' in their appearance" (p. 5). Additionally, while Guiterrez et al. (2023) found that the vast majority of interviewees (83%) indicated willingness to use an exoskeleton in their workplace, half considered these technologies to be for those who are physically weak (50%) or already injured (22%). In other words, use of the technology violated the workers' understanding of their identities because they did not view a medical device as relevant to their work or role in the organization.

In-group/out-group distinctions become accentuated and an individual's social identity becomes cognitively activated in certain situations such as the introduction of new technologies. When a worker's identity becomes salient, they will associate themselves with social group prototypes, which embody characteristics that become representative of the group (Hogg & Terry, 2000). Group prototypes are shared and agreed-upon views within the group that become group norms, and social groups influence the task–technology and identity–technology fit. Thus, when considering the introduction of new technologies such as exoskeletons into work contexts, it is important to look beyond simply the technology's fit to the task, and also consider the technology's fit with workers' identities (Kirkwood et al., 2021). If a technology is perceived not to fit a certain identity, then workers will be less willing to adopt it to accomplish a task that goes against the expression of that identity (Leonardi, 2011). It is important to realize that an individual's feelings of empowerment or weakness with the exoskeleton are highly related to their understanding of their in-group's norms, values, and characteristics, and are relevant to how it affects their relationships to others in their organization (Kirkwood et al., 2022).

24.3 Application of Interpersonal Theoretical Lenses

This section will review two interpersonal communication theories that have been utilized to understand the dynamics between humans and exoskeletons: Dyadic Power Theory (DPT; e.g., Dunbar, 2004) and Interaction Adaptation Theory (Burgoon et al., 1995). Though distinct theoretical lenses, they are fundamental to understanding the dynamics between humans and exoskeletons as, due to the wearability factors discussed in the previous section, successfully operating these technologies requires a high level of synchronized movement and a clear understanding of the delineation of control. However, before diving into the theories themselves, it is important to briefly discuss why we apply theories previously reserved for human–human interactions to technologies such as exoskeletons. We approach our research on technologies by conceptualizing exoskeletons as having agency.

Though there are many nuanced definitions, broadly, agency refers to the capacity of an entity, whether it be a person, group, or object, to act and make choices that can influence the world around them. It is the ability to exert power and make things happen, to have an impact on the environment and other entities, and is often associated with concepts such as autonomy, responsibility, and intentionality (Fortunati et al., 2018; Gibbs et al., 2021). The concept of agency involves the capacity to both sustain and transform structures, and is applicable to both humans and non-humans (Boudreau & Robey, 2005).

When considering more complex and even intelligent machines, Fortunati et al. (2018) defined agency as something *conferred* on machines by humans rather than being an inherent property of a technology. The authors argue that despite humans having the primary power in the human–robot relationship, humans oblige robots to shape their expectations and behavior in the interaction 'as if' they were fellow humans (p. 10). Agency is conferred on exoskeletons as they are anthropomorphic in physical design, and interact with human physiology (i.e., they move with a user or constrain the user's movement).

In general, communication scholarship acknowledges that technological artifacts have material agency and can impact human behavior and organizational processes —for example, a table placed in the middle of a room will influence how people walk through and use the room (Leonardi, 2011). Material agency has been defined as "the capacity for nonhuman entities to act on their own, apart from human intervention" (Leonardi, 2011, p. 148). More recently, machine agency has been used to refer to the capacity of technologies to perform specific tasks given its material features (Gibbs et al., 2021). Both humans and machines are influenced by the social and material contexts in which they operate, and their actions and behaviors are co-constructed through interactions with each other and their environment.

When considering wearable technologies such as exoskeletons, it is especially useful to view human and machine agency as co-constitutive, with the actions and behaviors of each influencing and being influenced by the other (Kirkwood et al., 2022).

Kirkwood et al. (2022) argue that the way humans interact with exoskeletons (and vice versa) is influenced by a complex interplay of social and material factors. Social factors may include the societal norms, beliefs, and attitudes towards exoskeletons, for example the social acceptability of wearing an exoskeleton, the level of knowledge or education about exoskeletons, and societal attitudes towards disability or enhancement. Material factors refer to tangible elements such as the design, functionality, and operation of the exoskeleton itself, and the physical environment in which it is used. These social and material factors are deeply interconnected and cannot be easily separated; they constantly influence each other and shape the way humans and exoskeletons interact (Leonardi, 2011). For instance, societal attitudes towards exoskeletons could influence the design of these devices, such as when people want an exoskeleton to look like a superhero and possess superhero strength, making designers want to push the boundaries of what is practical (social influencing material). Conversely, the capabilities and limitations of exoskeletons could affect societal perceptions and use of these technologies, for example exoskeleton wearers could be perceived as being less physically capable (material influencing social). Therefore, in understanding human–exoskeleton interactions, it is essential to take into account this intricate interplay of social and material factors, as to ignore one aspect would likely result in an incomplete or skewed understanding of the situation.

This way of considering exoskeleton agency allows for the application of interpersonal theoretical lenses to understand how humans and exoskeletons may function together as equal partners, becoming a singular entity. It also provides frameworks for understanding challenges faced by users, as well as the potential long-term impacts these technologies could have on individuals, industries, and broader society. First, we discuss interactional adaptation and synchrony, as they are the most critical to successfully function in an exoskeleton suit.

24.3.1 Interactional adaptation/synchrony

One of the fundamental components of collaborating with another agent, be it a machine or another human, is behavioral adaptation. Adaptation is considered both a biological and sociological process, and an imperative for a functioning society. It is through adaptation that we learn to interact with one another and form relationships, identities, and societal groups (Burgoon et al., 2017). Burgoon et al. (1995) argue that adaptation is often bi-directional, and is, in reality, 'co-adaptation,' as adaptation occurs on both sides of the interaction. Adaptation occurs in human–machine interactions as well. For example, human users of intelligent assistants must learn to ask questions a certain way in order to get a suitable response.

Exoskeleton technologies provide a unique opportunity for researchers to understand bi-directional adaptation between humans and machines as they use "shared control" (Ronsse et al., 2011, p. 1001) to achieve a common task goal, such as lifting a heavy object, towards whichboth the human user and exoskeleton contribute. Ideally, each agent needs to adapt to the other to coordinate, or synchronize, and complete the task: the human adjusts their movements to potentially leverage the strengths of the exoskeleton, and the robot typically responds to human movement (and intent) by adjusting its torque and kinematics.

In order to achieve the optimal performance, or for exoskeletons to become fully embodied machines, humans and exoskeletons must have a high level of interaction coordination. Interaction coordination is an intrinsic component of human behavior and can be considered the fundamental building block upon which all relationships rely (Bernieri & Rosenthal, 1991; Burgoon et al., 1995). Interactional synchrony, a type of interaction coordination, is generally defined as the degree to which separate behavioral patterns match, and comprises three subcomponents: rhythm, simultaneous movement, and the smooth meshing of interaction (Bernieri & Rosenthal, 1991, p. 403). Not only is our body programmed to synchronize with our own verbal utterances in the form of head movements and hand movements, but humans also tend to coordinate with nonverbal and verbal behaviors of interaction partners. For example, one partner may nod their head while listening to another partner who is saying something affirming or nodding as well.

In verbal communication the ability to coordinate or synchronize with another individual's speech rhythms is a necessity for learning language and interacting. Research on interactional synchrony between human dyads has shown a variety of prosocial outcomes including developing rapport, enhancing feelings of togetherness or similarity, fostering cooperation in groups (Wiltermuth & Heath, 2009), and society building (Burgoon et al., 1995, 2014). Indeed, moving synchronously with others can lead to feelings of creating the same enacted effect, such as a musical piece, and the blurring of lines between self and others (Wiltermuth & Heath, 2009).

Recent research has found that humans attempt to coordinate their behaviors and synchronize with non-human agents such as computers and robots with the same mechanisms and outcomes as with other humans (Fujiwara et al., 2022; Lehmann et al., 2015). This drive to synchronize is increased when human–robot teams are tasked with accomplishing a common goal. Much of this research has focused on performance or affective outcomes of synchrony with non-human agents. For example, Fujiwara et al. (2022) conducted a study on human–avatar synchrony in negotiation scenarios, and found that participants synchronized their movements with non-human counterparts and reported greater affiliation with the avatars when doing so. In an earlier study, Lehmann et al. (2015) found that participants watching a robot synchronize with another human rated the moving robot more likeable and intelligent than the non-moving robot, and the synchronously moving robot as the most likeable and intelligent.

While much early work on nonverbal synchrony explored body posture similarities in dyads, technological advancements have enabled researchers to make deeper exploration of the rhythm and timing components, which, many scholars agree, are the 'essence of synchrony' (Fujiwara & Daibo, 2016). Two overarching types of nonverbal synchrony are *interaction rhythms* (e.g., clapping using the same rhythm) during the course of an interaction, and *behavioral meshing* (e.g., clapping together with the same rhythm) that creates a larger whole (Bernieri & Rosenthal, 1991). Oft-cited examples of these conceptualizations of synchrony are jazz bands and sports teams whose members may differ in their posture and movements, but synchronize through rhythm and timing to form a larger, singular entity (Bernieri & Rosenthal, 1991; Fujiwara & Daibo, 2016; Lorenz et al., 2016).

Synchrony is measured by degree and pattern, with the degree of synchrony indicating the level of rhythmic convergence or temporal coordination (Fujiwara et al., 2023). Synchrony patterns consider movements in cycles, with interactants falling on a spectrum between in-phase and antiphase synchrony. In-phase synchrony refers to partners moving at the same part of the cycle at a given time, for example people clapping together in rhythm. Antiphase synchrony occurs when movements are made at opposite parts of the cycle at a given time, for example two individuals clapping alternatively in rhythm.

In-phase synchrony is particularly important for humans working with powered exoskeletons as the human and machine need to coordinate to become one entity (Kirkwood et al, 2021). While some human–exoskeleton dyads may be able to synchronize in one or two subcomponent areas such as body movements or timing, the challenge for synchronizing with a wearable robot is that it must be achieved across all threin refse domains: body movements, rhythm, and timing (Ronsse et al., 2022. The interface between a human and exoskeleton is physical movement by the wearer. Sensors in the suit detect motion and send information to the robot controller, which then triggers a mechanical response in the machine to anticipate and match that of the human motion (Ronsse, 2022). However, this technology is still developing and humans vary not only physically, such as height, strength, and range of motion, but also in learning and adaptability (Di Natali et al., 2020; Ronsse, 2022).

Wilkenfeld et al. (2023) found that participants relied on frameworks such as previous experiences with technologies and fellow humans in order to understand the machine itself, how to relate to it, and ultimately how to adapt and coordinate movements with it. Specifically, some participants used analogies to compare their experience to other technologies, comparing wearing the exoskeleton to other types of wearable items such as backpacks and ski boots; others thought of the exoskeleton like operating a larger machine such as a motorcycle. Participants also anthropomorphized the exoskeleton, ascribing it agency as a separate entity from themselves. The different frameworks for understanding the exoskeleton likely played a role in how participants perceived their own agency as relative to the agency of the exoskeleton, and which agent—the human or the machine—should be in control of movement.

Understanding collaboration and synchrony between two agents requires fundamental understanding of each agent and the factors that influence the relationship between them. As exoskeletons are wearable technologies and designed to augment human performance, wearers must relinquish some control over their own bodies in order to function in the suit. The delineation of responsibilities—including initiating movements and calibrating lifting strength—will influence how a user understands their agency as relative to the machine. To further unpack this idea, we turn to a theory on relational power.

24.3.2 Dyadic power theory

Dyadic Power Theory (Dunbar, 2004; Dunbar et al., 2016) is, as the name implies, a theoretic lens that explains the "dynamic interplay of conversational partners as they manage control of an interaction through dominance behaviors" (Dunbar et al., 2016, p. 76). Dunbar (2004) argues that power is inherent in all interactions, stemming from a combination of factors including structural forces and relationship history, and becomes more or less relevant depending on the context and goals of the interaction. Scholars define power as "the capacity to produce intended effects, and in particular, the ability to influence the behavior of another person" (Dunbar, 2004, p. 236). There are three different types of power across interactions: manifest power, latent power, and invisible power (Kirkwood et al., 2022). Manifest power is about the visible outcomes of power such as verbal and nonverbal strategies used to achieve certain ends. Latent power is a power that is not overtly expressed, for example when a more powerful person will not be challenged by a less powerful person because negative results will be feared. Invisible power is the result of social or psychological mechanisms that do not necessarily surface in overt behavior or latent grievances but may be manifest in systematic differences between men and women, parents and children, white people and non-white people. A powerful person might not be aware of their power because it is based in the relationship between two people at the dyadic level, but it is also influenced by cultural norms and societal-level systematic influences.

Since DPT was first introduced, several scholars have extended its application beyond dyadic relationships (i.e., marital partners). For instance, Dunbar and others have broadened DPT to consider how power dynamics are influenced by societal and cultural factors in organizations and other collective entities (Dunbar et al., 2016). Kirkwood et al. (2022) then extended DPT to relationships between humans and machines by theorizing the impacts of introducing exoskeleton wearable technologies into industrial applications at several levels, including individual role, team dynamics, and organizational power relationships.

24.3.3 Power

Kirkwood and colleagues pointed out that all three types of power covered in DPT can appear in dynamics between humans and exoskeletons, and will vary depending on the design of the suit, and other factors such as user trust in machines, work contexts, and specific tasks to be accomplished. For example, an exoskeleton may manifest power through the structural design limiting user motion, and employ visual or auditory warning signals (such as flashing lights or beeps) if a user is attempting to do something unsafe or move in a way that could cause injury.

Latent and invisible power of machines can be influenced by the creators of the technologies and the company leadership deploying these machines into their labor pool. For example, human drivers for Uber and Lyft are completely managed by an algorithmic authority (Lee et al., 2015). If a driver fails to meet certain requirements, there are consequences such as not getting access to higher-paying rides or even being terminated (Lee et al., 2015). So too may exoskeletons be designed to make the human move in specific ways and carry out tasks in a certain order. Industrial workers may be told they must use an exoskeleton in specific ways and even defer to the suit's standard of operation and safety protocols.

One of the key points Kirkwood and colleagues made was that exoskeleton technologies lie at the intersection of material and symbolic power. Material power is generally conceptualized as access to resources such as information, money, and valuable items, and can also include technologies that enhance performance or improve efficiency. As exoskeletons are designed to augment human strength, they can certainly be considered an object of material power. Symbolic power is power derived from the ability to influence meaning or identity, and exoskeletons as wearable technologies can influence both how individuals view themselves and how they are viewed by others. For example, when interviewing agricultural workers about their opinions after wearing passive exoskeletons, Upasani et al. (2019) found that many participants felt threats to their identities as strong and able-bodied, considering the suit to be a crutch for supporting weakness. In other words, the exoskeleton was a symbol of weakness.

We are able to understand power in relationships through the expressions of power, known as dominance (Dunbar, 2004). Dominance refers to the behavioral or communicative acts through which power is exerted and influence is achieved (Dunbar, 2004). Power is the *potential* to influence, or an *ability* to act, whereas dominance is the *actual* behavioral act through which an individual asserts their influence (Dunbar, 2004).

Dunbar and Burgoon (2005) argue that dominance is a profile of behavior reflecting a combination of individual temperament and situational features that encourage dominant behavior. Humans express dominance through nonverbal behaviors and verbal indicators such as body position, facial expressions, eye gaze, and tone (Dunbar & Burgoon, 2005), and these same indicators can apply to machines to make them seem dominant (Li et al., 2019). Even robots that have few human-like cues may be perceived as having higher status as compared with other robots when they make high-status motions (Li et al., 2019).

It is important to note this conceptualization of dominance requires an interaction to be considered within a specific context, such as a negotiation over task delineation. Power, in this case, is seen as relative, referring to the differences between partners in relation to one another. In other words, an individual cannot be dominant without an interaction partner who is submissive. Therefore, dominance is both behavioral and relational. Additionally, it is neither necessarily positive nor negative, but simply an action or reaction based on the dynamic and fluid conception of an individual's power in a specific moment. For example, when a human worker is wearing an exoskeleton with strength-related adaptation, the suit may unilaterally adjust the force or torque needed to successfully complete the task, thus expressing dominance over total lifting power. Additionally, an exoskeleton constraining a user by not allowing the user to, for example, sit down, has dominance over this range of user motion. Researchers of exoskeleton technologies recognize that different kinds of exoskeletons have varying amounts of power. Active exoskeletons have been considered more powerful than passive exoskeletons as the rigid links and powerful motors enable a rigid exoskeleton to dominate the gait of the wearer and to enforce a walking pattern, while soft wearable robots allow both the wearer and the robot to contribute to the movement (Piao et al., 2023).

24.3.4 Phases of power

DPT was originally proposed by Dunbar (2004) with eight propositions, then later expanded to include five more propositions that considered more nuanced influences on power dynamics such as domain-specific (i.e. spheres of activity over which one partner may have more control in a negotiation), broader cultural and structural influences (e.g. a culture in which age influences authority), and lastly extending DPT to collectivities such as organizations and organizational actors (Dunbar et al., 2016). The propositions are organized into pre-interactional factors that affect the relationship, the interactive process that occurs within the conversation, and finally the post-interactional implications of that process.

The first three propositions discuss pre-interactional factors: relative authority, relative resources, and relative power. Dunbar (2004) proposes that there is a positive correlation between authority and resources, with increases in relative authority being related to increases in relative resources (P1). The second and third propositions reflect a causal relationship with increases in relative resources producing increases in relative power (P2), and increases in relative authority producing increases in relative power (P3). Relative authority has both a direct and indirect influence on relative power. It acts as a power base from which an individual can have the potential for greater influence, and more relative authority can mean more access to resources, which in turn can increase relative power.

In a human–machine interaction context, relative authority and relative power can be considered as the delineation of tasks and decision making between a human user and the machine counterpart. For example, an exoskeleton by sheer design has authority over human movement such that it constrains certain motions such as twisting and sitting, and limits the speed at which wearers can make these motions (Piao et al., 2023). As such, human users must yield some of the power over their own bodies to the exoskeleton in order to function in the suit, thus increasing the relative power of the exoskeleton itself.

Relative resources from a machine perspective can mean information such as direct access to internet or underlying operations that may or may not be hidden from the user. For example, a powered exoskeleton can be designed with adaptive assistance mechanisms that increase or decrease strength augmentation based on the user's operating environment and task (Herr, 2009). Exoskeletons guided by intelligent, adaptive algorithms may have access to a much wider set of information, possess the capability to process data much faster than a human, and therefore have more power over decision making for certain task functions. A caveat to these ideas is the fact that often boundaries of access to resources and authority of machines are predetermined by the manufacturers of the machine, or the company management prior to implementation (Hinds et al., 2004). However, humans can also be limited by the bounds of management determining power, authority, and access to resources according to the structure and roles in an organization. For this reason, the first four propositions of the theory remain the most relevant to understanding the relationships between humans and exoskeleton technologies.

24.3.5 Interaction phase

The propositions in the interaction phase (P4–P6) describe behaviors between two individuals and consider how power dynamics influence dominance moves and attempts for control. Perhaps counterintuitively, DPT assumes that *larger* power imbalances between agents will result in *fewer* control attempts than if power is close to equal, and empirical studies have indeed shown evidence that more powerful individuals have less need to make dominance moves as their power is already understood (P4; Dunbar, 2004, p. 241). Propositions five and six address the dynamic nature of power in an interaction arguing that more control attempts increase the likelihood of gaining more control (P5), and the perception of increasing relative power will increase the amount of control attempts (P6).

An individual's perception of their own power as compared with a machine will likely vary according to the context and the individual's opinion of intelligent machines. In general, we tend to assume that humans should be the ultimate authority over machines—however, there are clear examples of cases in which humans defer to machine expertise. Examples of this are humans using a calculator for large computations, or relying on a navigation algorithm for driving directions. Those examples may not challenge our sense of authority—however, alternatively, if a warehouse employee is required to wear an exoskeleton that not only provides physical assistance but also directions for picking items off shelves in order to maximize shipping efficiency, a human could feel less power over their own performance.

We would expect more conflict to arise between a human user and a powered exoskeleton machine as they have equal power and must share control over one body. Wilkenfeld et al. (2023) noticed this conflict as users learned how to work with the machine to move—they not only thought of it as agentic, but also seemed to experience a power struggle. As one of the participants stated, the exoskeleton, "felt hard to control . . . it definitely felt like I was fighting it, a lot" (Participant #8). Another participant acknowledged they had to relinquish some control to the exoskeleton to perform successfully, saying "I think I learned to try and . . . not sort of fight against what it's doing" (Participant #5). These quotes indicate that participants consider the exoskeleton a fellow agent, perhaps with authority at or near their own level, and demonstrate how relative power dynamics impact user experience and successful performance.

23.4.6 Post-Interaction Phase

The final propositions cover the outcomes of interactions, including amount of control and levels of satisfaction. More (counter)control attempts increase the likelihood of gaining control. Additionally, another curvilinear relationship is expected between power difference and satisfaction, in that individuals who perceive their power as higher or lower than their partner's will be less satisfied with the interaction and the relationship overall than those in more equal partnerships (Dunbar, 2004).

In the human–exoskeleton relationship, it is important to recognize that these technologies impact a wearer's sense of their own agency and power as relative to the exoskeleton, and potentially even beyond the dyad to role and occupational identity. Previous research has shown that humans experience a change in their understanding of their own agency when working with a partner, whether it be human or robot (Ciardo et al., 2020). However, further work is needed to determine the connections between how a user understands the exoskeleton itself (e.g., is it a wearable, a tool, or an anthropomorphized agent?) and the impacts on agency and identity.

24.4 Looking to the Future

After Wilkenfeld and colleagues analyzed *users'* perspective on control and agency, we were then curious how *designers* perceived exoskeleton technologies in terms of power, agency, and control. We conducted 14 interviews with researchers and manufacturers of different types of exoskeletons (powered and passive). We found that designers are wrestling with how to create an adaptable exoskeleton that will result in human–machine synchrony, and embedded within the goal of synchrony is understanding how each agent will adapt to the other. All noted how current technologies require the human operator to adapt to the limitations of the exoskeleton and learn to function in it—to adjust how they move, think, and act in order to use it most effectively.

When asked about the power dynamics inherent in the human–exoskeleton interactions in terms of which entity should have more control over which tasks, most expressed an ideal situation being shared, adaptive control between the human and the machine. For example, one designer said, "I feel like the interface itself should be comfortable, but the way in which the I guess the . . . way in which the device is controlled should be . . . 'intuitive' . . . meaning that the control system has the human's intent embedded in it and so it doesn't need to be pushed, it's not reactive, it's proactive" (Designer #5). In this case, the designer believes the machine's movements should mimic a human's as much as possible and be able to anticipate the next movement without direct human intervention. However, many of the designers we interviewed also pointed out that while that's the ideal, our current technology simply has not reached that capacity and essentially the responsibility and power of control still fall to human wearers:

> I think the weakest point of any exoskeleton that we've developed, so far, is still the controller. You can watch a ballerina or a soccer player, somebody like Ronaldo or Messi, dribble a soccer ball through amazing defenses and then score. We cannot build a robot that can do that. And we can build hardware that's efficient, but we really can't build the controller that knows how to move your body with them. (Designer #2)

This designer indicated that we still have much to learn about how humans move in such a nuanced way, and then how to translate that into a machine that can not only make those same movements, but can also make them at the same time a human wearer makes them. As such, powered exoskeleton technologies are still emerging and there remain design challenges to accomplishing the goal of augmenting human strength without increasing physical, and financial, costs. For example, as one designer indicated, "there's still this kind of battle in the field between knowing what to do to achieve a performance outcome, like how to optimize a torque pattern for example, versus codifying that into something that's cheap and comfortable to wear and easy to maintain" (Designer #6).

24.5 Conclusion

Since General Electric's first attempt in 1965 to build the Hardiman, a large full-body exoskeleton, engineers and industrial designers have spent decades trying to develop devices that can augment human strength, increase performance, and decrease injuries from repetitive tasks and physical exertion. Like many recent technological innovations in the industrial sector, exoskeletons can serve as an extension of human capabilities rather than a substitution. This partnership between humans and machines combines human strengths in creativity, flexibility, empathy, and complex decision making, while allowing machines to process vast amounts of data, perform repetitive tasks, and execute tasks with high precision and speed (Kirkwood et al., 2021). This new collaborative dynamic requires a new perspective on the human–machine relationship—moving from simply human–machine interaction, which often assumes humans have full control and overlooks the agentic role that machines can play, to human–machine cooperation, which considers humans and machines as equal partners (Hoc, 2000).

In our ongoing studies, we have found the concepts of synchrony, agency, and power dynamics to be important to human–exoskeleton cooperation much as they are in human–human collaborations. For example, the processes and challenges of synchronizing with another occurred in interactions between humans and the exoskeletons they were wearing as though they were attempting to synchronize with a fellow human. Additionally, as exoskeletons are perceived by users as having agency, this can in turn impact a wearer's sense of their own agency which has implications for power in the user's relationships. Not only can we examine the power dynamics between the user and the machine, but exoskeleton technologies will also potentially influence power relations between the worker and their organization, power dynamics between social groups, dynamics between labor and capital, and relationships between humans and technologies in general.

Our preliminary research has shed light on the potential implications of exoskeleton technologies as wearable agentic machines, highlighting the importance of understanding the specific human–machine dynamics for the future design and implementation of such systems. It is clear these wearable technologies have the potential to significantly alter the landscape of work, health, and safety while challenging our traditional views of labor, agency, and identity. In essence, the future of work with exoskeleton technologies promises a fascinating new avenue for understanding human and machine cooperation.

References

BackX (2021). SuitX. https://www.suitx.com/backX.

Bernieri, F. J., & Rosenthal, R. (1991). Interpersonal coordination: Behavior matching and interactional synchrony. In R. S. Feldman & B. Rimé (Eds.), *Fundamentals of Nonverbal Behavior* (pp. 401–431). Cambridge University Press.

Boudreau, M. C., & Robey, D. (2005). Enacting integrated information technology: A human agency perspective. *Organization Science, 16*(1), 3–18. https://doi.org/10.1287/orsc.1040.0103.

Burgoon, J. K., Dunbar, N. E., & Giles, H. (2017). Interaction coordination and adaptation. In J. K. Burgoon, N. Magnenat-Thalmann, M. Pantic, & A. Vinciarelli (Eds.), *Social Signal Processing* (pp. 78–96). Cambridge University Press. https://doi.org/https://doi.org/10.1017/9781316676202.008.

Burgoon, J. K., Dunbar, N. E., & White, C. H. (2014). 10. Interpersonal adaptation. In C. Berger (Ed.), *Interpersonal Communication* (pp. 225–248). De Gruyter Mouton. https://doi.org/10.1515/9783110276794.225.

Burgoon, J. K., Stern, L. A., & Dillman, L. (1995). *Interpersonal Adaptation: Dyadic Interaction Patterns*. Cambridge University Press.

Chen, B., Ma, H., Qin, L. Y., Gao, F., Chan, K. M., Law, S. W., Qin, L., & Liao, W. H. (2016). Recent developments and challenges of lower extremity exoskeletons. *Journal of Orthopaedic Translation*, *5*, 26–37. https://doi.org/10.1016/J.JOT.2015.09.007.

Ciardo, F., Beyer, F., de Tommaso, D., & Wykowska, A. (2020). Attribution of intentional agency towards robots reduces one's own sense of agency. *Cognition*, *194*, 104109. https://doi.org/10.1016/J.COGNITION.2019.104109.

De Looze, M. P., Bosch, T., Krause, F., Stadler, K. S., & O'Sullivan, L. W. (2016). Exoskeletons for industrial application and their potential effects on physical workload. *Ergonomics*, *59*(5), 671–681. https://doi.org/10.1080/00140139.2015.1081988.

Dunbar, N. E. (2004). Dyadic Power Theory: Constructing a communication-based theory of relational power. *Journal of Family Communication*, *4*, 235–248. https://doi.org/10.1080/15267431.2004.9670133.

Dunbar, N. E., & Burgoon, J. K. (2005). Perceptions of power and interactional dominance in interpersonal relationships. *Journal of Social and Personal Relationships*, *22*, 207–233. https://doi.org/10.1177/0265407505050944.

Dunbar, N. E., Lane, B. L., & Abra, G. (2016). Power in close relationships: A dyadic power theory perspective. In J. A. Samp (Ed.), *Communicating Interpersonal Conflict in Close Relationships: Contexts, Challenges and Opportunities* (pp. 75–93). Routledge. https://doi.org/10.4324/9781315774237.

Fortunati, L., Sarrica, M., Ferrin, G., Brondi, S., & Honsell, F. (2018). Social robots as cultural objects: The sixth dimension of dynamicity? *Information Society*, *34*(3), 141–152. https://doi.org/10.1080/01972243.2018.1444253.

Fournier, É., Kilgus, D., Landry, A., Hmedan, B., Pellier, D., Fiorino, H., & Jeoffrion, C. (2022). The impacts of human–cobot collaboration on perceived cognitive load and usability during an industrial task: An exploratory experiment. *IISE Transactions on Occupational Ergonomics and Human Factors*, *10*(2), 83–90. https://doi.org/10.1080/24725838.2022.2072021.

Fujiwara, K., & Daibo, I. (2016, April). Evaluating interpersonal synchrony: Wavelet transform toward an unstructured conversation. *Frontiers in Psychology*, *7*. https://doi.org/10.3389/fpsyg.2016.00516.

Fujiwara, K., Hoegen, R., Gratch, J., & Dunbar, N. E. (2022). Synchrony facilitates altruistic decision making for non-human avatars. *Computers in Human Behavior*, *128*, 107079. https://doi.org/10.1016/J.CHB.2021.107079.

Fujiwara, K., Nomura, K., & Eto, M. (2023). Antiphase synchrony increases perceived entitativity and uniqueness: A joint hand-clapping task. *Frontiers in Psychology*, *14*. https://doi.org/10.3389/fpsyg.2023.1069660.

Gibbs, J., Kirkwood, G., Fang, C., & Wilkenfeld, J. N. (2021). Negotiating agency and control: Theorizing human–machine communication from a structurational perspective. *Human–Machine Communication*, *2*, 153–171. https://doi.org/10.30658/hmc.2.8.

Gutierrez, N., Ojelade, A., Kim, S., Barr, A., Akanmu, A., Nussbaum, M. A., & Harris-Adamson, C. (2023, February 15). Perceived benefits, barriers, perceptions, and readiness to use exoskeletons in the construction industry: Differences by demographic characteristics. *SSRN Electronic Journal*. Available at: https://doi.org/10.2139/SSRN.4361395.

Herr, H. (2009). Exoskeletons and orthoses: Classification, design challenges and future directions. *Journal of NeuroEngineering and Rehabilitation*, *6*(1), 1–9. https://doi.org/10.1186/1743-0003-6-21.

Hinds, P. J., Roberts, T. L., & Jones, H. (2004). Human–computer interaction: Whose job is it anyway? A study of human–robot interaction in a collaborative task. *Human–Computer Interaction*, *19*(1–2), 151–181. https://doi.org/10.1080/07370024.2004.9667343.

Hoc, J. M. (2000). From human–machine interaction to human–machine cooperation. *Ergonomics, 43*(7), 833–843. https://doi.org/10.1080/001401300409044.

Hogg, M. A., & Terry, D. J. (2000). Social identity and self-categorization processes in organizational contexts. *Academy of Management Review, 25*(1), 121–140.

Kapeller, A., Felzmann, H., Fosch-Villaronga, E., & Hughes, A.-M. (2020). A taxonomy of ethical, legal and social implications of wearable robots: An Expert perspective. *Science and Engineering Ethics, 26*, 3229–3247. https://doi.org/10.1007/s11948-020-00268-4.

Kim, S., Moore, A., Srinivasan, D., Akanmu, A., Barr, A., Harris-Adamson, C., Rempel, D. M., & Nussbaum, M. A. (2019). Potential of exoskeleton technologies to enhance safety, health, and performance in construction: Industry perspectives and future research directions. *IISE Transactions on Occupational Ergonomics and Human Factors, 7*(3–4), 185–191. https://doi.org/10.1080/24725838.2018.1561557.

Kirkwood, G. L., Otmar, C. D., & Hansia, M. (2021). Who's leading this dance? Theorizing automatic and strategic synchrony in human–exoskeleton interactions. *Frontiers in Psychology, 12*. https://doi.org/10.3389/fpsyg.2021.624108.

Kirkwood, G. K., Wilkenfeld, J. N., & Dunbar, N. (2022). Exoskeletons and the future of work: Envisioning power and control in a workforce without limits. *Human–Machine Communication, 4*, 187–206. https://doi.org/10.30658/hmc.4.10.

Knight, J. F., & Baber, C. (2005). A tool to assess the comfort of wearable computers. *Human Factors: Journal of the Human Factors and Ergonomics Society, 47*, 77–91. https://doi.org/10.1518/0018720053653875.

Lee, M. K., Kusbit, D., Metsky, E., & Dabbish, L. (2015). Working with machines: The impact of algorithmic and data-driven management on human workers. In *Conference on Human Factors in Computing Systems—Proceedings, 2015—April* (pp. 1603–1612). https://doi.org/10.1145/2702123.2702548.

Lehmann, H., Saez-Pons, J., Syrdal, D. S., & Dautenhahn, K. (2015). In good company? Perception of movement synchrony of a non-anthropomorphic robot. *PLoS ONE, 10*(5), 1–16. https://doi.org/10.1371/journal.pone.0127747.

Leonardi, P. M. (2011). When flexible routines meet flexible technologies: Affordance, constraint, and the imbrication of human and material agencies. *MIS Quarterly: Management Information Systems, 35*(1), 147–167. https://doi.org/10.2307/23043493.

Li, J., Cuadra, A., Mok, B., Reeves, B., Kaye, J., & Ju, W. (2019). Communicating dominance in a nonanthropomorphic robot using locomotion. *ACM Transactions on Human–Robot Interaction, 8*(1). https://doi.org/10.1145/3310357.

Lorenz, T., Weiss, A., & Hirche, S. (2016). Synchrony and Reciprocity: Key Mechanisms for Social Companion Robots in Therapy and Care. *International Journal of Social Robotics, 8*(1), 125–143. https://doi.org/10.1007/s12369-015-0325-8.

Lucas, K. (2011). Blue-collar discourses of workplace dignity: Using outgroup comparisons to construct positive identities. *Management Communication Quarterly, 25*(2), 353–374. https://doi.org/10.1177/0893318910386445.

Piao, J., Kim, M., Kim, J., Kim, C., Han, S., Back, I., Koh, J. S., & Koo, S. (2023). Development of a comfort suit-type soft-wearable robot with flexible artificial muscles for walking assistance. *Scientific Reports, 13*(1), 1–13. https://doi.org/10.1038/s41598-023-32117-2.

Ronsse, R. (2022). Adaptive Oscillators as Template for Modeling and Assisting Rhythmic Movements. In D. Torricelli, M. Akay, & J. L. Pons (Eds.), *Converging Clinical and Engineering Research on Neurorehabilitation IV* (Vol. 28, pp. 271–275). Springer International Publishing. https://doi.org/10.1007/978-3-030-70316-5_44.

Ronsse, R., Vitiello, N., Lenzi, T., van den Kieboom, J., Carrozza, M. C., & Ijspeert, A. J. (2011). Human–robot synchrony: Flexible assistance using adaptive oscillators. *IEEE Transactions on Biomedical Engineering, 58*(4), 1001–1012. https://doi.org/10.1109/TBME.2010.2089629.

Sarcos Corp (2019). Guardian XO full-body powered exoskeleton. https://www.sarcos.com/products/guardian-xo-powered-exoskeleton.

Smelik, A., Toussaint, L., & Van Dongen, P. (2016). Solar fashion: An embodied approach to wearable technology. *International Journal of Fashion Studies*, *3*(2), 287–303. https://doi.org/10.1386/infs.3.2.287_1.

Sposito, M., Di Natali, C., Toxiri, S., Caldwell, D. G., De Momi, E., & Ortiz, J. (2020). Exoskeleton kinematic design robustness: An assessment method to account for human variability. *Wearable Technologies*, *1*. https://doi.org/10.1017/wtc.2020.7.

Stirling, L., Siu, H. C., Jones, E., & Duda, K. (2019). Human factors considerations for enabling functional use of exosystems in operational environments. *IEEE Systems Journal*, *13*(1), 1072–1083. https://doi org/10.1109/JSYST.2018.2821689.

Tajfel, H., & Turner, J. C. (1979). An integrative theory of intergroup conflict. In W. G. Austin, & S. Worchel (Eds.), The social psychology of intergroup relations (pp. 33–37). Monterey, CA: Brooks/Cole.

Thoits, P. A., & Virshup, L. K. (1997). Me's and we's: Forms and functions of social identities. In R. D. Ashmore & L. J. Jussim (Eds.), *Self and Identity: Fundamental Issues* (pp. 106–133). Oxford University Press.

Upasani, S., Franco, R., Niewolny, K., & Srinivasan, D. (2019). The Potential for Exoskeletons to Improve Health and Safety in Agriculture—Perspectives from Service Providers. *IISE Transactions on Occupational Ergonomics and Human Factors*, *7*(3–4), 222–229. https://doi.org/10.1080/24725838.2019.1575930.

Wilkenfeld, J. N., Kim, S., Upasani, S., Kirkwood, G. L., Dunbar, N. E., & Srinivasan, D. (2023). Sensemaking, adaptation and agency in human–exoskeleton synchrony. *Frontiers in Robotics and AI*, *10*. https://doi.org/10.3389/frobt.2023.1207052.

Wiltermuth, S. S., & Heath, C. (2009). Synchrony and cooperation. *Psychological Science*, *20*(1), 1–5. https://doi.org/10.1111/j.1467-9280.2008.02253.x.

World Health Organization (2022, July 14). Musculoskeletal health. *World Health Organization*. https://www.who.int/news-room/fact-sheets/detail/musculoskeletal-conditions#:~:text=A%20recent%20analysis%20of%20Global,and%20rheumatoid%20arthritis%20(1).

Index

https://doi.org/10.1515/9783110792270-026